Medical Regulatory Affairs

Medical Regulatory Affairs
An International Handbook for Medical Devices and Healthcare Products

Third Edition

edited by
Jack Wong
Raymond K. Y. Tong

Jenny Stanford
PUBLISHING

Published by

Jenny Stanford Publishing Pte. Ltd.
Level 34, Centennial Tower
3 Temasek Avenue
Singapore 039190

Email: editorial@jennystanford.com
Web: www.jennystanford.com

British Library Cataloguing-in-Publication Data
A catalogue record for this book is available from the British Library.

Medical Regulatory Affairs: An International Handbook for Medical Devices and Healthcare Products (Third Edition)

Copyright © 2022 Jenny Stanford Publishing Pte. Ltd.

All rights reserved. This book, or parts thereof, may not be reproduced in any form or by any means, electronic or mechanical, including photocopying, recording or any information storage and retrieval system now known or to be invented, without written permission from the publisher.

For photocopying of material in this volume, please pay a copying fee through the Copyright Clearance Center, Inc., 222 Rosewood Drive, Danvers, MA 01923, USA. In this case permission to photocopy is not required from the publisher.

ISBN 978-981-4877-86-2 (Hardcover)
ISBN 978-1-003-20769-6 (eBook)

Contents

Preface	xxxvii

1. **How to Train University Students in Regulatory Affairs to Face the Medical Devices Market Growth after the COVID-19 Impact** — 1
   Raymond K. Y. Tong
   - 1.1 Introduction — 1
   - 1.2 A Sample of Regulatory Affairs Exercises for Students — 3
     - 1.2.1 Background — 3
     - 1.2.2 Lifelong Learning — 5

PART 1: INTRODUCTION

2. **The Evolution of the Regulatory Professional: Perspectives on the Skill Sets and Capabilities That Will Define the Next Generation of Regulatory Professionals** — 9
   David Martin and Neil Lesser
   - 2.1 Introduction — 9
   - 2.2 Drivers of Change — 10
   - 2.3 Historical Role and Skill Set of a Regulatory Professional — 12
   - 2.4 Changing Role and Skill Set of the Regulatory Professional — 14
   - 2.5 Develop as a Center of Intelligence — 14
   - 2.6 Advance Toward Strategic Relationship Management — 15
   - 2.7 Develop as a Strategic Business Partner — 15
   - 2.8 Conclusion: What Will It Take to Get There? — 16

3. The Role of the Asia Regulatory Affairs Team in Relation to the Commercial Team and Other Departments   19

Fredrik Dalborg

3.1	Introduction	19
3.2	Key Trends in the Asia Medical Device Industry	20
3.3	The Role of the Asia RA Team in a Global Medical Device Organization	21
3.4	Coordination between Commercial Teams and RA Teams	22
3.5	The Role of the RA Team during the Different Stages of a Product Life	23
	3.5.1 Product Development	23
	3.5.2 Market Introduction	23
	3.5.3 Product Maintenance	24
	3.5.4 Product Phase-Out	24
3.6	The RA Professional: A Trusted Advisor	25
3.7	Summary	26

4. Commercial Sense and What It Means for a Regulatory Manager or Executive   27

Annie Joseph

4.1	Know the Basics	28
	4.1.1 Marketing or Commercial Plan	28
	4.1.2 Priority Products and How They Are Ranked in the Company	29
	4.1.3 Know Your Commercial People	29
4.2	Maintain a Healthy Communication	29
4.3	Be Proactive and Part of the Solution	29
4.4	Remain the Expert and Provide Clarity to the Organization	31
4.5	Be a Champion for New Product Launches	31

5. Market Strategic Challenges for Medical Device (Asia/Pacific)   37

Sherwin Tan, Ki Eunyu, Kyser Tay, and David Lee

5.1	Strategic Changes	38

	5.1.1	Product Design and Innovation	38
	5.1.2	Technological Advancement	38
	5.1.3	Distribution Channels	39
	5.1.4	Geo-Political Changes	40
	5.1.5	Environmental and Health Pandemic	40
	5.1.6	Marketing Strategy and Tactics	41
5.2	Strategic Framework		41
	5.2.1	Market Analysis and Evaluation	42
	5.2.2	Competitive Advantage(s)	43
	5.2.3	Scope of Your Products and Markets	43
	5.2.4	Decide on Your Resources and Focus on Them	44
	5.2.5	Identify Changes, Prioritize and Implement Changes	44
	5.2.6	Continuously Monitor Performance and Review Strategy	45
5.3	Key Issues in the Market		45
	5.3.1	Internal Organizational Pressures	45
	5.3.2	Increasing International Competition	46
	5.3.3	Fast-Paced Innovations	46
	5.3.4	Organizational Restructuring and Merger and Acquisition	47
	5.3.5	Increased Quality Consciousness	47
5.4	Challenges Suppliers Are Facing		48
	5.4.1	Accessing Data across Platforms	48
	5.4.2	Identifying Customers across Channels	48
	5.4.3	Mapping the Entire Customer Journey	49
	5.4.4	Identifying New Potential Customers	49
	5.4.5	Maintaining a Consistent Customer Experience	49
	5.4.6	Maintaining Trust and Privacy	50
5.5	Six Growth Challenges		50
	5.5.1	Cash Flow Management	50
	5.5.2	Responding to Competition	51
	5.5.3	Nurturing a Great Company Culture	51

		5.5.4	Learning When to Delegate and When to Get It Done	52
		5.5.5	Keeping Up with Market Changes	52
		5.5.6	Deciding When to Abandon a Strategy	53
	5.6	Five Common Strategic Mistakes		53
		5.6.1	Poor Goal Setting	53
		5.6.2	Lack of Alignment	54
		5.6.3	People Not Being Connected to Common Goal	54
		5.6.4	Failure to Track Progress	54
		5.6.5	Not Utilizing Data Analytics	55

6. Regulatory Affairs as a Business Partner   **59**

Claudette Joyce C. Perilla

7. Introduction to Regulatory Affairs Professionals' Roles   **63**

Dacia Su

8. What It Means to Be a Medtech Regulatory Journalist   **67**

Amanda Maxwell

9. Accelerating Access to in vitro Diagnostics: Urgent Need for Increasing the Speed and Efficiency of Regulatory Review and Policy Development for in vitro Diagnostics for Antimicrobial Resistance and Epidemic Preparedness and Response   **73**

Rosanna W. Peeling, David Heymann, Noah Fongwen, Oliver Williams, Joanna Wiecek, Phil Packer, and Gabriela Juarez-Martinez

	9.1	Background	73
	9.2	The Need for Regulatory Harmonization, Convergence and Reliance for *in vitro* Diagnostics	76
	9.3	Need for a New Paradigm for Risk-Benefit Assessment for Diagnostics	77
	9.4	Next Steps	82

10. Regulatory Specialists in Medical Devices in Europe: Meeting the Challenge of Keeping Current in a Changing Environment—How TOPRA Supports Professionals in a Dynamic Industry   **85**

Lynda J. Wight

10.1	The European MedTech Environment	85
10.2	What Is TOPRA?	87
10.3	Why Be a TOPRA Member?	87
10.4	What Support Is Offered?	88
	10.4.1 Competency Frameworks	88
	10.4.2 Training and Development	89
	10.4.3 Education	90
	10.4.4 Networks and Collaborations	91
	10.4.5 Information	91
10.5	Looking to the Future	91

PART 2: MEDICAL DEVICE SAFETY AND RELATED ISO STANDARDS

11. Biomedical Devices: Overview   **95**

Piu Wong

11.1	Historic Aspect of Medical Devices	95
11.2	Biomedical Market Environment	97
11.3	Orthopedics	98
	11.3.1 Market	98
	11.3.2 Materials	99
	11.3.3 Biocompatibility	99
	11.3.4 Fabrication	100
	11.3.5 Polyethylene Fabrication	100
11.4	Vision Care	102
	11.4.1 Market	102
	11.4.2 Diagnostic Devices	102
	11.4.3 Treatment	103
11.5	Diabetics	104
11.6	Obesity	104

11.7	Vascular Disease	104
11.8	Concluding Remarks	105

12. Labeling, Label, and Language: A Truly Global Matter — 109

Evangeline D. Loh and Jaap L. Laufer

12.1	Introduction	109
12.2	Definition of Labeling	111
12.3	Elements of Labeling	113
12.4	Risk Management, Clinical Evaluation and Labeling: The Core Triangle for Safe and Effective Use of the Device	115
12.5	Labeling and Promotion	117
12.6	e-Labeling, Web Sites, Internet, and Social Media: A Brave New World for Labeling	117
12.7	Language, Language Level, and Intended User	119
12.8	Conclusion	121

13. Regulatory Affairs for Medical Device Clinical Trials in Asia Pacific — 123

Seow Li-Ping Geraldine

13.1	Introduction		123
13.2	Medical Device Clinical Trials versus Pharmaceutical Clinical Trials		124
13.3	Regulation of Clinical Trials		127
13.4	Country Regulations		133
	13.4.1	Australia	133
	13.4.2	China	135
	13.4.3	Hong Kong	136
	13.4.4	India	136
	13.4.5	Malaysia	137
	13.4.6	New Zealand	137
	13.4.7	Singapore	138
	13.4.8	South Korea	139
	13.4.9	Taiwan	140
	13.4.10	Thailand	141
13.5	Moving Ahead as Regulatory Affairs Professionals		141

14. Medical Device Classification Guide — **145**

Patricia Teysseyre

- 14.1 How to Carry Out Medical Device Classification — 145
  - 14.1.1 Scope — 145
  - 14.1.2 Definitions — 146
- 14.2 Main Classifications — 147
  - 14.2.1 Medical Devices — 147
  - 14.2.2 Active Devices — 147
  - 14.2.3 IVD Devices — 148
  - 14.2.4 IVD Case Study — 150
    - 14.2.4.1 US FDA — 151
    - 14.2.4.2 Canada — 151
    - 14.2.4.3 EU — 152
    - 14.2.4.4 Singapore — 153
- 14.3 Medical Device Classification: Practical Examples — 153

15. ISO 13485:2016: Medical Devices—Quality Management Systems—Requirements for Regulatory Purposes — **169**

Gert Bos

- 15.1 Introduction — 169
- 15.2 Background and Origins of ISO 13485:2016 — 170
- 15.3 Quality Management Systems — 172
- 15.4 ISO 9000 and ISO 13485 Quality Management System Family of Standards — 173
- 15.5 Global Regulatory Footprint of ISO 13485:2016 — 174
- 15.6 Implementing an ISO 13485:2016 Quality Management System — 175
- 15.7 Process Approach — 175
- 15.8 Planning the Implementation — 176
- 15.9 Scope, Exclusions and Non-Applicability — 178
- 15.10 Document Control — 178
- 15.11 Record Completion and Control — 178
- 15.12 Management Responsibility — 179

15.13	Resource Management	180
15.14	Product Realization	180
15.15	Risk Management	181
15.16	Design and Development	182
15.17	Purchasing and Supplier Control	183
15.18	Production and Service Provision	184
15.19	Monitoring and Measuring, Including Internal Audits and Management Review	184
15.20	Control of Non-Conforming Product	185
15.21	Analysis of Data	185
15.22	Improvement: Corrective Action and Preventive Action	186
15.23	Purpose and Goal of ISO 13485:2016 Certification	187
15.24	Achieving Certification and Continuing to Maintain Certification	187

16. ISO 14971: Application of Risk Management to Medical Devices — 191

Tony Chan and Raymond K. Y. Tong

16.1	Introduction	191
16.2	The Foundation of a Risk Management Framework: Policy, Plan, Team, Process, and Documentation	194
	16.2.1 Policy	195
	16.2.2 Plan	195
	16.2.3 Team	195
	16.2.4 Process	195
	16.2.5 Documentation	196
16.3	The RM Process	196
	16.3.1 Analyze Risk	196
	16.3.2 Evaluate Risk	197
	16.3.3 Control Risk	197
	16.3.4 Feedback from Production and Post-Production Information	199

16.4	Conclusion	200
16.5	Case Study—An Example to Illustrate Risk Management on a Medical Device: Functional Electrical Stimulation System for Walking	200
	16.5.1 Risk Management Process	202

17. Medical Devices—IEC International Standards and Conformity Assessment Services in Support of Medical Regulation and Governance   **209**

Gabriela Ehrlich

17.1	Introduction	209
17.2	What Is an IEC International Standard?	211
17.3	How Are IEC International Standards Developed?	211
17.4	IEC International Standards for Medical Devices	212
17.5	Conformity Assessment for Medical Devices	214
17.6	Conclusion	216

18. Good Submission Practice   **217**

Shinji Hatakeyama and Isao Sasaki

18.1	Preamble		217
18.2	Introduction		218
	18.2.1	Objective and Scope	218
	18.2.2	Background	219
	18.2.3	Definition	220
18.3	Principles of Good Submission		220
18.4	Management of Submission		222
	18.4.1	Planning for Submission	222
	18.4.2	Preparation and Submission of Application Dossier	224
	18.4.3	Quality Check	227
18.5	Communications		228
	18.5.1	Communications with the Review Authorities	228

	18.5.2	Communication within Applicants' Organization	231
18.6		Competencies and Training	231
	18.6.1	Core Competency of Applicants	232
	18.6.2	Training and Capacity Building	233
18.7		Conclusion	233

PART 3: MEDICAL DEVICE REGULATORY SYSTEM IN THE UNITED STATES, EUROPEAN UNION, SAUDI ARABIA, AND LATIN AMERICA

19. United States Medical Device Regulatory Framework — 237
Joshua Silverstein

19.1	Introduction		237
19.2	FDA's Center for Devices and Radiological Health		238
19.3	Legislation and Device Law		239
	19.3.1	FDA Law vs. FDA Regulations vs. FDA Guidance	240
19.4	The Regulatory Environment for Bringing a Medical Device to Market		241
19.5	Regulatory Considerations for Medical Devices		241
	19.5.1	Definition of Medical Devices	241
	19.5.2	Classification of Medical Devices	242
		19.5.2.1 General controls	246
		19.5.2.2 Special controls	246
		19.5.2.3 Premarket approval	247
	19.5.3	Convenience Kits	247
	19.5.4	Labeling	248
	19.5.5	Adulteration and Misbranding	248
	19.5.6	Establishment Registration and Medical Device Listing	249
	19.5.7	Quality System Regulation/Good Manufacturing Practices	249
	19.5.8	Medical Device Reporting	250
	19.5.9	Unique Device Identification	251

	19.5.10	User Fees	252
19.6	Premarket Submissions for Medical Devices		252
	19.6.1	eCopy Program for Medical Device Submissions	253
	19.6.2	Premarket Notification (510(k))	253
		19.6.2.1 Predicate device and substantial equivalence	253
	19.6.3	De novo Classification Request	255
	19.6.4	Premarket Approval Application	256
	19.6.5	Accessory Requests	257
	19.6.6	Investigational Device Exemption	258
	19.6.7	Q-Submissions: Requesting Feedback and Meeting with FDA	258
	19.6.8	Promoting Innovation: Breakthrough Devices	259
19.7	Summary		260

20. Regulation of Combination Products in the United States   263

John Barlow Weiner and Thinh X. Nguyen

20.1	What Products Are Considered Combination Products	264
20.2	The Standards for Determining If a Product Is a Combination Product	264
20.3	The Standards for Determining Which FDA Component Has Primary Responsibility for Regulating a Combination Product	266
20.4	Requests for Designation	267
20.5	Premarket Review Considerations	268
20.6	Post-Market Regulatory Considerations	269
20.7	Role of Office of Combination Products	270
20.8	International Harmonization and Coordination Activities with Foreign Counterparts	270
20.9	FDA Resources for Obtaining Additional Information	271

21. European Union Medical Device Regulatory System 273

Arkan Zwick and Gert Bos

- 21.1 Glossary of Terms 273
- 21.2 European Union: Medical Device Market and Structure 275
- 21.3 EU Regulations on Medical Devices 278
- 21.4 Harmonized Standards, Presumption of Conformity and State of the Art 278
- 21.5 European Associations 281
- 21.6 Overview of New Medical Devices Regulations 285
- 21.7 Guidelines MEDDEV/NB-MED 286
- 21.8 Guidelines MDCG and Common Specifications 287
- 21.9 Definitions 289
  - 21.9.1 Medical Device 289
  - 21.9.2 Devices without Medical Purpose 290
  - 21.9.3 CE Mark 290
  - 21.9.4 European Commission and MDCG 292
  - 21.9.5 Competent Authority
  - 21.9.6 Notified Body—Conformity Assessment Body 293
  - 21.9.7 The Manufacturer 295
  - 21.9.8 Authorized Representative 296
- 21.10 Classification 297
  - 21.10.1 Medical Devices 297
  - 21.10.2 Active Implantable Medical Devices 298
  - 21.10.3 In Vitro and Diagnostics Medical Devices 299
- 21.11 Conformity Assessment Procedures 300
- 21.12 General Safety and Performance Requirements 304
- 21.13 Labeling 305
- 21.14 Technical Documentation 305
- 21.15 Quality Management System 307
- 21.16 Risk Management 308
- 21.17 Clinical Evaluation 309

21.18 CE Mark Certificate and Declaration of
      Conformity   311
21.19 Post-Market Surveillance   312
21.20 From MDD to MDR   313
21.21 Transition and MDR Impact   314
21.22 General Safety and Performance Requirements   314
21.23 Stronger Notified Body Oversight   315
21.24 Economic Operators   315
21.25 Clinical Evidence, Postmarked Follow-Up and
      Risk Reviews   316
21.26 PSUR/PMCF Reports   316
21.27 UK Leaving the EU—BREXIT   317

22. Regulation of Combination Products in the European Union   321
Gert Bos

22.1 Introduction: Legal Basis   321
    22.1.1 Definitions   323
        22.1.1.1 Medical device   323
        22.1.1.2 Medicinal product 323
        22.1.1.3 Combination products: Principal mode of action   324
        22.1.1.4 Borderline products: MEDDEV 2.1/3   325
        22.1.1.5 Borderline products: Manual of decisions   326
22.2 Combination Products Regulated as Medicinal Products   326
    22.2.1 Examples of Combination Products Regulated as Medicinal Products   326
22.3 Combination Products Regulated as Medical Devices   327
    22.3.1 Examples of Combination Products Regulated as Drug-Delivery Devices   327
22.4 Combination Products Regulated as Devices Incorporating, as an Integral Part, an Ancillary Medicinal Substance   327

22.4.1 Examples of Devices Incorporating an Ancillary Medicinal Substance 328
22.4.2 Examples of Drug Substances Incorporated into Devices 328
22.4.3 Assessment of the Medicinal Substance Aspects of a Device Incorporating an Ancillary Medicinal Substance 329
22.5 The Consultation Process 330
22.6 Information to Be Provided on the Ancillary Medicinal Substance 330
22.6.1 General 330
22.6.1 Quality 330
22.6.2 Safety and Usefulness 331
22.6.3 Guidance 331
22.7 Combination Products Regulated as Drugs Incorporating, as an Integral Part, an Ancillary Medicinal Substance 332
22.8 Other Combination Products 332

23. Medical Device Regulatory Affairs in Latin America 335
Carolina Cera and Gladys Servia

23.1 Introduction 335
23.2 Latin America Market Analysis 336
23.2.1 Medical Device Market in Latin America 336
23.2.2 Trade Blocs 337
23.3 Overview of Medical Device Regulation in Latin America 338
23.3.1 Evolution of the Medical Device Regulation in Latin America 338
23.3.2 Regulatory Environment in Latin America 339
23.3.2.1 Finding local registration holder 339
23.3.2.2 Harmonization of medical device 340
23.3.2.3 Challenges in the region 341

23.4	Argentina and Brazil Medical Device System			342
	23.4.1	Definition of Medical Device		343
	23.4.2	Classification of Medical Devices		343
	23.4.3	Argentina Medical Device System		343
		23.4.3.1	Regulatory authority and medical device regulation	343
		23.4.3.2	Registration process of medical devices	344
		23.4.3.3	Documents and labeling requirements	344
		23.4.3.4	Official registration fee	345
		23.4.3.5	Timeline	346
		23.4.3.6	Regulatory action for changes and device modifications	346
		23.4.3.7	Post-market surveillance	346
	23.4.4	Brazil Medical Device System		347
		23.4.4.1	Registration process of medical devices in Brazil	347
		23.4.4.2	Documents and labeling requirements	349
		23.4.4.3	Official registration fee	350
		23.4.4.4	Timeline	350
		23.4.4.5	Regulatory action for changes and device modification	351
		23.4.4.6	Post-market surveillance	352
23.5	Colombia Medical Device System			352
	23.5.1	Regulatory Authority and Medical Device Regulation		352
	23.5.2	Definition of Medical Device		353
	23.5.3	Classification of Medical Devices		353
	23.5.4	Registration of Medical Devices		354
		23.5.4.1	Regulatory process	354
		23.5.4.2	Documents required	354
		23.5.4.3	Official registration fee	356

	23.5.4.4 Timeline	356
	23.5.4.5 Regulatory action for changes and device modifications	356
	23.5.4.6 Post-market surveillance	357
23.6	Mexico Medical Device System	357
	23.6.1 Regulatory Authority and Medical Device Regulation	357
	23.6.2 Definition of Medical Device	358
	23.6.3 Classification of Medical Devices	358
	23.6.4 Registration of Medical Devices	359
	23.6.4.1 Registration process	359
	23.6.4.2 Documents and labeling requirements	359
	23.6.4.3 Official registration fee	359
	23.6.4.4 Timeline	361
	23.6.4.5 Regulatory action for changes and device modifications	361
	23.6.5 Post-Market Surveillance	361

24. Saudi Arabia: Medical Device Regulation System — 363

Ali Aldalaan

24.1 Introduction	363
24.2 Legislative Responsibilities	364
24.3 Executive Responsibilities	364
24.4 Surveillance Responsibilities	365
24.5 Regulation Overview	366
24.6 Registration Requirements	367
24.7 Information to Be Provided to the SFDA	367
24.8 Medical Device Marketing Authorization	369
24.9 Medical Device Listing	370
24.10 Registration Fees	370
24.10.1 Medical Device Marketing Authorization	370
24.10.2 Medical Device Authorized Representative License	371

		24.10.3	Medical Device Establishment Licensing	371
	24.11	General Information and Documentary Evidence Need to Be Provided to the SFDA		371
	24.12	Labeling Requirement for Medical Device		372
	24.13	Clinical Investigation		374
	24.14	Post-Market Surveillance Requirement		376
		24.14.1	General	376
		24.14.2	Proactive Surveillance and Risk Assessment Activities	377
		24.14.3	Reactive Surveillance/Vigilance Activities	378
			24.14.3.1 Medical devices' management for incidents and adverse events	379
			24.14.3.2 Medical device field safety corrective actions	380
		24.14.4	Confidentiality of Information	380

Part 4: MEDICAL DEVICE REGULATORY SYSTEM IN ASIA-PACIFIC REGION

25. Australian Medical Device Regulations: An Overview 387

Petahn McKenna

	25.1	Medical Device Market in Australia		387
	25.2	Medical Device Regulations		388
		25.2.1	Overview	388
		25.2.2	Regulating Authority	389
		25.2.3	Legislation and Guidance	389
	25.3	Definition of Medical Device		389
	25.4	Classification of Medical Devices		390
		25.4.1	Classification of IVD Medical Devices	391
	25.5	Inclusion of Medical Devices on the ARTG		392
		25.5.1	Process for Supplying a Medical Device in Australia	392

		25.5.2	Process for Including Class 1 IVD Medical Devices in the ARTG	395
		25.5.3	Process for Including IVD Medical Devices in the ARTG	396
	25.6	Same Kind of Medical Device		398
	25.7	Unique Product Identifier		398
	25.8	In vitro Diagnostic UPIs		399
	25.9	Renewal		400
	25.10	Documentation Requirements		400
		25.10.1 Conformity Assessment Applications		400
	25.11	Application Audits		401
	25.12	Access to Unapproved Medical Devices		402

26. China: Medical Device Regulatory System — 403

Jack Wong

	26.1	Introduction			403
	26.2	Market Overview			404
	26.3	Overview of Regulatory Environment and What Laws/Regulations Govern the Medical Devices			404
		26.3.1	Guidance		406
		26.3.2	Standards		406
	26.4	Regulatory Body			407
	26.5	Regulatory Overview			408
		26.5.1	Definition of Medical Device		408
		26.5.2	Classification of Medical Devices		408
		26.5.3	Filing Process for Class I Medical Device		410
		26.5.4	Registration Process for Class II or III Medical Device		411
			26.5.4.1	Medical device registration certificate	411
			26.5.4.2	Documents needed for registration application	414

		26.5.4.3	Product technical requirements and registration testing	415
		26.5.4.4	Biocompatibility evaluation	415
		26.5.4.5	Clinical evaluation and clinical trials	416
		26.5.4.6	Enforcing GMPs	423
		26.5.4.7	Timeframes	424
		26.5.4.8	Registration alteration	424
		26.5.4.9	Registration renewal	426
		26.5.4.10	The registration fee	426
		26.5.4.11	Innovative medical devices	427
		26.5.4.12	Drug/device combination products	427
	26.6	Monitoring Adverse Events		428
		26.6.1	AE Reporting	428
		26.6.2	Periodic Risk Assessment Report	429
	26.7	Managing Recalls		429

27. Hong Kong: Medical Device Regulatory System 433

Jack Wong and Linda Chan

	27.1	Market Overview		433
		27.1.1	Market Environment	433
		27.1.2	Overview of Regulatory Environment and What Laws/Regulations Govern Medical Devices	434
		27.1.3	Regulatory Body	435
	27.2	Regulatory Overview		436
		27.2.1	The Definition of Medical Device	436
		27.2.2	Classification of Medical Devices	437
		27.2.3	The Role of Distributors or Local Subsidiaries	438
		27.2.4	Product Registration or Conformity Assessment Route and Time Required	439

		27.2.4.1	Suggested registration routes/steps	440

27.2.4.1 Suggested registration routes/steps 440
27.2.4.2 Technical material requirement 441
27.2.4.3 The labeling requirement of medical device 441
27.2.4.4 Post-marketing surveillance requirement 442
27.2.4.5 Manufacturing-related regulation 442
27.2.4.6 Clinical trial related regulation 442
27.2.4.7 Is there a procedure for mutual recognition of foreign marketing approval or international standards? 443
27.3 Commercial Aspects 443
27.4 Upcoming Events 443

28. India: Medical Device Regulatory System 445

Kulwant S. Saini

28.1 Market Overview 445
    28.1.1 Market Environment 445
    28.1.2 Overview of Regulatory Environment and What Laws/Regulations Govern Medical Devices 446
    28.1.3 Functions Undertaken by DCGI and Central Government 450
        28.1.3.1 Statutory functions 450
        28.1.3.2 Other functions 450
    28.1.4 Functions Undertaken by the FDA and State Governments 451
        28.1.4.1 Statutory functions 451
    28.1.5 Guidance Documents 452
    28.1.6 Indian Pharmacopoeial Commission 453
    28.1.7 Detail of Key Regulator(s) 453

28.2	Regulatory Overview		456
	28.2.1	Definition of Medical Device	456
	28.2.2	Classification of Medical Devices	457
	28.2.3	Role of Distributors or Local Subsidiaries	457
	28.2.4	Product Registration or Conformity Assessment Route and Time Required	458
	28.2.5	Quality System Regulation	460
	28.2.6	Product Registration and Quality System Regulation for Combined Device–Drug Product	460
	28.2.7	Registration Fee	461
	28.2.8	Technical Material Requirement	462
	28.2.9	The Labelling Requirement of Medical Device	462
	28.2.10	Post-Marketing Surveillance Requirement	463
	28.2.11	Manufacturing-Related Regulation	463
	28.2.12	Clinical Trial-Related Regulation	464
	28.2.13	Is There a Procedure for Mutual Recognition of Foreign Marketing Approval or International Standards?	465
28.3	Commercial Aspects		465
	28.3.1	Any Price Control of Medical Device	465
	28.3.2	Are Parallel Imports Allowed?	466
	28.3.3	Any Advertisement Regulation of Medical Device?	466
28.4	Upcoming Regulation Changes		466
28.5	Related Agencies/Departments and Ministries		468

29. Indonesia: Medical Device Regulatory System — 469

Mita Rosalina

29.1	Introduction	469
29.2	Regulating Authority	471
29.3	Definition of Medical Device	471

29.4	Classification of Medical Devices		472
29.5	Registration of Medical Devices		473
	29.5.1	Process	473
	29.5.2	Documents Required	474
	29.5.3	Official Registration Fee	474
	29.5.4	Time Line	474
	29.5.5	Validity of Product License	475
	29.5.6	Indonesian Labeling Requirement	475
	29.5.7	Regulatory Action for Changes and Device Modifications	475
29.6	Post-Market Surveillance System		476

30. Japan: Medical Device Regulatory System — 491

Atsushi Tamura and Keizo Matsukawa

30.1	Introduction		491
30.2	Regulatory Agency in Japan		492
	30.2.1	Ministry of Health Labour and Welfare/MHLW	492
	30.2.2	Pharmaceuticals and Medical Devices Agency	492
	30.2.3	PMDA Medical Device Unit	495
	30.2.4	Shared Responsibility of MHLW and PMDA on Medical Device Regulation	495
30.3	Legislation of Medical Devices		497
	30.3.1	Classification of Medical Devices	499
	30.3.2	Type of Product's Registration	500
		30.3.2.1 Notification of marketing	500
		30.3.2.2 Certification	500
		30.3.2.3 Approval	501
	30.3.3	Marketing Licenses	502
	30.3.4	Marketing Authorization Holder	503
	30.3.5	Accreditation of Manufacturing Businesses/Accreditation of Foreign Manufactures of Medical Devices	504
	30.3.6	Registered Certification Bodies	505

30.4	Quality Management System		506
	30.4.1	QMS Ordinance	507
		30.4.1.1 QMS conformity as an essential requirement	508
	30.4.2	QMS Organizational Structure and Personnel Requirements	509
	30.4.3	QMS Inspections	510
30.5	Post Market Safety Management		510
	30.5.1	GVP Ordinance	511
	30.5.2	Requirements for MAH	513
	30.5.3	Collection, Analysis and Reporting of Post-marketing Safety Information	513
		30.5.3.1 Obligations of MAH	513
	30.5.4	Recall	515
		30.5.4.1 Obligations of healthcare professional	516
	30.5.5	PMDA's Obligations during Post-Market Phase	516
		30.5.5.1 Various approaches	517
		30.5.5.2 Information services	517
30.6	New Regulatory Challenge		518
	30.6.1	SAKIGAKE Designation System	519
	30.6.2	Conditional Early Approval System	522
	30.6.3	Introduction of Approval System Based on the Characteristics of Medical Devices	522
		30.6.3.1 Early realization of change plan	525
		30.6.3.2 Improvement design within approval for timely evaluation and notice	525
	30.6.4	Electronic Labelling/Instructions for Use	525
	30.6.5	PMDA-ATC Medical Device Seminar	526
	30.6.6	Guidance for the Evaluation of Emerging Technology Medical Devices	527

		30.6.6.1	PMDA Science Board	527
		30.6.6.2	Publication of the guidance for the evaluation of next-generation medical devices	527
	30.6.7		Traceability	530
	30.6.8		Information Services	530
	30.6.9		Harmonization by Doing	532

31. Korea: Medical Device Regulatory System 535

Young Kim, Soo Kyeong Shin, and Jamie Noh

31.1	General Market Overview			535
	31.1.1	Key Healthcare Market Indicators of Korea		535
	31.1.2	Medical Device Market		538
	31.1.3	Import		540
	31.1.4	Domestic Production		544
	31.1.5	Export		551
31.2	Regulatory Approvals			551
	31.2.1	Responsible Authorities		551
	31.2.2	Qualifications for Medical Device Business		553
		31.2.2.1	Medical device business license	553
		31.2.2.2	Certification to medical device quality system management	554
		31.2.2.3	Medical device product approval or certification or notification	554
	31.2.3	Strategic Plan and Useful Tips for Efficient Registration		561
		31.2.3.1	Face-to-face meeting with reviewer	561
		31.2.3.2	Respond to reviewer's questions with respect	562

		31.2.3.3	Retain experienced regulatory affairs professionals	562
31.3	Reimbursement			563
	31.3.1	Overview of Reimbursement Scheme		563
	31.3.2	Medical Supplies		565
		31.3.2.1	Submission	565
		31.3.2.2	Pricing options	566
		31.3.2.3	Timeframes	567
		31.3.2.4	Stakeholders	568
	31.3.3	Health Technology Assessment		570

32. Overview of Medical Device Regulation in Malaysia — 573

Ir. Sasikala Devi Thangavelu

32.1	Medical Device Industry in Malaysia		573
32.2	Medical Device Regulatory Program		574
32.3	Introduction to Medical Device Act 2012		575
	32.3.1	Definition of Medical Device	576
32.3	Establishment license		577
32.4	Medical Device Registration		580
	32.4.1	Classification and Grouping for the Purpose of Medical Device Registration	581
	32.4.2	Conformity Assessment	584
	32.4.6	Conformity Assessment through Verification Route	585
32.5	Combinational Medical Devices		587
	32.5.1	Definition of Combination Product	587
	32.5.2	Registration Process of Combination Product	588
	32.5.3	Drug-Medical Device Combination Product Registration Process	590
	32.5.4	Medical Device-Drug Combination Product Registration Process	590
	32.5.5	Changes/Variation to Particulars of a Registered Drug-Medical Device Combination Product	590

		32.5.6	Changes/Variation to Particulars of a Registered Medical Device-Drug Combination Product	591

32.7 Medical Device Labelling — 591
    32.7.1 General Contents of Labelling — 593
    32.7.2 Additional Information for in vitro Diagnostic Medical Devices — 593

32.8 Medical Device Order 2016 — 594

32.9 Post-Market Surveillance and Vigilance — 595
    32.9.1 Distribution Records — 595
    32.9.2 Records of Complaint Handling — 595
    32.9.3 Mandatory Problem Reporting — 595
    32.9.4 Field Corrective Preventive Action — 596
    32.9.5 Recall — 596
    32.9.6 Voluntary Recall — 596
    32.9.7 Mandatory Recall — 597

32.10 Regulatory Action for Changes and Device Modifications — 597
    32.10.1 Change Notification for Registered Medical Device — 597
    32.10.2 Refurbishment of Medical Device — 598

32.11 Regulatory Control on Usage, Operation, and Maintenance of Medical Devices — 599

33. The Philippine Medical Device Regulatory System — 601
Rhoel Laderas

33.1 The Medical Device Market Profile — 601
33.2 Introduction to the Philippines Regulatory System — 602
33.2 The Medical Device Regulatory System — 603
    33.2.1 The Licensing of Medical Device Establishment — 603
    32.2.2 The Authorizations for Medical Device Products — 607

34. Singapore Medical Device Regulation   **619**

May Ng, Ray Soh, Trish, Beatrice, Bing Kang, Yiyu, Xinyu, Ivy Lim, and Tiffany Hu

34.1	Definition of Medical Device	620
34.2	Classification of Medical Device	621
34.3	Singapore Medical Device Notification, Registration and Other Authorisations	623
	34.3.1 Steps to Access MEDICS	623
	34.3.2 Establishment or Dealer Licenses	624
	34.3.3 Class A Notification	626
	34.3.4 Class B, C and D Pre-Market Product registration	626
	34.3.5 Special Authorisation Routes in Singapore	633
34.4	Post-Market Surveillance	637
	34.4.1 Adverse Event	637
	34.4.2 Field Safety Corrective Action	638
34.5	Medical Device Advertisement	639
34.6	Clinical Trial in Singapore	640
34.7	Price Control on Medical Device	641

35. Taiwan: Medical Device Regulatory System Introduction   **643**

Pei-Weng Tu

35.1	Market Overview	643
	35.1.1 Overview of Structure and Funding of Local Healthcare System	643
	35.1.2 Overview of Regulatory Environment and Laws/Regulations Governing Medical Devices	645
	35.1.3 Detail of Key Regulator(s)	646
35.2	Regulatory Overview	646
	35.2.1 Definition of Medical Device	646
	35.2.2 Classification of Medical Device	648

		35.2.3	Role of Distributors or Local Subsidiaries	648
		35.2.4	Product Registration, Technical Material Requirement, and Time Required	649
	35.3	Quality System Regulation		650
	35.4	Combined Device–Drug Product		650
	35.5	Registration Fee		650
	35.6	Labeling Requirements of Medical Devices		650
	35.7	Post-Marketing Surveillance Requirement		650
	35.8	Manufacturing-Related Regulation		651
	35.9	Clinical Trial–Related Regulation		651
	35.10	International Cooperation		651
	35.11	Commercial Aspects		653
		35.11.1	Price Control of Medical Device	653
		35.11.2	Parallel Imports	653
		35.11.3	Advertisement Regulation of Medical Devices	653
	35.12	Upcoming Events		653

36. Thailand: Medical Device Control and Regulation — 655

Kanokorn Pulsiri, Sirinmas Katchamart, Sansanee Pinthong, and Korrapat Trisansri

	36.1	Market Overview		655
	36.2	Medical Device Regulations		656
	36.3	Definition of Medical Device		657
	36.4	Medical Device Classification		658
	36.5	Pre-Marketing Control		659
		36.5.1	Documents Required for Medical Device Registration	659
			36.5.1.1 Licensed medical device and notified medical device	659
			36.5.1.2 Listing	661
		36.5.2	Labeling	662
		36.5.3	Advertising	663
	36.6	Post-Market Controls		663

37. Vietnam — 667
Nguyen Minh Tuan

    37.1 Market Environment — 667
    37.2 Roadmap for the Implementation of the Decree on Medical Device Management — 668
    37.3 License Fees — 671
    37.4 Labelling According to New Rules — 671
    37.5 Technical Requirements for Raw Materials — 672
    37.6 Domestic Production — 672
    37.7 Clinical Trial Evaluation — 672

PART 5: HOT TOPICS

38. A Strong Regulatory Strategy Is a Competitive Advantage to a Medical Device Company — 677
Jacky Devergne

    38.1 Competitive Advantage — 678
        38.1.1 Time to Market — 678
        38.1.2 Barriers of Entry — 679
        38.1.3 Continuity of Business — 679
        38.1.4 Best Use of Resources — 680
    38.2 The Right Organization — 681
        38.2.1 Investing in an Effective RA Organization — 681
        38.2.2 Integrating RA in Business Planning — 681
    38.3 Conclusion — 682

39. Regulatory Strategy: An Overview — 683
Pakhi Rusia

    39.1 Introduction — 684
    39.2 The Significance of Regulatory Strategy — 685
    39.3 Key Steps to Develop Regulatory Strategy — 686
        39.3.1 Defining Project Scope — 686
        39.3.2 Defining Timelines and Milestones — 687
        39.3.3 Stakeholder Alignment — 687

- 39.4 Key Factors to Be Considered for Developing Regulatory Strategy — 687
  - 39.4.1 Identification of Project Attributes — 688
  - 39.4.2 Geography and Regulatory Landscape — 689
- 39.5 Implementing the Regulatory Strategy — 690
- 39.6 Conclusion — 691

40. Leading the New Normal by Accelerating Digital Transformation — 693

Virginia Chan

- 40.1 China Medical Technology Market in 2025 — 693
- 40.2 Patient-Centric Digital Healthcare Model — 694
- 40.3 Accelerate Digital Transformation for Agility, Flexibility, Efficiency, Reimagine the New Business Model — 695
- 40.4 Digital Health Regulation: Focus on Quality Across the Product Life Cycle — 695
- 40.5 Connected Care — 696
- 40.6 Intelligent Design Control to Shorten Time-to-Market with Compliance — 697
- 40.7 Digital Innovation for Design Excellence: Augment Virtual Evidence to Reduce Design Time and Cost with in silico Clinical Trials — 697
- 40.8 Turning Complexity to Your Competitive Advantage: Gain Digital Dividend with Vision 2025 — 699

41. An Overview of the Herbal Product Regulatory Classification in Asia and General Guidelines for Health Product Development — 701

Jacob Cheong

42. Overview of Health Supplements: Singapore — 707

Srilatha Sreepathy, Geeta Pradeep, and A. V. Rukmini

- 42.1 Introduction — 707
- 42.2 What Are Health Supplements? — 708

42.3	What Are Not Health Supplements?	709
42.4	How Are Health Supplements Regulated in Singapore?	709
42.5	Safety and Quality Standards	709
42.6	Quality Standards	710
42.7	Stability Study and Shelf-Life of Health Supplements	712
42.8	Storage Condition	712
42.9	Product Labelling Requirement	713
42.10	Health Supplement Claims	714
42.11	Advertisements and Promotions	715
42.12	Other Important Aspects	715

43. International Medical Device School Experience — 719
Encey Yao

44. Medtech Start-Up: Journey to First Product Approval — 723
Sing Wee, Joel Tan, and Trish, May Ng

44.1	Introduction	723
44.2	Intended Use	724
44.3	Technical Documentation	725
44.4	Quality Management System	726
44.5	Clinical Evaluation	727
44.6	Conclusion	728

45. Digital Transformation of Healthcare and Venture Capital's Role in It — 729
Mark Wang

46. A Regulatory Career in Asia — 737
Ambrose Chan

46.1	Hiring Landscape in Asia	737
46.2	Opportunities across Markets	738
46.3	Getting Ahead in your Search	740

47. A Former FDA Investigator's Views on Compliance
    with the Medical Device Regulations   743
    Ken Miles

Index   751

Preface

Medical device regulation in Asian markets has become important. Governments and regulatory bodies of countries across the region have placed new regulatory systems or refined the existing ones. Regulatory affairs (RA) is a science of how to get a medical product registered with different countries' health authorities. A registered product would demand a lot of technical documentation to prove its efficacy, safety, and quality. To successfully and smoothly register a product, many soft skills are required for dealing with various key stakeholders in governments, testing centers, hospitals, and medical doctors.

The handbook is the first to cover medical device regulatory affairs in Asia. It is enriched by contributions by authors working with several regulatory bodies, including the US Food and Drug Administration (FDA), UK Medicines and Healthcare Products Regulatory Agency (MHRA), Japan Pharmaceuticals and Medical Devices Agency (PMDA), Saudi Food and Drug Authority (SFDA), Taiwan FDA). Each chapter provides substantial background materials relevant to a particular area to provide the reader a better understanding of regulatory affairs. The text also presents in-depth discussion on requirements for medical device registration in China and India.

Government bodies will find this book useful to understand the global regulatory environment to help enhance their regulatory systems. The medical device industry can use it to better understand and access the Asian market. Academics and students will find this book very important for their careers in biomedical engineering and medical device–related fields. In research and development, with the help of this book, companies can plan their projects and ensure that the developed medical devices adhere to the global regulatory environment.

The chapters have been grouped into four main parts as follows:

- Part 1 explains what RA is, how to be a good RA professional, how a RA professional works with other team members, and some associated soft skills.
- Part 2 focuses on medical device fundamentals, such as history, labeling requirement, clinical trial requirement, how to do classification, and two important standards for medical device regulatory (ISO 13485 and ISO 14971), IEC standard and Good Submission Practice.
- Part 3 discusses in detail the regulatory systems in the United States, the European Union, Saudi Arabia, and Latin America. These are important markets outside Asia. Their experience will be very helpful for Asia.
- Part 4 is the core of this book. It describes the regulatory system in the Asian market, with contributions from regulatory authorities, testing laboratories, and industries.
- Part 5 shares the hot topics in the regulatory field, including regulatory strategy, digital transformation, herbal regulatory system, health supplements, International Medical Device School, medtech start-up journey, regulatory career in Asia, and a former FDA investigator's perspective.

This book would not have been possible without contributions from outstanding experts in various topics discussed in it. We wish to express our gratitude to all of them for their precious efforts and strong support.

Fifty percent of the revenue from this book will be donated to the Asia Regulatory Professionals Association and the remaining 50% to the Department of Biomedical Engineering, the Chinese University of Hong Kong—both are for the development of medical device regulatory affairs in Asia.

Finally, many thanks to our families (Jack Wong's mother, Cheung Shim Kuen, wife, Sherry Kwan, and son, Jay Wong; Raymond Tong's parents, Wai-chuen Tong and Lai-lin Tsui, wife, Wai-nga Lam, and daughter and sons, Lok-ching, Lok-tin, and Lok-ting), for their support, encouragement, and patience. They have been our driving forces.

Jack Wong
Asia Regulatory Professional Association

Raymond Kai-yu Tong
Professor, Department of Biomedical Engineering (BME),
The Chinese University of Hong Kong (CUHK)
Hong Kong Academy Chair, Asia Regulatory Professionals Association

Chapter 1

How to Train University Students in Regulatory Affairs to Face the Medical Devices Market Growth after the COVID-19 Impact

Raymond K. Y. Tong

Department of Biomedical Engineering,
The Chinese University of Hong Kong, Shatin, Hong Kong
Asia Regulatory Professional Association (ARPA)

kytong@cuhk.edu.hk

This chapter focuses on how to design a regulatory course to train university students in regulatory affairs.

1.1 Introduction

Medical device regulation comes with new framework, new principal and supportive responsibilities in Asian and European countries. The COVID-19 pandemic has created the largest disruption of healthcare and education systems in history. The demand of medical devices for vaccination, diagnosis, treatment and telehealth purposes has surged. The Medical Devices Market Size was valued at US$ 483,285.8 million in 2019 and is projected to reach US$ 767,684.9 million by 2027 [1]. The compound

---

Medical Regulatory Affairs: An International Handbook for Medical Devices and Healthcare Products (Third Edition)
Edited by Jack Wong and Raymond K. Y. Tong
Copyright © 2022 Jenny Stanford Publishing Pte. Ltd.
ISBN 978-981-4877-86-2 (Hardcover), 978-1-003-20769-6 (eBook)
www.jennystanford.com

annual growth rate (CAGR) is estimated to be 6.1% from 2020 to 2027. Asian markets have significant growth. The market in China reached US$ 78.81 billion in 2018, at a rate of 22% from 2017. The Indian medical device market reached US$ 6 billion in 2017 and is projected to reach US$ 50 billion by 2025 [1]. The Medical Device Regulation (MDR), which was adopted in April 2017, changes the European legal framework for medical devices and introduces new principal and supportive responsibilities for European Medicines Agency and for national competent authorities in the assessment of certain categories of products. The course should be suitable for students who are studying biomedical engineering, biomedical science, healthcare, or medical device engineering. These students should have some experience in a health- or engineering-related field and should wish to start working in the growing field of regulatory affairs. The course in regulatory affairs aims to give both factual and practical knowledge of what is regulatory requirement and how to handle future regulatory tasks. On completion of the course, students will be able to demonstrate their understanding of how to meet the standards and regulatory requirements, and they will be able to handle regulatory tasks, including classification, risk management, ISO standard, product registration, and commercial materials review.

Students will be trained to understand the global, regional, and local medical device regulatory requirements and trends. Besides conducting lectures, students will be arranged into a small group to work on a medical device and practice preparing a medical device registration submission to a competent authority. The competent authority is a body that has authority to act on behalf of the government to ensure that the requirements of the medical device directives are transposed into the national law and are applied. Guest lecturers with regulatory experience and network will be invited to share their experience. Students who finish the course will be eligible for attending related medical device regulatory examination conducted by external bodies, such as notified bodies and the Asia Regulatory Professional Association.

1.2 A Sample of Regulatory Affairs Exercises for Students

1.2.1 Background

Consider this case. You have just been recruited by a manufacturing company, Poly Technologies Corporation (PTC), as project manager under its newly found business section for medical and healthcare products. The company has little, if any, experience in designing and manufacturing medical devices. Your primary duty is to provide internal consultancy to the top management of the company on new business development analysis and management of medical device design and manufacturing projects.

Last month your company's top management signed a business agreement to collaborate with an overseas client, which is a global distributor of medical products. Under this collaboration agreement, your client and PTC will co-develop a series of medical devices for global markets. Your client has proposed to start a project with the sterile disposable hypodermic syringes with needles — for single-dose hypodermic injection (Fig. 1.1).

To begin with, your client has suggested designing and manufacturing the products for the US and EU markets and, if possible, extend the market to China in future.

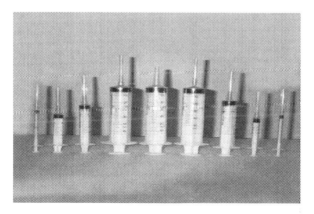

Figure 1.1 Various sizes of hypodermic syringes.

Task 1: Classification

Your first task is to brief your management about the product classifications of the device. For more details, see Chapter 14, "Medical Device Classification Guide."

Task 2: Risk Analysis

(For more details, see Chapter 16 on ISO14971.)

Target market	Device classification	What are the US FDA regulation numbers	Classification rules applied to the product under EU medical device directive	Notified body involvement for CE marking?
EU				
US				
China				

During the design phase of the device, suggest hazards for the product that the product design team has to include in the design input considerations.

Suggest the severity and occurrence for the suggested hazards. Ensure that you have provided your own table and definition of different levels of severity and occurrence.

Task 3: Risk Evaluation and Risk Control

(For more details, see Chapter 16 on ISO14971.)
You have to develop a risk graph (group for risk regions and decide the risk acceptability) and risk matrix (put all the above risks in your device).

Implement risk control measures for all of your hazards. After all risk control measures have been implemented and verified, the manufacturer shall decide whether the overall residual risk posed by the medical device is acceptable using the criteria defined in the risk management plan.

Task 4: Medical Device Registration Submission

Prepare a medical device registration submission to your competent authority.

For example, the student will represent the company to be the local responsible person in Hong Kong to submit an application for listing this product to the Medical Device Division (MDD), Department of Health in Hong Kong. The student will download the appropriate application form from the MDD website http://www.mdco.gov.hk/english/download/download.html and fill in the application form according to the information in the registration dossier provided. They will refer to the Overview of the Medical Device Administrative Control System (GN-01) and all guidance documents can be found at http://www.mdco.gov.hk/english/mdacs/mdacs_gn/mdacs_gn.html.

A mock up label will be required to show the Special Listing Information and provide the Essential Principles Declaration of Conformity in the submission. The sample of Essential Principles Declaration of Conformity is given in the Guidance Notes for Listing Class II/III/IV Medical Devices (GN-02) under the above-mentioned MDCO website.

1.2.2 Lifelong Learning

The above exercise is to equip students with lifelong learning skills to handle regulatory affairs and the requirements from the competent authority. Students should be encouraged to attend workshops/seminars which are organized by the competent authority, and key global (Asia-Pacific Economic Cooperation [APEC], Regulatory Affairs Professionals Society [RAPS]) and regional (Asian Harmonization Working Party [AHWP], Association of Southeast Asian Nations [ASEAN]) organizations.

References

1. Medical devices market growth sturdy at 6.1% CAGR to outstrip $767,684.9 million by 2027 — COVID-19 impact and global analysis, The Insight Partners, https://www.globenewswire.com/en/news-release/2021/04/26/2217087/0/en/Medical-Devices-Market-Growth-Sturdy-at-6-1-CAGR-to-Outstrip-767-684-9-Million-by-2027-COVID-19-Impact-and-Global-Analysis-by-TheInsightPartners-com.html.

2. ISO 13485 (Medical devices — Quality management systems — Requirements for regulatory purposes).
3. ISO 14971 (Medical devices — Application of risk management to medical devices).
4. Medical Device Division (MDD), https://www.mdd.gov.hk/en/home/index.html.
5. Medical Device Control Office (MDCO), http://www.mdco.gov.hk/eindex.html.
6. Asian Harmonization Working Party (AHWP) http://www.ahwp.info/.
7. International Medical Device Regulators Forum (IMDRF), http://www.imdrf.org/.
8. Asia Regulatory Professional Association (ARPA) https://www.linkedin.com/groups/3009071/.
9. International Organization for Standardization, http://www.iso.ch/iso/.

Part 1

Introduction

Chapter 2

The Evolution of the Regulatory Professional: Perspectives on the Skill Sets and Capabilities That Will Define the Next Generation of Regulatory Professionals

David Martin and Neil Lesser

Deloitte Consulting LLP
davidmartin@deloitte.com, nlesser@deloitte.com

2.1 Introduction

Historically, the role of the regulatory affairs function within companies that develop and sell medical devices has been more tactical than strategic. While this tactical focus has served the industry well in the past, regulatory authorities around the world are raising the bar for market access. Regulatory reforms as well as the increased availability of real-world safety and efficacy data continue to alter the path to approval and the underlying investment case for medical devices. Globalization will also play a key role in shifting regulatory requirements — New regulatory frameworks are evolving and regional partnerships will be the main driver

---

Medical Regulatory Affairs: An International Handbook for Medical Devices and Healthcare Products (Third Edition)
Edited by Jack Wong and Raymond K. Y. Tong
Copyright © 2022 Jenny Stanford Publishing Pte. Ltd.
ISBN 978-981-4877-86-2 (Hardcover), 978-1-003-20769-6 (eBook)
www.jennystanford.com

of harmonization going forward, especially in the Asia-Pacific market area.

In response, manufacturers are seeking a new type of regulatory professional that can continuously adapt to this more complex regulatory environment. The new regulatory professional will continue to execute tactical requirements and will also own strategic relationships with regulators and thought leaders that can help shape future policies. Guiding the regulatory professional will be a fluent understanding of the organization's strategy and active participation in product development and commercialization, making them a more robust business partner to the clinical and commercial functions.

2.2   Drivers of Change

The regulatory environment in the global medical device industry is and will continue to undergo major shifts. In December 2011, the Asia-Pacific Economic Cooperation group announced a new plan to harmonize regulations for medical devices by 2020. The proposal is based on the Global Harmonization Task Force (GHTF) — recently re-formed into the **International Medical Devices Regulators' Forum** (IMDRF) — and intends to create a predicable regulatory framework for device manufacturers. This large-scale initiative will require an unprecedented level of cooperation across national health authorities. If successful, the harmonization will eliminate some of the unknowns in the regulatory environment, but getting there will challenge manufacturers that must reconcile the harmonized promise of tomorrow against the fragmented regulatory landscape of the near term. In addition, regulations have a history of lagging behind the rate of product innovation, meaning there will be a continued need for sophistication in the regulatory affairs function to navigate the evolution of the national, regional, and global regulatory environment. Regulatory professionals in the medical device and diagnostics industry must become aware of the three main influencers to the environment: rapid change, harmonization on a global scale, and increased safety and effectiveness requirements.

First, the regulatory environment is changing rapidly and professionals must be able to gauge how new developments will affect the future environment. New regulations with far-reaching ramifications have emerged in only the last few months; in January of 2012, Chinese officials issued a set of rules to deal with the conflict of interest of healthcare government officials in connection with pharmaceutical and device manufacturers,[1] and released a new Five Year Plan for the pharmaceutical and medical device industries.[2]

Second, as firms increasingly develop a global footprint, the comparative global regulatory frameworks must be taken into account. These regulatory frameworks are embarking on a period of standardization, led predominantly by the Global Harmonization Task Force, founded in 1992 with the stated goal of achieving greater uniformity between national medical device regulatory systems. This agency is playing a key role in driving standardization and continues to leverage existing international organizations (such as the Association of Southeast Asian Nations (ASEAN)) to achieve greater coordination. For example, the Malaysian Medical Device Control Division has drafted guidelines for the in vitro diagnostics sector which are based on recommendations from GHTF and ASEAN guidelines.[3] In addition, these regulatory frameworks are also currently going through a period of change. The GHTF itself has recently announced a successor organization, the International Medical Devices Regulatory Forum (IMDRF) which will continue the goal of acceleration of harmonization, but will also assemble ad hoc working groups comprised of various stakeholders to discuss key issues.[4]

Third, there has been an increased emphasis placed on safety and effectiveness in the data required for clinical trials. Several developing nations are incorporating adverse event reporting capabilities and requirements into their regulatory frameworks, including India and China.[5] Singapore is also planning to introduce regulations for medical device clinical trials in 2012.[6] This trend is of increased importance, as clinical data will become more crucial for pre-market decisions due to the greater overall emphasis placed on patient safety recently. However, this trend will need to be monitored closely across the region as China has actually begun to relax clinical trial regulations for specific medical device classes.[7]

2.3 Historical Role and Skill Set of a Regulatory Professional

Historically, the regulatory professional has served an important operational role within a life sciences organization. The role could be described as process oriented in nature: managing the submission process, communicating to stakeholders and ensuring compliance with rules and policies to secure the smooth delivery of a product submission (see Fig. 2.1 below for an example of a job description for a traditional regulatory position).

- Compile and format departmental documentation for inclusion in regulatory submissions (IDE, 510(k), International Registrations).
- Participate in the complaint management process and ensure timely completion of adverse event reporting as required by the FDA and international regulatory agencies.
- Work closely with the Sr. Director of Regulatory and related departments in order to support Regulatory submission activities.
- Coordinate and consult with other departments, including Engineering, Manufacturing, Marketing and Quality, in order to support the collection of documentation, review and assembly of regulatory submissions.
- Ensure the quality, content and format of regulatory submissions to state, federal and international agencies.
- Ensure consistency, completeness and adherence of standards for all submissions. Submissions will include IDEs, 510(k)s, federal and state establishment licensing applications, international regulatory device licensing applications, etc.

Figure 2.1 Sample job description for a traditional regulatory professional role.[8]

The primary skill sets necessary for the completion of the above tasks are excellent organizational and managerial skills as well as a thorough understanding of standards and regulations. The requirements for this position are more focused on adherence to regulations and curating the appropriate compliance documentation, with less emphasis on analyzing the larger regulatory-industry environment to understand its far reaching implications.

While these process and organizational skills will remain essential, tectonic shifts in the environment have manufacturers seeking to unlock the strategic value of the regulatory function. They are looking for their regulatory leaders to contribute outside of their functional vertical and leverage their expertise across the organization. This enhanced position requires a broader skill set and leading manufacturers have already begun hiring for this evolved regulatory role as evidenced in the job description below (Fig. 2.2).

- Independently lead global regulatory strategy development, planning, and implementation for multiple complex programs and platforms. Participate in identification of risk areas and develop alternative courses of action including anticipation of regulators responses through scenario planning and development of contingency plans.
- Guide and influence technical groups in areas of product development and lifecycle enhancement. Participate in potential and established third part efforts (i.e. due diligence activities, joint ventures, etc.).
- Initiate and maintain appropriate communication within the RA function and represent Regulatory Affairs with business units and other functions. Implement policies to ensure ongoing compliance of regulatory requirements.
- Develop and implement regulatory strategy aligned with business strategy. Assess impact of new regulations and implement appropriate changes as well as lead development of company policy and positions on draft regulation and guidance.
- Responsible for negotiating and decision making with regulators and stakeholders with complex and high-risk projects.
- Provide direct supervision of individuals including mentoring, performance management and staffing decisions.
- Represent company externally at appropriate industry associations. May act as primary contact with regulatory authorities including the planning and leadership of meetings. May participate in management of budgets.

Figure 2.2 Example job description for the changing regulatory professional role.[9]

2.4 Changing Role and Skill Set of the Regulatory Professional

In light of the aforementioned shifts in the regulatory environment, it is clear that an expansion of capabilities will be required in the future within regulatory departments. These capabilities will go beyond the traditional realm of regulatory affairs and move the function closer to "heart of the business" issues that shape the strategic direction of the organization. Regulatory professionals that are able to make this leap will become trusted partners and contributors as they will have found ways to unlock value in the regulatory process. Described below are a few of the next-generation capabilities that will empower the regulatory professional.

2.5 Develop as a Center of Intelligence

To adjust to the dynamic regulatory environment found in both the Asia-Pacific region and the broader global environment, the regulatory function should strive to operate as a center of intelligence for the organization — proactively sensing signals of change in the external environment and disseminating the insights to the organization. Such a role requires that the professional be capable of synthesizing data and trends from a variety of sources and developing a recommendation that considers complex points of view. To do this successfully requires that regulatory professionals are able to capture intelligence from a variety of sources and detect signals from extraneous noise.

Regulatory intelligence is available from myriad sources: interactions with health authorities, conversations with key opinion leaders (KOLs), in the presentations of academics, through networking at conferences and summits, and from the outcome of product submissions. The efficiency and certainty of signal detection will improve with the collective experience of the regulatory professional and the level of rigor used to assess and document regulatory trends. In summary, the regulatory professional will need to develop a process to collect data, perform trend analysis, develop summary insights, and possess the ability to communicate the findings to various stakeholders within the organization.

2.6 Advance Toward Strategic Relationship Management

As the fundamental gatekeeper to market access, health authorities/regulatory agencies play a "make or break" role in the success of a product. For this reason, it is paramount that manufacturers consider means to maximize the value of interactions with key figures from a regulatory body. One such way to maximize value is to implement a more thoughtful approach to interacting with key figures in the approval process and those external stakeholders (e.g., KOLs) that might influence it. By formulating a clear understanding of its goals and needs, a manufacturer can align its outreach efforts to effect more purposeful and productive conversations with external figures. By cultivating advanced communication and negotiation skills the organization can help shape the dialogue around reforms and regulatory findings rather than being blocked from the conversation. While not necessarily a cure-all or a quick fix, an organization with a deep understanding of the landscape of influencers and a network to match is more prepared to respond in times of crisis or more able to pick up the phone and get an answer to a burning question.

2.7 Develop as a Strategic Business Partner

Foundational to the aforementioned capabilities will be the expansion of the regulatory professional as a strategic minded individual. With primary access to regulators and a keen understanding of the levers that guide their review and lines of inquiry, the regulatory professional is in a unique position to add value to the organization. By understanding where the key interests and points of debate are for a health authority (e.g., efficacy, safety, cost) the regulatory professional can help guide the direction of a clinical program or a commercial campaign toward those areas most likely to satisfy a regulators priorities while still serving the organization's needs. Part of this role will include actively participating in the strategic business planning process to lend a perspective on feasibility and any prior precedent. It will be critical for the internal expert to communicate across the organization into both commercial and clinical functions and serve as a strategic business partner that can help decipher the "noise" to guide informed decision making for commercial and clinical investment. Delivering in this expanded role will require flexible strategic thinking, complex stakeholder management and a firm understanding of the organization's goals and plans.

2.8   Conclusion: What Will It Take to Get There?

Having laid out a few of the capabilities and skills that the regulatory professional of the future will need, what follows is a discussion of how one can attain the skills and experiences to best position themselves for future success.

Primarily, one must gain the foundational strategic, analytical and communication skills that will be required. This is best accomplished through coursework in Strategy, Operations and Communications. Further coursework in the fields of Negotiation and Marketing would provide the necessary "big picture" view required of the new regulatory role, a skill set in synthesizing information and an understanding of how to manage multiple shareholders and their diverse incentives.

To complement this coursework, one should pursue opportunities to network with a wide array of students and professionals from across life sciences, health care, and the government in order to enrich one's understanding of the needs and motivations of various stakeholders in the system. Memberships in professional societies and industry groups and attending summits and conferences should be at the forefront of any career development strategy, but in this case these associations are particularly important in order to gain a viewpoint on the changing environment of regulation and develop a network of colleagues throughout the industry.

Further, a rotational role within a life sciences organization would be especially valuable. Such a role would provide exposure to a broad spectrum of functions and stakeholders and their diverse incentives and viewpoints. This position represents an ideal experience for a professional in this industry seeking to become fluent in the internal workings of an organization and competent in a strategic partner role.

Ultimately, success in this field requires that professionals recognize that the regulatory role is evolving and that this evolution must be matched by a new, broader focus in their education and career development. Fortunately, the resources exist to prepare for this evolved position and to build a career in this challenging, growing field — One might even say there has never been a more exciting time to be on the regulatory side of the medical device and diagnostics industry.

Acknowledgment

Contributions to this chapter have been made by Chris Nuesch, Bill Chiodetti, and Riddhi Roy, all from Deloitte Consulting LLP.

References

1. http://www.sidley.com/newsresources/newsandpress/Detail.aspx?news=5061.
2. http://www.pacificbridgemedical.com/newsletter/article.php?id=567.
3. http://www.rajpharma.com/productsector/medicaldevices/Malaysia-consults-on-four-draft-guidelines-for-IVD-sector-324052?autnID=/contentstore/rajpharma/codex/4072005d-15cc-11e1-bbe6-4d8e6a53eb99.xml.
4. http://www.emergogroup.com/blog/2011/11/ghtf-unveils-successor-organization.
5. "China Drug and Device Regulatory Update 2009", Pacific Bridge Medical Group, May 2009.
6. http://www.rajpharma.com/productsector/medicaldevices/Singapore-plans to introduce-new-framework-for-medtech-trials-in-2012-323092?autnID=/contentstore/rajpharma/codex/b5848548-046e-11e1-bbe6-4d8e6a53eb99.xml.
7. http://www.sidley.com/sidleyupdates/Detail.aspx?news=5019.
8. http://regulatorycareers.raps.org/jobs/4660783/regulatory-affairs-specialist.
9. http://regulatorycareers.raps.org/jobs/4699968/sr-manager-regulatory-affairs-cmc-biologics.

Chapter 3

The Role of the Asia Regulatory Affairs Team in Relation to the Commercial Team and Other Departments

Fredrik Dalborg

Terumo BCT Asia Pacific, Singapore
f.dalborg@gmail.com

3.1 Introduction

The role of the Regulatory Affairs (RA) department is evolving from that of a subject matter expert to becoming a business partner working closely together with the commercial team and other departments in order to ensure the successful implementation of the company's strategies.

With this collaborative approach, the RA team will work closely with the commercial team during all the stages of product life from product development to market introduction and eventual phase-out at the end of the product life.

Regulatory matters can sometimes be seen as hurdles and challenges delaying the launch of new products. When the regulatory team takes on the business partner role, then regulatory matters can evolve from being seen as a hurdle and instead, managed in the right way, they can become an opportunity

---

Medical Regulatory Affairs: An International Handbook for Medical Devices and Healthcare Products (Third Edition)
Edited by Jack Wong and Raymond K. Y. Tong
Copyright © 2022 Jenny Stanford Publishing Pte. Ltd.
ISBN 978-981-4877-86-2 (Hardcover), 978-1-003-20769-6 (eBook)
www.jennystanford.com

and a way to differentiate from competitors. With this, a strong regulatory team working in close collaboration with other departments using efficient processes can be a strategic asset and a key competitive advantage for the entire company.

3.2 Key Trends in the Asia Medical Device Industry

For many medical device products, the North American and Western European markets are slow growing, while the competition is fierce, customers are increasingly cost conscious when healthcare spending is under scrutiny, and markets are saturated. In this context, Asia represents an increasingly important growth opportunity for the global medical device companies.

Asian markets present many opportunities for medical device companies, including underlying market growth, improving national healthcare system coverage, increased ability to spend money on healthcare, increased healthcare sophistication, and an aging population.

Opportunities	Challenges
Strong underlying market growth	Limited healthcare spending per capita compared to the western world
Government ambition to improve national healthcare system coverage, gradually increasing spending	Diverse and complex markets and customer segments, requiring different approaches to serve the needs efficiently
Growing middle class with ability to pay for more advanced medical care	Underdeveloped medical infrastructure and workforce, limiting adoption of new technology
Increasingly sophisticated healthcare system capabilities, and the ambition to evolve the standard of care	Diverse reimbursement systems
Aging population, requiring more healthcare	Fragmented and inconsistent regulatory systems

At the same time, the Asian markets pose challenges, including limited healthcare spending, diverse and complex markets, underdeveloped medical infrastructure, diverse reimbursement systems, and fragmented and inconsistent regulatory systems.

In order to improve patient care and healthcare access in Asia and to realize the growth potential, these opportunities must be seized and the challenges be addressed. From a broad industry perspective, this will require improved collaboration between all stakeholders in the field, including regulators, health ministries, academics, healthcare providers, business executives, and patient groups. In this setting, the RA team has a key role to play in the overall effort to improve patient care and healthcare access in Asia.

Within the medical device companies, seizing these opportunities and overcoming the challenges will require detailed understanding of local and segment-specific conditions and strong collaboration between all functional teams, in particular commercial and RA teams.

3.3 The Role of the Asia RA Team in a Global Medical Device Organization

The medical devices industry is increasingly becoming a global business, with local companies adding global growth ambitions and multinational companies looking to ensure they balance established positions in mature markets with growth opportunities in emerging markets. To a very large extent, global companies are looking to Asia to find growth opportunities. As Asia becomes more important for global companies, the need for a better understanding of the Asian markets grows. The Asian regulatory environment is complex and less harmonized compared with other continents such as Europe. Due to this increased company focus on Asia, the importance of Asia RA matters increases, and together with that the need to effectively communicate the complexities and requirements of the Asian regulatory environment grows.

It is a key task for the Asia RA team to increase the awareness and understanding of the Asian regulatory environment, not only within the central corporate RA function but also within other key corporate functions such as commercial, legal, medical affairs, logistics, and R&D.

By facilitating this increased understanding of Asian RA matters, the Asian RA team can improve the abilities of the organization to succeed in its Asian market strategies.

3.4 Coordination between Commercial Teams and RA Teams

In larger organizations that manage broad product ranges and many geographical markets, the number of product registration activities to manage quickly becomes significant. The tasks involve not only managing new registrations but also maintaining current registrations, managing effects of changes to the products or manufacturing location, while at the same time keeping up with evolving medical device regulation. Obviously, the amount of ongoing activities quickly becomes significant, and the coordination becomes complicated and demanding. It is the task of the RA team to manage all these registration projects for various products in various geographical markets and to keep other departments updated on the development. Important aspects in this context include the following:

- Create formats for clear and simple status summaries, including key information such as expected dates for submitting the required documentation, expected time for regulatory approval, and the first possible date for the sale of the product. The status summaries should also include any identified risks with respect to the targeted timelines.
- Work with commercial teams to develop agreed methods of prioritization. The complexity and amount of ongoing activities will quickly force the RA team to prioritize. The framework for prioritization should be developed in close collaboration with the commercial teams and should include factors such as patient needs, size of the business potential associated with the registration, strategic and competitive considerations, the cost, complexity, and time required for the regulatory process.

A key aspect of this coordination and communication task is to adapt the message to the audience for the communication. In most cases, this will require the RA team to simplify the communication, leaving out some less important details in order to focus on the key information and to keep the information clear and consistent.

The commercial teams need to make sure to include the RA team early in their commercial planning process to ensure good coordination. This will allow for realistic and fact-based registration timelines and considerations to be included in the commercial plans. In order to support the joint prioritization efforts, the commercial teams need to share the information on the business potential and strategic considerations with the RA team.

3.5 The Role of the RA Team during the Different Stages of a Product Life

The role of the RA team is not limited to the stage in the product life when a new product is registered and introduced in the market. The RA team can and should play an important role during product development, market introduction, product maintenance, and the eventual phase-out of a product as well.

3.5.1 Product Development

During the product development phase, the RA team can play an important role by providing inputs and insight regarding how different designs and features, claims and supporting evidence, choice of manufacturing location, etc., can influence classification, requirements for clinical trials, and the resulting registration timelines and costs in different countries.

3.5.2 Market Introduction

As the product is being prepared for launch, the detailed planning of the registration work starts. At this stage, detailed assessments of registration timelines are required, including resource requirements for testing and trials, documentation, etc.

These inputs will be key for the commercial organization in the planning of launch events, advertising, potential requirements for expanding the field organization, training of the teams, etc.

A product launch often involves a whole family of products including different versions, accessories, disposable kits, etc. These different products can also have different classification. Coordination of the registration of the entire family of products is very important and can be complex.

The market introduction is a critical phase, and delays or changes can have significant revenue and profitability effects as well as manufacturing planning and competitive positioning effects; so the dialogue needs to be frequent and well coordinated between all involved departments, including marketing and sales, clinical and technical service, medical affairs, manufacturing, and supply chain.

3.5.3 Product Maintenance

When a product has been successfully introduced, the intensity of the RA workload related to that product is reduced, but not eliminated. At this stage, tasks can include

- renewal of registrations and licenses;
- new registrations or re-registrations required by changes in the regulatory environment;
- managing registration changes triggered by new product versions, new accessories, new claims, changes to components or manufacturing sites, etc.

This stage also requires close and proactive communication with other departments, so that the regulatory impact of changes of any kind can be fully understood upfront, and negative impacts can be minimized. For example, a change in product specification or manufacturing site may trigger significant regulatory impact, and these effects need to be fully understood before the decision about a change is made.

3.5.4 Product Phase-Out

When a product approaches the end of its life span, a new set of questions will arise:

- How long will it take to register and introduce a new replacement product in the market? This will be a key factor in what is often a multi-year process of a product generation change that includes notifying customers of the change well in advance (often several years), developing a replacement product, scientific studies, documentation, training field teams and customers on the new product, etc.
- Can registration timelines be reduced by minimizing certain changes compared with previous product generations?
- Will registration timelines be influenced by different product name and version number approaches?
- How will the different regulatory situations in the countries in question impact the sequence of the product generation change by country? A company would seldom phase out a product in all countries at once; a phased country-by-country approach is often used.

At this stage, a close collaboration with commercial teams, product development teams, and manufacturing teams is crucial.

3.6 The RA Professional: A Trusted Advisor

A key task for the RA team is to ensure compliance with all relevant regulations and laws. This is an increasingly important and complex task in the current environment:

- In general, the laws and regulations are becoming more and more stringent in all countries.
- The enforcement of regulation is becoming stricter.
- New regulation is introduced that covers multiple jurisdictions. Examples include the US Foreign Corrupt Practices Act and the UK Bribery Act.

Another key task is to ensure that all interaction with regulatory officials in all countries is handled in a professional and appropriate manner.

In this context, the role of the regulatory team also becomes that of a trusted and competent advisor similar to the role of the legal department. This includes to be very knowledgeable about the rules and regulations that apply in each situation and to provide advice to the decision makers in the organization. The first and foremost task is to keep the company, employees, and customers in compliance with rules and regulations and out of trouble. When that has been secured, the task becomes finding practical and creative ways within that framework to bring the products to market as efficiently as possible to the benefit of patients, caregivers, and the company.

3.7 Summary

The importance of the Asian markets is increasing for all global medical device companies. While the opportunities are significant, the challenges are many and complex. In this context, the Asia RA team plays an increasingly important role in relation to the commercial team and other departments, and through all the different stages of the product life. In this integrated role as a business partner and trusted advisor, the regulatory team becomes a strategic asset for the company, critical to the success of the company strategies.

Chapter 4

Commercial Sense and What It Means for a Regulatory Manager or Executive

Annie Joseph

Vital Signs Pte Ltd, 10 Anson Road #12-14 International Plaza, Singapore
079903annie@vitalsignsbc.com

Commercial in its simplest sense means "being in the business of making a profit." Hence, sales and marketing are often coined as commercial, i.e., the "arm" of a particular company that markets the product and brings in the revenue.

Adam Smith [1], the father of modern economics, noted that the essence of marketing is the *consumer*, i.e., the identification and satisfaction of a consumer's requirements form the basis of modern marketing. Marketing combines both the philosophy of business and its practice. As a result, the commercial function forms the interface with a company's existing and potential customers [2].

For commercial folks, regulatory managers tend to be associated with getting products registered with the local health authority. In many companies, you see a typical to-and-fro blame game between commercial and regulatory when new product

---

Medical Regulatory Affairs: An International Handbook for Medical Devices and Healthcare Products (Third Edition)
Edited by Jack Wong and Raymond K. Y. Tong
Copyright © 2022 Jenny Stanford Publishing Pte. Ltd.
ISBN 978-981-4877-86-2 (Hardcover), 978-1-003-20769-6 (eBook)
www.jennystanford.com

registrations are delayed, rejected or falls behind the competition. It is fair to say that there is always a long list of tasks to do as a regulatory manager, hence knowing what the commercial priorities are will help you become a successful regulatory partner. Here are some tips and suggestions to make you a commercial savvy regulatory executive.

4.1 Know the Basics

In any company you work for, there will be a process known as the commercial, marketing, or budget planning process. You will need to find out what this process is and where you should play a role. This is important so that you truly understand the issues facing the company.

4.1.1 Marketing or Commercial Plan

Marketers spend a considerable amount of time in developing a strategic marketing plan, which essentially maps out the strategy for the product or portfolio over a period of time (usually 3–5 years). While it is not necessary to go into the details of marketing plan for products that you are responsible for, it is imperative to comprehend the strategy of the product/the portfolio. Anyone would be impressed if a regulatory executive can clearly articulate the strategy that is spelt out in the marketing plan. The strategy is typically described in the following manner [3]:

- (a) the target audience/market
- (b) the core positioning
- (c) price strategy and positioning
- (d) value proposition
- (e) distribution strategy
- (f) communication strategy

Note: Refer to Appendix 4.A for an explanation of the terms above and Appendix 4.B for what a typical marketing plan is.

In Exhibit 4.1, you will find a useful checklist of questions you can ask in your meeting with a marketing or commercial colleague.

4.1.2 Priority Products and How They Are Ranked in the Company

While as a regulatory executive, sales is not part of your remit, it is pertinent to know sales by products as it is important to know the priority products in the organization. Products are typically prioritized according to how much profits they bring; however, many companies may have other criteria, for example, strategic products to the organization. This should be a question you ask your commercial colleagues.

4.1.3 Know Your Commercial People

As you understand the commercial priorities of the company, ensure you meet and know the different product or marketing manager that is responsible for the product/portfolio. Where possible, get involved in some of their marketing discussions to familiarize yourself with the issues or opportunities in the business.

4.2 Maintain a Healthy Communication

You should strive to build a partnership with commercial to an extent where they reach out to you proactively for solutions. The company benefits significantly when commercial and regulatory functions have a tight working relationship. To set this up for success, you should ensure that you do or develop the following:

(a) List of goals for the year: This should be agreed with your manager and aligned to the commercial priorities.
(b) Make your list known to commercial: Reinforce your goals in your communication with commercial so that they also have clarity on what is on your priority list.
(c) Routine discussion: Set up a recurring meeting with a clear set of meeting objectives.

4.3 Be Proactive and Part of the Solution

The regulatory function plays an important role in helping the company achieve its business objectives in a compliant manner.

However, they are various pressures within the commercial organization and interpretations of compliance. Within the different levels of sales and marketing, the level of understanding of what this means varies considerably. It is critical and important for regulatory executives to help the commercial team identify the areas of risk along with mitigation steps for that risk.

This does not mean that the regulatory team should remain in the archaic box known as the function that always says "No" or "Not allowed as it is off label." Instead, a commercial-wise regulatory executive will adopt the following stance:

(a) A solution-orientated approach

It is fair to assume that there will be many obstacles that will hinder what you would like to deliver to the commercial team. Proactively think of potential issues and manage difficult situations with strategic, flexible options that allow you to move forward. Provide realistic costs as well and it is likely that the least expensive, low risk option will be pursued.

(b) Essence of time

The criticality of time is often underestimated as different functions get lost under the pile of their individual "to-do" list. The focus on "time to market" is one of the pivotal concepts in commercial organizations and a proactive regulatory executive who understands this will be lauded as a strategic partner.

(c) Concept of commercially viable

The person in charge of strategic marketing typically is adamant about sticking to his or her demands when it comes to indications, label, or claims to maintain a competitive advantage. Hence, what is commercially acceptable or viable becomes paramount in finding alternative approaches in navigating through a regulatory roadblock. You can begin to find alternatives if you understand and appreciate the marketing plan or strategy (refer to Know the basics), the people and spend time with them in discussions. This concept is the provision of an alternative approach, usually different to the one defined by marketing, yet achieves the same end result of the strategy.

4.4 Remain the Expert and Provide Clarity to the Organization

Unlike regulatory executives who have to pour through detailed documents and dossiers, marketers have little patience in scouring through pages of documentation or legislation. If this is understood by the regulatory executive, that he or she can be a great partner and expert to ***simplify and interpret***.

In an increasingly complex regulatory environment, there is a need to think of new strategies. The commercial team depends on the acumen of the regulatory function to carry this responsibility effectively. Here are some of the expectations of a commercial head:

(i) established networks with regulatory and relevant government bodies

(ii) ability to effectively negotiate based on sound understanding of the local and regional (as relevant) legislation

(iii) clarify ambiguities in a legislation based on best available information and knowhow through networks/advocates

(iv) provide clear, concise information to senior commercial leaders in the organization confidently

4.5 Be a Champion for New Product Launches

New product development and launches are an important barometer for any organization, closely watched by the investment community. Underpinning this is the role of regulatory as a torchbearer for new products.

For many companies, new launches are the "life blood" of the organization and various efforts, teams, and processes will be placed to get it RIGHT, FIRST TIME. This is because, like all life forms, products too have finite lives (refer to Fig. 4.1). You will often hear the theme, You Get One Shot at Goal when it comes to launching successfully.

32 | Commercial Sense

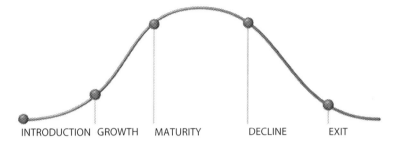

Figure 4.1 Product life cycle.

You will likely find a launch planning team in your organization. The following guide will help you understand the basic process of product launches, but more important, the active role you should play in driving a successful launch.

New Product Launches Guide for Regulatory Managers

1. Understanding the basic process	Typical launch planning process consists of the following: (a) Launch preparation (b) Launch readiness (c) Launch optimization
Launch preparation	Kicks off once a decision is made by the organization to conduct a Phase III or clinical study for registration purpose. This is a critical stage of the launch planning process and often very poorly managed. Medical and market access leads drive this stage; however, it is imperative that the regulatory manager clearly stipulates the evidence generation needed to obtain the desired label or indication.
Launch readiness	The stage between a decision to file in the first country and the actual launch. This can take anywhere from 1 to 3 years depending on the complexity of the registration, data/dossiers needs, and preparatory work needed by market access, commercial or market research.

Launch optimization	Once a product is launched, the commercial team will work to optimize the sales post launch. Early feedback, including how patients respond to the new product, is vital at this stage of the launch process.
2. Know where you will play a role	• Launch preparation is a critical stage for regulatory managers to have a very active role as they are the owners of the label and indication. • Know and clarify which processes you own and will lead. • Establish your goals and timelines with the launch planning team and your manager. • Keep a close working relationship with the launch planning team leader so that you can discuss emerging issues.
3. Develop a system for tracking and management	• Establish clear milestones and update the relevant launch planning team • Develop (and have it ready) a simple presentation in the event you are asked to provide an update at short notice
4. Develop an agreed KPI with the right stakeholders	• Your key stakeholders at the launch preparation stage are R&D and medical and market access. Establish a good working relationship with this group to ensure that as the team enters the launch readiness stage, you have the relevant resources and evidence to deliver the best for the organization. • KPIs or key performance indicators are important to keep all stakeholders accountable. You may also want to suggest the team use a RACI system.

A strong alliance between regulatory and commercial will lay the foundation for commercial success. A proactive, bold regulatory manager who seeks to understand the commercial needs and is able to provide viable solutions will go a long way in any organization.

Exhibit 4.1

Checklist

Product related

What are the products that are being marketed or about to launch?	☐
What are the sales by product/therapy/portfolio?	☐
List of priority products/portfolio	☐
What products are currently actively promoted by the sales force?	☐
Are there plans to discontinue any product line?	☐
Are there any quality related issues related to our products?	☐

Promotion related

For the existing product portfolio, what new indications are being planned?	☐
For the priority products, how are our indications/label different from that of our competitors?	☐
What are some of the adverse events or concerns related to efficacy/safety that we get from customers?	☐
How are these managed currently?	☐
Is the standard operating procedure adhered to?	☐
How are the sales force trained on managing these issues?	☐

Appendix 4.A

Target market	A well-defined set of customers you have identified to focus on
Core positioning	Refers to the position of the brand in the mind of the target customer
Price strategy	Closely linked to positioning; marketers may choose a price premium strategy, where the product is priced higher than the nearest competitor or a price penetration strategy where the product is offered at a discount

Value proposition	A clear statement that is persuasive to the question of "Why should I buy from you?"
Distribution strategy	Describes the business' distribution strategy for reaching its target market
Communication strategy	A marketing or product manager's decision on how the funds should be allocated to promotion, sales force, direct marketing, etc.

Appendix 4.B

At a minimum, every marketing plan should contain the following sections [3]:

Situation analysis	A SWOT (strengths, weaknesses, opportunities, threats) analysis, a description of the current situation and issues facing the business
Marketing objectives and Goals	What are the objectives/ measurable goals of the product in the planning period?
Marketing strategy	Defined in Appendix 4.A
Action plan	Translating the goals and strategies into concrete action plans with timelines
Measurement/control	Typically, key performance indicators against the defined action plan to meet the goals

References

1. Smith A, The Wealth of Nations, Random House, 1937.
2. Lancaster G, Massingham L, Essentials of Marketing, 1994.
3. Kotler P, Kotler on Marketing, 1999.

Chapter 5

Market Strategic Challenges for Medical Device (Asia/Pacific)

Sherwin Tan, Ki Eunyu, Kyser Tay, and David Lee

Jack.wong@arpaedu.com

This chapter begins by describing the strategic changes in recent years and moves on to introduce a strategic framework for businesses. The key issues in the market will be raised and challenges that suppliers are facing will be heightened. Lastly, the chapter will end by describing 6 growth challenges and 5 common strategic mistakes. The contents of the chapter are as follows:

1. Strategic changes
2. Strategic framework
3. Key issues in the market
4. Challenges that suppliers are facing
5. Six growth challenges
6. Five common strategic mistakes

Medical Regulatory Affairs: An International Handbook for Medical Devices and Healthcare Products (Third Edition)
Edited by Jack Wong and Raymond K. Y. Tong
Copyright © 2022 Jenny Stanford Publishing Pte. Ltd.
ISBN 978-981-4877-86-2 (Hardcover), 978-1-003-20769-6 (eBook)
www.jennystanford.com

5.1 Strategic Changes

"Given one hour to save the world, I would spend 55 minutes defining the problem and 5 minutes finding the solution."

—Albert Einstein

A huge aspect in managing the strategic challenges of this age is the ability to correctly identify the problem. In this section, the top 6 most common key strategic changes will be highlighted.

5.1.1 Product Design and Innovation

As times are changing, the needs and demands of the patients are also evolving. It is inevitable that in such times of change, there needs to be new alternatives, new ideas as existing solutions become obsolete. Businesses must come up with innovative ideas to tackle the new patient needs. This is the only way they will be able to adapt and stay relevant in the market.

When coming up with new innovations and product designs, it is important that businesses keep 3 things in mind, which is to consider the balance between desirability, economic viability and technical feasibility. Most importantly, they have to remember that design is human-centred, where it starts with what humans need. They must keep in mind that the ultimate goal of innovation in medical devices is to find the most effective way to meet the patient's needs[1]. With the focus and a good understanding of the patient's needs, businesses will be able to see the areas for innovation to allow them to remain competitive in the fast changing times.

5.1.2 Technological Advancement

Technology in the medical field is rapidly evolving. Almost every day, a new technology is being announced. It is crucial for businesses to keep abreast with such revolutionary technological advancements in the market to remain competitive. Some examples of technology that is rapidly evolving include Telemedicine and Virtual Reality (VR). Both technologies have big potential for growth, with global telemedicine market to be worth $113.1 billion by 2025[2] and the global healthcare market size for VR to grow to $30.4 billion by 2026[3]. These examples are just 2 out of the countless markets that are growing. To improve the quality of lives of people through better healthcare, technology will continue to evolve, and in return the market sizes will continue to grow. For businesses to make full use of this growing market, it is vital that they stay updated with the recent technological advancements.

5.1.3 Distribution Channels

A few decades ago, local distributors and dealers were commonly used by multinational medical device companies to penetrate foreign markets or to perform the "heavy lifting" on their behalf. However, in an increasingly globalized world with well-established trade relations between countries, multinational companies (MNCs) are experiencing a decrease in barrier to entry to foreign markets. In countries where business is expected to be profitable, MNCs of medical devices or drugs generally prefer to have their own "in-country" representation in those foreign countries instead of using local distributors or dealers in order to capture a larger share of the profit margin. Alternatively, it is a growing trend for MNCs to buy over their distributors that are successful to maximize their profits. This change in distribution channel is worth taking note of for distributors and dealers in the medical device or drug industry.

Another important point worth mentioning would be the increasing use of third parties by MNCs to become their license holder or authorised representative for their products. Such a move is strategic and rational to MNCs to cut off noncompliant or unsuccessful distributors or dealers. In the past where distributors or dealers themselves held the license, it became more complicated when MNCs decided to switch their distributorships and/or terminate their services. Hence, we are witnessing a situation whereby distributors/dealers are losing their bargaining power. In an increasingly competitive market, distributors/dealers will have to demonstrate their capability in order to remain viable.

5.1.4 Geo-Political Changes

Since China's economic reform and trade liberalization about 40 years ago, it has now become the second biggest economy in the world after the United States. America has been a traditional economic superpower and lauded by many as a land of opportunity. However, such a trend has seen significant changes in recent years with markets such as China and India emerging strongly. In the last 2 decades, multinational medical device companies have made large investments in emerging markets such as China and India as the economies of these countries grew and developed, the medical device companies benefited significantly as well[4]. With China and India economies developing into more mature ones, medical device companies are now exploring other potential Asian countries where the healthcare market is relatively untapped. In an increasingly interconnected and globalized world, it is imperative for businesses to keep abreast of such geo-political changes to remain viable and better anticipate the future.

5.1.5 Environmental and Health Pandemic

No matter how prepared a business is towards strategic challenges in the market, there are bound to be unforeseen circumstances. A great example would be the change in policies that comes with unprecedented environmental change and health pandemic. These factors will drastically affect the performance of businesses and it is crucial that they adapt their business models to ensure that they can remain competitive even in such unfavourable times.

In such an interconnected world. It is important for businesses to align their values to progress towards sustainable development of the world. A way to do so will be by tackling the Sustainable Development Goals set by the United Nations, which deals with issues such as environmental degradation, climate change, inequality and poverty[5]. Meeting these goals will not only create a sustainable world for all businesses to succeed, but they also outline new markets and opportunities for companies[6].

In a health pandemic, businesses will have to re-imagine and re-align their business models to adapt to the "new normal". It is inevitable that consumers will turn to online services and reduce face-to-face interactions as it is contactless and safe. Businesses hence must think of ways to integrate such online services, such as e-Commerce, digital marketing and other Social-Media platform, in their model[7]. The speed of setting up such services is also crucial for rapid revenue recovery.

5.1.6 Marketing Strategy and Tactics

With advancements in technology, the marketing landscape has evolved and experienced key changes. To remain on top of the changes and remain competitive, it is imperative to be aware of these 2 technologies in contemporary marketing: automation, robotics and artificial intelligence (AI)[8]. Studies have shown that automation in marketing has been on the rise, with almost half of the marketing efforts being automated. It was also discovered that for companies that outperformed their counterparts, 63% were adopting automation in their marketing strategies. Another evolving area in marketing is the increasing use of AI to automate customer interactions instead of the traditional methods where humans themselves manage these interactions.

With the rise of digital and social media platforms as well as an increasingly digital-literate consumer base, it is unsurprising that digital marketing will be given much attention. However, a study has shown that traditional marketing such as traditional print and television marketing is not completely obsolete and in fact still makes up 52% of the marketing budgets[9]. To maximize marketing avenues, businesses will need to strike a balance between digital channels and traditional marketing.

5.2 Strategic Framework

With the strategic challenges in the market identified, it is crucial for businesses to come up with a framework to overcome such challenges in the most effective and efficient way. In this section,

6 steps of a strategic framework to handle the challenges will be discussed.

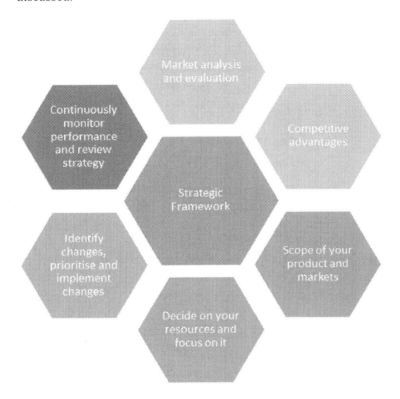

5.2.1 Market Analysis and Evaluation

Effective market analysis and evaluation are done in a comprehensive and critical manner. Businesses must continuously ask questions about the market, regarding the potential customers, the existing customers' expenditure habits, the target market, the competitor's strengths and weaknesses[10]. Every question must be broken down in order to fully understand the market.

Effective market analysis and evaluation is the first step to a successful business. Market analysis provides the businesses with valuable insights into shifts in the economy, their competitors, current trends, demographics and characteristics of the customers' spending[11]. It is only with this information that businesses are able to make informed and wise decisions for their plan ahead.

5.2.2 Competitive Advantage(s)

For the success of the business, they must be able to reflect on themselves and think whether their products and services have a competitive advantage. To do so, they must have a good analysis of themselves as well as their competitors. Upon identification of potential competitors, a thorough analysis must be done. Analysing the competitor includes identifying their market positioning, their marketing strategies, their reviews on various platforms, their way of handling customers and their design of their products[12]. Businesses must see which areas the competitors are excelling, their reason for success and come up with an idea to overtake them in this area. To do so, they must first have a deep understanding of their products and services.

Competitive analysis allows the businesses to get inspiration and insight[13]. It is through competitive analysis that businesses are able to generate ideas that distinguishes them from their competitors. Competitive analysis should not be underestimated as it plays a big role in helping the business be unique to stand out in the market.

5.2.3 Scope of Your Products and Markets

Product and market scope refers to the products and markets which the business will concentrate on[14]. In this process, businesses decide what kind of product they will offer and which markets the business will target these products. When defining the scope, businesses must analyse their current resources and set a realistic scope that is economically viable. For small businesses, setting a narrower and more focused scope will be necessary with the limit on their resources. For large businesses, with enough resources, they will be able to handle broad product lines and wider marketing campaigns. Setting a wrong scope will only dilute the business' core competency. The most important factor that businesses must remember is that the features of the product and the needs of the market must be aligned for the success of the business. Only the products that meet the market's needs will be able to stay appealing and competitive in the market.

5.2.4 Decide on Your Resources and Focus on Them

For the optimum use of business' resources, resource allocation is essential. Resource allocation refers to the process of planning, managing and assigning resources in order to meet the business' strategic goals[15]. For effective resource allocation, the project must first be broken down into tasks. For each task, it is important to know what resources are essential for its success and allocate them accordingly. Types of resources can include labour, equipment, materials, facilities and others. After the initial allocation, resource levelling, the process of inspecting the resources to allow the tasks to flow smoothly, and re allocation of resources may be needed[16].

Resource allocation is important as not only does it save the waste of resources, it also improves the business' productivity, time management and staff morale[17]. Successful allocation of resources minimises the risk that the business has to face, allowing them to stay competitive despite the ever-changing markets.

5.2.5 Identify Changes, Prioritize and Implement Changes

Demands and expectations of consumers undoubtedly change over time. Markets are always constantly shifting and the businesses that are able to identify the need for a change and adapt to the volatile markets will be able to stand out from their competitors. The need for a change can be identified by several factors: unsatisfactory performance, regular unpleasant surprises, competitors doing better and rise of new opportunities[18]. For each of these factors, the business must analyse what are the essential factors that are hindering the business and prioritise on implementing changes in these areas.

Implementing change in the strategy is a form of preparing the business for the future. With the ability to implement change, the business' internal capabilities are built up and they will be able to stay competent among the rest of the businesses, regardless of any unprecedented changes the market is bound to have.

5.2.6 Continuously Monitor Performance and Review Strategy

Businesses should not have a fixed mindset towards their business strategy and must have a mindset to change when necessary. The only way to do so is to stay proactive, continuously monitor their performance and review their strategy. Monitoring the performance includes analysing the current strengths and weaknesses, the results that are produced from certain strategies and the potential for growth and development[19].

The analysis not only becomes a method of the businesses to get an idea of how they are performing, but it also becomes an opportunity for them to check whether their current strategies align to their overall vision. Setting long term goals, rather than focusing too much on short term goals, is important in this process so that the strategies remain aligned to the overall vision. Staying committed to the overall vision will allow the business to have a strong sense of identity and drive towards their success.

5.3 Key Issues in the Market

For a business to continue thriving, familiarity with the market is essential and important. This section will expound upon several key issues in the market that businesses should be aware of.

5.3.1 Internal Organizational Pressures

Organizations are continuously subjected to external or environmental forces that pressure them to make changes internally in order to remain viable. One salient example of internal organizational pressures resulting from an external factor is the COVID-19 pandemic. The pandemic forced many businesses to deal with the sudden transformation of shifting their businesses from the traditional methods to online modes. Another recent example would be the transformation of the conventional hardcopy documents to electronic records in many industries. These organizational changes require expertise, time and money to execute and according to a research from McKinsey & Company,

70% of organizational transformations do not succeed and they often cost more and require more effort than initially expected[20].

Some ways to cope with organizational pressures include remaining positive and rising to the challenge by viewing the situation as an opportunity to challenge oneself instead of an insurmountable obstacle[21].

5.3.2 Increasing International Competition

Companies are operating in an era of increased international competition and these could be due to a myriad of factors such as reduction of trade barriers, enhancement in communications, advancement in information technology, etc. In the face of increased competition, it becomes paramount for businesses to identify their competitive advantage. Competitive advantage can be simply defined as the points that make one better than one's competitors in the minds of customers[22]. The competitive advantage could be in the form of a unique feature or better quality product or services. Importantly, the top management of the company must be able to effectively position the company and carve out a niche for their products or services that appeal to the needs of consumers.

5.3.3 Fast-Paced Innovations

Many big organizations continue adopting an incremental innovation framework that is to search for the next version of a product or the evolution of an area that is within their comfort zone[23]. While incremental innovation remains important and viable, there is also a need for another type of innovation—fast-paced exploratory innovations. Such innovations are opposite of the usual incremental innovation and will enable organizations to better leverage on the opportunities presented by the advancements and developments in the world.

Notable examples of innovations that are fast-paced and exploratory in nature are how some manufacturing sites rapidly transformed themselves to produce Personal Protective Equipments (PPEs) during the COVID-19 pandemic where these supplies faced a global shortage. This allows businesses to obtain immediate revenues and to stay relevant.

5.3.4 Organizational Restructuring and Merger and Acquisition

Mergers and acquisitions (M&A) are fairly common occurrences. It is therefore an important issue in the market that business leaders should have knowledge about. Merger refers to the combination of companies whereas acquisition refers to a company purchasing a majority stake in another chosen company[24]. Such activities are typically with the goal of enlarging market share, obtaining competitive advantages, achieving synergies, enhancing growth, or influencing supply chains.

In the medical device industry, there has been a lot of consolidation in recent years and such a trend is expected to continue[25]. M&A activities in the medical device industry allows companies to strengthen and diversify their product range as well as reduce competition. An example that highlights this would be Abbott's acquisition of St. Jude's medical which resulted in Abbott occupying close to 20% of the global cardiovascular market[26]. Especially in the medical device industry where the regulatory processes are known to be long and costly, M&A is one of the ways medical device companies can stay profitable.

5.3.5 Increased Quality Consciousness

In a fast-paced world where time is highly valued, it is not uncommon for the quality of products to be compromised at the expense of speed. However, with the easy access to information enabled by information technology and the availability of alternatives, consumers are now more sophisticated and increasingly quality-oriented. As such, certifications or standards start to take on a more important role.

The International Organization for Standardization (ISO) develops and publishes international standards[27] and companies who adopt standards by ISO demonstrates their commitment to the highest-quality standards available. In fact, manufacturers in many industries such as medical are required to have certain certifications. With a more informed customer base, manufacturers without certifications will inevitably lose out to those who have compiled with quality guidelines or parameters. Though there are many other issues in the market, this section highlighted some of the key issues that companies should take note of. Awareness of these market issues will enable companies to be better prepared to tackle these challenges.

5.4 Challenges Suppliers Are Facing

As the demand for suppliers increases with the ageing population and healthcare demands, it is crucial that they identify the challenges in order to overcome them and succeed as a supplier. In this chapter, such challenges will be identified and discussed.

5.4.1 Accessing Data across Platforms

Collecting data about the customers should be the utmost priority when it comes to understanding the customers. It is through data collected across various platforms that allows the businesses to fully understand the user needs for targeted marketing, which increases the chances of their success.

Although data collection may appear tough in today's age where customers are so concerned with privacy, it was found that more than 60% of customers were willing to share their information when it came to personalisation or discounts[28]. Therefore, businesses must be able to come up with necessary tactics in this process of data collection so that they will be placed at a competitive advantage with the data that they have.

5.4.2 Identifying Customers across Channels

In today's world, there are endless number of ways the suppliers are able to reach out to the customer, such as through Social Media or E-Commerce platforms. With the increased number of channels, the supplier must be able to identify the target audience of the channels so that they will be able to project the product to the most appropriate audience. An accurate and early recognition of the customer base is the foundation for the success as a supplier. With the identified target audience, the supplier will be able to set a clear focus of whom the business will serve and why the customers need the products[29] to market their product in the most appealing way.

5.4.3 Mapping the Entire Customer Journey

Customer journey mapping refers to the process of creating a visual map of all the possible customer interactions with the brand. An example of mapping would be to identify all the customer touchpoints of the business and creating a persona of the customer in each of the touch points. This will ensure that the businesses realise the perspective of the customers. This strategic method will also help the businesses in understanding the customer expectations and hence the essential components needed to optimise the customer experience[30]. With 80% of consumers now considering the experience of the company as important as its products, mapping the customer journey is crucial for the success of the business.

5.4.4 Identifying New Potential Customers

New potential customers may arise in various ways. It may happen when businesses expand their product range or when they begin to market the product in a new way or to a new group of buyers[31]. These potential new customers are essential for the business' future growth. Not only do they bring in more revenue, but in the process of looking for new potential customers, businesses are also encouraged to be proactive in the ways they promote their product, which is an important characteristic of a successful business. New potential customers may also act as a plan B when the current pool of customer's demands change. Even if the business has a stable pool of customers, they should always remain driven to find new potential customers to ensure their progress as a business.

5.4.5 Maintaining a Consistent Customer Experience

Maintaining a consistent customer experience is extremely important for any business. When potential customers choose a certain business, one of the first things they look at is undoubtedly the customer experiences of the previous customers. Potential customers rely heavily on reviews or recommendations. In fact, in a study done by Nielsen, it was found out that about 92% of customers rely on recommendations from their friends or families[32]. With such high reliance on customer experience, it is important that businesses maintain a consistent and pleasant customer experience.

Other than acquiring more customers, a pleasant customer experience ensures customer retention. With the foundation of loyalty built between the business and the company, the customer will choose to patronize the business even if there are some disadvantages to it compared to the competitors. Therefore, businesses must work to build this relationship with the customers for their continuous success.

5.4.6 Maintaining Trust and Privacy

Customer privacy is a basic right that any customer has. Every customer believes in the importance of privacy, especially in today's world where so much personal information is shared online[33]. Maintaining privacy plays a big role in determining the trustworthiness of a business and hence, the business that takes the effort to maintain the trust will be at a competitive advantage. Most businesses that are seen as trustworthy have very specific privacy policies in place. Therefore, for businesses to succeed, it is crucial that they come up with their own privacy policies to gain the trust of the consumers for their success.

5.5 Six Growth Challenges

The growth of a company is essential to its survival. It has been estimated that two-thirds of businesses make it through the 2-year mark of operations, half survive the 5-year mark and just one-third make it through the 10-year mark[34]. While companies generally aim towards growing their businesses and increasing their overall sales and profits, there are many factors that can affect or slow down the growth of a company. In this section, common challenges to growth will be discussed.

5.5.1 Cash Flow Management

Cash flow management refers to the process of tracking the amount of money coming in and going out of the business[35]. This allows businesses to anticipate the money that will be available and identify the money needed for expenses such as payment to suppliers and staff salary. Having a good cash flow management enables companies to better prepare for the future and resolve cash flow issues in a timely manner.

The top reason for failure for many small businesses is cash flow problems and 82% of small businesses fail due to cash flow problems[36]. It is recommended for companies to have a cash flow buffer of at least 6 months. This will help businesses to tide through challenging or unexpected situations. For a company to be successful and continue their growth, it is critical for them to have robust cash flow management system in place.

5.5.2 Responding to Competition

One major challenge to a company's growth would be the rise of competition. The awareness of companies to their competitors and their ability to analyse the treat they pose is the first step to responding to competition. A case in point would be Apple's iPod and Creative Technology's Nomad Jukebox[37]. Before the development of the iPod which skyrocketed the popularity of MP3 players, Creative was already one of the leading pioneers in the market. However, the size of the iPod which is similar to a deck of cards was revolutionary and it soon became the world's top choice shortly after its launch in 2001. In this technologically advanced world where innovation and new breakthroughs are constantly taking place, businesses need to recognise competition at an early stage to better prepare their responses to it.

5.5.3 Nurturing a Great Company Culture

In a broad sense, company culture is the behaviours and attitudes of the company and the people working inside[38]. It includes various elements such as the company mission, leadership style, work environment, values, etc.[39]. A great company culture is important as employees are more likely to find joy in their work, forge good relationships with fellow colleagues and finally become more productive. Companies who are seeking to grow their business should invest in building a company culture where there is trust in the company and their products or services. A positive company culture can promote a working environment where people are highly motivated and united, thereby increasing efficiency and laying a strong foundation for growth.

5.5.4 Learning When to Delegate and When to Get It Done

While managers have their own tasks to perform, one of their primary roles is delegation. Delegation allows managers to have time to focus on planning, collaborating with other managers and monitoring the performance of their employees to ensure that the overall mission and goal of the organization can be achieved.

One big failure in organizations is the failure of managers to delegate effectively[40]. Effective managers in the company should know the responsibilities to delegate to develop employees. Additionally, they should provide constructive feedback and opportunities for development for the employees. However, in actual practise these hardly take place due to various reasons such as the need to make one's self indispensable or the belief that the employees are unable to perform the task as well as the manager. For a company's growth potential to be maximised, effective delegation should take place so that employees can be well-trained and future leaders can be identified.

5.5.5 Keeping Up with Market Changes

As highlighted in detail in the section on "Key issues in the market", businesses should be familiar with certain key issues such as the increase in international competition, merger and acquisition activities, increased quality consciousness etc in order to better navigate their businesses in the competitive and volatile global market. However, keeping up with market trends is never easy while running the business on a day-to-day basis. Resources, time and energy are limited after all. Therefore, the following smart ways can be employed to keep up with market changes in a resourced and time-pressed environment: Using social media to stay up-to-date with industry's influencers, leveraging on latest publications of industry report and research, gaining insights from surveys of existing customers and observing competitors through avenues such as their website[41].

5.5.6 Deciding When to Abandon a Strategy

Companies invest a lot of time into forging their business plans and strategies. Therefore, not many would be willing to abandon or radically change an existing strategy where much effort has gone into. Common reasons for this include the thought that too much has been invested in the current strategy that the value to cancel the project becomes very low as well as the greed of leaders who were taught to maximise their returns on projects that they often refuse to believe that a project has failed[42]. However, most successful companies do in fact abandon their strategies when they do not work out. It is said that the hallmark of a successful company is not their brilliant original strategy but rather how they reacted when confronting the reality of their strategies[43]. For a company's continued growth, a certain degree of fluidity and flexibility is required of their strategies. As the markets where businesses operate in are not static, neither should companies' strategies be.

5.6 Five Common Strategic Mistakes

"It's good to learn from your mistakes. It's better to learn from other people's mistakes."

—Warren Buffett

As pointed out by one of the most successful investors of all time, mistakes of others are lessons in which we do not have to pay any price. This section will raise 5 common strategic mistakes of businesses.

5.6.1 Poor Goal Setting

Goal setting is a positive and powerful tool when it is able to provide clear direction and ignite enthusiasm[44]. The same is true of the converse. Poor goal setting creates confusion and makes people doubtful and discouraged. A common strategic mistake of organizations, especially new ones is that the goal setting was not given due importance and thoughtlessly created. Other common mistakes are goals that were made with the intention only to impress but not to provide clear guidance on efforts, goals that are unrealistic that it demoralises the staff on the ground and goals that are too broad that it lacks focus. When executed poorly, goal setting can negatively impact an organization and hence should not be taken lightly.

5.6.2   Lack of Alignment

After goals have been set in place, the next critical step is to ensure that everyone is aligned to the goal. Organization alignment is the idea that the entire organization from the entry-level staff to the Chief Executive Officer share a common goal[45]. It is this common goal that spurs collaboration and synchronises efforts for the attainment of the goal. However, according to research, only 40% of employees across organizations are aware of the goals of their company[46]. This means that more than half of the staff are working aimlessly without having clear direction on the organization's focus. The lack of alignment is a pitfall that organizations should avoid.

5.6.3   People Not Being Connected to Common Goal

A huge obstacle about goals is the lack of connectedness to it. Leaders of corporations should not take it for granted that everyone will work towards the goals once they are set. As leaders, it is important to inspire and convince the employees to believe in the goals of the organization. This can be achieved through activities such as team building events where team spirit can be cultivated and social interaction where the staff can be influenced and inspired towards the common goal.

5.6.4   Failure to Track Progress

One of the reasons for failure in projects is due to the absence of effective progress tracking[47]. In an effective progress-tracking management, the progress or milestones set must be measurable or trackable or there will be no way to actually measure the progress of the project. The milestones decided should also be realistic to be able to successfully drive employees as well as to minimise disruptions or changes to the overall progress chart. By implementing an effective system to track progress, accountability is promoted in the organization and deadlines can be better monitored.

5.6.5 Not Utilizing Data Analytics

Data analytics indicates the process of analysing datasets to obtain information about the information they contain[48]. Data analytics have been given much attention and with the rise of data scientists and analysts, companies that fail to take advantage of this technology will naturally be at a loss compared to their counterparts. In the context of businesses, the datasets may include data from the past and new information collected from a particular project. Such analysis allows companies to better track their successes and failures and even to identify their competitive advantage. In relation to the previous point on tracking progress, data analytics can possibly help to measure performance and serve as a progress tracking tool. In this technology-advanced society, the ability to fully utilize such technologies will contribute to the success of businesses.

References

1. https://www.consultantsreview.com/cxoinsights/importance-of-Innovation-in-medical-devices-industry-vid-761.html#:~:text=Innovation%20can%20help%20in%20getting,compliance%20to%20various%20regulatory%20requirements.
2. https://www.proclinical.com/blogs/2019-2/top-10-new-medical-technologies-of-2019.
3. https://www.fortunebusinessinsights.com/industry-reports/virtual-reality-vr-in-healthcare-market-101679.
4. https://www.asiabiotech.com/19/1909/19090021x.html#gsc.tab=0.
5. https://www.un.org/sustainabledevelopment/sustainable-development-goals/.

6. https://www.unglobalcompact.org/what-is-gc/our-work/sustainable-development.
7. https://www.mckinsey.com/business-functions/marketing-and-sales/our-insights/reimagining-marketing-in-the-next-normal.
8. https://online.maryville.edu/blog/4-trends-changing-the-marketing-landscape/.
9. https://www.targetmarketingmag.com/article/the-evolution-marketing-then-now/all/.
10. https://www.patriotsoftware.com/blog/accounting/how-to-conduct-a-market-analysis/.
11. https://www.businesswire.com/news/home/20180801005423/en/Importance-Market-Analysis-Improving-Business-Growth--#:~:text=Market%20analysis%20is%20one%20of,and%20making%20wise%20business%20decisions.&text=Effective%20market%20analysis%20can%20help,the%20traits%20of%20customers'%20expenditure.
12. https://www.bigcommerce.com/blog/how-perform-competitive-analysis/.
13. https://cerealentrepreneur.academy/the-importance-of-competitor-analysis/.
14. https://the-definition.com/term/product-market-scope#:~:text=Definition%20(1)%3A,will%20target%20with%20these%20products.
15. https://www.orangescrum.com/blog/top-10-reasons-why-you-need-effective-resource-allocation.html#:~:text=Resource%20allocation%20is%20a%20process,vital%20in%20delivering%20project%20efficiently.
16. https://www.projectengineer.net/the-6-steps-of-resource-allocation/.
17. https://www.orangescrum.com/blog/top-10-reasons-why-you-need-effective-resource-allocation.html#:~:text=Resource%20allocation%20is%20a%20process,vital%20in%20delivering%20project%20efficiently.
18. https://www.shmula.com/determining-the-need-for-change-in-an-organization/22516/.
19. https://www.clearreview.com/why-performance-management-important/.
20. https://www.forbes.com/sites/brentgleeson/2018/03/23/organizational-change-can-suck-3-ways-to-manage-fear-and-stay-energized/#67e37d47609d.

21. https://psychcentral.com/lib/tips-for-coping-with-organizational-change/.
22. https://velocityglobal.com/blog/get-ahead-international-competition/.
23. https://www.industryweek.com/technology-and-iiot/article/22028234/feats-of-fastpaced-exploratory-innovation.
24. https://www.investopedia.com/ask/answers/why-do-companies-merge-or-acquire-other-companies/.
25. https://www.massdevice.com/category/business_financial_news/mergers_acquisitions/.
26. https://medical-technology.nridigital.com/medical_technology_jan18/medical_device_mergers_and_acquisitions_of_2017.
27. https://www.iso.org/home.html.
28. https://www.leightoninteractive.com/blog/why-collecting-customer-data-is-crucial#:~:text=Collecting%20customer%20data%20is%20the,your%20target%20audience%20or%20persona.&text=Social%20media%2C%20email%20sign%2Dups,who%20you're%20marketing%20to.
29. https://smallbusiness.chron.com/importance-target-audience-consumers-37173.html#:~:text=Identifying%20a%20target%20audience%20provides,audience%20at%20a%20manageable%20level.
30. https://www.salesforce.com/uk/blog/2016/03/customer-journey-mapping-explained.html#:~:text=Customer%20journey%20mapping%20helps%20businesses,business%20from%20the%20customer's%20perspective.&text=A%20map%20helps%20reveal%20issues,customers%20interact%20with%20your%20business
31. https://www.thebalancesmb.com/identifying-opportunity-in-new-potential-markets-4043634.
32. https://www.stratfordmanagers.com/customer-experience-advantages/#:~:text=Exceptional%20customer%20experience%20builds%20a,providing%20extra%20value%20through%20interactions.
33. http://clocktowerinsight.com/customer-privacy-why-its-more-important-than-ever/#:~:text=Protecting%20user%20privacy%20can%20enable,practices%20by%20a%20small%20margin.
34. https://blog.hubspot.com/sales/growth-strategy.

35. https://www.xero.com/us/resources/accounting-glossary/s/what-is-cash-flow-management/#:~:text=Cash%20flow%20management%20is%20the,going%20out%20of%20your%20business.&text=Cash%20flow%20is%20the%20term,analyzing%20any%20changes%20to%20it.
36. https://www.score.org/blog/1-reason-small-businesses-fail-and-how-avoid-it#:~:text=In%20fact%2C%2082%25%20of%20small,will%20%E2%80%93%20of%20several%20underlying%20causes.
37. https://www.forbes.com/2004/12/06/cx_ah_1206mondaymatchup.html#456f67d91134.
38. https://www.thebalancecareers.com/what-is-company-culture-2062000#:~:text=Company%20culture%20is%20important%20to,coworkers%20and%20be%20more%20productive.
39. https://hbr.org/2018/01/the-culture-factor.
40. https://www.shrm.org/resourcesandtools/hr-topics/organizational-and-employee-development/pages/delegateeffectively.aspx.
41. https://www.bl.uk/business-and-ip-centre/articles/how-to-identify-market-trends-for-long-term-business-planning.
42. https://www.marketleadership.net/abandon-your-strategy/.
43. https://www.foxwoodassociates.com/2019/05/01/when-to-abandon-a-strategy/.
44. https://www.thebalancecareers.com/the-darker-side-of-goal-setting-why-goal-setting-fails-1916826#:~:text=Poor%20goal%20setting%20makes%20people,your%20progress%20in%20achieving%20them.
45. https://www.martechadvisor.com/articles/digital-transformation/the-importance-of-organizational-alignment-and-how-to-achieve-it/.
46. https://hbr.org/podcast/2012/03/good-strategys-non-negotiables.html.
47. https://blog.kintone.com/business-with-heart/blog/why-projects-fail-poor-progress-tracking-and-management/.
48. https://www.lotame.com/what-is-data-analytics/#:~:text=Data%20Scientists%20and%20Analysts%20use,content%20strategies%20and%20develop%20products.

Chapter 6

Regulatory Affairs as a Business Partner

Claudette Joyce C. Perilla

Asia Regulatory Professional Association

cjcperilla@gmail.com

If I receive a dollar for every time I hear, "*our RA Department is like the police, they won't let me do anything,*" I would have enough money to spend most of my days frolicking at the beach in the French Riviera or partying it up in Ibiza. Those words, like a knife through your back, will haunt you throughout your stint as a regulatory affairs officer. BUT don't fret! It is in your powers to change this mindset.

So, what does it take to be a business partner?

Know and understand your role. Making sure that your product registrations and licenses are in compliance to the applicable regulations are your bread and butter. But more than this, you have to remember that you are part of a team. You are an integral part of the whole business ecosystem and have equal opportunities as the other departments to ensure profit and business success.

Medical Regulatory Affairs: An International Handbook for Medical Devices and Healthcare Products (Third Edition)
Edited by Jack Wong and Raymond K. Y. Tong
Copyright © 2022 Jenny Stanford Publishing Pte. Ltd.
ISBN 978-981-4877-86-2 (Hardcover), 978-1-003-20769-6 (eBook)
www.jennystanford.com

Regulatory affairs management is hard work. For the most part, we get so caught up in ensuring regulatory compliance that we tend to forget that we were brought in to support the business. We are here to help maximize business benefits while being able to follow the regulations applicable to us. We are here to make sure that we find ways to be able to launch early, providing early and continuous patient access to our products with strong compliance mindset.

Build relationships. One of the most common complaints from RA officers that I hear is that the other departments do not understand your job. Trust me, they also feel the same way.

Being caught in our own bubble is common not just to RA but to everyone else in the team. The demands of our job may prevent us from reaching out and talking. We tend to forget the human side of this job. The people whom we work with still have emotions and needs, and they still need to trust you and ensure that you will not lead them to their professional demise.

While this should not be an issue at a professional level, it can still play a role in your day-to-day activities. To put it simply, imagine if your friend comes to you for help. You don't immediately shut them down and tell them you can't do it because it's not your job to do so, right? They have their own families that should help them, not you, right? But because they are your friend, you understand that they came to you for a reason, you tend to listen and understand what they need, and see what you can do. They trust you and you trust them that they will do right by you.

This sort of relationship works in the business setting, for both internally and working with your authorities. There are a lot of times when people come to you last minute or similarly, you may require their help in a short amount of time. In my experience, I would most likely get help from somebody who knows me and understands that I am not there to just add on to their workloads, that what I need is his expertise and that's why I have reached out and not just passing the responsibility to them.

Do not get me wrong, though. It will not be easy. Trust will never come easy.

However, as with anything else, start small. Small conversations. Baby steps. It will all work out in the end.

Stay curious. Change, as we all say, is the only constant thing in this world. And this, in fact, is also constant in the RA world.

The global regulations have been fast-evolving. More and more countries have seen the need for mandatory regulations for therapeutic products and are set to implement these, if they haven't yet. And if you are not quick on your feet, you might get left out.

Regulators understand that the industry will always ask for information. They do also understand that it will also be beneficial to them if the industry is up-to-date and understands the requirements. The better understanding of the industry has on the regulations, the better the quality of the dossiers that they submit, the faster the regulators can do their evaluations.

Read. Listen. Ask. Seek new information.

It does not hurt to be information-hungry.

Keep an open mind. Regulatory affairs provide a very good opportunity to work our creative juices and think out of the box. Yes, all these while still within the bounds of regulations/legislations.

Just because a certain strategy has not been done, it does not mean it cannot be done. And similarly, just because a comment on regulations did not come from Regulatory Affairs, does not mean that it may not work. Remember, there are different angles to what we are doing and we are only privy to one side of it.

As mentioned earlier, the demands of our job are so high, we can easily get caught in our RA bubble. Fresh eyes and fresh minds always help when you're stuck. Also, it doesn't hurt to get help from the outside.

You have to remember that the people whom you work with have the same ultimate goal as you: patient access to your products, compliance to applicable regulations, but with right product and right profile being registered to maximize business needs.

Don't be too quick to shut down ideas.

Listen. Understand. And then act upon it.

Chapter 7

Introduction to Regulatory Affairs Professionals' Roles

Dacia Su

Regulatory Affairs Manager, Asia Pacific,
Terumo BCT Asia Pte. Ltd.

dacia.su@terumobct.com

Regulatory affairs professionals can be found in all sorts of companies, for example, biotech, pharmaceuticals, medical devices, diagnostics, and even nutritional products. They play critical roles in the health product lifecycle, from development through post-market approval. The usual responsibilities would be in the following general areas:

1. To keep track of the ever-changing regulations in the countries in which the company wishes to distribute its products.
2. To advise companies on the regulatory changes that would affect proposed activities where this is often one of the important factors for a commercial product launch. It often enables the commercial team to be able to launch the product effectively, by being prepared once product approval is given.

Medical Regulatory Affairs: An International Handbook for Medical Devices and Healthcare Products (Third Edition)
Edited by Jack Wong and Raymond K. Y. Tong
Copyright © 2022 Jenny Stanford Publishing Pte. Ltd.
ISBN 978-981-4877-86-2 (Hardcover), 978-1-003-20769-6 (eBook)
www.jennystanford.com

3. To ensure that their companies comply with all of the regulations and laws.
4. To identify risk and develop contingency plans. For medical products, risk associated in the medical device would often translate to harm to people.
5. To present registration documents to regulatory agencies and carry out all subsequent negotiations necessary to obtain and maintain the right to market the products. It is not a simple process as it often involves collating and evaluating scientific data.

Required competencies of regulatory affairs professionals

It is crucial for a regulatory affairs professional to be adaptive and believe in continuous learning. As the regulatory landscape is constantly changing with new technology, new risk potential, or reimbursement, an organization with an understanding of regulatory landscape is more prepared to respond to regulators in times of crisis.

The regulatory affairs professional must also be able to capture intelligence from a variety of sources (for example, interactions with health authorities, conversations with key opinion leaders, networking at conferences). This should translate to the ability of critical thinking as well, as one is now able to anticipate the changes in the external environment and disseminate internally to the organization as a value-add.

An all-rounder is often required to understand science, business, and government language and procedures. This is seen from the need to learn product knowledge from the scientific and technical team and then translate the safety reports to the regulators.

Strong negotiation and communication skills are also essential with working multiple stakeholders (commercial colleagues and government agencies). For example, if given a minute to justify a recommendation with a busy CEO, the important message should be communicated in the brief time in order to make progress in the project.

One should be detailed to process the numerous data for product submissions. In addition, project management skills should be employed as there is currently a complex global regulatory framework and projects have to be managed from the launch stage to the ready-to-market stage.

Operating in an ethical manner would allow both the professional and the organization to respond quickly to recalls and conduct business in a fair manner, as this is a role that comes with heavy responsibility.

Who should join the regulatory affairs profession?

A person who wishes to shape healthcare decisions and policy, with the belief that the work you are doing is to help people, many of whom have life-threatening diseases, should consider entering the profession. A believer of continuous education would thrive on the job, as you never stop learning on the changes to regulatory policies and procedures.

Chapter 8

What It Means to Be a Medtech Regulatory Journalist

Amanda Maxwell

Medtech Regulatory Affairs Editor, Clinica Medtech Intelligence
Amanda.Maxwell@informa.com

The medtech regulatory world is a relatively small community. Being a journalist in this community brings with it a sense of belonging and of responsibility to accurately report developments and inform those in medtech regulation.

Writing about the rules that will help ensure products are safe enough to give patients the best possible chance of survival, recovery, or improvement is a vital and important job.

This is a complex area, and I am equally aware of the importance of achieving a balance to reflect the need for appropriate rule-making; overly strict rules could be as damaging to patient chances as lax rules if they mean that products do not make it on time to market for some patients.

This is something I felt passionate back in 1984, when I joined Clinica Medtech Intelligence, then known simply as Clinica, and I still feel passionate about it now.

Medical Regulatory Affairs: An International Handbook for Medical Devices and Healthcare Products (Third Edition)
Edited by Jack Wong and Raymond K. Y. Tong
Copyright © 2022 Jenny Stanford Publishing Pte. Ltd.
ISBN 978-981-4877-86-2 (Hardcover), 978-1-003-20769-6 (eBook)
www.jennystanford.com

Many people ask me if I become bored after so long in this sector. The answer is a definite "No"! Medtech regulations need constant interpreting and reinterpreting as medical technology develops, attitudes to risk change, and patients become more informed.

Being a journalist means offering a canvas to the sector on which its problems with the regulations and difficulties in their interpretation can be fairly represented for all to view and consider.

I am often the messenger interpreting what the lawmakers actually mean and aim to achieve behind their frequently complex jargon, so that those who are going to apply these rules understand clearly what is required. My aim is to achieve accuracy so that an informed discussion can take place at the necessary levels to achieve a balanced outcome.

It is by analyzing the impact of current rules and proposed changes and seeking different views, as well as through international comparisons in rule-making that we can communicate where appropriate regulation lies and the possible means to achieve this, and the impact of different choices.

Highlighting these issues is part of the chain of motivation for change.

It is not uncommon to hear that Clinica articles have been taken into key high-level meetings, in Europe included, and cited as demonstration of how the status quo is simply not working.

How I became a medtech journalist and why I stayed

I am a career journalist. I joined Clinica fresh out of university with a degree in languages. My mission back then was to investigate the French and Italian medtech markets.

Soon after I joined in 1984, the medtech industry began working with the European Commission on the EU's medical device directives. It was my first foray into regulations and an uphill journey getting to grips with the regulatory issues; the complexities nearly overwhelmed me in those early days. However, I soon discovered I had found my niche and was fascinated with the subject. I have loved my job ever since.

Since then, and following an early career traveling throughout Europe to key regulatory meetings, I developed a strong network of medtech regulatory contacts. Many of those who I met in the early days are still active in this sector. Some, of course, have moved on or retired having left their powerful mark on the sector; some have sadly died but will be remembered as having played a key role. And there are also many dynamic new faces on the conference circuits, many keen to make their mark. Although, nowadays, people seem to move on into other sectors more frequently too.

Highlights

My career highlights have come in many shapes and forms.

Perhaps one of my most memorable moments was swimming in the Aegean Sea in Greece in the 1990s, at a conference which was effectively a first step on the way to developing the European medical device database, Eudamed ,which was attended by many of the EU's key regulators.

At the time, the then head of devices at the European Commission was also swimming nearby and clearly having an interesting conversation about devices with another regulator. I think that must have been my one and only water-based tip-off for a story that later emerged!

Another highlight was attending a meeting in the European Parliament on the revision of the Medical Device Directives and having the honor of seeing and hearing the formidable Dagmar Roth Behrendt, a lead European Parliament Member of Parliament at that time, speak passionately in person about how she thought high-risk devices should be regulated—akin to drugs.

My sources

People frequently ask me where I get my information from. It comes mostly from talking to contacts and reading documents circulating the industry—between the lines. So often material is put out, but the news will lie in what is not said, rather than what is said. It is questions that are left deliberately unanswered that can lead to the real issues.

Meetings are particularly valuable—from the point of view of listening to opinion and for helpful context to the issues that are out there. Updates on websites are often valuable leads into issues that turn out to have a wider significance.

Changing technology

One of the most stimulating challenges in this industry is the rapid technological change and the wide variety of different products and ways in which they work. eHealth is revolutionizing our sector, and the medtech and pharmaceutical sectors have found ways of making use of each other's products to optimize the drug delivery and device efficacy. Nanotechnology is enabling the miniaturization of so many treatments, and human engineering and bioprinting are creating treatments and solutions barely dreamt of back in the 1980s when I started at Clinica. Regulators are challenged to keep up.

If technology has changed so much in this timeframe, where will it lead to—buoyed by big data—over the next three decades? Where indeed? Aesthetic products—such as fillers and implants—are increasingly being regulated around the world as medical devices, and we are approaching a point where human enhancement devices—for example, some brain stimulation devices—are needing to be considered within the context of medical device rules. We are likely to have not only technological and regulatory issues ahead but some big ethical questions too. And all these matters will continue to pose a tough challenge to those involved.

Changing global outlook

While technology is changing, so is the world. With the Internet available almost globally, information is becoming so much easier to find. When I started at Clinica, it was unimaginable to have the breadth and frequency of stories we now have in 2016 on countries in Asia Pacific, for instance, and in South America.

Of course, many of these countries lacked regulations back then too, or had very primitive and skeleton arrangements, but the intervening time has since witnesses a sharp rise in the number of countries that have joined regulatory groups, such as the International Medical Device Regulators Forum and the Asian

Harmonization Working Party, as a way to interpret and adopt a regulatory medtech pathway.

Changing world of journalism

Of course, it is not only the medtech world that has changed over these 30 years or so. The world of journalism has also changed—from paper copies to online only, from a focus on news to a focus on intelligence and analysis, from national to increasingly international and global news. In my early days, we had to write articles by hand, but had a much bigger editorial team and more junior staff.

We have now moved to a smaller number of experienced staff, supported by a team of mature experienced writers with deeper regulatory knowledge.

The future

With 30 years behind me, how many more ahead? Well, the passion is still there, the contacts remain dynamic, and all over the world new regulatory patterns are emerging—new regulations, groups of countries working together, global initiatives.

In the EU, in particular, we are awaiting a new set of regulations. Hopefully, they will be agreed to and adopted in 2016. They will bring with them fresh challenges as manufacturers, notified bodies, and authorities have to adapt considerably. New acts to support these regulations will need to be drafted and then will follow the gradual realization of what works well and what does not work so well. And no doubt the EU experience—just as it did some 20 years or so ago when the Global Harmonization Task Force was set up—will again influence the global regulatory stage.

Once again, my job will be to highlight the issues and what they could mean to the sector in the future so stakeholders can have the necessary insights to prepare as best as possible.

When will I become bored? Not ever. This is too dynamic and important a sector. It is a privilege to cover it and to work with committed and interesting people.

Chapter 9

Accelerating Access to in vitro Diagnostics: Urgent Need for Increasing the Speed and Efficiency of Regulatory Review and Policy Development for in vitro Diagnostics for Antimicrobial Resistance and Epidemic Preparedness and Response

Rosanna W. Peeling,[a] David Heymann,[a] Noah Fongwen,[a] Oliver Williams,[b] Joanna Wiecek,[b] Phil Packer,[c] and Gabriela Juarez-Martinez[d]

[a]*London School for Hygiene and Tropical Medicine, UK*
[b]*Wellcome Trust, UK*
[c]*InnovateUK, UK*
[d]*Knowledge Transfer Network, UK*

9.1 Background

In 2015, the US National Academy of Medicine convened a workshop entitled, Global Health Risk Framework for Research and Development of Medical Products [1]. The following are excerpts from the Workshop Session on "Convergence of Regulatory

---

Medical Regulatory Affairs: An International Handbook for Medical Devices and Healthcare Products (Third Edition)
Edited by Jack Wong and Raymond K. Y. Tong
Copyright © 2022 Jenny Stanford Publishing Pte. Ltd.
ISBN 978-981-4877-86-2 (Hardcover), 978-1-003-20769-6 (eBook)
www.jennystanford.com

Expectations, Review, and Approval: Coordination, Harmonization, and Convergence."

According to Hans-Georg Eichler, European Medicines Agency, "Even though regulators have the same overall goals, there are many different actors and often a divergence of outcomes. There are costs and risk associated with nonconvergence among regulators and other decision makers. Different evidence standards, for example, lead to opportunity costs (e.g., developing a product for a global market might require one study for regulator A and a different study for regulator B, etc., which ties up capital that could be used for other innovations). Multiple assessments of priority or probability of success can lead to uncoordinated competition for clinical trial participants and trial sites, resulting in delays in bringing products to market. Differing regulatory procedures also affect timelines and opportunity costs. Different outcomes (e.g., market authorization in one country or region, but denial in another; product licensed, but not funded or reimbursed) can lead to delayed or unequal patient access to products, as well as political tensions."

Margaret Hamburg, National Academy of Medicine, emphasized the need for "streamlining processes, including decreasing bureaucratic and logistical complexities, better sharing of data, aligning across jurisdictions regulatory expectations of companies and the scientific community in terms of the burden of proof for product review and approval, moving toward common data and evidence standards, and advancing regulatory science (i.e., identifying gaps and developing the knowledge and regulatory tools necessary for more streamlined oversight)."

By working together, Hamburg continued, global regulatory authorities will be better able to meet expectations in terms of scientific review, more effectively share information about advances in science and emerging technologies, and develop more flexible approaches to assessing complex products and defining acceptable levels of risk in public health emergency situations. "Regulatory coordination is vital in a public health emergency to ensure meaningful action, especially given the often limited population to study or time sensitivity. Working together more effectively also helps to ensure quality and safety." The establishment of the International Coalition of Medicines

Regulatory Authorities (ICMRA) in 2015 is an important activity geared toward enhancing global regulatory convergence and capacity.

In moving forward, Hans-Georg Eichler, European Medicines Agency, suggested that "regulators can help minimize the opportunity costs through regulatory convergence on both content and process. Convergence on content includes, for example, alignment on how to balance speed, feasibility of studies, and affordability of product with quality, validity, and robustness of information. In order to have content convergence there is a need for agreement on the kind of evidence needed. Other content areas include agreement on outlines of study protocols; development of bridging authorizations; flexible regulatory tools to give legal certainty to manufacturers to deploy products in countries of need; and building clinical trial infrastructure in country. Convergence on process involves aligning how interactions with product sponsors are handled from beginning to end, and across authorities." Eichler also suggested that timelines could be shortened by undertaking some processes in parallel rather than serially (such as product assessment by the European Medicines Agency (EMA), the World Health Organization (WHO) prequalification, and local authorization from the country where the product will be used). In summary, following were the key points raised during the workshop:

- The harmonization of regulatory processes and regulatory science standards among global regulatory authorities is key to more effective and rapid assessment of complex products and define acceptable levels of risk in global public health emergency situations.
- Regulatory coordination is a core component of meaningful action in a public health emergency, given the often limited clinical trial capacity (population to study, and clinical sites) and limited window of time for study.
- Action needs to be taken now, before the next crisis emerges, to enhance regulatory convergence, strengthen preapproval of clinical trial designs and protocols, the development and approval of prototype or plug-and-play platforms and capabilities, the building of clinical trial infrastructure, and negotiating agreements for data sharing among authorities.

9.2 The Need for Regulatory Harmonization, Convergence and Reliance for in vitro Diagnostics

Accelerating access to in vitro diagnostics (IVDs) is a global priority as outbreaks of infectious diseases are occurring with increased frequency and intensity. Antimicrobial resistance is now a global crisis with rapid emergence of "super-bugs" that threaten the safety of our healthcare institutions and risk a return to the pre-antibiotic era when people can die from minor infections, and surgery and child birth can become high risk. New and improved diagnostics, introduced rapidly and effectively where they are required, are needed to combat these global health priorities.

Despite progress with regulatory harmonization and convergence through the work of the Global Harmonization Working Party (former Asia Harmonization Working Party), the Pan-African Harmonization Working Party (now part of the Africa Medical Devices Forum) and the Latin America IVD Alliance (ALDDIV) [2–4], the current process of bringing a new point-of-care diagnostic to the clinic remain costly and can take up to ten years, as the licensing procedures alone can take up to five years to obtain the necessary clinical risk assessments and approvals by regulatory authorities. Harmonization or convergence of regulatory requirements in each region or group of countries could substantially reduce the duplication of efforts and delays to access. Although significant progress has been made to reduce the duplication in inspections of manufacturing quality through the establishment of a medical device single audits programme (MDSAP) by the International Medical Device Regulatory Forum (IMDRF), harmonization and convergence of other areas of regulatory review has been slow [5]. Once a diagnostic is successfully licensed, further work by policy makers is needed to evaluate the clinical benefit to patients and the cost-effectiveness of introducing a new test into the health system. The lengthy health technology assessments (HTAs) are followed by the development of policies for implementation and further guidance which can extend this process for an additional three to five years (Fig. 9.1). The length and the complexity of this process substantially increases developer costs, lowers the attractiveness of this sector to product developers and delay access to life saving commodities.

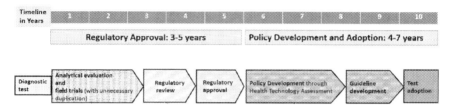

Figure 9.1 Bench to bedside pathway for access to quality-assured diagnostic tests with current timelines.

9.3 Need for a New Paradigm for Risk-Benefit Assessment for Diagnostics

In 2016, member states of the United Nations committed to a set of Sustainable Development Goals [6] with which countries pledged to make health care available for all and to leave no one behind. The lack of access to diagnostics has been a major reason why health care services fail in much of the developing world. Accurate diagnostic tests for infectious diseases are traditionally performed in laboratories with trained personnel and sophisticated equipment. In low-and middle-income countries (LMICs) where laboratories are underequipped and laboratory infrastructures are often limited, lack of access to diagnostics has led to misdiagnosis or presumptive use of antibiotics, exacerbating the global crisis of antimicrobial resistance. Recent advances in the development of rapid diagnostic tests (RDTs) that can be performed at the point-of-care (POC) or point-of-need have transformed the diagnosis of major infectious diseases such as HIV, tuberculosis and malaria. These diagnostics are not as accurate as laboratory-based tests but allow greater number of individuals to be tested, i.e. improving access. Figure 9.2 shows a simple mathematical relationship between diagnostic access and accuracy. Let us take an example of a test that is highly accurate (e.g. 100% sensitive) but requires a laboratory with highly trained personnel is thus only accessible to 30% of the population. This laboratory-based test can at best correctly identifying 30% of those who are infected. On the other hand, a RDT with reduced accuracy (e.g. 90% sensitivity compared to the laboratory test) but can be accessed by 90% of the population at all levels of the health care system, this RDT can now correctly identify 81% of those who are infected. On a theoretical basis, it may be easy to conclude that the benefit of having a test that can correctly identify 81% of those who are truly infected far outweighs the scenario in which only 30% of those infected can be identified. This increased benefit has to be balanced against the risk of having 10% false negative results.

	Sensitivity			
Access	100	90	80	70
100	100	90	80	70
90	90	81	72	63
80	80	72	64	56
70	70	63	56	49
60	60	54	48	42
50	50	45	40	35
40	40	36	32	28
30	30	27	24	21
20	20	18	16	14
10	10	9	8	7

Figure 9.2 Trade-offs between test accuracy and accessibility.

In reality, these trade-offs between accuracy and accessibility are often quite complex. The assessment of risks and benefits depends on the nature of the disease, its transmissibility, the harm of false positive and negative results and whether there are ways to mitigate these risks. Inequity of access represents more than physical inaccessibility because of geographic location. Accessibility of a diagnostic test includes ease of use outside of laboratory settings, timeliness of results, affordability and in the absence of constraints such as stigma. These trade-offs often require regulators and policy makers to collaborate in balancing the level of risks that are acceptable against the incremental benefits that a new test can bring. The following examples serve to illustrate the need for these decisions:

Case study 1: FDA approval of HIV self-tests [7]

In the US in 2010, the Centres for Disease Control and Prevention estimated that 18% of HIV-infected individuals did not know their status, mainly because they were marginalised from the healthcare system or were reluctant to get tested due to stigma. The availability of a rapid HIV self-test that can be performed in the privacy of one's own home would offer a potential solution. The performance of the first self-test to be submitted for market authorization showed that the sensitivity of this test when performed by lay persons did not meet the FDA performance requirements. However, a modelling study by the Centers for Disease Control and Prevention showed that if rapid HIV self-tests were available for sale, more than 2.7 million people per year in the US would self-test, from which more than 45,000 new HIV infections would be identified. The FDA approval of HIV self-tests was based on the estimated public health benefit of interrupting the transmission of 4,000 HIV cases outweighing the risk of 1,100 false negative results. The FDA mitigated the harm of potential false negative results by ensuring that the instruction for use contain messages informing the user of the potential for false negative results and the importance of re-testing at frequent intervals.

Case study 2: Diagnosis of HIV in infants

The diagnosis of HIV born to HIV-infected mothers requires molecular testing to detect virus in the blood of infants as maternal HIV antibodies tend to stay in the blood of infants up to 18 months of age. In many countries where molecular testing for HIV is not widely available, blood collected from the infants are often spotted onto a piece of filter paper, air dried and transported to a central laboratory for testing. In settings where the demand for molecular testing is high, results are often returned to the health provider several months after sample collection. According to UNAIDS and UNICEF, "many infants across the world are dying needlessly because they are not being tested early enough for HIV and treated if they have the virus. Without treatment, half of all HIV-positive babies will not live long enough to see their second birthday; a third will not see their first" [8]. In Mozambique, an estimated 62% of HIV-exposed infants received the results of their diagnosis more than 1 month after sample collection. This likely contributed to increased loss to follow-up, morbidity and mortality. In a study modelling two ways of diagnosing infants with HIV, Meggi et al. showed that the current scenario was that blood samples were collected from infants born to HIV positive mothers and sent to a central laboratory where they were tested using a molecular test with >95% sensitivity. However, as results often came back very late, many babies were lost to follow-up. This current scenario resulted in only 375/1000+ babies being identified as HIV positive and treated 28 days after specimen collection. The model showed that if a rapid POC test which detects HIV p24 antigen in 15 minutes and which had a sensitivity of 72% compared to the molecular test, 683/1000+ would have been identified 15 minutes after blood collection, a gain of 81% more HIV+ infants being identified early and getting appropriately treated. The model shows the life-saving potential of using a less sensitive but more accessible test compared to the status quo [9]. This rapid POC test has not been approved for use in any country as the WHO and most regulatory authorities considered that a sensitivity of 72% was not acceptable. This study was conducted by the National Institute of Health in Mozambique to demonstrate that diagnostic tests should not be approved based on accuracy alone. Consultation of regulatory authorities with policy makers on the trade-offs between test accuracy and accessibility is critical to a balanced risk-benefit analysis.

Case study 3: Ebola diagnosis

During the Ebola outbreaks in West Africa, patients suspected of having Ebola had a blood sample taken in the triage tents for molecular testing. However, waiting times for disease confirmation using molecular tests can sometimes be days, depending on how many people present for testing. The World Health Organization decided to give emergency use authorization to a rapid test which was 85% sensitive compared to the molecular test but could give a result in 15 minutes. However, this test was not widely adopted because policy makers in the affected countries considered the risk of 15% of those who were truly infected being sent back to the communities where they would continue to transmit the infection as not acceptable. The trade-offs in this case are quite complex, as the nature of the disease, the number of people presenting for triage, the stage of their disease by the time they presented and the risk of those who were not infected being infected while waiting in the triage tent all had to be taken into consideration. There were no right or wrong answers and lives would be lost no matter which option was chosen.

Case study 4: COVID-19 diagnostic tests

A similar risk-benefit analysis to case study 3 was required in the current COVID-19 pandemic response. On April 8 2020, the WHO issued a scientific brief saying, "In response to the growing COVID-19 pandemic and shortages of laboratory-based molecular testing capacity and reagents, multiple diagnostic test manufacturers have developed and begun selling rapid and easy-to-use devices to facilitate testing outside of laboratory settings. These simple test kits are based either on detection of proteins from the COVID-19 virus in respiratory samples (e.g. sputum, throat swab) or detection, in blood or serum, of human antibodies generated in response to infection. The WHO applauds the efforts of test developers to innovate and respond to the needs of the population. However, before these tests can be recommended, they must be validated in the appropriate populations and settings. Inadequate tests may miss patients with active infection or falsely categorize patients as having the disease when they do not, further hampering disease control efforts" [10].

In the meantime, despite the best efforts of many countries to scale up molecular testing, hospital emergency rooms were overwhelmed with people presenting for care with symptoms consistent with COVID-19. The demand led to long delays in obtaining results to allow the patients to be triaged to COVID areas or non-COVID wards. Access to testing was also limited outside of major urban centres.

On September 11, 2020, the WHO published an Interim Guidance on Antigen Detection for the diagnosis of SARS-CoV-2 using rapid immunoassays, in which the WHO recommended the use of rapid antigen tests with sensitivities >80% and specificity > 97% compared to a molecular test [11]. The consensus on these sensitivity and specificity thresholds took many discussions involving a panel of experts, regulatory authorities and policy makers. This consensus might have been reached sooner for a more effective pandemic response if a mechanism for these collaborative discussions and models for consensus building among major stakeholders were already in place.

Whereas regulatory approval of medical products such as drugs, vaccines and diagnostics have traditionally been based on non-inferiority (i.e. only approving a product that is equal or better than the last approved product), for diagnostics, the comparison to the last approved product has always been based on measurements of accuracy. However, now access has to be taken into consideration. While accuracy can be measured in terms of sensitivity and specificity, access cannot be so easily quantified. Assessing the risk of false positive and negative results from a more accessible but less accurate test versus the benefit that wider access can bring requires a new paradigm in which regulators and policy makers work together with subject matter experts and statisticians to ensure maximum safe and effective use of available diagnostics for public health.

9.4 Next Steps

To address this problem, Wellcome Trust and Innovate UK are funding an Accelerated Diagnostics Access Project (ADAP) focused on accelerating the uptake of diagnostics in low and middle-income countries (LMICs). This work, led by the London School of Hygiene and Tropical Medicine (LSHTM) and Chatham House, will involve conducting a series of regional consultations to explore whether there are opportunities to decrease the time that it takes for licensing and introduction to use, and thereby accelerate access to diagnostics in LMICs. Through these consultations, the project will gather the perspectives of different stakeholders and build awareness of the bottlenecks, barriers, gaps, duplications and fragmentation that impede rapid and timely licensing and introduction to use of new diagnostic tests and Identify possible solutions to overcome the identified challenges. Based on the outcomes of the consultation process, a global meeting will be held to develop recommendations for accelerated access to diagnostics. The final recommendations for national policy makers and regulators will be made public through a series of policy briefings, produced by Chatham House.

It is hoped that this work will lay the foundations for a paradigm change that will ensure greater and faster access to quality-assured diagnostics, and thereby improve patient care, decrease antibiotic use to preserve antimicrobial drugs for future generations, and support more rapid and effective responses to infectious disease outbreaks.

References

1. National Academy of Medicine. Global Health Risk Framework for Research and Development of Medical Products. Workshop Summary. 2016. National Academy Press, ISBN 978-0-309-38099-7| DOI: 10.17226/21853. http://www.nap.edu/21853.
2. Asia Harmonization Working Group. AHWP Transformation into GHWP with Unchanged Position on Regulatory Authorities-Industry-Partnership. http://www.ahwp.info/index.php/news.
3. Pan-African Harmonization Working Party. http://www.pahwp.org/index.html.
4. Latin America Alliance for the Development of IVDs. https://www.aladdiv.org.br/?lang=en.
5. International Medical Devices Regulators Forum. Medical Device Single Audit Program. http://www.imdrf.org/workitems/wi-mdsap.asp.
6. World Health Organization. Sustainable Development Goals. http://www.who.int/sdg/targets/en/.
7. Peeling RW. Accessible diagnostics. In (Wong J, ed) *Handbook of Regulatory Affairs in Asia*. First edition.

8. UNAIDS. Early diagnosis and treatment save babies from AIDS-related death. https://www.unaids.org/en/resources/presscentre/featurestories/2009/may/20090527unicef.
9. Meggi B, Bollinger T, Mabunda N, Vubil A, Tobaiwa O, Quevedo JI, et al. (2017) Point-of-Care p24 Infant testing for HIV may increase patient identification despite low sensitivity. *PLoS ONE* 12(1): e0169497. doi:10.1371/journal.pone.0169497.
10. World Health Organization. Advice on the use of point-of-care immunodiagnostic tests for COVID-19. April 8 2020.
11. World Health Organization. Interim advice on antigen detection for diagnosis of COVID-19 using rapid immunoassays. September 11 2020.

Chapter 10

Regulatory Specialists in Medical Devices in Europe: Meeting the Challenge of Keeping Current in a Changing Environment—How TOPRA Supports Professionals in a Dynamic Industry

Lynda J. Wight

The Organisation for Professionals in Regulatory Affairs (TOPRA)

Lynda@topra.org

10.1 The European MedTech Environment

Europe is an important market for medical devices and the medical technologies industry in Europe has an annual turnover of €110 billion, employing almost 600,000 people in more than 22,000 companies (see www.medtecheurope.org/). The sector is driven mainly by small and medium-sized enterprises (SMEs), which make up almost 95% of the industry although there is an increasing involvement of many larger pharmaceutical companies which have developed combination products and companion diagnostics. The sector in the EU has been growing on average by 4% a year over the past six years making it a

---

Medical Regulatory Affairs: An International Handbook for Medical Devices and Healthcare Products (Third Edition)
Edited by Jack Wong and Raymond K. Y. Tong
Copyright © 2022 Jenny Stanford Publishing Pte. Ltd.
ISBN 978-981-4877-86-2 (Hardcover), 978-1-003-20769-6 (eBook)
www.jennystanford.com

significant contributor both to the EU economy and to the improvement of health outcomes in the population.

The Medical Device Directives are the core legal framework in the EU and were harmonized in the 1990s. They form the foundation of Europe's regulatory framework but are currently being revised in a process that began in 2008. This revision had several purposes:

- to ensure a consistently high level of health and safety protection for EU citizens using these products
- to maintain the free movement of the products throughout the EU
- to ensure the legislative framework is suitable for products developed under the significant technological and scientific progress in this sector

The process to review the legislation has been protracted and the implications for regulatory professionals in the industry are not yet fully known. Meanwhile new aspects and technologies such as healthcare software continue to challenge both the industry and regulators.

For the individual regulatory professional, often working in a small team in an SME environment, this presents a number of challenges and raises several questions:

- How can I be aware of the legislation and procedures that currently apply to my products and ensure my company is compliant?
- How can I keep track of how new legislation is unfolding and can I have any influence on the process?
- Who can I turn to if I have issues that cannot be answered by colleagues in my company?
- How can I build relationships with other professionals in this sector which can assist me in my day-to-day work?
- How can I ensure I have the skills and competencies that will support me now and in my future career as it develops?

TOPRA exists to assist with these challenges for regulatory professionals in the medtech as well as the pharmaceutical and veterinary medicines sectors.

10.2   What Is TOPRA?

TOPRA is the professional membership organisation for individuals working in healthcare regulatory affairs. The members are individuals working in regulatory affairs all over the world—in more than 55 countries. They are actively involved in shaping and delivering high-quality regulatory services in a range of healthcare sectors including regulatory agencies, notified bodies, academia, regulatory affairs consultancies, clinical research organisations and many others.

TOPRA represents and promotes the global healthcare regulatory profession so that the value our members bring to the delivery of safe and effective healthcare products is appreciated.

TOPRA provides members with top-quality, relevant support with a particular focus on European matters. This support is tailored to match career stages to help members perform to the highest level and to help retain the brightest and the best within the profession.

10.3   Why Be a TOPRA Member?

The regulatory role in the medical technologies industry is a constantly evolving one, but there is an increasing need for a professional approach and a vehicle for those working in this sector to have a defined *professional identity*.

Our Statement of Values joins all members together under a united *framework of good practice*.

Our *community* of membership gives each member a voice and a means to contribute to debate and to influence the direction of regulatory development.

TOPRA is an independent, non-political organisation that acts as a *"safe space"* for the exchange of ideas and for debate on current issues.

All this is backed up with *information and services to support members* in their daily work.

10.4 What Support Is Offered?

- We provide independent space in which the members from different sectors can network, talk freely and share expertise—both face-to-face and online.
- Our membership structure allows individuals to signal their commitment, experience and currency by the use of MTOPRA or FTOPRA as a post-nominal.
- We host round-table debates on topical issues so that we can hear the views of our members and then publish information on the subject to invite wider debate.
- Our Special Interest Networks provide an opportunity for networking and information-sharing in a variety of specialist areas.
- We have regular meetings and discussion with legislators and other opinion formers to share the expertise of our members, and help shape the future of regulation.
- We produce guidance on standards for our members and encourage them to participate in continuing professional development
- We run our own training and development programmes at all levels, including a formal MSc postgraduate programme in medical devices, with a broad range of high-calibre speakers
- We keep our members informed of developments in regulatory matters by drawing together and publishing useful information in an easily accessible and digestible form, including through our monthly journal *Regulatory Rapporteur*.

10.4.1 Competency Frameworks

The profession of "regulatory affairs" is a relatively new one with key medicines Directives being enacted in Europe in the 1960s. It evolves as new legislation is enacted, and the pace of evolution has increased in recent years as medicine and technology have developed. The roles performed by regulatory specialists in companies are wide-ranging and varied, and in the devices sector, in particular, the job description in one company can be wildly different from the other.

Nonetheless, there are core competencies that all individuals, whether working in agencies or in a commercial setting, must attain and develop as their level of responsibility grows throughout their career. TOPRA has developed a competency framework to help employers identify the skills their employees must have, and for those employees to follow as they pursue their career goals.

10.4.2 Training and Development

The TOPRA training programme for regulatory professionals is geared to key career stages as shown in the following figure.

For each of these stages, there are training programmes to provide students with the knowledge and skills required through a blended approach of teaching, practical case studies and exchange of experiences.

Exploring: The "**Basics of Medical Device Regulatory Affairs**" is a one-day programme for those thinking of moving into the profession, but is also for those in SME organisations who need to understand the regulatory opportunities and constraints in order to perform their roles (for example, in marketing) more effectively.

Establishing: Based on a successful model that has been in place for over 35 years, the **Medical Devices Introductory Course** is a three-day intensive residential training programme in which students are taught by and interact with industry experts, notified bodies, regulatory agency representatives and consultants. This course gives a firm foundation to anyone looking to make a career in medical technology regulatory affairs.

Consolidating: Regulatory professionals who are moving into new areas or taking on additional responsibilities need to refresh, update or deepen their understanding. Courses in the TOPRA **CRED** (Continuing Regulatory Education and Development) series are designed to meet these needs. In 2016 , new Devices courses, including one on software, will be added to the selection already available. The content of these programmes is developed by active TOPRA members to ensure the material is relevant, up to date and, above all, practical. Classic teaching is complemented by group case study work.

Driving: Regulatory affairs specialists are increasingly taking senior strategic roles and need to have a comprehensive understanding not only of the technical and legal issues but also of management and strategic planning. The TOPRA **Masterclasses** provide immersion into the topic with in-depth information on the latest regulatory initiatives and issues, case studies and interaction with leading figures from industry, regulatory authorities, notified bodies and the academic sector.

Influencing: At the most senior level, a regulatory professional may expect to shape future legislation and the best way to do this is to be involved in TOPRA's **Horizon** programme of independent conferences, including the flagship **Annual Devices Symposium.** Here there is an opportunity to dialogue with European Commission representatives and a global faculty at the highest level. Each year, in a unique model for the devices sector, the conference is staged in partnership with a different EU regulatory agency giving opportunities to establish new relationships and develop a deeper understanding of the issues across the continent. This unique model means the Symposium is seen by many as the leading event in Europe.

10.4.3   Education

TOPRA has developed a world-leading MSc qualification in Medical Technologies Regulatory Affairs (MTRA), which is delivered in partnership with Cranfield University in the UK. This university is solely for post-graduate students and has a global reputation in engineering and management, making it an ideal choice to accredit the MTRA MSc. The course is part-time, based on eight face-to-face modules designed to fit in with a busy work schedule. Once again, the training is delivered by experienced regulatory professionals and a key part is the interaction between students and experts. The MSc dissertation is a chance for the student to undertake original research which could benefit their employer and adds to the body of knowledge in this emerging profession.

10.4.4 Networks and Collaborations

In any regulatory role, access to colleagues who can assist, inform and support is crucial—especially for those whose working environment can make them feel isolated. The TOPRA Medical Devices Special Interest (SPIN) group is an international network of TOPRA members who can communicate online in a closed environment and support each other. The management team of the group also arrange webinars and teleconferences on topics of immediate importance so that SPIN members can access new information at their desks and get answers to questions as they come up in their day to day work.

10.4.5 Information

All TOPRA members receive the peer-reviewed journal *Regulatory Rapporteur*. Internationally valued as a definitive source of regulatory intelligence and analysed information, this journal regularly features medical device topics. The members also have access to an online version as soon as it is published and a searchable archive of past issues.

10.5 Looking to the Future

There is no doubt that the regulatory environment in the EU is entering a phase of seismic change. At the same time, the medical devices industry is producing ever more complex products which defy classical regulatory approaches. A partnership between manufacturers and those charged with developing and implementing the legislative framework is more important than ever and can only be achieved through mutual trust and open debate in a neutral environment.

TOPRA provides the independent safe space where this partnership can develop and underpins this with the tools and resources to make regulatory professionals in the medical devices sector the best they can be.

For more information, see www.topra.org.

Part 2

Medical Device Safety and Related ISO Standards

Chapter 11

Biomedical Devices: Overview

Piu Wong

The Whitehead Institute for Biomedical Research, Cambridge, USA
pwong0610@gmail.com

11.1 Historic Aspect of Medical Devices

Medical devices are part of our daily life. They could range from simple equipment such as a thermometer to a highly sophisticated machine such as a pacemaker for a person with heart problem. They all serve to meet the medical needs. In contrast to pharmaceutical products, their modes of action are not pharmacological, metabolic, or immunological. Some of the landmarks in medical device development are illustrated in Table 11.1 [1–2].

Between 1980 and 2000, high-resolution imaging devices, in particular, radiographic and fluoroscopic units, became more commonly used in hospitals and clinics. Devices designed for the real-time monitoring of cardiovascular parameters such as heart rate and blood pressure were part of the standard equipment. Moreover, there was no effective treatment for some of the

---

Medical Regulatory Affairs: An International Handbook for Medical Devices and Healthcare Products (Third Edition)
Edited by Jack Wong and Raymond K. Y. Tong
Copyright © 2022 Jenny Stanford Publishing Pte. Ltd.
ISBN 978-981-4877-86-2 (Hardcover), 978-1-003-20769-6 (eBook)
www.jennystanford.com

diseases such as chronic kidney failure until dialysis machines were used to greatly improve their quality of life. Computerized axial scanners (CT) and magnetic resonance imaging (MRI) have become crucial tools in medical imaging to supplement X-rays and ultrasound for screening disease such as patients with a high risk of colon cancer.

Table 11.1 Landmarks in medical devices

Year	Device	Inventor
1943	Dialyzer	Dutch physician, Willem Kolff
1950	Artificial hip replacement	Performed by English surgeon Sir John Charnley
1951	Artificial heart valve	US team led by electrical engineer Miles Edwards
1952	External cardiac pacemaker	US cardiologist Paul Zoll
1960	Internal pacemaker	US electrical engineer Wilson Greatbatch
1970	Computerized axial tomography (CT) scanner	Nobel Prize winners Godfrey Hounsfield and Allen Cormack
1977	Magnetic resonance imaging	US team led by physician Raymond Damadian
1982	Artificial heart	Willem Kolff
1985	Implantable cardioverter defibrillator	Polish cardiologist Michel Mirowski
2000–2010	Robotic medical devices, such as NeuroArm and CyberKnife	Various inventors (http://www.mritechnicianschools.org/twelve-amazing-robots-that-are-revolutionizing-medicine/)

It is estimated that approximately 50,000 people died because of the human error [3]. From 2000 onward, robotics became a reality and began to be used in clinical procedures where extreme precision is critical. For example, a minor error in neurosurgery will lead to paralysis. NeuroArm combines MRI and a surgical robot to perform microsurgery and biopsy-stereotaxy with high precision [4].

Another advantage is that remote-controlled surgery is possible if the doctor is not next to the patient. Da Vinci Surgical System has developed over 100 devices so that surgeons can remotely access the machine and perform surgery such as double vessel bypass surgery in the human heart [5].

11.2 Biomedical Market Environment

Aging population, enhanced awareness of the patients, and improved quality of the medical system are the driving forces for the continued demand and growth of the medical device industry in Asia. HSBC's "The Future of Retirement" report estimates that the number of people aged 65 or older will reach 1.4 billion by 2050. The United Nations also predicts that the number of people aged 60 or older will reach 2 billion by 2050, representing 32% of the world's population. For example, in China, 25% of the population will be over 65 by 2050. In Japan, 35.6% of the population will be over 65 by 2050. Standard & Poor's (S&P) reported that 80% of the older individuals suffer from at least one of the chronic health conditions, such as arthritis and cataracts [6, 7]. Therefore, there is a strong need and market to develop technologies to better diagnose disease and improve the treatment for the aging population.

The global medical devices industry has generated revenues in excess of $300 billion in 2011 worldwide [8]. Despite the fact that Asia has more than 60% of the world's population, the healthcare expenses constitute about 15% of the global healthcare expenditure. This suggests that there is an enormous upside potential in the Asian market to improve the healthcare system possibly by better diagnosis and treatment using advanced medical devices. For example, China's import of US medical devices has increased steadily, and it is the second largest market for US medical devices after Japan, with 1.5 billion US medical devices imported in 2008 [9].

The United States is the largest producer of medical devices worldwide. S&P reports that US medical device manufacturers receive 40–50% of their revenues from the foreign markets. The largest US-based medical device companies based on S&P Healthcare products and supplies survey, 2010, are Johnson & Johnson, GE Healthcare, Medtronic, Boston Scientific, Abbott Labs, Becton Dickinson, Stryker, Zimmer, and St. Jude Medical. Companies based outside the United States, such as Siemens (Germany), Hitachi Medical Corporation, and Toshiba (Japan), are major players in the biomedical device industry [10]. US firms are building manufacturing and marketing centers in Asia to streamline manufacturing and distribution efficiencies. High-technology medical products that address the prevalent health issues, especially for the aging population, have a very positive outlook in the future. Here I will illustrate the importance of medical devices by focusing on only diagnosis and treatment of age-related diseases such as arthritis and vision degeneration.

11.3 Orthopedics

11.3.1 Market

Approximately 1 million patients worldwide are treated annually for total replacement of arthritic hips and knees joints. One in forty persons requires orthopedic surgery because of the common joint disorder, osteoarthritis. The number is increasing because of the aging population and enhanced awareness among patients. Stryker Corporation reported that the 2009 worldwide market for reconstructive devices was in the order of $12 billion in size [11]. Knee reconstruction is the most common surgery, constituting 50% of the total joint replacement market. For example, DePuy Rotating Platform Knee mimics the human knee's natural range of motion (Fig. 11.1). Spinal surgery such as thoracolumbar (relating to thoracic and lumbar region of the vertebral cord) implant generates about $5.5 billion worldwide revenue. The United States contributes about 60% of the global orthopedic device market. Major players include Johnson & Johnson (DePuy), Stryker, Zimmer Inc, Medtronics, and Biomet [12].

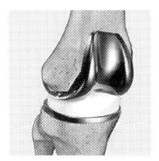

Figure 11.1 The DePuy Rotating Platform Knee mimics human knee's natural range of motion.

11.3.2 Materials

Lightweight, strength, and biocompatibility are the key requirements for the material to be chosen for medical implants. The most commonly used orthopedic implant materials are stainless steels, cobalt–chromium alloys, and titanium alloys [13]. Stainless steel is often preferred for making bone plates, screws, pins, and rods because of its resistance to chemicals and environmental conditions found inside the human body. Cobalt–chromium alloys are used in joint replacement and fracture repair implants. Titanium alloys have a significantly higher strength-to-weight ratio than stainless steels. The higher flexibility offered by titanium alloys allows the design of shape-memory-based implants to aid in bone in-growth for better grip. Trabecular metal, which is made from tantalum over carbon, is also strong, flexible, and biocompatible and is porous enough to allow tissue in-growth. Ceramic materials are also used to make implant surfaces that rub each other but do not require flexibility, such as the surface of the hip joint [14].

11.3.3 Biocompatibility

The evaluation of medical device compatibility determines if the contacts of the device with the body produce any harmful local or systemic effect, e.g., degraded products are carcinogenic. The ISO 10993 guidelines have been developed to cover the testing of materials and devices that come into direct or indirect contact with the patient's body [15]. The very first consideration of the selection of materials shall be fitness for purpose with regard to characteristics of the materials such as chemical, toxicological, physical, electrical, morphological, and mechanical properties. The extent of the test depends on the existing pre-clinical and clinical safety data. More stringent tests will be carried out for a novel material. Some of the critical issues include characterizing the material for genotoxicity, carcinogenicity, and reproductive toxicity, irritation and skin sensitization in vitro cytotoxicity, and identification and quantification of degradation products from polymeric medical devices ceramics and metals. Importantly, testing should be done on the sterile final product, or representative samples processed in the same manner as the final products [16].

11.3.4 Fabrication

The characteristics of the metal or the biomaterial that I described earlier can be changed depending on the methods to shape the material into the implant, a process called fabrication. In other words, some fabrication methods can strengthen the material, while others can weaken them [17]. The most common fabrication methods used for metal implants are machining, investment casting, and hot and cold forging.

Machining of implant materials is usually performed with computer to allow high precision and reproducibility. For example, cobalt–chrome stem caps for hip transplant are often made using automated machining process. Investment casting is often employed for more complex shapes such as knee implants. Metal is melted and poured into a ceramic mold to create the desired shape of the implant. Forging process presses materials into shape between two molds. This often involves heat treating for molding (hot forging), or the metals are shaped at room temperature (cold forging) [18]. Implants made by forging are often stronger than similar parts made by casing.

11.3.5 Polyethylene Fabrication

Turning powdery polyethylene into a solid piece of polyethylene is called consolidation. Different ways of consolidation can influence the characteristics of the final implants. One of the consolidation processes is called ram extrusion. The polyethylene powder is heated and forced out through a tube. The combination of heat and pressure deposits the powder into the solid bar for further machining. Another way to fabricate polyethylene implants is called compression molding, where the raw powder is heated and then pressed into sheets that can be cut and shaped to create the implant. A slight variation of this process is called net-shape compression molding. The powder is directly deposited into the final implant shape, instead of some intermediate product. This method improves wear reduction of the finished material. After fabrication, heat-treating the metal implant and then slow cooling it reduces brittleness [19, 20]. Because of its durability and performance, metal-on-polyethylene has been the leading artificial hip component material chosen by surgeons since the approval by the US Food and Drug Administration (FDA) 30 years ago. The metal ball is cobalt chrome molybdenum alloy and the liner is polyethylene. This ability to adapt and customize during the surgical procedure is an important attribute of polyethylene (Fig. 11.2).

Figure 11.2 This artificial hip is composed of cobalt chromemolybdenum alloy made metal ball and polyethylene liner. Polyethylene liner provide flexibility during surgical procedure. (http://bonesmart.org/hip/hip-replacement-implant-materials/).

11.4 Vision Care

11.4.1 Market

Age-related vision disorders such as cataract, muscular degeneration, and glaucoma have boosted the growth of eye-care industry in the area of cosmetic surgery such as the implantation of refractive lens. BCC Research reported that the global market for ophthalmic devices, diagnostics, and surgical equipment was valued at $15.2 billion in 2009, excluding consumer eye-care products [21]. The market is estimated to reach $19 billion by 2014. Major manufacturers include Advanced Medical Optics, Bausch & Lomb, Canon, Nidek, Topcon, and Zeiss Meditec.

11.4.2 Diagnostic Devices

Diagnostic instruments are the key tools to identify diseases by high-resolution inspection of the cornea and the macula. One of the most commonly used equipment among ophthalmologist is slit lamp (Fig. 11.3). The most important field of application is the examination of the anterior segment of the eye, including the crystalline lens and the anterior vitreous body. The illumination system produces a slit image that is as bright as

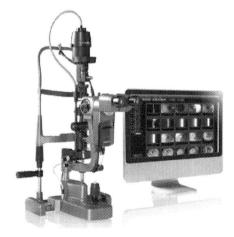

Figure 11.3 Digital slit imaging used for the detection of cataracts, corneal injury, and other eye diseases (http://www.in.all.biz/g182258/).

possible, at a pre-defined distance from the instrument. The slit imaging system is used for detecting cataracts, corneal injury, muscular degeneration, and retinal detachment. Another commonly used procedure is tonometry, which is used to determine the intraocular pressure, the fluid pressure inside the eye. It is a critical test to evaluate the risk of glaucoma [22].

11.4.3 Treatment

Cataracts lead to blurred and cloudy vision. Cataracts grow very slowly and become denser over time. More than 90% people have a cataract and half of them will lose part of the vision by age 75 [23]. Cataract surgery is the most effective treatment. Further advancement of technology in intraocular lens implants such as multifocal refractive lens has significantly improved the vision of the cataracts patients [24], e.g., the implant of artificial lens such as ReZoom IOL, marketed by Abbott restore vision (Fig. 11.4), after the removal of patients' clouded native lens.

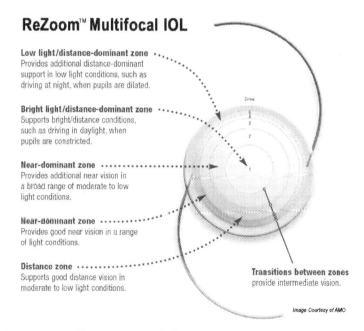

Figure 11.4 ReZoom is a multifocal refractive intraocular lens that scatters light over five optical zones, as indicated, to provide near, intermediate, and far vision.

11.5 Diabetics

Glucose monitoring is essential for diabetes management. LifeScan, J&J company One Touch Ultralink blood glucose monitoring system allows accurate estimation of blood glucose [25]. In combination of an insulin pump, proper dosing can be administered in response to blood glucose level before or after meals. To streamline these processes, an artificial pancreas system helps reduce the severity of a drop in glucose levels by automatically adjusting insulin flow. The system combines a continuous glucose monitor, an insulin infusion pump, and a glucose meter [26].

11.6 Obesity

Obesity is a major global problem affecting billions of people. Unfortunately, there is no single FDA-approved pill that can effectively stop this pandemic. In addition to the traditional obesity surgery to reduce the size of the stomach, there are a few less invasive devices that slow down digestion, making people feel fuller longer. EndoBarrier, is a flexible tube inserted into the small intestine, forming a barrier between food and intestinal walls. It has been approved in Europe and Australia and clinical trials are being run in the United States [27]. SatisSphere is a device that can be placed in the small intestine, mimicking the shape of duodenum. Therefore, the food journey is delayed and possibly gives the body wrong perception that plenty of food has been taken [28].

11.7 Vascular Disease

One of the major vascular diseases, abdominal aortic aneurysms (AAA) is a localized swelling (ballooning) of the abdominal aorta, caused by the weakness of the aortic wall. Untreated condition is likely to cause rupture and associated death in one-third of patients. Although there are about 27 million people worldwide suffering from AAA, only 20% are detected [29]. Ultrasound screening for AAA can be adopted to improve the accuracy of the detection. Endovascular aneurysm repair (EVAR) is a minimally invasive procedure that is also used in the treatment of AAA. In an EVAR, a manufactured device known as a stent-graft is deployed under x-ray guidance inside the aneurysm to exclude it from the circulation (30). The principle is illustrated in Fig. 11.5.

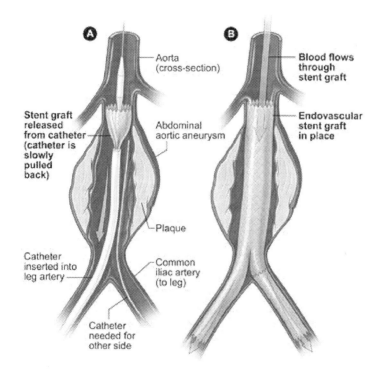

Figure 11.5 Reconstruction of a stent graft in AAA. (A) A catheter is inserted into the weakened artery. The stent graft is then released. (B) The stent supports the artery walls and divert the blood to pass through the aneurysm, therefore preventing it from rupture (http://www.hearthealthywomen.org/treatment-and-recovery/pvd-treatment-and-recovery/aortic-aneurysm-repair/page-3.html).

11.8 Concluding Remarks

Biomedical devices are diverse and rapidly evolving. This brief introduction aims to highlight representative examples, and we hope the readers will gain an appreciation of medical devices in everyday life. Recent trends have indicated that various health care-related disciplines such as biological sciences, nanotechnology, cognitive sciences, information technology, and material sciences converge into the development of medical devices. This will ultimately improve the quality of medical devices.

References

1. Health Technologies Timeline. Retrieved October 10, 2011, from http://www.greatachievements.org/?id=3824.
2. Reiser SJ. (1978). *Medicine and the Reign of Technology*, Cambridge University Press, New York.
3. Kohn LT, Corrigan JM, Donaldson MS (1999). *To Err Is Human: Building a Safer Health System,* National Academy Press, Washington, D.C.
4. Neuroarm. Retrieved October 11, 2011, from: http://www.neuroarm.org/.
5. The *da Vinci*® Surgical System. Retrieved October 11, 2011, from http://www.davincisurgery.com/davinci-surgery/davinci-surgical-system/.
6. *World Health Statistics 2009.* (2009). Geneva, World Health Organization. Retrieved October 20, 2011, from *www.who.int/entity/.../**world_health_statistics**/EN_WHS09_Full.pdf.*
7. Department of Economic and Social Affairs, Population Division (2002). *World Population Ageing, 1950–2050.* United Nations, New York.
8. Medical Device Revenue to Top $300 Billion This Year: Kalorama. Retrieved February 25, 2012, from http://www.kaloramainformation.com/about/release.asp?id=1826.
9. U.S. Department of Commerce and the U.S. International Trade Commission (2009). *Medical Devices Industry Assessment.* US Trade Department, USA.
10. Industry Surveys Healthcare Product and Supplies. Retrieved October 19, 2011, from http://issuu.com/alanahealthcare/docs/s-p-healthcare-product-and-supplie_20110819_142411.
11. Handout on Health: Osteoarthritis. Retrieved October 30, 2011, from http://www.niams.nih.gov/health_info/osteoarthritis/default.asp.
12. Tanna S (2004). *Osteoarthritis—Opportunities to Address Pharmaceutical Gaps.* Geneva, World Health Organization, Switerland.

13. Breme HJ, Helsen JA (1998). *Metals as Biomaterials*, Chichester, John Wiley & Sons, United Kingdom.
14. Alireza N, Peter DH, Cui'e W (2007). Biomimetic Porous Titanium Scaffolds for Orthopedic and Dental Applications, Institute for Technology Research and Innovation, Deakin University, Australia.
15. Practical Guide to ISO 10993. Retrieved October 30, 2011, from www.distrupol.com/images/ISO_10993.pdf.
16. Use of International Standard ISO-10993, "Biological Evaluation of Medical Devices Part 1: Evaluation and Testing" Retrieved October 28, 2011, from http://www.fda.gov/MedicalDevices/DeviceRegulationandGuidance/GuidanceDocuments/ucm080735.htm.
17. Nakajima H (2007). Fabrication, properties and application of porous metals with directional pores. *Progress in Materials Science*, **52**, 1091–1173.
18. Ryan G, Pandit A, Apatsidis DP (2006). Fabrication methods of porous metals for use in orthopaedic applications. *Biomaterials*, **27**, 2651–2670.
19. Avitzur B (1987). Metal forming, *Encyclopedia of Physical Science & Technology*, **8**, 80–109.
20. Drozda T; Wick C, Bakerjian, Ramon V, Raymond F, Petro L (1984). *Tool and Manufacturing Engineers Handbook: Forming, SME.*
21. Global Markets for Ophthalmic Devices, Diagnostics and Surgical Equipment. Retrieved October 30, 2011, from http://www.bccresearch.com/report/ophthalmic-devices-markets-hlc083a.html.
22. Amm M, Hedderich J (2005). Transpalpebral tonometry with a digital tonometer in healthy eyes and after penetrating keratoplasty. *Ophthalmologe*, **102**(1), 70–76.
23. Cataract. Retrieved October 27, 2011, from: http://www.who.int/topics/cataract/en/.
24. ReZoom multifocal lens. Retrieved 25 October 2011 from http://www.rezoomiol.com/rezoom_multifocal_lens.html.
25. One touch. Retrieved October 25, 2011, from http://www.onetouch.com/our_products.
26. Artificial pancreas system. Retrieved October 25, 2011, from http://www.fda.gov/MedicalDevices/ProductsandMedicalProcedures/HomeHealthandConsumer/ConsumerProducts/ArtificialPancreas/default.htm.
27. Endobarrier. Retrieved October 24, 2011, from: http://www.endobarrier.com/.

28. Columbus may be the home to the next big craze in weight loss. Retrieved October 22, 2011, from http://www.examiner.com/extreme-weight-loss-in-columbus/columbus-may-be-the-home-to-the-next-big-craze-weight-loss.
29. First patients enrolled in Cordis Trial of New Stent Graft System to Treat Abdominal Aortic Aneurysm. Retrieved October 23, 2011, from http://www.investor.jnj.com/textonly/releasedetail.cfm?ReleaseID=457755.
30. Greenhalgh RM, Powell JT (2008). Endovascular repair of abdominal aortic aneurysm. *N. Engl. J. Med.*, **358**(5): 494–501.

Chapter 12

Labeling, Label, and Language: A Truly Global Matter

Evangeline D. Loh and Jaap L. Laufer

Emergo Group

evangeline@emergogroup.com, Jaap.Laufer@dekra.com

12.1 Introduction

For most medical devices, the label on the product is the primary interface between the user and the product. As an integral part of the product proper, it provides a visual way to verify that the product in the packaging is what the user intends to use and ensures the packaging is used as intended (e.g., for a sterile, double packed product). But labeling is a lot more—it encompasses the whole interface between the user and the manufacturer. Labeling tells the reader what the product may be used for and when not, what benefits and risks it is expected to have, and which precautions the user should take or consider when used as intended. Some legal labeling definitions include any and all promotional utterances, while others restrict its remit to the printed materials

---

Medical Regulatory Affairs: An International Handbook for Medical Devices and Healthcare Products (Third Edition)
Edited by Jack Wong and Raymond K. Y. Tong
Copyright © 2022 Jenny Stanford Publishing Pte. Ltd.
ISBN 978-981-4877-86-2 (Hardcover), 978-1-003-20769-6 (eBook)
www.jennystanford.com

that accompany the product. Labeling also establishes the use intended by the manufacturer and depending on the classification of the product, may have been reviewed in the regulatory submission.

But all labeling must be based on the risk management performed by the manufacturer during design and development, no exceptions. This pivotal document dictates not only in essence the labeling content but also the fundamentals upon which the clinical evaluation of the product should be based. So, as the product is on the market, the experience gained in the "post-market" phase will dictate through its feedback on the risk management an evolution of the labeling content commensurate with the stage in the life cycle of the device.

Labeling and languages are often mentioned in one breath. However, languages required by the respective countries or economies are just a subsection of the need to adapt the language of the instruction for use to the intended user or users. Complicated use of even absence of the native language of a user may defy the purpose of communicating the essential elements of the information required to use the device safely and for its intended use. So, labeling language should be interpreted in those two distinct ways. It constitutes a critical aspect of Human Factors in Design and other core considerations when creating a new product or adapting an existing one.

An aspect not always appreciated is that "world" languages such as English, French, Spanish, and increasingly, Mandarin Chinese, have an (often unintended) impact beyond the national borders of the jurisdiction for which they were intended. This may lead to conflicting versions of label content in different countries and, thus, cause confusion among the readers. This calls for a global harmonization of labeling content—a lofty but often elusive goal. Note that label content is distinguished from the indications for use, which is also conveyed by the labeling. There can often be different indications for use based on what the manufacturer was able to substantiate in their regulatory submission in a particular market. (Or stated another way, what the regulatory authorities in a particular market were comfortable to accept as indications for use based on the clinical data.)

Finally, the revolution in information has impacted the requirements put forward by notably hospitals and other electronically advanced users. While we included the information available at this time, this is most certainly the aspect that will need to be updated in the next five years.

12.2 Definition of Labeling

Labeling means different things in different countries. The European Medical Devices Directive 93/42/EEC as amended by Directive 2007/47/EC (the MDD) does not define labeling per se or its constituent components. Apart from the ubiquitously present stipulations in the MDD when and where the CE-marking must be affixed, par. 13 of Annex I (Essential Requirements) spells out in great detail which information must be supplied by the manufacturer. The elements of labeling mentioned there are limited to the information accompanying the device (i.e., package insert or instruction for use, which includes a manual for large or complex devices) and the label(s) proper for individual units either on the product itself or on the unit packaging, on the "sales packaging." Terms such as "shipping carton," labeling for double packaging, and other commonly used designations have not been codified in the law.

One of the key elements of labeling in the near future will be the use of a UDI, a unique device identifier. While the idea itself makes eminent sense, it must be feared that in absence of a well-defined electronic structure allowing for the proper handling and management of the huge quantity of data influx, the objective of the electronic tracing of individual products from the manufacturer to the user will stay an illusion for many years to come. To this end, the Global Harmonization Task Force (GHTF) has issued guidance on UDI [1] (UDI System, September 16, 2011) which implores countries to consider the GHTF guidance in promulgating regulations on UDI. It would appear the USA FDA will be the first in line to publish a draft rule (expected in the fall of 2011) [2]. The FDA has published this draft UDI as a proposed rule (Federal Register, Vol. 77, No. 132, 10 July 2012). http://www.gpo.gov/fdsys/pkg/FR-2012-07-10/

html/2012-16621.htm; however, the EU, with their revision of the medical devices directives, will similarly seem to require UDI [3]. The draft Medical Device Regulation introduces the use of UDI. http://www.gpo.gov/fdsys/pkg/FR-2012-07-10/html/2012-16621.htm.

As the concept of "labeling" in Europe means different things in the respective EU Member States, vastly different regulatory regimens exist. France, Germany, the United Kingdom, and even Italy and Spain show an increasingly strong enforcement of their respective laws concerning promotion, including the Internet. The enforcement of a harmonized label content (especially concerning intended uses, indications, and claims as well as contraindications) is just beginning, enhanced by the COEN (Committee of Enforcement), and lately by the CMC (Central Management Committee), two Member State-only committees that promotes a coordinated approach toward interpretation and implementation of laws and regulations for devices.

Based on Title 21, Section 820.1 of the Code of Federal Regulations (CFR), the United States has had a long and historically driven tradition of strong and sometimes even heavy-handed enforcement of labeling compliance. Labeling in the United States means any written or verbal expression of information accompanying or concerning the device and controlled in any way by the manu-facturer, including but not limited to labels in the strict sense, promotional materials, any verbal statement by company representatives, the Internet, and even information about products not yet approved but shown to the public at trade fairs. The FDA is strictly monitoring the compliance of labeling content with the claims approved. Any proof that a manufacturer is intentionally or commercially exploiting the often gray area of "off-label" use can lead to draconian fines and/or criminal prosecution.

In Korea, a medical device label is one attached onto the outer most packaging though there does not appear to be an official reference to the definition of labeling. In China, NMPA (National Medical Products Administration) Order No. 10 regulates the labeling (Instructions, Labels, and Packaging of Medical Devices). There are 23 Articles in the Order. Countries such as Singapore and Hong Kong, which use more GHTF (or

AHWP)-like regulatory systems, have similarly defined labeling as in the EU.

12.3   Elements of Labeling

Labeling may consist of the following:

- The label proper (see above) increasingly including the UDI code.
- A manual, instruction for use, package insert, etc., intended for the intended user(s).
- Where applicable or demanded by laws or regulations, promotional materials where they relate to the "approved" claims or intended uses. For instance, where an insulin pump must not be used when the user is swimming, the manu-facturer or distributor cannot show a lady on the beach wearing that pump. (Historic example! The case was whistle-blown by a competitor and the manufacturer cancelled the ad.)
- Materials placed by a manufacturer or a distributor of the product on the Internet or spread by e-mail, social media, involuntary pop-ups, etc. This is such a growing and complicated subject that we will treat it separately further in this chapter. Suffice it to say here that electronic utterances that are in perfect compliance with the laws in one country may be offensive in another, even when the national language of the latter country is not used. A historic example springs to mind: When a representative of a UK-based company showed a training video to illustrate the use of a new invasive surgical device (hardly glamorous) to physicians, he was arrested in Libya for "pornography." The video had used a female patient.

A good source for labeling content guidance is the GHTF document "SG1-N70:2011 Label and Instructions for Use for Medical Devices," issued on September 22, 2011. It is a bit general but gives a solid basis for which aspects of labeling should absolutely be considered before placing a device on the market. Also, while there are country differences and deviations, there is generally a core set of elements in labeling, which is captured in the GHTF guidance.

In many countries in Asia, South America, and even North America (Mexico specifically), there is a requirement that the medical device registration number be included on the labeling.

In Korea, labeling is generally a sticker that is attached onto every product intended for the Korean market, which must include the information about the following: KFDA registered model number, manufacturer information, Korean license holder information, KFDA registration number, manufacturing date/lot number, etc. In China in particular, the NMPA Order No. 10, Article 8, stipulates the number of the Import Medical Device Registration Certificate (IMDRC) number as well as the reference to the relevant technical standards.

The Australian system leverages the EU system, and generally labeling developed for the EU can be used in Australia with the addition of the Australian Sponsor. In New Zealand, the Medicines Regulations 1984, Regulation 12(4) [4], establishes the requirements for the labeling of medical devices. The regulation states, "No person shall sell any medical device that does not bear the name of the manufacturer of the medical device or the name of the manufacturer's distributor in New Zealand."

In Hong Kong, the Medical Device Administrative Control System (MDACS) is modeled on the recommendations of the GHTF, and, thus, compliance should not be particularly burdensome for products that have already met the criteria in the EU. The labeling requirements are delineated in Appendix 3, Additional Medical Device Labelling Requirements of the guidance, GN-01. Again, the device's Listing Number ("HKMD No. ####") needs to be on the label. There is guidance on labeling in the form of the Singapore Health Sciences Authority (HSA) GN-23 Guidance on Labelling for Medical Devices, August 2009; and, again, the medical device regulatory system in Singapore is largely based on the GHTF guidance documents.

The applicable Brazilian labeling requirements are described in Resolution RDC No. 185, Annex III.B, information on the labeling and IFU for medical devices. The name and address of the Brazilian importer should be included if appropriate (2.1), but more specifically, the Brazilian Registration Holder, technical expert certified by ANVISA, technical expert's name must be delineated on the label (2.11). Also, once ANVISA has accepted the application, the ANVISA registration number must be included on the product label preceded by "ANVISA" (2.12). For the most part, Resolution RDC No. 185, Annex III.B, resembles the MDD, Annex I, ER, Section 13. Information supplied by the manufacturer.

In Chile, at the writing of this section, only a limited number of products are subject to mandatory control by the ISP: examination gloves, surgical gloves, condoms, and sterile hypodermic needles and syringes for single use. The regulatory framework for medical devices is based on Law 19,497 and Products Control Regulations and Medical Use Elements (DS No. 825/98), and Article 26 discusses the labels. Again, as observed in many countries, the Chilean device registration number issued by the ISP must be included as well as all the standard elements.

12.4 Risk Management, Clinical Evaluation and Labeling: The Core Triangle for Safe and Effective Use of the Device

Few risk analyses are concluded without specific inclusion of risk mitigating statements in the labeling, mostly in the instructions for use. These statements are commonly divided into contra-indications, warnings when used as intended, precautions to be taken before or after device use, and which side effects may be expected, as well as other considerations. Most countries do not define under which header a particular statement must be placed; however, this aspect is usually emphasized by a regulatory authority whenever a risk mitigating statement is insufficiently complied with by the user, sometimes replete with the instruction to print bold or similar.

Clinical evaluation and investigation are the tool to validate critical labeling content, e.g., under what conditions the device should be used, what performance specifications may be expected, what precautions need to be taken, etc. Very important, the specific nature, relevance and incidence of adverse events or incidents must be critically evaluated, as it directly relates to the expectations expressed in the initial risk assessment. The final conclusion about the risk/benefit ratio is frequently determined by the capability to address the residual risks adequately by explicitly stated warnings in the instruction for use or the manual. Any risk that turns out to have been underestimated in any way will have to be reassessed, especially whether it can be mitigated by improved design and/or labeling content, or poses an insurmountable block to further use.

So, even though the labeling content is drafted in the course of the design development, the adequacy of the instruction for use can only be verified when it is applied in practice, i.e., in a clinical investigation or in the post-market phase. (Note also, the regulatory authority review of the regulatory submission can lead to changes in the labeling content as well.) Consequently, the post-market surveillance (PMS) is the single most important feedback phase in the life cycle of a label. This is where refinements of the actual handling, expanded indications (if they do not require further corroboration), more detailed warnings, precautions, etc., as well as government-dictated insertions occur. For instance, in a case not too long ago, the United Kingdom demanded that for a relatively low-risk device, a cautionary statement be included about the lack of data for pediatric use, and consequently, the restricted use in patients under 16. In another case, the French agency determined that the package insert of mammary implants must display a warning that such products may interfere with the effectiveness of classic mammography.

This leads to one more consideration: In this global age, any intervention at the government level of one country will lead to inquiries by other regulatory authorities, especially if the initial intervention originated in one of the GHTF founding countries, but increasingly also in emerging economies. This puts a big stake in the proper formulation of label content.

In conclusion, any significant deviations between the risk assessment, the results of clinical evaluation, and the labeling will likely lead to problems for the manufacturer. Both competitors and regulatory authorities are watching this matter closely!

12.5 Labeling and Promotion

A general rule is that promotion must be consistent with labeling. A medical device cannot be promoted beyond the comments delineated in the labeling for which the manufacturer obtained the regulatory approval, clearance, or registration. It is not common for manufacturers of medical devices to be marketing professional-use medical devices to consumers.

Widely different regulatory regimens exist for promotional materials for devices. For example, in the EU, promotion and advertising are not explicitly discussed in the MDD, nor in the proposed Draft Revision of the Directives.

Promotion directed to patients or lay users may also be regulated differently [5]. This is specifically the case in Australia, where a code exists: the Therapeutic Goods Advertising Code 2007 [6]. Advertisements related to medical devices directed to consumers must comply with this code.

Many lesser developed economies do not have specific regulations for promotional materials, or they may be regulated by pharmaceutical laws.

12.6 e-Labeling, Web Sites, Internet, and Social Media: A Brave New World for Labeling

While the term e-labeling has been widely accepted to mean electronic labeling, the official reference to the source was not readily apparent, of course, an adoption from the moniker electronic mail, "email." For the purposes of this discussion, electronic should mean non-paper though this is a moniker that each regulatory environment will need to define.

At the time of this writing, Europe appeared to have made the greatest accomplishment with e-labeling. A European guidance, MEDDEV, was published in 2007 on e-labeling of IVDs [7]. The guidance established the use of the other-than-paper format IFUs (different media) by different means of supply for certain categories of in vitro diagnostic devices. There were, of course, very stringent provisions that needed to have been addressed in order to supply IFUs by different media and through different means of supply, not withstanding a toll-free telephone number. A toll-free number within the 27 EU member states as well as Norway, Iceland, Lichtenstein, and Switzerland is quite a feat! In the summer of 2011, the European Commission issued draft regulations on electronic labeling (e-labeling) and shared these publically as part of their WTO Technical Barriers Trade activities [8]. At that time, the publication date was December 14, 2011. While this date has passed, this legislation is expected in Q1 2012. There is now published an EU Commission Regulation (Regulation (EU) No 207/2012) on electronic instructions for use of medical devices. http://eur-lex.europa.eu/LexUriServ/LexUriServ.do?uri=OJ:L:2012:072:0028:0031:EN:PDF.

While recitals are in essence background comment without legal merit, Recital (3), of the draft indicates: "In order to reduce potential risks as far as possible, the appropriateness of the provision of instructions for use in electronic form should be subject to a specific risk assessment by the manufacturer." Ultimately, provided the medical device meets the provisions of the legislation, this still needs to be considered. This, of course, is similar to the recommendation for labeling, provided the device in the EU is a category that qualifies for e-labeling; ultimately, the decision is risk management related.

> *"The draft Regulation sets out conditions according to which instructions for use in paper form may be replaced by electronic instructions for use. It limits the possibility of providing instructions for use in electronic form to defined medical devices and accessories intended to be used in specific conditions. Furthermore, it contains a range of procedural safeguards. Thus instructions for use have to be provided in paper form on request, and a specific risk assessment by the manufacturer and information on how to access to the instructions for use is needed."*

The draft Regulation also sets up a few basic safety requirements for the following:
- *instructions for use in electronic form which are provided in addition to complete instructions for use in paper form*
- *Web sites containing such instructions for use*

Article 1 of the draft legislation states the premise that information supplied by the manufacturer may be in "electronic form instead of in paper form." Article 2(2) describes electronic

IFUs: electronic form by the device, electronic storage medical with the device, and on a Web site. Only manufacturers of explicit categories of medical devices can consider e-labeling (Article 3(1)): fixed installation, implanted medical devices, devices built-in system visually displaying the IFUs, software, and professional use.

As it stands, many of the AHWP members do not accept e-labeling. In Korea, the package insert including intended use, instructions for use, etc., can be provided in an electronic format such as a CD ROM [9]. In China, e-labeling is not permitted, *per se*, but can certainly be provided in addition to the paper version.

At this time, one can comfortably assert that there is nothing to preclude provision of hard copy labeling with e-labeling. Whether e-labeling alone is accepted is an entirely different matter and will likely be increasingly permitted over time.

Early 2012, social media have yet to make a visible impact on the sale or promotion of medical devices. It is to be expected that we will see an increased use of sale channels especially for "OTC" devices. This is addressed cursorily in the proposed European Draft Regulation to become enforceable later in this decade.

12.7 Language, Language Level, and Intended User

One of the more critical aspects of the labeling content is the language. The famous non-verbal IKEA® "manuals" and labels do not use any language predicated on the assumption that every individual will recognize certain pictures and symbols. The same is true for the universally recognized symbol for "exit." So, the problem posed by the limited space (as well the high cost) on most devices relative to the need for translation in national languages has been partially resolved by the use of symbols. Insofar these have been included in the "Harmonized Standard" European standard EN 980, there is a "presumption of compliance" with the respective subsections of the ER 13 of Annex I of the MDD. Any other symbols, including but not limited to those in ISO 15223, must be explained in the instruction for use in all applicable languages. That said, EN 980:2008 at the time of publication of this text will have been withdrawn, and the EN harmonized standard EN ISO 15223-1:2012.

In the United States, the use of symbols is much less appreciated. There, as well as in many other mainly Anglosaxon jurisdictions, culture dictates a graphic description of details in the instruction for use, which in the United States is also driven by the fear of liability for anything not explicitly expressed in the labeling. Health Canada does not explicitly state that symbols are accepted and Health Canada does not recognize EN 980 or ISO 15223 though Health Canada appears to accept symbols. While this attitude is most outspoken in the Anglosaxon cultures, it tends to become also increasingly prominent in other ones. The ultimate result is an instruction for use or manual that is for most intended readers, at best difficult to access or comprehend.

Brazilian RDC No. 185, Annex III.B, Section 1.4, permits the use of symbols; however, symbols must comply with regulations and technical standards. If there are no regulations or technical standards applicable, symbols used must be described in the IFU. The standards that are published by Associação Brasileira de Normas Técnicas (ABNT) include ABNT NBR ISO 15223-1:2010 titled *Productos para a saúde—Símbolos a serem utilizados em rótulos, rotulagem e informações a serem fornecidas de produtos para saúde*.

This dilemma between readability and completeness has been partially resolved by demanding that manufacturers develop and enclose instruction for use that represents a language comprehension of a 12-year-old individual, taking into account that the language used is native. The use of a language other than the native language(s) is in most countries of the world restricted to higher professional users or specialists. For instance, an insulin pump might be used by an array of users from a highly educated physician diabetologist to a semi-literate elderly person. An implantable left ventricular assist device (LVAD) may have up to four different manuals for the respective users of the device and its accessories.

The EU requires that the knowledge and experience of the user be considered (MDD Annex I, Section 1). In the United States, there has been greater emphasis on human use factors [10]. One should not dismiss the standard of IEC 62366:2007, Medical Devices—Application of Usability Engineering to Medical Devices. Human use factors have implications on the labeling, which should also be considered.

Then there is the language proper. Especially in multi-country or -language areas (most of the world, actually….), it is very difficult for a manufacturer to develop an instruction for use with sufficient coverage of languages to satisfy the users' needs. Oddly enough, it is here that e-labeling (see elsewhere in this article) accessible by smartphones could provide a very legitimate solution. This appears to be a very cost-effective, environmentally friendly, and practical solution for many sparsely populated or developing regions of the world, with the only limitation being expressed by the saying "you can lead the horse to the well, but you cannot make it drink": The user—even an educated one—should at least a few times, read the instructions for use and not discard them unread when opening the package.

12.8 Conclusion

Labels and labeling are among the most important parts of a medical device. While cultures, educational levels, reading proficiency, and familiarity with a product do vary, manufacturers must pay utmost attention to the formulation of content for the respective components of labeling to ensure the safe use of the device when used as intended. The use of language commensurate with the various intended users is critical, as is the effort by the manufacturer to keep the labeling contents current to the knowledge gained by the use of the product in practice.

References

1. GHTF Guidance on UDI. Retrieved December 2011 from http://www.ghtf.org/documents/ahwg/AHWG-UDI-N2R3.pdf.
2. US FDA Draft Rule on UDI. Retrieved December 2011 from http://www.fda.gov/MedicalDevices/DeviceRegulationandGuidance/UniqueDeviceIdentification/default.htm. http://www.gpo.gov/fdsys/pkg/FR-2012-07-10/html/2012-16621.htm.

3. EU UDI. Retrieved December 2011 from http://ec.europa.eu/health/medical-devices/specific-areas-development/udi/index_en.htm.
4. New Zealand Regulations. Retrieved December 2011 from http://www.legislation.govt.nz/regulation/public/1984/0143/latest/DLM 96155.html.
5. TGA Online. Retrieved December 2011 from http://www.tga.gov.au/industry/legislation-tgac.htm.
6. Australia Regulation. Retrieved December 2011 from http://www.comlaw.gov.au/Details/F2007L00576.
7. E-Labeling of IVD. Retrieved December 2011 from http://ec.europa.eu/health/medical-devices/files/meddev/2_14_3_rev1_ifu_final_en.pdf.
8. EU Draft regulation on e-Labeling. Retrieved December 2011 from http://ec.europa.eu/enterprise/tbt/index.cfm?fuseaction=Search.viewDetail&Country_ID=EEC&num=381&dspLang=en&nextpage=1&basdatedeb=&basdatefin=&baspays=&baspays2=&basnotifnum=&basnotifnum2=&bastypepays=ANY&baskeywords=electronic%20instructions&fromform=viewKeyword".
9. Korean Ministry of Health and Welfare (2011) *Korea Medical Device Act No. 0564, Amended as of April 7, 2011 effective from October 08, 2011*.
10. US FDA. Retrieved December 2011 from http://www.fda.gov/downloads/MedicalDevices/DeviceRegulationandGuidance/GuidanceDocuments/UCM095300.pdf.
11. Council of the European Parliament (July 12, 1993), Council Directive of 14 June 1993 concerning Medical Devices (MDD 93/42/EEC), as amended by Directive 2007/47/EC of the European Parliament and of the Council of 5 September 2007, *Official Journal of the European Communities (OJ)* L169, p1.
12. CEN/CENELEC EUROPEAN STANDARD (May 2008), *Symbols for use in the labelling of medical devices*, EN 980, ICS 01.080.20; 11.120.01.
13. CEN/CENELEC EUROPEAN STANDARD (October 2008), *Information supplied by the manufacturer of medical devices*, EN 1041.
14. Code of Federal Regulations, Title 21, Volume 8 (Revised as of April 1, 2011), Title 21-Food and Drugs, Chapter I-Food And Drug Administration, Department of Health and Human Services, Subchapter H—Medical Devices, *Part 820—Quality System Regulation*, CITE: 21CFR820.1.

Chapter 13

Regulatory Affairs for Medical Device Clinical Trials in Asia Pacific

Seow Li-Ping Geraldine

Johnson & Johnson Medical Asia Pacific,
No. 2 International Business Park,
#07-01, Tower One, The Strategy, 609930 Singapore
lseow@its.jnj.com

13.1 Introduction

Over the past decade, the Asia-Pacific region has grown to become an important player in global clinical research development and has contributed to the provision of quality clinical data to gain market access for new products. Even though the most remarkable growth has been observed mainly in pharmaceutical clinical trials, industry stakeholders have recognised the high potential of medical device clinical trials which have hitherto been unregulated in most Asia-Pacific countries. In recent years, more regional regulators have too been turning their focus on the regulation controls required for medical device clinical trials. As both regulators and industry stakeholders seek for direction in this changing landscape, there is a generated demand for regulatory

---

Medical Regulatory Affairs: An International Handbook for Medical Devices and Healthcare Products (Third Edition)
Edited by Jack Wong and Raymond K. Y. Tong
Copyright © 2022 Jenny Stanford Publishing Pte. Ltd.
ISBN 978-981-4877-86-2 (Hardcover), 978-1-003-20769-6 (eBook)
www.jennystanford.com

affairs professionals to participate actively in this quest, and work synergistically with their clinical research colleagues, to ensure that any additional regulatory controls would be better managed without impeding the path to market access for new medical devices. It is therefore pertinent for regulatory affairs professionals to gain a good understanding of medical device clinical trials and develop expertise in related regulatory requirements.

13.2 Medical Device Clinical Trials versus Pharmaceutical Clinical Trials

For those who are new to the world of medical device clinical trials, it would be fair to assume that there are differences between medical device clinical trials and pharmaceutical clinical trials, whereas they share a similar overall clinical trial process (Fig. 13.1).

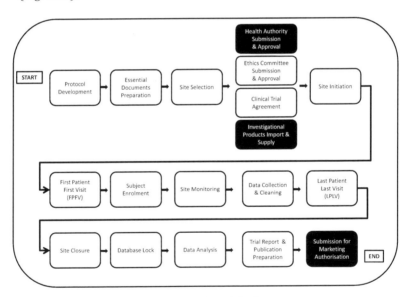

Figure 13.1 The clinical trial process (simplified).

The most obvious difference would be the product types involved in the trial. Medical devices vary in complexity and would include a wide range of health or medical instruments used in the treatment, mitigation, diagnosis or prevention of disease or abnormal physical condition. Medical devices with pharmaceutical components may be categorised as a combination product or a medical device or a pharmaceutical product. Interestingly, some medical devices are classified and regulated as drugs in a few countries. The classification of the investigational medical device should be clarified upfront by the sponsor with the relevant regulatory bodies as this would influence the requirements for conducting the clinical trial.

Trials are designed to answer medical questions. The purpose of a trial is to test out a hypothesis about the effect of a particular intervention upon a disease or condition. This effect could be tested using a single-arm (uncontrolled) design or a comparative (controlled) design (Fig. 13.2). The comparator could be (1) no intervention, (2) a sham device/procedure (placebo in pharmaceutical studies), or (3) existing therapies. The use of placebo or a blinded product may not always be ethical or possible in medical device trials. A comparator could be an existing product which has either been first in the market or has proven efficacy and safety, or an existing procedure which uses different products and instrumentation but has similarities to the investigational procedure, or existing standard of care which may not be linked to a product or a procedure. The added clinical value compared to an existing medical device is of interest to healthcare providers, patients and payers. Due to the innovative nature of medical devices, this "gold standard" may not always exist. For first in class medical devices, it is also a challenge to identify the comparator to be used for the study. Alternatively some trials are conducted with a single arm design, only to compare the results with those from a historical trial involving a predecessor.

The focus in a pharmaceutical clinical trial is solely on the investigational drug. In the medical device context, the investigational device is but part of a complex system comprising the device, the instrumentation or accessory devices and the procedure for using the device. Sometimes all components are unregistered products and the indicated procedure has an investigational status. The clinical trial protocol needs to address how the device is the only investigational component without being influenced by the variability from the other components of this system, for example, through standardised certified surgical technique training before using the device in the trial. Another relevant consequence would be the scope of safety and complaints reporting which should cover all components and the expectedness of events, unless otherwise specified in the protocol.

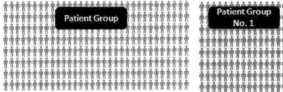

Trial Design - Uncontrolled Patient Group
When there is just one treatment or intervention used during the clinical trial and all participants receive the same treatment, the clinical trial is referred to as a **single arm** study. Single arm studies ae also called "**uncontrolled**" – since the outcome of the experimental intervention is not compared against a control group.

Trial Design - Controlled Patient Group
It is generally considered better to design a study with more than one arm - so that the experimental intervention can be compared to
1) no intervention, 2) a sham device/procedure (or placebo), or 3) existing therapies, etc.
This is called a controlled trial.

Figure 13.2 Uncontrolled versus controlled trial design.

The pharmaceutical phasing of trials from phase 1 to 4 does not apply to medical device clinical trials. Instead the 4Ps are commonly used: Preclinical trials including in vitro/bench-top testing and animal studies, Pilot trials, Pivotal trials and Post-marketing studies (Table 13.1). Generally, the clinical development of a medical device tends to move at a faster pace in order to attain the innovative leader position for the technology involved and to ensure that the customers can access to the latest device. Some devices, depending on their classification and the selected regulatory path, do not even require clinical data for registration which is based on literature evaluation of data from predicate or previously approved devices, thus some regulatory bodies warrant the need for post-marketing studies to monitor long-term effects and to gain additional safety and effectiveness clinical data for the new device.

Many similarities exist between medical device clinical trials and pharmaceutical clinical trials. Those with experience in pharmaceutical clinical trials could leverage on this experience and apply their knowledge with an open mindset and adaptability to the different scenarios in medical device clinical trials.

The principles of the Good Clinical Practice Guideline by the International Conference on Harmonisation of Technical Requirements for Registration of Pharmaceuticals for Human Use (ICH) Guideline, or commonly referred to as ICH GCP [1], that form the ethical foundation for pharmaceutical clinical trials, do apply to medical device clinical trials.

Table 13.1 Phases of clinical trials involving medical devices

	Phase	Description
Preclinical trials	In vitro (benchtop) testing	In vitro research takes place in the laboratory. Laboratory researchers attempt to simulate conditions of the human body, conducting multiple tests to establish whether the device is effective, safe, and functioning properly.
	Animal studies	Initial tests to *establish safety* of the product or materials in question will be conducted *on animals*.
Clinical trials	Feasibility (pilot) studies	A feasibility, or pilot, study will be conducted to *test safety and performance* of the product in question, usually a multi-centre study, involving 30 to about 200 subjects and lasts for 30 days to 6 months.
	Pivotal trials	Pivotal trials are conducted to *gain evidence of safety and effectiveness*, also usually multi-centre studies, with 100 to over 1,000 subjects and lasts for 30 days to several years. Used to obtain regulatory approval from the relevant approval body.
	Post-marketing studies	After the launch of a product, post-marketing studies will be conducted to *monitor long-term effects and to gain additional safety and effectiveness data*.

13.3 Regulation of Clinical Trials

Basically, regulation of clinical trials is necessary to ensure

(i) The safety, well-being and rights of the clinical trial subjects
(ii) The scientific conduct of the clinical trial
(iii) The credibility of the clinical trial data

This is in coherence with a Regulatory Authority's role in the governance of public health and safety. When a clinical trial application is submitted to the regulatory authorities for approval, the reviewers would typically evaluate the protocol, the informed consent form as well as product-related essential documents like the Investigator's Brochure to establish an expert opinion on whether the trial should be conducted from a scientific perspective and whether there are measures in place to ensure the safety, well-being and rights of the subjects. Credibility of clinical trial data, which would be analysed and collated into a clinical trial report, is reviewed as part of the evaluation of a product registration dossier. In some countries, these aspects are further verified through inspections performed by the health authority either during or after the trial. Other basic aspects of regulation would also cover standards for the conduct of the clinical trial, reporting of safety-related events, product defects and regular trial status updates, and labelling, import, accountability and traceability of investigational medical devices which should be manufactured according to relevant standards. A risk-based approach to the scope of control is adopted by most regulatory authorities. Observations of latest regional trends also reveal that Regulatory Authorities are going beyond merely enforcing regulation by playing a more contributing role, to enhance the local clinical research infrastructure and competency of clinical trial personnel, through organising training programmes in medical device clinical trials for clinical trial personnel, provisioning for the registration of trials in a public clinical trial registry, and even revising their local version of GCP to keep up with the changing clinical trial environment (Table 13.2).

Apart from the Regulatory Authority, the responsibility of upholding the above-mentioned three ethical pillars of a clinical trial is also shared by the other major stakeholders in a clinical trial albeit from different perspectives: the Investigator, the Sponsor and the Institutional Review Board (IRB), which is sometimes also referred to as the Ethics Committee (EC) (Fig. 13.3). In fact, although ICH GCP, being a guideline to be followed when generating clinical trial data that are intended to be submitted to regulatory authorities, is commonly used as a basis to regulate clinical trials, it does not have a section specifically for "Regulatory Authority". The Declaration of Helsinki, which describes the ethical principles for medical research involving human subjects, is addressed primarily to physicians, i.e. the Investigator [2]. The other common reference for medical device clinical trials, ISO14155:2011 (E) [3], does not describe the role and responsibilities of the regulatory authority. Hence, while regulatory affairs professionals from the Sponsor organisations have guidance on the scope of their responsibility, those from the regulatory authorities need to establish their specific role and responsibilities within their own clinical research environment.

Table 13.2 Possible areas of regulation

New clinical trial application	Local GCP
Protocol & amendment review	Reporting of trial status & deviations
Informed consent form review	Reporting of safety-related events
Investigator's brochure review	Reporting of product defects
Surgical technique review	Clinical trial report review
Label & packaging review	Inspection-sponsor, site, CRO, warehouse
Instructions for use review	Clinical personnel qualification & training
Product import into country	Public clinical trial registry
Product accountability	Certification of competent sites

Figure 13.3 Major stakeholders in a clinical trial.

The harmonisation of regulation within the region and the opportunity to learn about the implementation of regulation in other countries including outside the Asia-Pacific region will definitely help local regulators and regulatory affairs professionals to formulate regulation that would effectively meet the needs of their country as well as to increase acceptance of local clinical trial data in overseas submissions. The US Food and Drug Administration (US FDA) would be a forerunner in medical device clinical trial regulations since implementing medical device amendments to the Federal Food, Drug & Cosmetic Act on 28 May 1976. The European Medical Directives, which were implemented in the last decade, adopt a different regulatory approach through a combination of regional and local laws, ISO standards and notified bodies. Within the region, within the last decade, the regulatory authorities in Japan, South Korea and Taiwan have implemented changes to their organisations and clinical trial regulations to address the needs of the growing number of local medical device clinical trials.

From an operational perspective, using the same clinical trial reviewer team who has been evaluating pharmaceutical clinical trials may seem like an easily adaptable and convenient approach to undertake the task of evaluating medical device clinical trials. However, a copy and paste success should not be expected without investing in equipping and/or building up the team with knowledge and experience in medical device clinical development process, investigational methodology and the greater emphasis on technology expertise, diagnostic approach and surgical techniques. A similar consideration should be made for the team who traditionally reviews safety reports and product complaints for pharmaceutical products. This scaling up of clinical trial and safety vigilance resources should be part of the medical device regulation development strategy. Due to the rapidness and specificity of the technology and innovation involved in medical device clinical development, the engagement of independent experts to perform the evaluation could help to augment the required resources to support this effort.

Industry stakeholders who often take on the role of the Sponsor or clinical research organisations are watching these developments closely. Changes to clinical trial regulations have an impact on clinical trial operations and more crucially affect the clinical development pathway in terms of speed and complexity. Proper change management coupled with effective organisational strategy for regulatory affairs and clinical research will help to cushion the impact of these changes and prepare the industry to take on the challenges of a changing regulatory landscape.

Interestingly, there is a heavy reliance on the IRB to be the only ethical and scientific "third party" gatekeeper to evaluate and approve medical device clinical trials when the regulatory authority's approval is not required. Inadvertently, most of these regulatory authorities would have guidelines regarding IRB's organisation and procedures. In 2001, the Association for the Accreditation of Human Research Protection Programs, Inc. (AAHRPP), an independent human research protection accreditation program, was established in the United States and has defined domains of responsibility: Organization, Institutional Review Board (IRB) or Ethics Committee (EC), and Researchers and Research Staff. AAHRPP has even accredited eight organisations outside the United States among which there are three each from India and South Korea and one each from China and Singapore.

In the United States, clinical evaluation of devices that have ot been cleared for marketing requires an Investigational Device Exemption (IDE); if the clinical trial involves a significant risk device, the IDE must also be approved by the US FDA. The scope and strategy for regulation of medical device clinical trials varies within the Asia-Pacific region. This could be influenced by the maturity of medical device regulation, the adoption of medical device clinical trial regulation from other countries, the available infrastructure, regulation and resources for pharmaceutical clinical trials, the role and responsibilities of the IRBs, and the local demand for regulation of medical device clinical trials, among others. Most countries require compliance to ICH GCP or a localised version of GCP in their regulations, noting that localised GCP or local regulation will generally prevail over ICH GCP requirements where there is a conflict between the local and international references. Some implement more stringent controls on clinical trials involving unapproved products than those involving approved registered products. Countries with well established regulatory organisations for evaluation of pharmaceutical clinical trials, for example Australia, Japan, South Korea and Taiwan, have

Table 13.3 Overview of the current regulation of medical device clinical trials in some Asia-Pacific countries

Country	Regulatory authority	Submission	Notification	Approval	Import licence	Safety reporting	Inspection
Australia	Therapeutic Goods Administration (TGA)	Sequential	Within 10 business days (CTN scheme)	Within 50 days (CTX scheme for review of safety of device)	Part of CTN/CTX	Local SUADEs/SUSARs within 7–15 days	NA
China	Center of Medical Device Evaluation, State Food and Drug Administration (CMDE, NMPA)	Sequential (Before IRB submission)	Yes (protocol only)	NA	NA	SAE within 24 hours	Site-specific
Hong Kong	Medical Device Division, Department of Health	NA	NA	NA	NA	Voluntary	NA
India	Medical Device Division, Central Drugs Standard Control Organization, Drugs Controller General (India) (DCGI)	Parallel	NA	About 12–16 weeks. Prior IRB approval required	Yes	Local SUADEs/SUSARs within 14 days	NA
Malaysia	Medical Device Control Division, Ministry of Health (MOH)	NA	NA	NA	NA	Yes	NA
New Zealand	New Zealand Medicines and Medical Devices Safety Authority (Medsafe)	Sequential	Within 45 business days.	NA	NA	Local SUADEs/SUSARs within 7–30 days	NA
Singapore	Clinical Trials Branch & Medical Device Branch, Health Sciences Authority (HSA)	NA	NA	NA	Yes	Local SUADEs/SUSARs within 48 hours or 10–30 days	NA
South Korea	Medical Device Safety Bureau, Korean Food & Drug Administration (KFDA)	Parallel	NA	Officially 30 days	Per import	Local SUADEs/SUSARs within 7–15 days	NA
Taiwan	Division of Medical Device and Cosmetics, Taiwan Food and Drug Administration (TFDA)	Parallel	NA	1–3 months	Yes	Local SUADEs/SUSARs within 7–14 days	GCP inspection by TFDA
Thailand	Thailand Food and Drug Administration	NA	NA	NA	Yes	NA	NA

leveraged on their experience and allocated dedicated teams for evaluation and consultation for medical device clinical trials. In Southeast Asia, Singapore is taking the lead in the implementation of medical device clinical trial regulation from 2013. In a few countries where the specific regulations are not implemented, certain regulations for pharmaceutical clinical trials could sometimes apply to medical device clinical trials on a case-by-case basis; for these countries, prior consultation with the relevant authorities would be advised. As of the end of 2011, Hong Kong and Malaysia currently do not have any regulation on medical device clinical trials and import of investigational medical device for clinical trial use.

13.4 Country Regulations

Figure 13.4 describes the typical regulatory activities that are relevant to the clinical trial process.

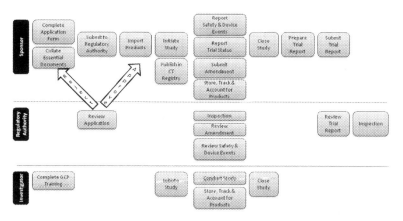

Figure 13.4 Typical regulatory activities in the clinical trial process.

Table 13.3 shows an overview of the current regulation of medical device clinical trials in some of the Asia-Pacific countries.

13.4.1 Australia

Medical devices that are not in the Australian Register of Therapeutic Goods (ARTG) can be legally supplied in a clinical trial through the Clinical Trial exemptions mechanism (Clinical Trial Notification (CTN) Scheme or the Clinical Trial Exemption (CTX) Scheme), which is managed by the Therapeutic Goods Administration or TGA. In addition, prior quarantine clearance must be obtained to import any material of biological origin (human, animal, plant or bacterial) and questions about requirements for an import permit should be directed to the Australian Quarantine & Inspection Service (AQIS). The responsibility for monitoring a clinical trial rests with the Sponsor, Institution in which the trial is being conducted, the Ethics Committee and the Investigator. Clinical trials must be approved by a Human Research Ethics Committee (HREC) which must be constituted and operating in accordance with the NHRMC's National Statement on Ethical Conduct in Human Research. The HREC is responsible for considering the scientific and ethical issues of the proposed clinical trial protocol. Approval of a trial is required from a HREC with jurisdiction over the specific site where the trial is conducted. The Sponsor decides whether to use the CTN CTX Scheme when the trial involves any device which is not included in the ARTG or used beyond the conditions of its marketing approval. The Sponsor in this scenario must be an Australian clinical trial Sponsor. Clinical trials in which the medical devices are used within the conditions of their marketing approval are not subject to CTN or CTX requirements but still require approval by a HREC before the start of the trial.

The CTN process involves a notification only to the TGA with a nominal notification fee, thus no approval or decision is made by the TGA. CTN trials cannot commence until the appropriately completed CTN scheme form and the notification fee is submitted to the TGA who will send the clinical trial sponsor an acknowledgement letter. The notification of the CTN form and the associated fee automatically creates the exemption necessary to allow lawful supply of the unapproved medical devices for the clinical trial.

In the CTX approval process, TGA reviews the safety of the device by assessing the summary data and usage guidelines for a proposed clinical development programme within 50 working days, and if approval if granted, subsequent trials must be carried out under the terms of the approval and be notified to the TGA.

The CTX application does not require the submission of the clinical trial protocol which should be reviewed by the HREC. A CTX trial cannot commence until TGA has provided written advice that the application has been approved, and a HREC and the institution at which the trial will be conducted have provided approval for the conduct of the trial.

The clinical trial Sponsor needs to notify TGA of the data of trial completion and the reason for completion. TGA does not require notification of protocol amendments which clarify the use of, and/or monitoring of treatment but may require if there is a major change to the protocol and the HREC requires a change to the conditions of their approval.

In terms of adverse event reporting, the clinical trial Sponsor is required to report to TGA about serious and unexpected adverse device events. Fatal or life-threatening adverse device events must be reported initially within 7 calendar days of knowledge, followed by a full report within 8 additional calendar days. Other serious unanticipated device events must be reported within 15 calendar days of knowledge. The Investigator is responsible for adverse event reporting to the HREC as required by the HREC and to the Sponsor as required by the protocol [4–7].

13.4.2 China

Submission of a clinical trial application in China is currently limited to Class III medical devices implanted into human body that are not yet available in the market, or those medical devices that are based on the theory of traditional Chinese medicines only [8]. The clinical trial protocol should be filed with the Center of Medical Device Evaluation (CMDE) of the National Medical Products Administration (NMPA). Medical devices to be used in a clinical trial must fulfill certain conditions including measuring up to the registered product or industrial standard, possession of a testing report from the Sponsor and a qualified type testing report from a NMPA-recognised testing centre, as well as data from animal studies. Approval requirements are set on a case by case basis after consultation with NMPA since the need for a NMPA approval before commencing a clinical trial is not clearly stated in the regulations. Once the protocol has been filed with the CMDE, the Sponsor can proceed with the submission to the IRB for review and approval. All serious adverse effects must be reported to NMPA within 24 hours; however, there is no clear definition for "serious adverse effects" in the regulations. NMPA may conduct site inspections during or after the end of the trial. It is not mandatory to obtain local clinical data if global clinical data is available for imported medical devices. In addition, there is new regulation with regards to clinical data exemption for some class II medical devices [9]. As of the fourth quarter of 2012, China is in the process of revising its Medical Device GCP.

13.4.3 Hong Kong

Hong Kong has not implemented regulations for medical device clinical trials. Although reporting of serious adverse events in a medical device clinical trial is still done on a voluntary basis, the Medical Device Division would prefer to be notified of local reportable adverse events which have occurred at sites in Hong Kong.

13.4.4 India

Generally an approval from the Drugs Controller General (India) (DCGI) is required for first-in-man and pivotal clinical trials; this is also dependent on whether the investigational device is included in a list of notified devices which are classified by the DCGI as drugs. The medical device advisory committee (MDAC) reviews the submitted clinical trial application for completeness and, in the case of early phase of first-in-man trials, may refer the application to the Indian Council for Medical Research for scientific review. The application dossier is rather comprehensive and consists of many sections requesting for most essential documents including the insurance certificate and indemnity agreement but excluding the clinical trial agreement with each site. Both DCGI approval and IRB approval for the institution(s) where the trial will be conducted are required before the trial can commence. An import licence is required for importing investigational medical devices into India.

The application for an import licence for the investigational medical devices is submitted together with the clinical trial application. Any unexpected serious adverse event (SAE), as defined in the Indian GCP Guidelines, occurring during a clinical trial should be communicated promptly (within 14 calendar days) by the Sponsor to the Licensing Authority and to the other Investigator(s) participating in the study [10–12].

13.4.5 Malaysia

As of end of 2011, regulation of medical device clinical trials has not officially implemented in Malaysia. Future guidelines based on the recently passed Medical Device Bill will provide more information on the regulation of medical device clinical trials in Malaysia.

13.4.6 New Zealand

Clinical trials of medical devices do not require approval under New Zealand legislation, but the New Zealand Medicines and Medical Devices Safety Authority or Medsafe would like to be informed by email of such trials. However, Health and Disability Ethics Committee approval should be obtained for medical device clinical trials. If the medical device used in a clinical trial contains a hazardous substance or a new organism, an approval under the requirements of the Hazardous Substances and New Organisms Act 1986 (HSNO Act) is required. Any medical device imported for use in a clinical trial is exempted from mandatory notification to the Web Assisted Notification of Devices (WAND) database [13].

All clinical trials conducted in New Zealand are expected to be conducted in accordance with the CPMP Note for Guidance on Good Clinical Practice, regardless whether an approval under the Medicines Act is required for the trial. The Sponsor, who must be a person in New Zealand, assumes responsibility, including legal liability, for the trial in New Zealand in terms of preservation of records, reporting adverse events, submit 6 monthly progress reports, notifying protocol changes and ensuring that the trial is conducted in accordance with both New Zealand law and Good Research Practice standards. Expedited reports of all fatal or life-threatening suspected unexpected serious adverse reactions (SUSARs) occurring in New Zealand trial participants must be sent to Medsafe by the Sponsor within 7 days of knowledge. The Sponsor is required to hold reports of all worldwide SUSARs which may be requested by Medsafe. Adverse events that cause injury and that are associated with medical devices should also be reported to Medsafe, initial reports within 7 calendar days for actual or potential death or serious injury and within 30 calendar days for those with or without actual or potential injury [14].

Medsafe also administers a self-certification scheme for clinical trial sites that have patients in residence, and maintains a list of sites for which it has received evidence of compliance with GCP requirements.

13.4.7 Singapore

Import of unregistered medical devices into Singapore for use in clinical trials requires the approval of the Health Sciences Authority (HSA). An application form for Clinical Trial Test Materials (CTM for Medical Devices) must be submitted by the the Sponsor [15, 16]. Upon importation of the investigational medical devices into Singapore, the Sponsor is responsible for complying with safety and defects reporting, notification of device recalls, and device labelling, accountability and traceability requirements. Although currently medical device trials do not require regulatory approval, all adverse events or defects in medical device trials must be reported. Adverse events that represent a serious threat to public health must be reported within 48 hours while those that have led to the death, or a serious deterioration in the state of health, of a patient, a user of the medical device or any other person must be reported within 10 days. Events where a recurrence of which might lead to the death or a serious deterioration in the state of health, of a patient, a user of the medical device or any other person must be reported within 30 days. Initial reports should be followed by a final report within 30 days of the initial report, detailing the investigation into the adverse event.

During the final phase of implementation of the medical device regulatory framework in Singapore in the fourth quarter of 2012, the new Health Products (Clinical Trials) Regulations will introduce regulations for medical device clinical trials in Singapore using a risk-based approach. Proposed changes, which affects clinical trials to determine safety and performance of all Class C and D registered and unregistered medical devices, will include a new Clinical Trial Notification route for listed investigational devices versus a Clinical Trial Approval route for unregistered investigational devices. Clinical trials involving Class A and B medical devices, non-invasive and non-confirmatory in vitro diagnostic products as well as observational trials are excluded from this scope of regulation. Changes to current safety reporting requirements are also expected. Another proposed change of interest is the allowance of a trial to have more than one Sponsor in order to address various industry business and outsourcing models where the Sponsors may have joint or allocated responsibilities. All unregistered medical devices which are imported for non-investigational purpose in the conduct of a clinical trial, regardless of product risk class, will be regulated under the Health Products (Medical Device) Regulations instead.

13.4.8 South Korea

From 2007 to 2011, the Korean Food and Drug Administration (KFDA) has approved 120 medical device clinical trials. Applications for clinical trial protocol and amendment approval and reporting for adverse events, product defects, trial status updates and import quantities are submitted online via the KiFDA Electronic Filing System. For trials where the essential documents are prepared in English, a Korean translation of the protocol and participant-facing written material must be prepared and submitted in addition to the English versions. Otherwise all remaining essential documentation can be submitted in English. Submission to KFDA can be done in parallel with the submission to the IRB, however KFDA requires an IRB approval for the site where the trial is conducted before KFDA issues the approval. Although the official review time for a clinical trial application is 30 days, this timeline could be delayed by, for example, IRB-required changes to the informed consent form.

Device-related death or life threatening adverse events must be reported within 7 days for the initial report and a follow-up detailed report within the next 8 days; all other types of SAEs must be reported within 15 days of knowledge of the event [17–19].

KFDA has even taken one step further by creating a new system requiring hospitals and medical doctors to be qualified specifically as medical device clinical centres. To be qualified, hospitals must have standard operating procedures according to GCP and medical doctors must have training in medical device clinical development and GCP. By the end of September 2011, a total of 94 university hospitals were registered by KFDA as medical clinical device centres nationwide.

13.4.9 Taiwan

The regulatory system for investigational medical devices in clinical trials consists of regulation for GMP (Good Manufacturing Practice) or QSD (Quality System Design), determination of device classification and regulation for import of these devices. The documents required for the import of investigational medical devices used in a clinical trial and the clinical trial approval include the Certificate of Free Sale in marketed countries, the protocol and the informed consent form as well as most other clinical trial essential documents including the clinical trial agreement, insurance certificate and indemnity agreement. If the Certificate of Free Sale in marketed countries is not available, technical data including preclinical safety and functional tests plus the IRB approval letter must be provided. The clinical trial application can be submitted in parallel both to the IRB and the Division of Medical Device and Cosmetics, Taiwan Food and Drug Administration (TFDA) for review. After the clinical trial results are submitted as reference for product registration, a TFDA review team will perform a GCP inspection at the site before approving the use of these results as reference for product registration. To shorten the review time for medical device clinical trials, the review process has recently been simplified for clinical trials which have obtained US FDA approval and for those with minor changes to the protocol (excluding safety and design changes).

In addition, TFDA has implemented a local version of GCP as well as established a list of good clinical research centres, mandated attendance of compulsory GCP training programme for clinical trial personnel and organised training on clinical trials involving medical devices and technology for regulators, industry and site personnel.

Reporting of adverse medical device reactions (ADR) and product defects can be done online, using the National Reporting System of Adverse Medical Device Reactions and the Medical Product Defect Reporting System respectively, by consumers, medical personnel and manufacturers. For SAEs, manufacturers are expected to provide an oral or initial report is expected within 7 days of knowledge of the event, followed by a full report within 14 days [20].

13.4.10 Thailand

To date, the Thailand Food and Drug Administration (Thai FDA) does not have any regulation on the conduct of medical device clinical trials in Thailand. Only the import of unapproved medical devices used in clinical trial must be pre-approved by Thai FDA who will issue an import licence that specifies the approved quantity before the products can be shipped into Thailand.

13.5 Moving Ahead as Regulatory Affairs Professionals

Existing regulations will change as the learning curve progresses. The clinical research environment and the regulations for medical device clinical trials will evolve to meet the demands and challenges in the conduct of medical device clinical trials in the Asia-Pacific region. A careful effort to streamline regulations governing clinical trials could reduce redundancy in the system while ensuring ethical conduct [21]. In order for Asia Pacific to contribute even more significantly to the globalisation of evidence based medicine, the crucial role of regulatory affairs professionals in this evolution could be enhanced by exploring various possibilities of continuous personal development of regulatory affairs and clinical research knowledge and trends, and through collaborations with their clinical research colleagues in the same quest using a synergistic goal approach. Implementation of new or revised regulations should be coupled with motivated considerations for appropriate change management measures within the organisations. Resources could be dedicated to extend the organisation's knowledge database, outside their area of jurisdiction or countries, about sites, investigators, participants, other key stakeholders, clinical research practices and the quality of trial data. The continuation of using a risk-based approach may not be disadvantageous, though more long-term benefits could be derived from a proactive strategy than a reactive one. Meanwhile, where regulations are not yet available, the principles of globally recognised good clinical practice guidelines, like ICH GCP, should be used as foundation beacons to guide the management and conduct of medical device clinical trials.

References

1. International Conference On Harmonisation Of Technical Requirements For Registration Of Pharmaceuticals For Human Use (10 June 1996). International Conference on Harmonisation (ICH) Tripartite Guideline, Guideline for Good Clinical Practice E6 (R1). Retrieved 10 January 2011, from http://www.ich.org/fileadmin/Public_Web_Site/ICH_Products/Guidelines/Efficacy/E6_R1/Step4/E6_R1__Guideline.pdf.
2. World Medical Association (2008). Declaration of Helsinki: Ethical Principles for Medical Research Involving Human Subjects. Retrieved 10 January 2011, from http://www.wma.net/en/30publications/10policies/b3/17c.pdf.
3. International Organisation for Standardisation (2011). ISO 14155:2011: Clinical Investigation of Medical Devices for Human Subjects- Good Clinical Practice.
4. Therapeutic Goods Administration, Australia *Government* (March 2006). The Australian Clinical Trial Handbook. Retrieved 10 January 2011, from http://www.tga.gov.au/industry/clinical-trials.htm.
5. Therapeutic Goods Administration, Australia *Government* (May 2011). Australian Regulatory Guidelines for Medical Devices version 1.1. Retrieved 10 January 2011, from http://www.tga.gov.au/industry/devices-argmd.htm.

6. Therapeutic Goods Administration, Australia Government (October 2004). Access to Unapproved Therapeutic Goods: Clinical Trials in Australia. Retrieved 10 January 2011, from http://www.tga.gov.au/industry/clinical-trials.htm.
7. National Health and Medical Research Council. Australian Government (October 2007). National Statement on Ethical Conduct in Human Research. Retrieved 10 January 2011, from http://www.nhmrc.gov.au/health-ethics/human-research-ethics-committees-hrecs/human-research-ethics-committees-hrecs/national.
8. State Food and Drug Administration, P.R. China (17 January 2004). Provisions of Clinical Trials for Medical Devices, SFDA Order No.5. Retrieved 10 January 2011, from http://eng.sfda.gov.cn/WS03/CL0768/61644.html.
9. State Food and Drug Administration, P.R. China (24 November 2011). Notification about Clinical Data Exemption for Certain Class II Medical Devices (In Chinese). Retrieved 10 January 2011, from http://www.sda.gov.cn/WS01/CL0845/67233.html.
10. Central Drugs Standard Control Organisation, Ministry of Health, India (December 2008). Guidance for Industry on Submission of Clinical Trial Application for Evaluating Safety and Efficacy, CT/71108, Version 1.1. Retrieved 10 January 2011, from http://cdsco.nic.in/CDSCO-GuidanceForIndustry.pdf.
11. Central Drugs Standard Control Organisation, Ministry of Health, India (04 August 2010). Requirements for Conducting Clinical Trial(s) of Medical Devices in India - Draft. Retrieved 10 January 2011, from http://cdsco.nic.in/Requirements%20for%20Conducting%20Clinical%20Trial(s)%20of%20Medical%20Devices%20in%20India.PDF.
12. Ministry of Health and Family Welfare, Government of India (20 April 2010). List of Notified Medical Devices. Retrieved 10 January 2011, from http://cdsco.nic.in/Medical_div/medical_device_division.htm.
13. New Zealand Medicines and Medical Devices Safety Authority (May 2011). Guideline on the Regulation of Therapeutic Products in New Zealand Part 11: Clinical Trials- Regulatory Approval and Good Clinical Practice Requirements, Edition 1.1. Retrieved 10 January 2011, from http://www.medsafe.govt.nz/regulatory/clinicaltrials.asp.
14. New Zealand Medicines and Medical Devices Safety Authority (10 May 2011). Information for Medical Device Suppliers: Medical Device Adverse Event Reporting. In *MedSafe Regulatory Information*. Retrieved 10 January 2011, from http://www.medsafe.govt.nz/regulatory/devicesnew/9adverse-event.asp.

15. Health Sciences Authority, Singapore (01 September 2010). Health Products Act (Chapter 122D): Health Products (Medical Devices) Regulations 2010. Retrieved 10 January 2011, from http://www.hsa.gov.sg/publish/etc/medialib/hsa_library/health_products_regulation/legislation/health_products_act.Par.69129.File.dat/HEALTH%20PRODUCTS%20(MEDICAL%20DEVICES)%20REGULATIONS%202010.pdf.

16. Health Sciences Authority, Singapore (01 September 2010). Health Products Act (Chapter 122D): Health Products (Medical Devices) (Exemption) Order 2010. Retrieved 10 January 2011, from http://www.hsa.gov.sg/publish/etc/medialib/hsa_library/health_products_regulation/legislation/health_products_act.Par.94239.File.dat/HEALTH%20PRODUCTS%20(MEDICAL%20DEVICES)(EXEMPTION)%20ORDER%202010.pdf.

17. Korean Food and Drug Administration, South Korea (25 November 2011). Guideline for Medical Device Korean Good Clinical Practice (In Korean). Retrieved 10 January 2011, from http://www.kfda.go.kr/index.kfda?mid=90.

18. Korean Food and Drug Administration, South Korea (28 November 2010). Enforcement Regulations of the Medical Device Act. Retrieved 10 January 2011, from http://www.kfda.go.kr/eng/eng/index.do;jsessionid=VAnDsf9zaa5M8jZvOtuhIQi8TiICe6RPiZA5PK1o3UtxBONueQ7AEtlQxPaNsgTE?nMenuCode=46&searchKeyCode=125&page=1&mode=view&boardSeq=66026.

19. Korean Food and Drug Administration, South Korea (01 September 2010). Enforcement Decree of the Medical Device Act. Retrieved 10 January 2011, from http://www.kfda.go.kr/eng/eng/index.do;jsessionid=VAnDsf9zaa5M8jZvOtuhIQi8TiICe6RPiZA5PK1o3UtxBONueQ7AEtlQxPaNsgTE?nMenuCode=46&searchKeyCode=125&page=1&mode=view&boardSeq=66026.

20. Taiwan Food and Drug Administration, Taiwan (12 April 2006). Guidelines for the Registration of Medical Devices. Retrieved 10 January 2011, from http://www.fda.gov.tw/eng/people_laws_list.aspx.

21. Seth W. Glickman, *et al* (2009). *Ethical and Scientific Implications of the Globalization of Clinical Research*. N Engl J Med 360:816–823.

Chapter 14

Medical Device Classification Guide

Patricia Teysseyre

Johnson & Johnson Medical Asia Pacific,
No. 2, International Business Park,
#07-01, Tower One, The Strategy, 609930 Singapore

pteyssey@its.jnj.com

14.1 How to Carry Out Medical Device Classification

14.1.1 Scope

It is the responsibility of the manufacturer to determine the classification of its medical device as early as possible in the device development. Its determination should be based on rules derived from the potential of a medical device to cause harm to a patient or user and thereby on its intended use and the technology it utilizes.

The manufacturer should first clearly define the intended use of the device [1]. The next step is to consider the applicable classification regulations in force in the country where the product is going to be registered before being marketed. Most countries have their own classification scheme, such as Japan, People's Republic of China [5], and India. The classification schemes of Canada and Australia/New Zealand resemble the European one, whereas the one in the United States is different. Efforts are under way to effect a global harmonization of

---

Medical Regulatory Affairs: An International Handbook for Medical Devices
and Healthcare Products (Third Edition)
Edited by Jack Wong and Raymond K. Y. Tong
Copyright © 2022 Jenny Stanford Publishing Pte. Ltd.
ISBN 978-981-4877-86-2 (Hardcover), 978-1-003-20769-6 (eBook)
www.jennystanford.com

classification. For instance, Study Group 1 of the Global Harmonization Task Force (GHTF/SG1/N15:2006) [8] has made progress and has been taken as a reference in some Asian countries such as Singapore, but it can still be noticed that systems differ and it is the up to the manufacturer to check and adapt its medical device class to the appropriate local classification scheme. Although different in many jurisdictions, they carry remarkable resemblance in most cases, having 4 categories of risks identified and treating in vitro diagnostics in their own right [3]. A large percentage of the products would be in the same class across the globe, but care should be taken to examine individual product classifications carefully. GHTF consolidates the classes as A to D, with A being the lowest risk, and D the highest.

In accordance with the classification rules, the manufacturer should document and justify the class of the device.

The class determination can be influenced by several factors, including the duration of device contact with the body, the degree of invasiveness, whether the device delivers medicinal products or energy to the patient.

In general, when two or more classification rules may apply to the device, the device is allocated to the highest level of classification. In case of doubts, the appropriate Notified Body or Regulator should be consulted.

14.1.2 Definitions

Central circulatory system: the major internal blood vessels including the following: arteriae pulmonales, aorta ascendens, arcus aorta, aorta descendens to the bifurcatio aortae, arteriae coronariae, arteria carotis communis, arteria carotis externa, arteria carotis interna, arteriae cerebrales, truncus brachio-cephalicus, venae cordis, venae pulmonales, vena cava superior, vena cava inferior.

Central nervous system: means brain, meninges, and spinal cord.

Duration

Transient: Normally intended for continuous use for less than 60 minutes.

Short term: Normally intended for continuous use for not more than 30 days.

Long term: Normally intended for continuous use for more than 30 days.

Intended use/intended purpose: means the use for which the device is intended according to the data supplied by the manufacturer on the labelling, in the instructions and/or in promotional materials.

Invasiveness

Invasive devices: A device which, in whole or in part, penetrates the body, either through a body orifice or through the surface of the body.

Body orifice: Any natural opening in the body, as well as the external surface of the eyeball, or any permanent artificial opening, such as a stoma.

Surgically invasive device: An invasive device which penetrates inside the body through the surface of the body, with the aid of or in the context of a surgical operation.

Implantable device

Any device which is intended

(i) to be totally introduced into the human body or
(ii) to replace an epithelial surface or the surface of the eye by surgical intervention which is intended to remain in place after the procedure.

Any device intended to be partially introduced into the human body through surgical intervention and intended to remain in place after the procedure for at least 30 days is also considered an implantable device.

14.2 Main Classifications

14.2.1 Medical Devices

The flow chart in Fig. 14.1 shows the main steps to be taken and questions to be addressed when carrying out a medical device classification. This decision tree is based on the GHTF (GHTF/SG1/N15:2006) [8] and MEDDEV 2.4/1 Rev. 9 Guidance [6].

14.2.2 Active Devices

The classification of rules for active medical devices in some jurisdictions differ from the medical devices ones and are shown in Fig. 14.2, as they are defined in the European Medical Devices Directive 93/42/EEC [2]. Figure 14.2 is an extract from the MEDDEV 2.4/1 Rev. 9 Guidance [6]. It should be noted that these are additional classification rules for active devices. The general rules 1–8, and the higher-numbered special rules may be applicable as well.

14.2.3 IVD Devices

Regarding in vitro diagnostic devices, their classification is well defined in each local regulation, which usually provides the class corresponding to the assay or reagent (and the associated equipment).

For instance, GHTF has a separate risk based classification. Europe uses risk assessment from 10 years ago. Health Canada has a separate classification for IVD but bringing them individually into the generic 4 risk classes.

In vitro Diagnostic (IVD) product means any reagent product, calibrator, control material kit, instrument, apparatus, equipment or

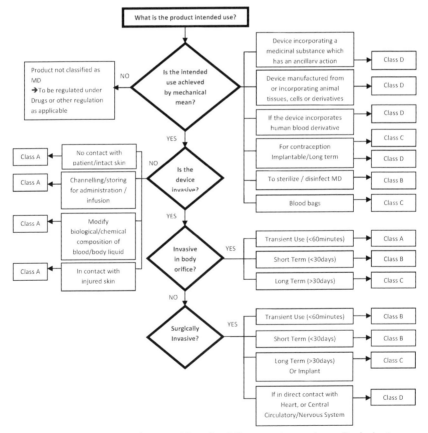

Figure 14.1 Steps to be considered while carrying out medical device classification.

Main Classifications | 149

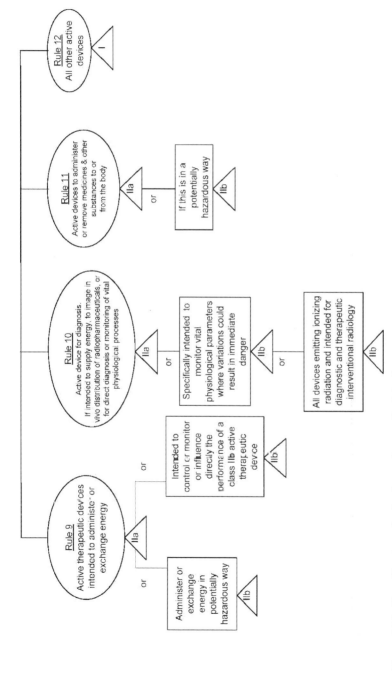

Figure 14.2 Classification of rules for active medical devices.

system, whether used alone or in combination, that is intended by its product owner to be used in vitro for the examination of specimens, including blood and tissue donations, derived from the human body, solely or principally for

(i) providing information concerning a physiological or pathological state,
(ii) providing information concerning a congenital abnormality,
(iii) determining the safety and compatibility of donations, including blood and tissue donations, with potential recipients, or
(iv) monitoring therapeutic measures

and includes a specimen receptacle but not a product for general laboratory use, unless that product, in view of its characteristics, is specifically intended by its product owner to be used for in vitro diagnostic examination.

According to GHTF standards [7], IVD medical devices are classified into four classes, based on the individual risk and public health risk level (Table 14.1).

Table 14.1 Classification system for IVD Medical Devices

Class	Risk level	Device examples
A	Low individual risk and low public health risk	Clinical chemistry analyser, prepared selective culture media
B	Moderate individual risk and/or low public health risk	Vitamin B_{12}, pregnancy self-testing, anti-nuclear antibody. urine test strip
C	High individual risk and/or moderate public health risk	Blood glucose self testing, HLA typing, PSA screening. Rubella
D	High individual and high public health risk	HIV blood donor screening, HIV blood diagnostic

14.2.4 IVD Case Study

IVDs are medical devices intended to perform diagnoses from assays in hematology, immunology, microbiology, etc. Table 14.2

Table 14.2 Classification overview of Chlamydia serological reagents in different legislations

Country/region	US	Canada	EU	Singapore
Class	Class I	Class III	List B	Class C
Rule	21CFR866.3120	Rule 2(b) (ii) of SOR/98-282) SCHEDULE 1 (Section 6)	Directive 98/79/EC— Annex. II	GN-14 Guidance on the Risk Classification of in vitro Diagnostic Medical Devices

illustrates the Chlamydia serological reagents. The exercise conducted around it aimed to determine its classification within different regulations; the rules which served as a reference to determine the class are highlighted.

14.2.4.1   US FDA

- **Rule: 21CFR866.3120 [10]:**

Title 21—Food And Drugs: Chapter I—Food And Drug Administration Department Of Health and Human Services Subchapter H—Medical Devices Part 866—Immunology and Microbiology Devices

Subpart D—Serological Reagents

Sec. 866.3120 Chlamydia serological reagents.

(a) *Identification.* Chlamydia serological reagents are devices that consist of antigens and antisera used in serological tests to identify antibodies to chlamydia in serum. Additionally, some of these reagents consist of chlamydia antisera conjugated with a fluorescent dye used to identify chlamydia directly from clinical specimens or cultured isolates derived from clinical specimens. The identification aids in the diagnosis of disease caused by bacteria belonging to the genus Chlamydia and provides epidemiological information on these diseases. Chlamydia are the causative agents of psittacosis (a form of pneumonia), lymphogranuloma venereum (a venereal disease), and trachoma (a chronic disease of the eye and eyelid).

(b) *Classification.* Class I (general controls).

14.2.4.2   Canada

- Medical Devices Regulations (SOR/98-282) [9] SCHEDULE 1 (Section 6) Classification Rules for Medical Devices Part 2, in vitro Diagnostic Devices Use With Respect to Transmissible Agents
- **Rule 2(b) (ii):**

An IVDD that is intended to be used to detect the presence of, or exposure to, a transmissible agent is classified as Class II, unless

(a) it is intended to be used to detect the presence of, or exposure to, a transmissible agent that causes a life-threatening disease if there is a risk of propagation in the Canadian population, in which case it is classified as Class IV; or

(b) it falls into one of the following categories, in which case it is classified as Class III:

(i) it is intended to be used to detect the presence of, or exposure to, a transmissible agent that causes a serious disease where there is a risk of propagation in the Canadian population,

(ii) **it is intended to be used to detect the presence of, or exposure to, a sexually transmitted agent,**

(iii) it is intended to be used to detect the presence of an infectious agent in cerebrospinal fluid or blood,

or

(iv) there is a risk that an erroneous result would cause death or severe disability to the individual being tested, or to the individual's offspring.

14.2.4.3 EU

Directive 98/79/EC [3]—Annex II—List of Devices Referred to in Article 9(2) and (3)

- **Rule: List B**
  - Reagents and reagent products, including related calibrators and control materials, for determining the following blood groups: anti-Duffy and anti-Kidd,
  - reagents and reagent products, including related calibrators and control materials, for determining irregular anti-erythrocytic antibodies,
  - reagents and reagent products, including related calibrators and control materials, for the detection and quantification in human samples of the following congenital infections: rubella, toxoplasmosis,
  - reagents and reagent products, including related calibrators and control materials, for diagnosing the following hereditary disease: phenylketonuria,
  - ⇒ **reagents and reagent products, including related calibrators and control materials, for determining the following human infections: cytomegalovirus, chlamydia,**
  - reagents and reagent products, including related calibrators and control materials, for determining the following HLA tissue groups: DR, A, B,
  - reagents and reagent products, including related calibrators and control materials, for determining the following tumoral marker: PSA,
  - reagents and reagent products, including related calibrators, control materials and software, designed specifically for evaluating the risk of trisomy 21,

— the following device for self-diagnosis, including its related calibrators and control materials: device for the measurement of blood sugar.

14.2.4.4 Singapore

Health Sciences Authority GN-14 Guidance [4] on the risk classification of in vitro Diagnostic Medical Devices

- **Rule: 2**

Tests to detect infection by HIV, HCV, HBV, HTLV. This Rule applies to first-line assays, confirmatory assays and supplemental assays.
Rule 2: IVD medical devices intended to be used for blood grouping, or tissue typing to ensure the immunological compatibility of blood, blood components, cells, tissue or organs that are intended for transfusion or transplantation , are classified as **Class C,** except for ABO system [A (ABO1), B (ABO2), AB (ABO3)], rhesus system [RH1 (D), RH2 (C), RH3 (E), RH4 (C), RH5 (e)], Kell system [Kel 1 (K)], Kidd system [JK 1 (Jka), JK2 (Jkb0)] and Duffy system [FY1 (Fya), FY2 (Fyb)] determination which are classified as Class D.

Rationale: The application of this rule as defined above should be in accordance with the following rationale: A high individual risk, where an erroneous result would put the patient in imminent life-threatening situation places the device into Class D. The rule divides blood-grouping IVD medical devices into two subsets, Class C or Class D, depending on the nature of the blood group antigens the IVD medical device, is designed to detect, and its importance in a transfusion setting.

14.3 Medical Device Classification: Practical Examples

Figures 14.3–14.8 provide examples of the way to use the classification decision trees provided in the various guidelines when classifying a medical device. These examples focus on one type of medical device (i.e. a syringe). Depending on its intended use, a simple syringe can be classified in each of the categories, from the lowest one (lower risk) to the highest one (higher risk). The rules taken as reference for this case study are the ones from the European regulation as laid down in Appendix IX of the Medical Devices Directive 93/42/EEC with the supporting charts from the MEDDEV 2.4/1 Rev. 9 Guidance [6].

Syringe without needle

↳ Intended use: **channel or store substances which will eventually be administered to the body; typically, it can be used in transfusion, infusion, extracorporeal circulation**

  ↳ Invasive?
    ↳ NO
      ↳ Rule 2: Channelling or storing for eventual administration
        ↳ **Class I**

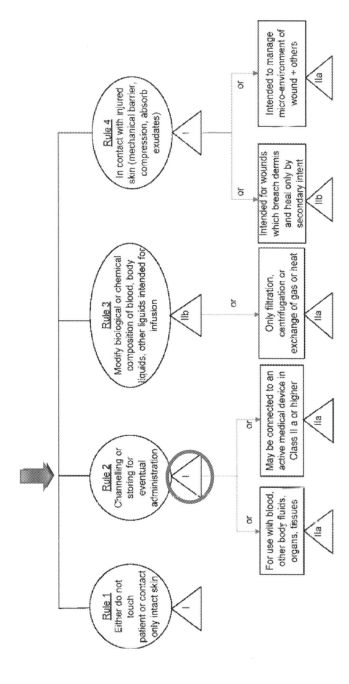

Figure 14.3 MEDDEV 2.4/1 Rev. 9—Non-Invasive Devices.

Syringe without needle

⇧ Intended use: channel or store substances which will eventually be administered to the body; to be used on infusion pumps

    Invasive?
    ⇧ NO

⇧ Rule 2: Channelling or storing for eventual administration

It may be connected to an active medical device (infusion pump)

⇧ **Class IIa**

Medical Device Classification | 157

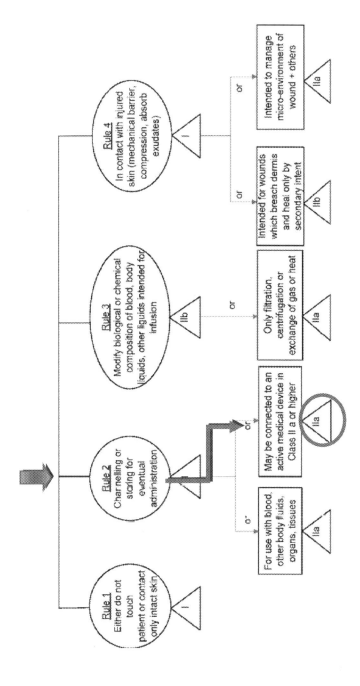

Figure 14.4 MEDDEV 2.4/1 Rev. 9—Non-Invasive Devices

Syringe without needle
↳ Intended use: **deliver oral medication**

   Invasive?
   ↳ YES

      Invasive in body orifice?
      ↳ YES

         Transient Use?
         ↳ YES (the the syringe is aimed to be used for less than 60 minutes)

            Class I

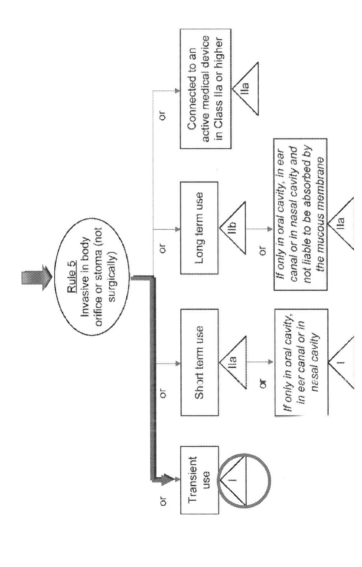

Figure 14.5 MEDDEV 2.4/1 Rev. 9—Invasive Devices.

160 | Medical Device Classification Guide

Syringe with needle

↳ Intended use: **to administer substances to the body**

    Invasive?

    ↳ YES Invasive in body orifice?

        ↳ NO

          Surgically invasive?

          ↳ YES

            Transient Use?

            ↳ YES (the syringe is aimed to inject medicine for less than 60 minutes)

              ↳ Class IIa

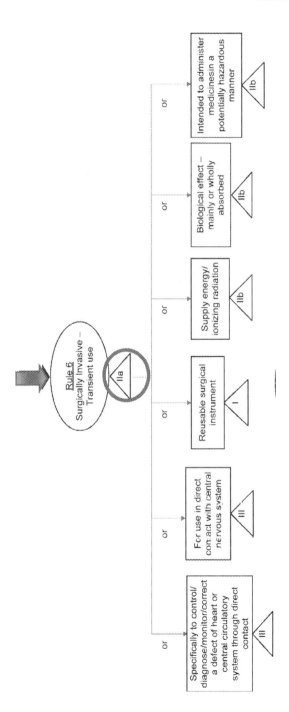

Figure 14.6 MEDDEV 2.4/1 Rev 9—Surgically Invasive Devices.

Syringe with needle

↳ Intended use: **to administer or withdraw substances in the spine**

    Invasive?

    ↳ YES

        Invasive in body orifice?

        ↳ NO

            Surgically invasive?

            ↳ YES

                Transient Use?

                ↳ YES (the syringe is aimed to be in contact with the body in less than 60 minutes)

                    For use in direct contact with the Central Nervous System

                    ↳ **Class III**

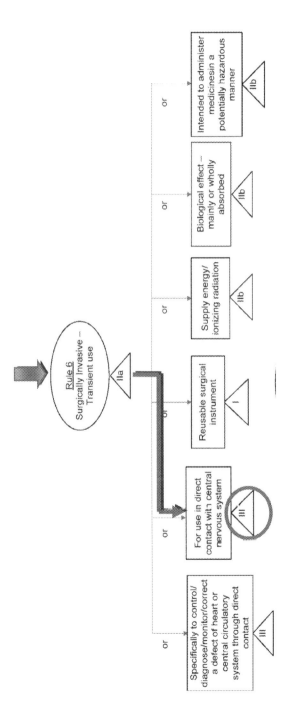

Figure 14.7 MEDDEV 2.4/1 Rev. 9 – Surgically Invasive Devices.

Syringe without needle, pre-filled with Sodium Chloride Solution

↳ Intended use: **to rinse vascular access devices**

    Mechanical?

    ↳ YES (rinsing, flushing catheters)

        ↳ Device incorporating a substance, which, if used separately, can be considered to be a medicinal product, and which acts on the human body with an ancillary action

            Class III

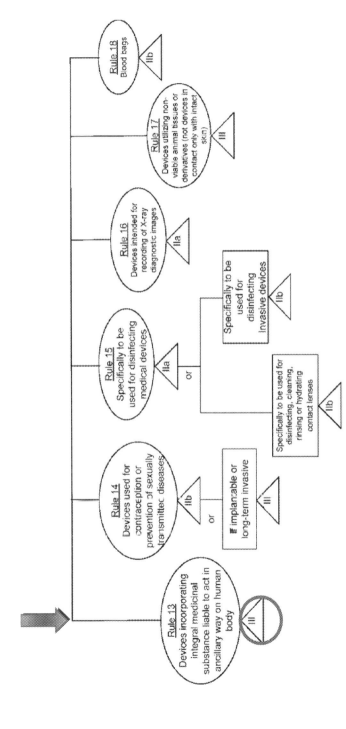

Figure 14.8 MEDDEV 2.4/1 Rev. 9—Special Rules.

Note: If the syringe is prefilled with a drug, the entire combination is reviewed under the pharmaceutical legislation in Europe. If it would be filled with degradable hyaluronic acid for injections in knees as lubricant, they would be class III, based on rule 8; filled with bovine collagen they would be class III under rule 17 and animal tissue directive would be applicable as well.

References

1. Classification of Medical Devices in Europe—An Overview Dr Jaap Laufer, President, European Medical Device Consultants (EMDC).
2. Council of the European Parliament (5 September 2007) Directive 2007/47/EC of the European Parliament and of the Council of 5 September 2007 amending Council Directive 90/385/EEC on the approximation of the laws of the Member States relating to active implantable medical devices, Council Directive 93/42/EEC concerning medical devices and Directive 98/8/EC concerning the placing of biocidal products on the market, *Official Journal of the European Communities* OJ L247 p. 21.
3. Council of the European Parliament (27 October 1998), Directive 98/79/EC of The European Parliament and of The Council of 27 October 1998 on *in vitro* diagnostic medical devices, *Official Journal of the European Communities* OJ L331 p. 1.
4. GN-14 Guidance on the Risk Classification of *In Vitro* Diagnostic Medical Devices Revision 1 (2008). Medical Device Guidance, Health Sciences Authority, Singapore.
5. Guidance Notes: GN-01 Overview of the Medical Device Administrative Control System Government of the Hong Kong Special Administrative Region, The People's Republic of China Date of Issue: 1 September 2005, Department of Health.
6. European Commission—DG Health and Consumer—Directorate B, Unit B2 "Cosmetics and medical devices"—June 2010 MEDDEV 2. 4/1 Rev. 9—Guidelines Relating to the Application of the Council Directive 93/42/EEC on Medical Devices, OJ L 169, 12.7.1993, p. 1.
7. Principles of IVD Medical Devices Classification SG1 Final Document GHTF/SG1/N045:2008, Global Harmonization Task Force.
8. SG1(PD)/N77R4 Global Harmonization Task Force—(Revision of GHTF/SG1/N15:2006) Title: Principles of Medical Devices Classification—Authoring Group: Study Group 1 of the Global Harmonization Task Force—Date: October 3, 2011.

9. Medical Devices Regulations (SOR/98-282) Schedule 1 (Section 6) Classification Rules for Medical Devices Part 2, *in vitro* Diagnostic Devices Use With Respect To Transmissible Agents. Retrieved 06 April 2012, from: http://laws.justice.gc.ca/eng/regulations/SOR-98-282/page-23.html#h-68.

10. Title 21—Food And Drugs: Chapter I—Food And Drug Administration Department Of Health And Human Services Subchapter H—Medical Devices Part 866—Immunology And Microbiology Devices Subpart D—Serological Reagents Sec. 866.3120 Chlamydia Serological Reagents. Code of Federal Regulations, Revised as of April 1, 2011, CITE: 21CFR866.3120. Retrieved 06 April 2012, from: http://www.accessdata.fda.gov/scripts/cdrh/cfdocs/cfcfr/CFRSearch.cfm?fr=866.3120.

Chapter 15

ISO 13485:2016: Medical Devices—Quality Management Systems—Requirements for Regulatory Purposes

Gert Bos

Qserve Group, Utrechtseweg 310, building B42, 6812 AR Arnhem, The Netherlands
Gert.Bos@Qservegroup.com

This chapter describes the origin, implementation and use of the quality management system requirements in medical device globally based on a commonly used standard ISO 13485.

15.1 Introduction

A standard is, in essence, an agreed repeatable way of doing something. It is a document, published after collective work from one or more committees, which contains a technical specification or other precise criteria designed to be used consistently as a rule, guideline, or definition. Standards help to make life simpler and to increase the reliability and the effectiveness of many goods and services we use. Standards are created by bringing together the experience and expertise of all interested parties such as the producers, sellers, buyers, users and regulators of a particular material, product, process or service.

Medical Regulatory Affairs: An International Handbook for Medical Devices and Healthcare Products (Third Edition)
Edited by Jack Wong and Raymond K. Y. Tong
Copyright © 2022 Jenny Stanford Publishing Pte. Ltd.
ISBN 978-981-4877-86-2 (Hardcover), 978-1-003-20769-6 (eBook)
www.jennystanford.com

Standards are designed for voluntary use and do not impose any regulations. However, laws and regulations may refer to certain standards and make compliance with them compulsory.

15.2 Background and Origins of ISO 13485:2016

ISO 13485:2016 is a Quality Management System (QMS) for medical devices specifically for regulatory purposes and, whilst based upon the foundation of ISO 9001:2000 (quality management system standard), is a standalone standard.

ISO 13485:2016 is published by various organizations and when it is published by those organizations this is recognized by letters being placed before the ISO 13485:2016. For example, in the Netherlands it is published by Nederlands Normalisatie Instituut (NEN) and this is denoted by the Standard being NEN ISO 13485:2016 and as it is also a European Norm it is published as NEN EN ISO 13485:2016.

Some of the changes that had been introduced with the publication of ISO 9001:2000, including the reduction in the number of required documented procedures and the inclusion of concepts such as customer satisfaction and continual improvement were not in alignment with the regulatory requirements for medical devices.

The standard ISO 13485:2016 Medical devices—Quality Management systems–requirements for regulatory purposes has grown to be the certification standard of choice for medical device manufacturers. All the requirements are specific to organizations that are providing medical devices or services to the medical device industry and this does not depend on the size or type of the organization. To facilitate comparison to the underlying ISO 9001 standard, a comparison table has been provided in one of the annexes of the standard, including some explanatory notes.

The ISO 13485:2016 standard is built on the historic structure for the quality management standards, and did not follow the general restructuring of most quality management system standards. The key objectives in the latest update focused around getting the requirements closer to the state of art expectations in the global regulatory environment. Many elements from the US Quality System Regulations (QSR) that did not make it into the previous version of the standard have now made it in. Also, key concepts from the global audit model MDSAP (the IMDRF Medical Device Single Audit Program) have been integrated, as well as the changes in the new EU Medical Device Regulations (EU-MDR).

These additional and new elements compared to the previous version of the standard (ISO 13485:2003) include the following:

- Increased focus on compliance with regulatory requirements
- Process controls based on risk management
- Increased requirements for design and development, including considerations of usability, use of standards, verification and validation, design and development transfer and design records
- Increased controls for outsourced processes and suppliers
- Increased requirements for process validation
- Increased requirements for 'feedback', including complaint handling
- Introduction of statistical techniques for data analysis

The main difference between ISO 13485:2003 and ISO 9001:2015 besides its structure is related to customer satisfaction and continuous improvement, elements that are not included in ISO 13485. In ISO 13485 these items are replaced by processes for ensuring continuing effectiveness of the quality system to meet customer and regulatory requirements. Second, it promotes active systems for customer feedback related to if the organizations need to meet customer and regulatory requirements.

In addition, the requirements of ISO 13485:2016 include many more requirements for documented procedures, documented requirements and records, supporting its use in meeting regulatory requirements.

A last diverging element is the focus in ISO 9001 on the company risk assessment and strategy, whereas ISO 13485 remains focused on the risks and safety of the medical devices a company produces.

15.3 Quality Management Systems

Quality Management Systems consist of common elements that are expressed as the organizational structure, processes, procedures, work instructions and resources needed to implement quality management (Fig 15.1).

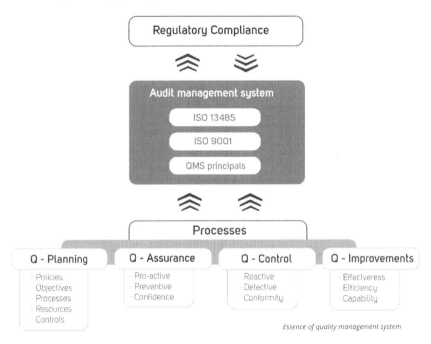

Essence of quality management system

Figure 15.1 Quality management system essence.

Shared elements of organizational structure, authorities and responsibilities, methods and processes, data management, resources, training, maintenance, customer satisfaction/requirements, product quality and continuous improvement are based on the principles of quality management:

(i) **Customer and regulatory focus**: By understanding current and future customer needs, organizations can meet their requirements whilst ensuring that all known applicable regulations are applied.

(ii) **Leadership on purpose and direction**: Leaders, by establishing unity of purpose and direction of the organization,

should create and maintain the internal environment in which people can become fully involved in achieving the organizations objectives.
(iii) **Involvement of people at all levels**: People are the essence of an organization and their full involvement enables best performance in the organization.
(iv) **Process approach to resources and activities**: This principle commonly this leads to the PDCA approach of Plan–Do–Check–Act.
(v) **Systems approach to management**: Core in the system's smooth running is identifying, understanding and managing inter-related processes of a system as it contributes to the organizations effectiveness and efficiency in achieving its objectives and reducing overall process risks.
(vi) **Factual approach to decision making**: Decisions based on analysis of data and information are typically more effective. Monitoring and measuring will allow an organization to understand its ability to supply a safe and effective product and service.
(vii) **Mutually beneficial supplier relationships**: An organization is responsible for ensuring control of its outsourced processes, and with increasing regulatory pressure, supply chain management truly becomes one of the keys to success.
(viii) **Improvement as an ongoing objective**: Long withheld in medical device industry, legislation is now changing to stimulate product improvement in a continuous effort.

15.4 ISO 9000 and ISO 13485 Quality Management System Family of Standards

ISO 13485:2016 is commonly seen as part of the 9000 series of standards, despite it being a standalone standard. That means that effective implementation can best be achieved by combining the requirement of ISO 13485 with content, views and guidance of a series of adjacent standards, representing further detail and in some cases national variations:

(i) ISO 9000:2005 describes the fundamentals & vocabulary.

(ii) ISO 9001:2015 provides the basic QMS requirements.
(iii) ISO 9004:2018 provides guidelines for performance improvement beyond general QMS requirements.
(iv) ISO 19011:2018 provides guidelines for quality and/or environmental management systems auditing, including internal audits.
(v) ISO/TR14969:2004 provides guidelines for application of ISO 13485:2003, the older version of the medical device QMS standard; but still contains some useful concepts.

In addition, there is a standard used to supervise third party assessment bodies that audit against, e.g. ISO 13485:2003:

(i) ISO/IEC 17021-1:2015 provides requirements for bodies providing audit and certification of management systems.

15.5 Global Regulatory Footprint of ISO 13485:2016

With the increasing use of ISO 13485 by regulatory authorities worldwide the use of the standard by manufacturers and other organizations is increasing. ISO 13485:2016 supports the regulatory compliance in several countries including the European Union for CE marking, Australia, Canada and Taiwan. Japanese Ministerial Ordinance (MO) No. 169 is similar to ISO 13485:2003. In Europe, it is a harmonized standard for the three medical devices directives and for the new EU MDR and EU IVDR regulations, which means that adhering to it for the quality management systems aspects identified in these directives there is a presumption of conformity, details of which are specified in the so-called annexes Z found at the back of the European versions of the standards. Current annexes are still for directives; annexes for regulations are being drafted.

For the USA ISO 13485:2016 is optional and may typically be the basis for the QMS that is used by many manufacturers, but it is not a distinct requirement. It can also be used for low risk devices manufacturers that are exempt from QSR's 21 CFR 820. The US FDA has been cooperating closely with the development of ISO 13485:2016, in the attempt to further align ISO 13485 and US QSR (Quality System Regulations) requirements. Some of the concepts from ISO 134854:2016 form also the basis of the IMDRF MDSAP program on Medical Device Single Audits.

With the growing interest in regulatory regimes in Asia and other parts of the world, many more regulatory requirements are added to the equation. ISO 13485:2016 certification is a great tool to harmonize the various QMS requirements from all these existing and developing legislations, as it is very easy for national regulators to add some country specific requirements to the general framework provided in the global standard.

15.6 Implementing an ISO 13485:2016 Quality Management System

There are various reasons why an organization implements a certified ISO 13485 quality management system and this primarily includes meeting regulatory requirements when registering medical devices worldwide. In addition, it might help improve customer confidence and increase the organization's competitive edge. As such it makes good business sense and forms an excellent basis for an efficient and effective business to be certified against ISO 13485.

A well-implemented ISO 13485-based QMS supports regulatory compliance and improves customer confidence. It improves the consistency and stability of the processes used by the organization. It can reduce waste and defects, improve employee motivation and participation and is a basis for monitoring, managing or potentially improving the performance of suppliers.

15.7 Process Approach

A process is a set of interrelated or interacting activities that uses resources to transform inputs into outputs. The process approach systematically identifies and manages the linkage, combination, and interaction of a system of processes within an organization, as depicted in Fig. 15.2 linking the various parts of ISO 13485 in a generic process flow. The QMS documentation will reflect the actual process flows of the particular organization.

The process approach considers the importance in understanding and meeting requirements by looking at processes in terms of the value that they can add, in measuring the performance of processes, determining if that process is effective and using the objective measurement of processes to improve that process.

The process model of Plan–Do–Check–Act (PDCA cycle) is used in the ISO 13485 standard to link the clauses of the standard. It is used to show how the requirements of customers and regulatory authorities can be met (Fig 15.3).

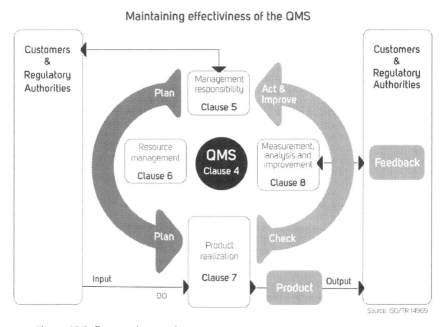

Figure 15.2 Process interactions.

15.8 Planning the Implementation

The implementation process begins with the assumption that the decision to implement has already been taken. Best practice traces back the origin of the decision to implement and the underlying causes or views, such as supply chain requirement, regulatory requirement, or desire to improve performance, reduce waste, etc.

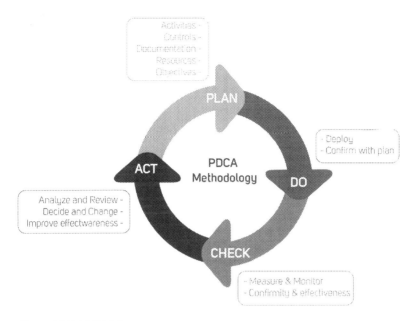

Figure 15.3 NPDCA for your processes.

Some initial general steps for a successful implementation effort would include knowing how to interpret the ISO 13485:2016 requirements for your system, ensuring team members are knowledgeable on the current system, arranging management commitment, and developing a comprehensive project plan which includes monitoring steps. Where needed training should be provided to staff to enhance their knowledge and to boost their commitment which is crucial to the successful implementation and continuing compliance to the requirements of the standard.

Key elements would include promoting awareness, performing a critical gap analysis, and following effective implementation of course getting ready for continually improving the system, where regulatory requirements allow.

The ultimate assessment and verification of the first implementation is in a round of internal audits followed by a management review concluding the effectiveness of the QMS.

With 2016 as basis, it might be considered to claim also continued compliance to ISO 13485:2003 version for those countries where that version (and EN ISO 13485:2012 for Europe) is still used by the legislators as semi-legislative requirement.

15.9 Scope, Exclusions and Non-Applicability

ISO 13485 can be applied to any organization who wishes to demonstrate the ability to provide medical devices and/or related services that meet customer and regulatory requirements ISO 13485. ISO 13485 can be applied irrespective of type or size. Providing it is permitted in the regulatory requirements that apply to the organization that is implementing ISO 13485 exclusions are permitted. Organizations can exclude design and development controls, Clause 7.3 but the justification for the exclusions must be clearly documented in the organizations quality system and must be an appropriate exclusion for any regulatory system being followed. Other clauses can be excluded if they are not applicable, e.g. sterilization if the manufacturer does not produce sterile medical devices.

In the 2016 version, the scope is expanded to include broader supporting stakeholders such as consultancy firms, authorized representatives, etc.

15.10 Document Control

To be efficient and effective an organization must manage how to ensure that those working within that organization carry out process in a consistent way. This includes both large and small organizations. In large organizations, this means that tasks are completed in the same systematic way. Large organizations or those with complicated processes will benefit significantly with the adoption and implementation of an appropriate quality management system.

Section 4.2 of ISO 13485 describes the requirements for documents and records that are used to support the quality management system, which very often is divided into four tiers with their own types of documentation (Fig. 15.4).

15.11 Record Completion and Control

Quality records are an important part of any quality management system and especially in ISO 13485 as they are the evidence that activities that are required as part of the quality management system have been completed. Requirements for records are referenced in at least 50 places within the standard but many organizations will have the need for records for other aspects of the Quality Management system due to the nature of the business. To meet regulatory compliance demands, the number of records needed is continuously increasing.

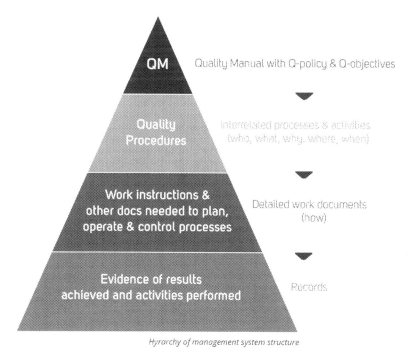

Hyrarchy of management system structure

Figure 15.4 Typical four-tier quality management system setup.

15.12 Management Responsibility

The emphasis in Clause 5 is on the role top management has within the quality management system in ensuring that the system is maintained and remains effective, and that customer and regulatory requirements are met by the organization.

Top management has to ensure that the organizations quality policy is defined and that this is linked to the quality objectives in the organization. A management representative is appointed by top management.

Reviews are conducted at planned intervals and in the standard the specific aspects of the quality management system that should be considered at these reviews are defined.

Since the last revision, an increased focus on compliance with regulatory requirements is required, and key objectives will need to be redefined. The focus is globally shifting to a regulatory life cycle management, with much emphasis on post market surveillance and post market clinical follow up; elements feeding into continuous product improvements, risk reduction and verification on continuation of meeting state of art compliance.

15.13   Resource Management

It is important for any organization to have adequate resources to be able to meet the customer and regulatory requirements and these needs to ensure that the requirements of the products are met. These should include the effective initiation of the resources and their continued maintenance to support the processes of the quality management system. It is the role of top management to ensure that the appropriate resources are available.

There are many aspects to resources they can be people, infrastructure, work environment, information, suppliers and partners, natural and financial resources. Even if these resources are outsourced, they have to be managed by the organization like the processes that are within the organization.

15.14   Product Realization

Product realization includes all the processes that bring the product or service to the customer. These include planning, understanding customer requirements, communicating with customer, the design and development of the product or service, purchasing of the materials and services, production/manufacturing, delivery of the products/service and calibration of any equipment that is used.

15.15 Risk Management

The only place where risk management is identified within ISO 13485 is under the planning of product realization. Risk management is a requirement throughout product realization. The requirements for risk management must be documented and records of the risk management must be retained. It ultimately forms the basis of the decision to place a product on the market or to refrain from it, depending on the outcome of the final risk to clinical benefit evaluation.

There is a standard which gives guidance on risk management ISO 14971:2019 Medical devices—Application of risk management to medical devices. Further guidance on how to integrate risk management throughout the entire QMS can be found in the Global Harmonization Task Force (GHTF) guidance from Study Group SG3.

The 2016 version envisages all process controls be based on risk management (Fig. 15.5).

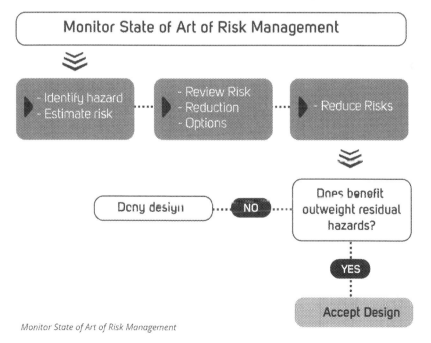

Monitor State of Art of Risk Management

Figure 15.5 Essential steps in risk management.

15.16 Design and Development

The design and development process includes all aspects of the process of bringing the product or service from the original concept of the idea and ensuring that is in line with the user and regulatory requirements through to that product or service that is available to customer and then how any changes to that design are managed.

In order to do this, there are various aspects that are identified within the requirements of ISO 13485:2019, including the respective design gates between the various phases:

 (i) Planning
 (ii) Inputs
 (iii) Outputs
 (iv) Review
 (v) Verification
 (vi) Validation
 (vii) Control of change

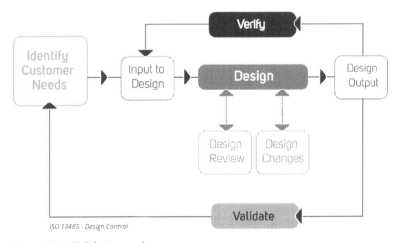

Figure 15.6 Validation cycle.

The latest version of the standard prescribes increased requirements for design and development, including considerations of usability, use of standards, verification and validation, design and development transfer and design records. It also envisages increased requirements for process validation (Fig. 15.6).

15.17 Purchasing and Supplier Control

The management of suppliers and the materials or services received from those suppliers is critical to the organization and the quality of the product and service supplied by the organization to their customers.

The GHTF has published guidance on the Control of Products and Services Obtained from Suppliers. The guidance describes the process of establishing controls for products and services obtained from suppliers. This typically comprises six phases, which include the following:

(i) Planning
(ii) Selection of potential supplier(s)
(iii) Supplier evaluation and acceptance
(iv) Finalization of controls
(v) Delivery, measurement and monitoring
(vi) Feedback and communication, including Corrective Action and Preventive Action process

When an organization outsources the process, they remain responsible for that process. They must ensure that the processes remain suitable and effect as part of the quality management system. They need to ensure that that the product requirements are not affected as a result of outsourcing that process as the manufacturer is responsible for the product throughout the time the product is being managed by the subcontractor. For example, if the design process is being outsourced the business that is outsourcing the process would need to have involvement in defining the design inputs to ensure that the customer and regulatory requirements are met. Other key processes that may be outsource or subcontracted are sterilization or distribution.

The 2016 version introduced increased controls for outsourced processes and suppliers. Stricter compliance reviews enhance the chance of regulatory oversight at subcontractors, e.g. unannounced visits from EU Notified Bodies, EU regulators, China's National Medical Products Administration (NMPA), etc., indicate that contracts will need to define regulatory as well as quality agreements in much more detail.

15.18 Production and Service Provision

The standard clearly calls out to plan and carry out provisions under controlled conditions, including availability of information describing product characteristics, availability of documented procedures, work instructions, and reference standards, use of suitable equipment and use of measuring and monitoring devices. As for controlled environments used, there is no clear guidance, but applied conditions should be found suitable in the process risk assessment and appropriate to the final use of the device.

Clearer requirements are provided on the validation need for the processes for production and service provision—usually following a master validation plan approach, identification and traceability, customer property including drawings and IP, and the preservation of the product.

For many medical devices ensuring that the device is stored and handled to ensure that the performance of the device is maintained is important. Preservation of products covers many perspectives including storage at the defined temperature, making sure the product remains sterile, protection of fragile products. In order to be able to do this the organization has to understand the requirements specific to the product, for example, what temperature range should the product be stored at, how should the product be protected if it is fragile. Any special storage conditions have to be controlled and recorded. What this means is that the area where the product is stored it need to be known to be capable of maintaining that temperature range throughout whether is it a refrigerator or freezer or a walk-in freezer and it has to have been qualified to show that it maintains the temperature range throughout.

15.19 Monitoring and Measuring, Including Internal Audits and Management Review

The quality management system needs to be implemented to a high level before internal audits are conducted. Of course, to conform to ISO 13485, an internal audit program must be operational, helping the organization investigate any quality problems and verify solutions are effectively implemented. Periodic audits also keep the quality awareness level of staff high and foster improved internal communications.

The internal audits are expected to cover all areas and all shifts, auditing some areas more often based on previous results, whilst increasing audit frequency for changed or high-risk areas. ISO 19011 can serve as guidance.

Based on the internal audits and further management information, top management arranges the management review to come to a supported conclusion towards the (continuing) effectiveness of the QMS as well as to ensure continuing suitability. The standard identifies minimum requirements to be taken into the review including also any new or revised regulatory requirements for any country the company sells products or provides service to. Records of such reviews are kept, including action lists and their follow up activities.

The 2016 version of the standard follows the global regulatory push to enhance the requirements for 'feedback', including complaint handling. Requirements for statistical techniques for data analysis are being introduced.

15.20 Control of Non-Conforming Product

Non-conforming product needs to be identified and controlled to prevent its unintended use or delivery. Controls and responsibilities need to be defined in a documented procedure that should arrange for action taken to eliminate the detected non-conformity and for the authorization of non-conforming product's use, release or acceptance by concession. Such acceptance of non conforming product by concession can only be signed off if regulatory requirements, including a dedicated risk assessment, are met and the identity of the person authorizing the concession is identified and recorded.

15.21 Analysis of Data

The intent of this section is to stimulate the manufacturer to collect and analyze data on the quality system, both from the inside (e.g. production inspections), as well as from external sources which include complaint handling and more active ways of post market surveillance. It further supports the demonstration of the suitability and effectiveness of the QMS. In many cases, a selection of the analyzed data might well be used to stimulate quality awareness and motivate staff to further improve quality and efficiency. The trending becomes also more relevant in regulatory compliance, as for example the new EU regulations heavily depend on trending and sampling in the post-market phase.

15.22 Improvement: Corrective Action and Preventive Action

Following the documented procedure for corrective action and recording the results of the action taken will allow quick identification of any findings in one area. This can potentially help the remaining areas to avoid a similar problem. The correction is the immediate remedy to a non-conformity, e.g. clearing an uncontrolled messy area. The corrective action, which follows the root cause analysis (e.g. using the five why method), is the sum of measures that will ensure the same non-confirming situation does not re-occur. A different part of the improvement system are the preventive actions, which in a similar way work to ensure a non-conforming situation does not occur, but based on rationales and other input, not stemming from an existing non-conformity.

Clear distinction should be made between cases where some non-conforming situation was found, where correction and or corrective action to prevent reoccurrence are warranted, and the cases where no non-conformity was found but where preventive action can prevent a non-conformity from happening at all.

Please note ISO 9001 is focusing on correction and corrective action, whilst ISO 13485 unchangedly includes also the need to define preventive actions.

15.23 Purpose and Goal of ISO 13485:2016 Certification

Formal certification is achieved only through external (third party) assessment by a Certification Body. Certification can only be achieved after correct preparation for the initial audit, which is a two-stage event comprising a readiness review followed by a complete in-depth audit of all applicable areas of the standard.

Benefits of certification include expanded access to world markets, the improved ability to bid for contracts, the use as a market differentiator. The display of the certification mark (e.g. a symbol granted by the Certification Body which usually is a combination of the registrar's logo and ISO 13485) and the evidence of independent audits by professional, independent certification bodies designated by an accreditation agency can help to provide a much greater trust in the company's ability to provide a consistent high quality product or service.

These benefits clearly outweigh the barriers to certification such as the difficulty to identify and create new processes for the system, development of the necessary documented procedures and instructions and the resistance by some employees to change (and process measurements) that are found in most organizations.

15.24 Achieving Certification and Continuing to Maintain Certification

For an organization to be ready for certification, it is important that they have fully implemented an ISO 13485:2016 QMS and to be using the system. When the QMS is being assessed for certification, there are two aspects the certification body are considering:

(i) Are all the requirements included in the QMS? This will be completed by the review of the QMS as it is documented and by review of the internal audits and management reviews that have taken place. This is often called a Stage 1 assessment or Document review.

(ii) The second stage of the certification process is to show the effectiveness of the processes that have been put in place to meet the requirements of the standard. During this assessment, the organization has to demonstrate the continual improvement of the quality management system. This is Stage 2 of the assessment.

An organization needs to be sure that they are ready to demonstrate both these aspects. Often organizations try to complete the process too quickly and do not have sufficient evidence of the working of the QMS to demonstrate compliance. Best practice is to work with people experienced in ISO 13485 QMS systems when developing or upgrading a QMS.

Other pitfalls are as follows:

(i) There is insufficient awareness throughout the organization.
(ii) Top management have not been sufficiently involved to show that they are taking top management responsibility seriously.
(iii) There has been insufficient training throughout the organization to understand what is required.
(iv) The requirements especially related to Clause 6 and 7 have not been related to the products or services that the organization is manufacturing or delivering.
(v) Internal audits have not been completed of the whole QMS.
(vi) Appropriate metrics and measures have not been put in place for the analysis of data.

Once an organization has received its assessment and achieved certification to ISO 1345:2016, this is not the end of the process. The continuing process of maintaining certification begins as part of the continual improvement process.

Becoming certified to ISO 13485:2016 is a significant achievement for an organization and is a cause for celebration. When the celebrations are over it is important for everyone in the company to realize that maintaining the effective implementation of the quality management system is just as important. This includes understanding that it is a continuous process that everyone is committed to and involved in to ensure that the requirements of the quality management system continue to be met.

References

1. GHTF SG3/N15R8/2005—Implementation of risk management principles and activities within a quality management system.
2. GHTF/SG3/N17:2008 Quality Management System—Medical devices—Guidance on the control of products and services obtained from suppliers.
3. ISO 13485:2003/2016: Medical devices—Quality management systems—Requirements for regulatory purposes.
4. ISO 14971:2019: Medical devices—Application of risk management to medical devices.
5. ISO/TC 210 N382: Explanation of differences between ISO 13485: 2003 and ISO 9001:2008.
6. ISO/TR 14969:2004: Medical devices—Quality management systems—Guidance on the application of ISO 13485: 2003.

Chapter 16

ISO 14971: Application of Risk Management to Medical Devices

Tony Chan[a,b,c] and Raymond K. Y. Tong[d,e]

[a]*Virginia Tech, 21520 Yorba Linda Blvd. Suite G288, Yorba Linda, CA 92887, USA*
[b]*US Co-Chair, ISO TC 210 JWG1 Risk Management*
[c]*US Expert Member, IEC TC 56 WG4 Dependability Management*
[d]*Department of Biomedical Engineering,*
The Chinese University of Hong Kong, Shatin, Hong Kong
[e]*Asia Regulatory Professional Association (ARPA)*

chant@vt.edu, tchan@agsm-inc.com, kytong@cuhk.edu.hk

16.1 Introduction

The development of ISO 14971 was not carried out in a vacuum but instead was shaped by precedent activities in the European Union (EU) to establish risk management standards quite early in the 1990s. The first step made by the EU was the publication in 1997 of a European Standard, EN (European Norm) 1441 on Risk Analysis for Medical Devices. The framework focused on using a specific set of methods called Failure Mode and Effect Analysis (FMEA), for medical device risk analysis. ISO adopted this European Standard in 1997, when it essentially duplicated its publication under the title "*Medical Device—Risk Analysis*," ISO 14971-1. This standard was useful for prioritizing activities but remained incomplete because of its focus on the particular use of a single approach,

---

Medical Regulatory Affairs: An International Handbook for Medical Devices and Healthcare Products (Third Edition)
Edited by Jack Wong and Raymond K. Y. Tong
Copyright © 2022 Jenny Stanford Publishing Pte. Ltd.
ISBN 978-981-4877-86-2 (Hardcover), 978-1-003-20769-6 (eBook)
www.jennystanford.com

namely FMEA. Thus, a working group was formed to incorporate risk management practices already defined in the IEC 60601 Standard Series, a set of standards focusing on medical electrical equipment, into its new international standard, designated as ISO 14971, in 2000. This standard then replaced the ISO/IEC 14971-1 standard and was recognized as both a European-harmonized standard (GHTF SG3/N15R8:2005 [1]) and an FDA-recognized standard. It emphasized a more complete risk management framework, shown in Fig. 16.1.

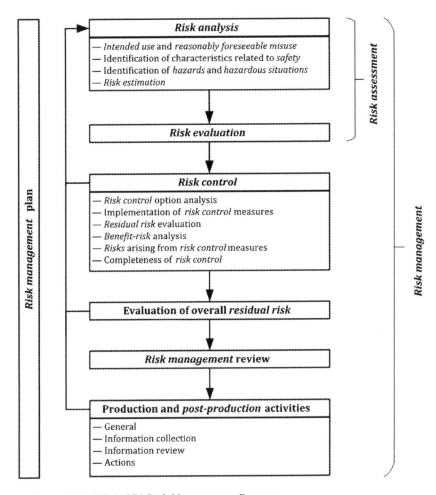

Figure 16.1 ISO 14971 Risk Management Process.

The ISO 14971 Standard includes all the basic principles of risk management that are applicable to medical devices. It specifies the steps required for identifying hazards, estimating risk, evaluating risk, and controlling risk. A medical device manufacturer's risk management process must include these steps. The new standard has two important aspects that deserve special mention. First, the standard expands the scope and potential approaches to the risk management process that enables manufacturers to satisfy the essential requirements of the medical device directives of the EU. These directives have similar authorities in the EU as the US Code of Federal Regulations in the United States. Essential requirements are necessary elements for protecting the public interest in the EU, are mandatory for product compliance before they are put into commerce and must be applied as a function of the hazards inherent to a given product.

Second, the standard deemphasized the use of FMEA, which is recommended only if it fulfills appropriately the particular risk analysis task under study; many other analytical approaches are acknowledged to potentially be suitable to achieve the same end. The expanded scope of the standard emphasizes management responsibilities. The standard stipulates that management responsibility is the initial and key requirement. More precisely, management must incorporate the following tasks:

(1) defining policies for determining acceptable risk
(2) ensuring the provision of adequate resources, including the identification of a risk management team
(3) ensuring the assignment of trained personnel to perform risk assessment and management activities
(4) reviewing the results of risk management activities at defined intervals to ensure the effectiveness of the risk management process

The expanded scope also mandates implementation of the following records:

(1) risk management plan
(2) risk management file
(3) risk management report
(4) production and post-production information

These requirements are interesting because they draw attention to the fact that risk management rests on more than having a framework, but also on a human dimension where the activities of people must be directed and organized. Currently, the ISO 14971 Standard has just gone through a second edition that incorporated only minor changes from the initial edition; however, the volume of explanation of the stipulated framework in the second edition has increased almost three-fold (from 22 pages to 65 pages). The ISO 14971:2019 *"Application of Risk Management to Medical Devices"* [2] stipulates an expectation of a risk management framework that considers the total lifecycle of the product. It includes a risk management process, and defines executive responsibilities, personnel qualifications for performing risk management activities, and the documentation needed to provide a record of risk management activities. Some limitations in the document are still perceived by some to exist. For example, it does not provide guidance with respect to clinical decisions that may be a significant part of the risk/benefit analysis. In fact, the EU is currently challenging the committee on the sufficiency of the guidance with regard to risk/benefit analysis. In addition, some readers of the standard can develop the false impression that it only addresses the design control elements of a quality management system because manufacturing only appears once in the text although it is in fact mentioned a further twenty times in the annexes of the document which are provided for explanations but not requirements.

In summary, the ISO Standard provides a framework to the medical device manufacturers for effective management of the risks associated with the use of their products and within which experience, insight, and judgment are applied systematically to manage these risks. It also helps practitioners to identify processes by which they can identify hazards associated with a medical device, estimate and evaluate the risks associated with these hazards, control these risks, and monitor the effectiveness of that control.

16.2 The Foundation of a Risk Management (RM) Framework: Policy, Plan, Team, Process, and Documentation

The foundation of an RM framework consists of five elements: (1) policy, (2) plan, (3) team, (4) process, and (5) Documentation.

16.2.1 Policy

A risk management policy is essential to a company that operates on RM. This policy should take into account of relevant international, national, or regional standards and regulations. It should define how to determine acceptable risk and include the review of results of risk management activities at defined intervals. The review ensures the continuing suitability and effectiveness of the RM process. Moreover, the policy should be established before or at the time of implementation of the RM process. The RM process should not begin without such policy in place.

16.2.2 Plan

The company should establish a risk management plan for every product based on its risk management process. This plan [1] should

(a) identify and describe the product
(b) define a scope
(c) state verification activities
(d) indicate roles and responsibilities of the allocated resources
(e) define intervals for review
(f) specify criteria for risk acceptability
(g) record the plan changes and its revision

16.2.3 Team

A description of each of the RM team members' roles and responsibilities should be defined and recorded. It is a part of the evidence to demonstrate sufficient resource is allocated for risk management activities.

16.2.4 Process

A medical device company should establish a process describing how, what, and when risk management activities are carried out during the entire product life cycle (See Fig. 16.1 and Section 16.3). At the broadest level, the RM process consists of a four-part, continuous-management process: (1) analyze risk, (2) evaluate risk, (3) control risk, and (4) get feedback from both production and post-production information. The next section describes the details of each major activity of the process.

16.2.5 Documentation

Documentation can be electronically stored records or paper records. It provides objective evidence for regulatory compliance as well as for legal purposes. The documentation should include policy, procedures, plans, and records of all risk management activities during the entire product life cycle.

16.3 The RM Process

16.3.1 Analyze Risk

This initial task involves the following steps:

(a) **Familiarize yourself with the product.**

It is important to understand thoroughly the requirements of developing a product before the start of any risk management activities. The background understanding of requirements, such as marketing, performance, functional, regulatory, legal, etc., would provide a firm foundation for those who are engaged in the risk management activities.

(b) **Understand intended purpose and use.**

In addition to aforementioned requirements, the intended purpose and use of a product provide a boundary and scope for the subsequently risk management activities. Understanding the medical purpose, anatomy and/or organs of body affected, patient profile, user profile, and application environment of the product forms the basis to initiate hazard identification and for carrying out further risk management activities.

(c) **Identify hazards and hazardous situations.**

In the framework of the Standard, hazard is defined as the potential source of harm; harm is physical injury or damage to the health of people, property, or the environment. Thus, one should focus on identifying the immediate source or a situation(s) that causes the hazard that results in harm, i.e., the hazard or hazardous situation. It is also beneficial to distinguish that a failure may not cause harm or directly cause harm.

(d) **Estimate risk.**

Once a hazard or hazardous situation is identified, its consequence or severity of harm and probability of occurrence of harm need to be estimated. The consequence or severity of harm of a hazard or hazardous situation could be assessed directly and immediately. However, the probability of occurrence of harm may be an indirect assessment since the harm is related to the hazard and the hazard is related to the product. The risk level of a hazard or a situation(s) that initiate the hazard would then be estimated based on the probability of occurrence and severity of that harm posed by the hazard or the hazardous situation.

16.3.2 Evaluate Risk

This subsequent activity after risk analysis is to make an acceptability decision. Consequently, there are many actions need to be taken depending upon the level of risk on each of the identified issues. Generally, there are three kinds of actions. First, the risk could be accepted without further actions. Second, the risk is so high that control measures must be taken to reduce it to an acceptable level. Lastly, the risk is at a marginally acceptable or unacceptable level. It requires further investigation, either conduct a risk/benefit analysis before making accept or reject decisions, or implement further risk control measure(s) to reduce the risk to an acceptable level.

16.3.3 Control Risk

After the initial risk analysis and evaluation, risk control measures may be deemed necessary to maintain, reduce or eliminate the risk. In the framework of the Standard, risk has two components: the probability of occurrence of harm and the severity of that harm. Therefore, a risk control measure could either affect the probability of occurrence of the harm or the severity of that harm. Generally, it is much easier to work on maintaining or minimizing the occurrence aspect of the harm. It would take relatively much more resources and efforts in reducing or eliminating the severity of harm. In fact, risk control is aiming at controlling the identified hazards and the hazardous situations that are causing harm.

The Standard is very specific about how risk should be controlled. It stipulates an Option Analysis [1] that requires a hierarchy of priority order when approach to risk control measures. Risk control measure(s) should be considered in the following priority order:

(a) inherent safety by design (the most preferred risk control measure)

(b) protective measures in the product or in the manufacturing process

(c) information for safety (the last resort for risk control measure)

It is indispensable to ensure the implementation of these risk control measures have taken place as well as to demonstrate their effectiveness after implementation. Objective evidence of implementation and demonstration of effectiveness of these measures should be documented.

In practice, risk analysis, risk evaluation, and risk control are steps of an iterative process. These activities may happen at least three times in a product development process. Initial risk is analyzed, evaluated for acceptance the first time and then risk control measures are considered as necessary for each hazard and hazardous situation. Then, residual risk needs to be analyzed and evaluated for acceptance as related to each individual hazard after the control measures are applied. This process continues until the residual risk is judged acceptable. Finally, a review of the overall residual risk, i.e., the accumulation of all the residual risks posed by each individual hazard after effective control measure(s) is applied, should take place. This is when a company decides whether a product is safe for use.

If an individual residual risk or the overall residual risk [1] is judged unacceptable using the policy or criteria pre-established by the company and further risk control is impractical, the company may gather and review data and literature on the medical benefits of the intended purpose and use of the product to determine if they outweigh the residual risk. If the evidence does not support the conclusion that the medical benefits outweigh the residual risk, then the risk remains unacceptable. The project should be cancelled or the product needs redesign. If the medical benefits outweigh the residual risk, then the company should document all the evidence to support the acceptability decision. This iterative process should continue until all individual residual risk and the overall residual risk are judged acceptable before the commercialization of a product.

16.3.4 Feedback from Production and Post-Production Information

In this part of the risk management process, the focus is on information received after all risk control measures are deployed. When considering the entire product life cycle, the company should establish and maintain a systematic process. It could be a well-developed information system, to review information gained about the product after development. This information should include information from design and process validation, product transfer, design to production, vendor to company, company to vendor, outsourced activities, production activities, field use, etc. The information should be evaluated for possible relevance to safety [1] especially the following:

(a) if previously unrecognized hazards are present
(b) if the estimated risk(s) arising from a hazard is no longer acceptable
(c) if the original assessment is otherwise invalidated

If any of the above conditions is satisfied, the results of the evaluation should be fed back as an input to the risk management process. A review of the appropriate steps of the risk management process for the product should be considered. If there is a potential that the residual risk(s) or its acceptability has changed, the impact on previously implemented risk control measures should be evaluated.

The above four risk management activities are integral parts of a continuous process improvement within a management system.

Certain activities may be initiated out of the aforementioned order. They are useful if applied appropriately, especially early in the product development process.

16.4 Conclusion

RM is about focusing on the number of safety issues to the vital few that will make a difference. A comprehensive RM framework helps a company gain confidence that it understands those risks that really matter to avoid unacceptable and unexpected crises or surprises. The framework also helps the company know how to avoid, eliminate, or contain the causes or drivers of those significant risks that may cripple the business if they were neglected. It also provides a basis for measuring, comparing, controlling, monitoring, and preventing those risks that have the potential to exhibit significant negative impact to the company.

In conclusion, RM is a disciplined, structured, and scientific approach that provides a proactive business strategy to enable a company to manage and prevent potential crisis and create a sustainable strategic advantage while maximizing shareholder value. Through the RM process, a company is better able to protect and secure its value, guard its growth potential, and hedge against market, capital, operational, and financial volatility.

16.5 Case Study—An Example to Illustrate Risk Management on a Medical Device: Functional Electrical Stimulation System for Walking

Neuromuscular Electrical Stimulation (NMES) is generally referred to as an artificial electrical stimulation of a muscle that has diminished nervous control. When the aim is to provide muscular contractions and produce functionally useful movements, it is commonly referred to as functional electrical stimulation (FES). The FES is a technique that uses bursts of short electric pulses to generate muscle contractions by stimulating motor-neurons or reflex pathways.

A FES system for walking consists of a main unit, a control sensor, and adhesive surface electrodes connected with wires

(Fig. 16.2). The surface electrodes are located on the surface of the skin, above a targeted muscle or nerve. With regard to the lower limbs, drop-foot is a common clinical feature of those who have suffered from stroke, which is characterized by the inability to dorsiflex the foot during the swing phase of gait. People affected tend to have a labored and unsafe gait, and compensatory movements such as hip hiking and limb circumduction may occur. The intended use of an FES drop-foot stimulator system is to correct the problem of drop-foot by stimulating the ankle dorsiflexors or the common peroneal nerve to achieve toe clearance (Fig. 16.3). The stimulation of the hemiplegic leg can be controlled by a heel-switch worn on the hemiplegic side. When the switch is on the hemiplegic side, stimulation is initiated by a heel off and terminated by a heel strike. The effectiveness of the system is to save effort in walking, increase the patient's ability to walk, and reduce spasticity.

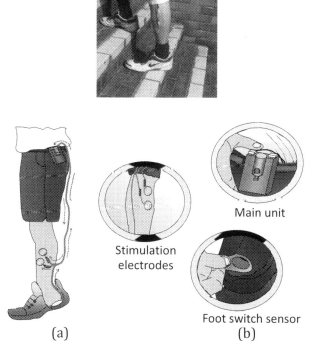

Figure 16.2 Using FES for walking.

(a)   (b)

Figure 16.3 Right hemiplegic stroke patient: (a) without FES and drop foot on the right side; (b) with FES system to correct drop foot problem.

16.5.1 Risk Management Process

Step 1: Intended use/intended purpose

- For the particular medical device or accessory being considered, the manufacturer shall describe the intended use/intended purpose and any reasonably foreseeable misuse. They should also have a list of all those qualitative and quantitative characteristics that could affect the safety of the medical device and, where appropriate, their defined limits.

The intended use of the FES drop-foot stimulator system is to correct the problem of drop-foot by stimulating the ankle dorsiflexors or the common peroneal nerve to achieve toe clearance. The timing of stimulation is controlled by a heel-switch worn on the hemiplegic side. Electrical stimulation is initiated by a heel off and terminated by a heel strike in every walking step on the hemiplegic side.

The FES system only can be applied on stroke patients with drop foot problem by a trained clinician/rehabilitation engineer. The patients must be trained by the clinician/rehabilitation engineer before they can take the device back home for training and daily use.

The FES stimulation parameters are 40 Hz with adjustable pulse width of the stimulation pulse (100–500 us) and the maximum stimulation intensity is 100 mA. These parameters are safe and they are currently use in clinical/hospital for rehabilitation with FES.

Step 2: Identification of known or foreseeable hazards

- The manufacturer shall compile a list of know or foreseeable hazards associated with the medical device in both normal conditions and fault conditions.

Example of Hazards

1. Surface stimulation electrodes are placed on the wrong position on the skin and cause injury.
2. There are three connection ports on the FES main unit and they are connected wrongly: one is connected to the stimulation electrodes, another one is for foot switch and last one is for computer data transfer.
3. The output electrical stimulation is too high and cause injury.
4. Patient pressed or turned wrong button to change the FES parameters, which causes injury.
5. There is skin allergy to the surface electrode.

Step 3: Estimation of risks for each hazard

- ISO14971: For each indentifies hazard, the risk(s) in both normal and fault conditions shall be estimated using available information or data. The hazard is combined with the hazard severity and the probability of occurrence (Tables 16.1 and 16.2).

    Estimated Risk = (Estimated hazard Severity) × (Estimated Probability of Occurrence)

Each manufacturer is free to define his own category system (Tables 16.1 and 16.2) and has to recorded these in the risk management file.

Table 16.1 Example of five qualitative severity levels (ISO 14971:2007 Table D.3)

Common terms	Possible description
Catastrophic	Results in patient death
Critical	Results in permanent impairment or life-threatening injury
Serious	Results in injury or impairment requiring professional medical intervention
Minor	Results in temporary injury or impairment not requiring professional medical intervention
Negligible	Inconvenience or temporary discomfort

Table 16.2 Example of semi-quantitative probability levels (ISO 14971: 2007 Table D.4)

Common terms	Examples of probability range
Frequent	$\geq 10^{-3}$
Probable	$<10^{-3}$ and $\geq 10^{-4}$
Occasional	$<10^{-4}$ and 10^{-5}
Remote	$<10^{-5}$ and $\geq 10^{-6}$
Improbable	$<10^{-6}$

Step 4: Risk Evaluation

- For each identified hazard, the manufacturer shall decide the criteria in the management file to consider whether the estimated risk(s) is needed to have risk reduction or not (Tables 16.3 and 16.4).

Table 16.3 Estimation of risks for each hazard

Hazards	Severity	Probability
R1. Surface stimulation electrodes are placed on the wrong position	Serious	Remote
R2. The three connection ports on the FES main unit are connected wrongly with wires	Minor	Probable
R3. The output electrical stimulation is too high	Serious	Remote
R4. Patient pressed or turned wrong button to change the FES parameters	Serious	Probable
R5. Skin allergy to the surface electrode.	Minor	Occasional

Table 16.4 Risk evaluation

		Qualitative severity levels				
		Negligible	Minor	Serious	Critical	Catastrophic
Semi-quantitative probability levels	Frequent					
	Probable		R2	R4		
	Occasional		R5			
	Remote			R1, R3		
	Improbable					

Key
- Unacceptable risk
- Investigate further risk reduction
- Insignificant risk

Step 5: Risk Control

- The manufacturer shall identify risk control measure(s) that are appropriate for reducing the risk(s) to an acceptable level. Risk control shall use the following in the priority order listed:
  (1) inherent safety by design
  (2) protective measures in the medical device itself or in the manufacturing process
  (3) information for safety

R2, R3, R4, and R5 are using inherent safety by design to reduce the risk and R1 are using information for safety.

Step 6: Implementation of risk control measure(s)

- The manufacturer shall implement the risk control measure(s) selected in step 5.

Implementation of risk control measure(s) (Table 16.5)

R1: A qualified trained clinician/rehabilitation engineer shall provide information sheet and taught the patients on how to place the stimulation electrode correctly.

Table 16.5 Table with risk mitigation measures

		Qualitative severity levels				
		Negligible	Minor	Serious	Critical	Catastrophic
Semi-quantitative Probability levels	Frequent					
	Probable					
	Occasional					
	Remote		R5	R4		
	Improbable		R2	R1, R3		

Key
- Unacceptable risk
- Investigate further risk reduction
- Insignificant risk

R2: The connection ports are designed only one plug can fit in one socket for all the three ports (Fig. 16.4).

R3: The system use standard 9 V battery and the circuit can only generate maximum 100 mA. And the stimulation pulse width can be changed by the clinician through an FES program software only.

R4: Patient can only change the stimulation amplitude in a limited range, and all the parameters are pre-programmed by the clinician.

R5: Biocompatible electrode for clinical use on surface electrical stimulation shall be used. They will receive the biocompatible standard.

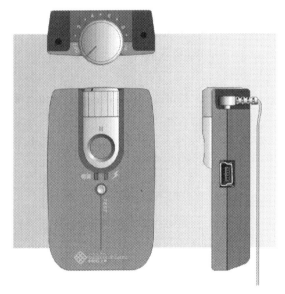

Figure 16.4 The connection ports are designed only one plug can fit in one socket. And the patient can only adjust the stimulation intensity in the limited range on a knob from 0 to 10.

Step 7: Residual risk evaluation

- Any residual risk that remains after the risk control measure(s) are applied shall be evaluated using the criteria defined in the risk management plan. If the residual risk does not meet the criteria, further risk control measures shall be applied.

  The residual risk is judged acceptable for the FES system.

References

1. GHTF (2005). SG3/N15R8:2005 Implementation of risk management principles and activities within a Quality Management System, Global Harmonization Task Force.
2. ISO 14971:2019—Application of Risk Management to Medical Devices.
3. ISO/IEC Guide 51:1999—Safety Aspects—Guidelines for their inclusion in standards.

Chapter 17

Medical Devices—IEC International Standards and Conformity Assessment Services in Support of Medical Regulation and Governance

Gabriela Ehrlich
International Electrotechnical Commission
geh@iec.ch

17.1 Introduction

Today, electronics and electrical, including medical devices, are more similar than ever before. Their parts and subassemblies transit through many countries before a device is built somewhere, shipped, installed, and then used—anywhere in the world. Electrical devices are generally no longer the industrial manufacture of a single country; more often than not, they are "made in the world".

Broad use of harmonized, globally agreed technical rules—generally IEC International Standards—helps increase the availability of safe, affordable, quality medical devices. At the global level, this results in broader access to better healthcare.

Medical Regulatory Affairs: An International Handbook for Medical Devices and Healthcare Products (Third Edition)
Edited by Jack Wong and Raymond K. Y. Tong
Copyright © 2022 Jenny Stanford Publishing Pte. Ltd.
ISBN 978-981-4877-86-2 (Hardcover), 978-1-003-20769-6 (eBook)
www.jennystanford.com

While the trade of medical electrical equipment is global, the regulatory environment is still local to each specific jurisdiction. This lends emphasis and importance to an international approach to standardization but also to greater regulatory convergence.

Medical electrotechnical equipment is of increasing importance in areas that were traditionally covered by non-electrical medical equipment.

Healthcare services and the application of medical electrotechnical equipment, healthcare software and IT networks are growing rapidly, largely driven by the following factors:

- The life expectancy and size of population is increasing.
- The impact of information technology is growing.
- Cost saving goals are gaining importance in medical practice.
- Developing countries are generating new equipment markets.

Growing demand can have a positive impact on cost and innovation, if manufacturers are able to access large markets that allow them to off-set often high development costs.

All medical devices need to meet a rather complicated set of safety requirements. Compliance regulations will heavily affect their design and manufacture, including mechanical, electrical, and software matters.

Clear, harmonized regulatory guidelines help encourage investment in R&D and future innovations and reassure manufacturers that regulatory requirements will remain sensible.

However, when regulation is too severe in terms of safety or ambiguous or confusing in its requirements, it can hinder investment and with it patient access to affordable diagnosis and treatment options.

Use of international standards with little or no national/regional variations is a guarantee for optimum patient access to reasonably prized and safe healthcare.

The International Electrotechnical Commission (IEC) is a globally trusted organization. It brings together 172 countries, 88 Members and 84 Affiliates, which are mostly developing countries (www.iec.ch).

17.2 What Is an IEC International Standard?

An international standard is, in essence, an agreed and repeatable way of accomplishing the same outcome time after time. It is a document that presents the combined expertise, know-how and consensus of many experts and is designed to provide technical specifications or other precise criteria as guidance or definition.

Standards, regulations, and certification are closely interlinked. Standardized technologies that are tested and verified for safety, quality and performance contribute to overall risk-management. They also help improve the understanding of device interactions.

With IEC 60601-1 and other international standards, the IEC offers the globally recognized technical foundation for a large range of medical devices. These international standards provide essential guidance for medical electrical equipment, electrical systems and software used in healthcare and their effects on patients, operators, other persons, and the environment.

17.3 How Are IEC International Standards Developed?

IEC International Standards are voluntary, and consensus driven. Many experts from many different countries participate in the standardization process on the global, neutral, and independent IEC platform. That is where they try to achieve consensus on solutions to challenges that are faced by multiple countries.

Only when at least a two-thirds majority of all experts have reached agreement; all major issues have been addressed and major opposition has been overcome, can a standard be sent for approval by the IEC members.

Each member country can invite all national stakeholders to comment on the proposed standard. This national input is then brought to the global level in the IEC. At that point, every member country can comment, accept or reject the proposed standard. Every country has one vote; this helps avoid that a big country can push its needs to the forefront to the detriment of smaller countries.

The result is an international standard that is near universally relevant, can be used by all and helps level the playing field.

IEC International Standards offer a double consensus at the expert and at the country level: they combine the know-how and consensus of thousands of technical experts, as well as the vote and approval of many countries.

17.4 IEC International Standards for Medical Devices

Several IEC technical committees and subcommittees (TCs and SCs) prepare international standards that concern medical devices. For example IEC TC 29 develops standards related to hearing aids and hearing instruments; IEC TC 87 covers the breadth of ultrasonic equipment for all diagnostic or therapeutic purposes; IEC TC 66 prepares international standards that cover the safety requirements of electrical equipment that is used for measurement, control and laboratory use; experimental devices are also part of its scope.

IEC TC 62, which was established in 1968, prepares the majority of international standards for medical devices. It focusses on general safety aspects, as well as product safety and performance. Together with its different subcommittees IEC TC 62 covers a vast field of product categories including for example diagnostic imaging, radiotherapy, nuclear medicine, radiation dosimetry, electromedicine, anaesthesia, critical care, surgery, artificial respiration or paediatrics.

When appropriate, and to ensure that international medical device standards fit seamlessly together, IEC TC 62 and its SCs cooperate with other committees of the IEC, ISO or other organisations, based on the expertise each organization embodies.

The most important publication of IEC TC 62 is the IEC 60601 family of standards, which is widely accepted for basic safety and essential performance of medical electrical equipment. Compliance with this standard series is a requirement for certification in many countries.

Medical Electrical Equipment (Reference IEC 60601-1, 3.63) is defined as "electrical equipment having an applied part or transferring energy to or from the patient or detecting such energy to or from the patient". That includes EEG monitors, IV pumps, imaging systems, ECG devices, vital signs monitors, and similar devices that connect directly to a patient.

The IEC 60601-1 series addresses basic safety, as well as electrical and mechanical safety; essential performance; operator protection; radiation, temperature and other hazards; programmable systems, transformers, batteries, mobile equipment, and so forth. The series contains more than 60 special requirements documents geared to specific devices.

In addition to safety design requirements, strict manufacturing rules are associated with the traceability of source materials and manufacturing procedures, as well as quality control.

Risk management is intrinsically a part of the philosophy of the IEC in general and IEC 60601-1 in particular. In a unique approach, IEC 60601-1 integrates mandatory risk management principles of ISO 14971 but applies them to device instead of process certification. IEC 60601-1 facilitates compliance with legal and regulatory requirements specific to the healthcare area. It guides manufacturers in the identification of all hazards associated with a medical device, including for all operation modes and all fault scenarios. One outcome is a risk matrix that is associated with the use of an individual device. Probability of occurrence, vs. severity of harm is classified. Anything significantly harmful and somewhat probable of occurring is addressed.

The IEC 60601-1 standard series also provides well-defined rules for isolation of both the patient and the operator of the device. This includes tight controls on leakage currents, voltages applied, and energy limits that can make it to the patient. Clear guidance is provided on how to design, manufacture or test external power supplies or power converters.

This is complemented by other IEC International Standards which help verify for example device stability when the power supply is briefly pulled out of the wall or during voltage fluctuations.

Several IEC International Standards address multiple types of outside interferences that can impact the ability of a medical device to function error free. Among other things, they provide guidance for tests that evaluate the device's ability to work correctly when faced with electromagnetic interference, radio frequency pulses, power frequency magnetic fields, etc. They help monitor and identify any type of error messages, system resets, component failures, changes to programmable parameters, changes of operating modes, false alarms, faulty patient information, etc. The aim is to ensure not only damage-free test survival of the medical device, but successful, error-free functioning beyond.

With the help of IEC International Standards, medical electrical equipment manufacturers are able to ensure that medical devices as well as complex systems and assemblies are safe and effective.

17.5 Conformity Assessment for Medical Devices

In addition to standardization, the IEC also offers a unique globally standardized approach to testing and certification for medical devices under IECEE (IEC System of Conformity Assessment Schemes for Electrotechnical Equipment and Components). Since 1986, IECEE members certify that medical devices comply with IEC International Standards.

The IECEE is the global testing and certification system for electrotechnical equipment. It has issued more than a million certificates that are recognized worldwide. IECEE administers third party conformity testing and certification schemes that address the safety, quality, efficiency and overall performance of components and goods that are used in the home, office, or health facilities.

IECEE certification is based on the principle of reciprocal acceptance. Members of the System issue test reports and certificates that are mutually accepted by the other members of the System as a basis for obtaining certification or approval at national level. This approach is essential in facilitating international trade and allowing direct access to the marketplace for vendors, retailers or buyers. It eliminates unnecessary duplicate testing and reduces the costs related to the certification process. IECEE is also an important partner for regulators in the verification of compliance. IECEE operates the CB Scheme and the Full

Certification Scheme. The latter includes factory inspections. More information on IECEE is available at www.iecee.org.

IECEE verifies that medical electrical and electronic devices and equipment are reliable and meet expectations in terms of performance, safety, reliability and other criteria. The scheme not only applies to the medical electrical equipment itself but also covers risks to patients, those who operate the equipment—doctors, nurses and technicians, for instance—and maintenance personnel.

Medical technology may be speeding ahead but IECEE is close behind in terms of managing the risks associated with new technologies. For example, the IECEE CTL Expert Task Force 03 "MEAS, MED" deals with technical matters of standards for medical electrical equipment. The sub-group Task Force "MED" specifically deals with the implementation of risk management requirements as set out in the third edition of IEC 60601-1.

The Expert Task Force consists of about 35 members who represent various interests in the field of medical electrical equipment (government agencies, testing laboratories, certification bodies, IEC technical committees). The group meets regularly and is responsible for

- developing guidelines and working instructions on how to implement the relevant clauses of IEC 60601-1 in helping manufacturers demonstrate compliance with the "risk management process" as defined in ISO 14971, Medical devices—Application of risk management to medical devices;
- establishing a consensus with methods that are acceptable for determining compliance with all the relevant clauses (in relation to ISO 14971) of IEC 60601-1;
- developing a checklist aimed at assisting the medical equipment industry, official authorities and stakeholders around the world to test in the appropriate manner;
- acting as an Advisory Group on the common understanding of ISO 14971 with respect to IEC 60601-1;
- considering technical enquires pertaining to testing requirements, test method and test; and
- developing decision sheets.

17.6 Conclusion

An international standard and a global Conformity Assessment System is at its most effective when regulatory requirements are harmonized. The ultimate aim is to achieve an outcome where one standard, one test is accepted everywhere. That is when it offers the biggest advantages for regulators, industry, and the patient.

Chapter 18

Good Submission Practice

Shinji Hatakeyama[a] and Isao Sasaki[b]

[a]*Asia Regulatory Affairs, Medicine Development Center, Eisai Co., Ltd.*
[b]*International and Greater China Regulatory Affairs,*
Regulatory Affairs, Astellas Pharma Inc.
s-hatakeyama@hhc.eisai.co.jp, isao-sasaki@astellas.com

18.1 Preamble

Good Submission Practice (GSubP) refers to the best practices by industries. It is aimed to enhance the quality and efficiency of the medical products registration process by improving the quality of submission as well as its management. Promotion of GSubP by the industries has been recognized as important in facilitating Good Review Practice (GRevP) by the regulatory authorities. Furthermore, synergic promotion of the two practices by the regulatory authority and the industries are recognized as Good Registration Management (GRM) under APEC umbrella. The two regulatory authorities, Taiwan FDA and PMDA in Japan, have led GRM for facilitating best practice of medical products registration in APEC.

Asia Partnership Conference of Pharmaceutical Association (APAC) was established in 2012 and is a platform built by

---

Medical Regulatory Affairs: An International Handbook for Medical Devices and Healthcare Products (Third Edition)
Edited by Jack Wong and Raymond K. Y. Tong
Copyright © 2022 Jenny Stanford Publishing Pte. Ltd.
ISBN 978-981-4877-86-2 (Hardcover), 978-1-003-20769-6 (eBook)
www.jennystanford.com

13 Asian pharmaceutical industry's associations to advocate industry proposals in order to expedite the launch of innovative medicines for the peoples in Asia. Regulations and Approvals Expert Working Group (RA-EWG), formed under APAC, has been pursing rolling out of GSubP to Asian economies through establishment of high-level guidance for GSubP, which has been endorsed by APEC in 2016, for sharing its concept widely, and collaboration with Taiwan TFDA and PMDA for establishing APEC GRM Center of Excellence (CoE) workshop for capacity building.

APEC GRM CoE Workshop has been held at Taipei since 2016. Total trainees from 2016 to 2019 are around 250 from 16 economies in APEC (around 110 of regulators from 12 economies and around 140 industry members from 11 economies). The workshop has adapted "Train the trainer" model that trainees at the work shop to be trainers for GRM/GRevP/GSubP at their economies. As results, local GRevP and/or GSubP workshops were held in Singapore, Chinese Taipei, Thailand, Malaysia, the Philippines, Indonesia and Malaysia from 2017 to 2019. Furthermore, pilot GRM CoE workshops were held in Mexico and Thailand in 2017 and 2019, respectively. Thai FDA has been also endorsed as official GRM CoE in 2020 by APEC after the pilot GRM CoE workshop at Bangkok.

APEC GRM CoE workshop consists of three types of session: a common session, the GRevP session and the GSubP session. The common session facilitates fruitful communications between reviewers and applicants. The GRevP session introduce concept of WHO GRevP guideline [1] to reviewers. This chapter intends to share APEC GSubP guideline, main content of the GSubP session, to gentle readers, and support best practice of regulatory submissions for the medical products in your field.

18.2 Introduction

18.2.1 Objective and Scope

The objective of this document is to provide general and high level guidance on Good Submission Practice (GSubP, see Section 18.2.3, Definition) principles and processes which applicants of medical products should keep in mind. The recommended processes are not intended to provide detailed instructions on how to conduct each submission or to serve as prerequisite for the medical product submissions.

The goal of GSubP is to enhance efficiency and quality of medical product registration process which leads to enhance early access to these products by patients. The GSubP principles and elements described in this document will help applicants to achieve the goal.

Regarding detailed procedures for submission preparation, applicants should consider to generate Standard Operating Procedures (SOPs) in their own organization considering the general guidance provided in this document and specific conditions and requirements in each country.

This document is envisioned as a companion document to Good Review Practice (GRevP, see Section 18.2.3, Definition) guidelines, and sufficiently expandable to accommodate additional annexes or ancillary documents in the future.

This document applies to any aspects related to the regulatory submission for medical product registration and its management by applicants. It also covers associated activities by applicants in planning, submission and review stages up to approval.

Although this document was written focusing on application submission for pharmaceutical products and biologicals and higher-risk medical devices for use in humans, the concepts may be applied to other types of medical products. Similarly, the concepts described here may also be applicable to the entire product lifecycle from investigational testing to new product applications, updates or variations to existing marketing authorizations and maintenance of the product.

18.2.2 Background

In general, registration submission of medical products is made in the final stage following time consuming product development process. It can be regarded as the compilation of all development program and activities of the product. In order to obtain early approval in the registration process, it is important for applicants to prepare and submit application with good quality dossier. It is also essential that applicants keep close communications with the review authorities by taking prompt and appropriate actions in ensuring of smooth registration process. Application submission with poor quality of dossier and management will lead to failure in getting final approval or cause significant delay due to a large number of inquiries and requests from the review authorities. Applicants should always seek ways to improve their submission in quality and efficiency.

18.2.3 Definition

Good Submission Practice (GSubP)
An industry practice for any aspect related to the process, format, contents and management of submission for registration of medical products by applicants. It is the practice to enhance the quality and efficiency of the product registration process by improving the quality of submission as well as its management.

To promote continuous improvement, all aspects of GSubP should be evaluated and updated on an ongoing basis.

Applicant (as defined in WHO GRevP guidelines [1])
The person or company who submits an application for marketing authorization of a new medical product, an update to an existing marketing authorization or a variation to an existing marketing authorization.

Good Review Practice (GRevP) (as defined in WHO GRevP guidelines [1])
Documented best practices for any aspect related to the process, format, content and management of a medical product review.

GRevP has been introduced and moved towards step-wise implementation by many regulatory authorities to enhance the timeliness, predictability, consistency, transparency, clarity, efficiency and quality of product review. The GRevP guideline document was endorsed by World Health Organization (WHO).

18.3 Principles of Good Submission

The objective of GSubP is to help applicants prepare good quality submission leading to successful registration. The 'principles' of a good submission describe the key elements of GSubP which applicants should follow in order to achieve successful product registration. The following five key principles of good submission are provided as a general guide. Applicants should keep them in mind when planning and managing submissions.

Key Principles of Good Submission

1. **Strong Scientific Rationale and Robust Data with Clarification of Benefit-Risk Profile:**

    A good submission should be based on strong scientific rationale and robust data in terms of integrity, relevance and completeness. The nature of the benefits and types of risks should be clarified with sound evidence.

2. **Compliance to Up-to-date Regulatory Requirements**

    A good submission is made in compliance with the up-to-date regulations. In addition, it should keep reasonable consistency with internationally harmonized regulatory standards.

3. **Well-Structured Submission Dossier with Appropriate Cross-references**

    A good submission will be made with well-structured dossier complying with the acceptable format by the review authorities. To ease the reviewing process, applicants are encouraged to use appropriate cross-references in the dossier.

4. **Reliability, Quality, Integrity and Traceability of Submission Documents and Source Data**

    A good submission is made ensuring the reliability, quality, integrity, and traceability of information and data described in submission documents including their sources.

5. **Effective and Efficient Communications**

    A good submission and timely review can only be achieved by keeping effective and efficient communications with the review authorities throughout the product development and registration process. In addition, good communications within the applicants' organization(s) are essential for successful submission as well as its management.

18.4 Management of Submission

The working scheme for applicants to prepare and manage the submission differs by the size of the applicant's organization. Appropriate resource management is also important, considering the submission work is time- and labor-intensive activity.

In case the applicant is a small to medium-sized organization, submission preparation is often managed and handled by a person or a small group of people. Therefore, the applicant may need to plan their submission with sufficient lead time and consult with the review authorities when necessary.

In case of large organization, preparation of an application submission is generally conducted by collaborative work among concerned parties. For example, a submission team consisting of clinical, non-clinical and quality experts, statisticians, medical writers, regulatory staffs, project managers and other relevant stakeholders is formed and work together for a submission. The individual roles and responsibilities should be defined in advance.

Sometimes local and international collaborations among multiple organizations are also required, e.g. local affiliate and headquarters, sponsor and co-development partners, originator and licensee companies.

In any of aforementioned working scheme, applicants should appropriately manage the whole process of product registration including submission.

The principles of project management and quality management are critical for well-organized submission preparation. The submission practices of careful planning, good communications and clearly-defined work instructions can maximize the quality and efficiency of submission process.

18.4.1 Planning for Submission

Preparation for application submission generally starts with planning phase. Often, submission for product registration takes place in the last stage of lengthy product development process. Even so, applicants need to initiate discussions on submission strategy from an early stage of product development and establish a clear strategy for submission. Clarification of product profile as well as its update according to ongoing development program is a critical part of such strategic discussions. For that purpose, some companies use a document so-called "Target Product Profile," a summary format of product development program described in terms of labeling concepts.

It is also important for applicants to conduct clinical and non-clinical studies as necessary in compliance with the up-to-date regulatory standards, guidelines and regulations. Applicants shall strive to obtain and understand the regulatory information necessary for product development and registration.

It should be noted that progress has been made in regulatory convergence and harmonization by international cooperation scheme among the regulatory authorities, e.g. the International Council for Harmonization of Technical Requirements for Pharmaceuticals for Human Use (ICH), Harmonization of Standards and Technical Requirements in The Association of Southeast Asian Nations (ASEAN) and The International Medical Device Regulators Forum (IMDRF). It is necessary that applicants keep abreast with not only local but also regional and international standards, guidelines and regulations, and update their own submission strategy accordingly.

In order to plan and manage an application submission efficiently, applicants are recommended to prepare and use the following tools:

Checklist

Applicants are encouraged to make a checklist to plan for every required component of submission dossier. The list may include name of each document with information such as responsible person/party, target date and status. Such list will be useful not only to check if there is any missing component but also to manage the whole process of submission preparation efficiently.

Glossary

It is important to keep consistency of terminology used throughout a submission dossier. Applicants are recommended to create a list of general glossary before initiating preparation of study reports and summaries.

Template

Template is a standard file format document containing pre-defined layout, styles, texts and graphics. Templates help authors to prepare each component document in structured and consistent manner complying with the required format and contents, e.g. ICH M4 and E3. It will also enhance efficiency of preparation. Submission with a unified format of study reports and summaries also enables reviewers to perform review smoothly.

Timeline table

Development and management of timeline is one of the most important tasks in submission planning phase especially when the submission is performed by collaborations among multiple parties of applicants. It is recommended that applicants generate and keep updating a timeline table or a Gantt chart including the role and responsibility of each person/party to manage the whole process of submission preparation.

If necessary, applicants shall also plan for pre-submission meeting with review authorities (see Section 18.5.1).

These activities in planning stage will enhance quality and efficiency of submission preparation and its management.

18.4.2 Preparation and Submission of Application Dossier

There are two main steps in preparation of an application dossier. One is preparation of each component, i.e. writing study reports and summaries, and preparing other required documents. The other is compilation and assembling of submission dossier.

In general, authors of reports and summaries are assigned from experts in each scientific field or medical writers, and overall handling of submission is conducted by regulatory function or professionals.

Writing study reports and summaries

Study reports and summaries are key components of technical documents in application dossier. The former corresponds to Module 3, 4 and 5, and the latter constitutes Module 2 in ICH-CTD.

The contents of study reports should be based on strong rationale and robust data with scientific evidence. Needless to say, applicants should ensure reliability, integrity and traceability of data described in the reports. Applicants also need to refer to the relevant guidelines on the format and contents of study reports which can be accepted by the review authorities, e.g. ICH M4 and E3.

Summary documents should be generated based on the contents of study reports to provide clear rationale with justification. It is also necessary to clarify the nature of benefits and risks of the product based on sound scientific evidence.

The contents of these documents shall comply with up-to-date standards and regulations at the time. In addition, it is essential to prepare these documents by taking into account alignment with current international standards and guidelines.

The authors of these documents should strive to write a concise and easy to read document. Sometimes peer review by competent third party or person is effective in order to check the validity of scientific contents before final draft.

Translation of original documents to other language is sometimes required in the process of submission preparation. In such case, applicants should pay careful attention to ensure accuracy and validity of translation.

Besides study reports and summaries, other types of documents are required at application submission by each regulatory authority as regional or country-specific requirements. They include, but are not limited to, application form, proposed labeling, letter of authorization, patent statement, certificates issued by the competent authority etc. Type of required documents differs depending on the type of medical products, category of application and the local regulations. Applicants should review the list of required documents provided by the national regulatory authorities.

Compilation and assembling of dossier

Before compiling and assembling submission dossier, applicants should review the structure and format of dossier accepted by the national regulatory authorities, e.g. ICH-CTD. Collection and review of each component document should be performed in reference to defined table of contents. A checklist will help applicants to manage the collection process efficiently. In compiling and assembling of submission dossier, applicants need to ensure that every document has been prepared consistently and placed in the correct location of the dossier.

Some regulatory authorities have been accepting application submission with electronic dossier. Applicants should review the local regulatory requirements and follow the relevant instructions when intending to submit their application electronically.

Submission of application

Each review authority has defined acceptable format, process and route of application submission, e.g. hard copy or electronic dossier, on-line, mailing or on-site submission. Sometimes, a pre-submission consultation with the review authorities is required to fix the date of submission.

Applicants are required to submit application dossier following the procedure and instructions provided by each authority. To avoid rejection of filing, applicants should ensure that the submission is made in proper category and contains all the required information and materials using appropriate format.

Standard operating procedure for submission preparation

Preparation of application dossier is a complicated and time-consuming process. It is often performed by collaborations among applicants' parties or group of organizations. It is therefore beneficial for applicants to generate SOPs and share them within the parties or organizations for proper management of the whole process of submission preparation.

SOPs may be structured to contain or refer to additional tools that could assist in performing the procedure for submission, e.g. template, standard format of checklist.

Additional working procedure documents may also be created to give more detailed instruction and structure in support of SOPs. These documents can describe in detail how a particular process is performed, e.g. procedure for drafting, reviewing and finalization of each study report and summary. SOPs may also outline the workflow processes which facilitate project management when multiple parties work on different parts of the application dossier.

These SOPs need to be updated depending on the change in applicant's working environment, e.g. change in organization, scheme of work-sharing etc.

18.4.3 Quality Check

Quality check (QC) of submission dossier and its components is critical and indispensable process in order to achieve a submission of good quality. The purpose of QC is to ensure that information and data described in submission dossier have sufficient quality in accuracy, integrity and traceability of scientific data/information, and to check compliance to pre-defined format, template and structure. Some regulatory authorities require submission of QC declaration by applicant.

The following types of QC can be conducted depending on the subject, timing and stage of submission preparation.

QC of study reports and summary documents

The main purpose of this QC is to ensure accuracy, integrity and traceability of scientific data and information. This type of QC is usually conducted just before or at finalizing of each report, summary document or any other document which refer to the contents of these documents, e.g. product labeling.

It should also be conducted when making revisions to the contents of these documents.

In case translation of these documents to other language is required, it is useful to review the accuracy and validity of translation. In addition, compliance to pre-defined format and template, e.g. ICH M4, E3, needs to be checked as a part of QC process.

QC of submission dossier

This is the QC process to be conducted at compilation and assembling of submission dossier. The purpose is to check if every required component is ready and placed in the correct section of the dossier.

QC of electronic dossier

This type of QC is required in case of application submission with electronic dossier, e.g. ICH eCTD or NeeS (non-eCTD electronic submissions). The purpose is to confirm if the dossier is compliant with the review authority's electronic dossier requirements, e.g. electronic bookmarks, cross references and hypertext links are correctly functioned.

A record should be created when conducting QC. Also, applicants are highly recommended to have a written SOP for QC procedure.

18.5 Communications

Effective communications will help applicants to improve quality and efficiency of the product development as well as registration process, thereby realize timely approval and earlier patient access to new products. Applicants should foster good communications with the review authorities and those within applicants' organization(s).

18.5.1 Communications with the Review Authorities

Communication with the review authorities can take place in various forms such as meetings, inquiry and response. Applicants should be aware of available communication mechanisms in pre- and post-submission stages and make effective use of them in product development and submission processes. Interactions with the review authorities throughout the processes are greatly facilitated by having a clearly defined contact point in applicants' organization. It is recommended that main communications with the review authorities are conducted consistently through regulatory professionals for all projects.

During the post-submission stage, applicants should make prompt and appropriate responses to the inquiries and/or requests from the review authorities. To do so, applicants need to track the progress of review in timely manner and adjust the schedule as well as internal resource accordingly. It is also important to ensure that the review authorities and applicants are able to share information about the timeline and progress of review.

Communications in pre-submission stage

Meeting with review authorities

Meetings with regulatory authorities in product development and pre-submission stages help applicants to fix design of planned clinical/non-clinical studies, clarify requirements and potential concerns of ongoing development program and envisage possible questions in the forthcoming application process. It enables applicants to progress their development program efficiently in compliance with regulatory requirements and prepare a good quality submission dossier by dealing with potential questions and concerns in advance. It will increase the probability of a positive outcome in the forthcoming application submission.

Applicants should proactively use pre-submission communication with the review authorities to make successful submission.

In order to hold effective and productive meeting, applicants should keep the following points in mind.

- Study and follow the defined rules and procedure for the meeting
- Clarify the purpose and discussion points
- Prepare good quality meeting materials
- Discuss based on reasonable scientific rationale
- Prepare and circulate meeting minutes/memo on discussion points and agreements
- Take appropriate follow-up measures on comments and advice received from the authorities

Communications in post-submission stage

Meeting with review authorities

Opportunity of meetings in post-submission review stage is useful not only to track progress of review but also to discuss and solve potential issues, questions and requests raised by reviewers. It also helps applicants to have clear understanding on the background of received inquiries and prepare appropriate response to the point.

Availability of these meetings in post-submission stage depends on the review process adopted by each review authority. Often such meeting is held according to a request from review authorities. Applicants should follow the instructions described in previous section (see Section 18.5.1) when having a meeting in post-submission stage.

Inquiry and response

Inquiry and response form the critical communication between reviewers and applicants. In general, inquiries from the review authorities are issued in two separate stages of the course of application review, i.e. in screening phase and in scientific review stage.

Screening or validation for application filing is usually performed by the review authorities at receipt of submission to ensure that the dossier is complete and of suitable quality for scientific review. In case any screening inquiry is received, applicants should correctly understand the contents and make prompt and appropriate action for response.

Once screening phase is finished, official scientific review starts. The review authorities may request additional information based on the outcome of scientific review.

At receipt of an inquiry in scientific review stage, it is important for applicants to clarify and understand the background as well as intention of the reviewer with that inquiry. To make it possible, the review authorities often allow applicants to ask for clarification. Applicants should make the best use of such opportunity.

When applicants received a critical inquiry which would require an additional study (or studies), they should have a consultation meeting with the review authority as much as possible to clarify details of the required study (or studies).

Proper management of the timeline for response preparation is another important element. It is advisable for applicants to confirm the deadline of response, set a reasonable timeline and appropriately manage prompt response preparation.

Preparation of response package needs to be conducted following the procedures and instructions described in section 18.4 Management of Submission.

Depending on the country, the review authorities may request applicants to confirm the contents of draft evaluation report in the last stage of the review process. In such case, applicants should confirm the contents carefully.

18.5.2 Communication within Applicants' Organization

In many cases, preparation of an application submission is performed by collaborative work among concerned persons or parties in applicants' organization. Sometimes collaborations among multiple organizations are required.

Applicants should understand that a good submission can be achieved only when concerned parties within or among applicants' organizations share a clear strategy and work collaboratively throughout the product registration process up to approval. Good communication within submission team is the key to successful submission.

It is highly recommended that the submission team clarifies and confirms its operation model as well as the role and responsibility of team members when they have a kick-off meeting. Good communications within the team will be facilitated by establishing and sharing standardized working procedure and having a platform of information sharing such as regular meetings.

In case of global product registration, collaboration among multiple regions with time and geographical differences are required for submission in each country. In such case, applicants should make an effort to achieve effective and efficient communications among the regions.

In the case that applicants have outsourced manufacturing, research and development or submission operations, it is applicants' responsibility to keep close communication with the contractor.

18.6 Competencies and Training

It is recommended that applicants possess general core competencies to properly manage and prepare submissions. Type of recommended competencies depends on the role and responsibility of each person or party in the submission team.

18.6.1 Core Competency of Applicants

A core set of recommended general competencies includes the following elements:

Scientific knowledge and expertise

Applicants should have professional knowledge and expertise that relate to the product safety, efficacy and quality. These knowledge and expertise are especially significant for the authors and reviewers of technical documents in submission dossier. Writing skills are also essential for the authors.

Logical application of each field of scientific knowledge, understanding on risk-benefit analysis, and critical thinking methodology to ensure compliance with regulatory standards and guidelines are also recommended competencies.

Good understanding of up-to-date regulations

Applicants should always keep abreast with the latest regulatory environment. This can be done by following the regulatory authorities' website and check updated news, notices or highlights. If available, applicants can also subscribe to a mail delivery service provided by the review authorities to allow applicants to receive updated regulatory information from the authorities' website periodically.

Applicants should carefully study published regulations, technical guidelines, notices, Q&A documents etc. Applicants can also attend training programs provided by the regulatory authorities, industry associations or other third parties to help understand the contents and background of these regulations.

Other hard and soft skills

It is recommended that applicants develop the following hard and soft skills and abilities as a part of their competencies to move forward submission and its management efficiently.
- Planning and project management
- Medical and technical writing
- Technical skills for electronic submission (as necessary)
- Problem-solving
- Communication

Integrity and reliability

Applicants should approach the process with honesty, integrity and reliability and should not jeopardize the confidence of the regulatory authorities and other stakeholders.

18.6.2 Training and Capacity Building

Training is essential for applicants to acquire sufficient core competencies and strengthen skills and capacity. For that purpose, applicants can make use of various opportunities of training programs provided by regulatory authorities, industry associations and other third parties.

For example, these parties often hold periodical educational programs, workshops and training sessions for applicants. Sometimes the authorities also provide briefing sessions for applicants when they release a new regulation or guideline, and prepare Q&A documents.

These external training programs are valuable for applicants to deepen their own scientific, technical and regulatory knowledge and expertise. Applicants should strive to participate in these training programs whenever possible.

In addition, it is essential for applicants to acquire necessary skills and competence through their own day-to-day operations. Opportunities of in-house training, self-training and on-the-job training in applicants' organization should be leveraged as a part of capacity building program.

It is also recommended for applicants to establish good documentation practice to document submission requirements and past applications for reference and to allow good practices sharing within their organization(s) so that they can make good use of these valuable experiences and competence for future submissions as well as capacity building. Archival can be in the form of manuals, SOP, database or any other appropriate tools.

18.7 Conclusion

The goal of GSubP is to enhance the quality and efficiency of the medical product registration process by improving the quality of submission as well as its management. The implementation and execution of GSubP will help applicants to reduce the risk of filing rejection and number of critical/major inquiries from review authorities and guide applicants to obtain early approval in the product registration process.

Acknowledgments

This chapter is referenced from the APEC Good Submission Practice Guideline developed by APAC. The authors wish to thank the Regulatory and Approvals Expert Working Group of APAC for their work.

Glossary

APEC: Asia Pacific Economic Cooperation
ASEAN: The Association of Southeast Asian Nations
CTD: The Common Technical Document
ICH: The International Council for Harmonization of Technical Requirements for Pharmaceuticals for Human Use
ICH-CTD: Common Technical Document agreed in ICH
ICH eCTD: Electronic Common Technical Document defined by ICH
IMDRF: The International Medical Device Regulators Forum
Inquiry: Questions or information requests made by the review authorities on submitted registration application
NeeS: Non-eCTD electronic Submissions
Q&A: Questions and Answers
QC: Quality Check
RA-EWG: Regulatory and Approval Expert Working Group established under APAC
SOP: Standard Operating Procedure
WHO: World Health Organization

Reference

1. Good Review Practices: Guidelines for National And Regional Regulatory Authorities. WHO Technical Report Series, No. 992, 2015, Annex 9. http://www.who.int/medicines/areas/quality_safety/quality_assurance/Annex9-TRS992.pdf?ua=1.

Part 3

Medical Device Regulatory System in the United States, European Union, Saudi Arabia, and Latin America

Chapter 19

United States Medical Device Regulatory Framework

Joshua Silverstein

US Food and Drug Administration, Center for Devices and Radiological Health, 10903 New Hampshire Ave., Silver Spring, Maryland, USA

joshua.silverstein@fda.hhs.gov

The regulatory information contained in this chapter is intended for market planning and is subject to change frequently. Explanations of laws and regulations are unofficial.

19.1 Introduction

The United States Food and Drug Administration, commonly known as the FDA, is an agency within the U.S. Department of Health and Human Services (HHS). The FDA is a science-based, regulatory, and public health agency. Its statutory mandate is to promote and protect the public health, including ensuring that there is a reasonable assurance of the safety and effectiveness of devices intended for human use and that the public health and safety are protected from electronic product radiation. Furthermore, the

---

Medical Regulatory Affairs: An International Handbook for Medical Devices and Healthcare Products (Third Edition)
Edited by Jack Wong and Raymond K. Y. Tong
Copyright © 2022 Jenny Stanford Publishing Pte. Ltd.
ISBN 978-981-4877-86-2 (Hardcover), 978-1-003-20769-6 (eBook)
www.jennystanford.com

FDA also regulates foods, drugs, biological products, cosmetics, and tobacco products. The FDA does not regulate the practice of medicine. State law regulates medical practice.

The FDA's regulatory authority extends to the fifty (50) States, the District of Columbia, Puerto Rico, Guam, the Virgin Islands, American Samoa, and other U.S. territories. Estimates by Select USA (https://www.selectusa.gov/medical-technology-industry-united-states), a government-wide program housed in the U.S. Department of Commerce, indicate the United States remains the largest medical device market in the world with a market size of approximately $156 billion, and it is expected to reach $208 billion by 2023. The U.S. market value represented about 40% of the global medical device market in 2017. U.S. exports of medical devices in key product categories identified by the Department of Commerce exceeded $43 billion in 2018. These estimates also indicate there are almost 2 million medical device jobs in the United States, including both direct and indirect employment. Medical device companies are mostly small and medium-sized enterprises, with more than 80% of these companies having fewer than 50 employees.

The learning objective for this chapter is to provide students and other interested individuals a concise overview of the U.S. regulatory framework for medical devices. More specifically, the goal is to introduce and focus on the general regulatory requirements, pre- and post-market, necessary to legally market medical devices in the United States.

19.2   FDA's Center for Devices and Radiological Health

In the United States, the FDA's Center for Devices and Radiological Health (CDRH) is responsible for regulating medical devices and radiation-emitting electronic products (both medical and non-medical). The CDRH's regulatory paradigm in regulating medical devices is based on risk. That is, the higher the risk of harm to humans, the greater the level of regulatory control. The CDRH evaluates most new medical devices for safety and effectiveness before they are marketed. It also monitors products already on the market, so that unsafe and/or ineffective products are corrected or removed from the market. In addition, it regulates companies that design, manufacture, repackage, relabel, reprocess, sterilize, and/or initially import medical devices into the United States. The CDRH also regulates radiation-emitting electronic products, including both for medical and non-medical purposes, such as lasers, diagnostic x-ray systems, ultrasound equipment, and microwave ovens.

The CDRH's Division of Industry and Consumer Education (DICE) is available to answer questions from both the medical device industry and consumers (https://www.fda.gov/medical-devices/device-advice-comprehensive-regulatory-assistance/contact-us-division-industry-and-consumer-education-dice).

To learn more about the CDRH's mission, vision, and shared values, as well as strategic planning, transparency and innovation initiatives, the following website provides comprehensive information: https://www.fda.gov/about-fda/center-devices-and-radiological-health/cdrh-mission-vision-and-shared-values.

19.3 Legislation and Device Law

The Radiation Control for Health and Safety Act of 1968 (RCHSA) required a radiation control program that included the development and administration of performance standards to control the emission of electronic product radiation from electronic products, such as microwave ovens, diagnostic x-ray equipment, and others. The goal of this law was to provide for the protection of the public health from radiation emissions from such products. A few years later, on May 28, 1976, President Gerald Ford signed into law the Medical Device Amendments to the U.S. Federal Food, Drug, and Cosmetic Act (FD&C Act). This was the result of a U.S. Senate finding that faulty medical devices had caused 10,000 injuries, including 731 deaths. The Medical Device Amendments established a comprehensive system for the regulation of medical devices intended for human use in the United States, including premarket review requirements. There have been numerous subsequent amendments to the FD&C Act, including notably the Safe Medical Devices Act of 1990,

FDA Modernization Act of 1997, FDA Amendments Act of 2007, FDA Safety and Innovation Act of 2012, and the 21st Century Cures Act of 2016.

19.3.1 FDA Law vs. FDA Regulations vs. FDA Guidance

Federal laws establish the legal framework within which the FDA operates. The FD&C Act is a federal law enacted by Congress in 1938 that gave authority to the FDA to oversee the safety of food, drugs, and cosmetics. The FD&C Act has been amended many times. For instance, it was amended in 1976 as described above to incorporate the Medical Device Amendments into law.

Based on the laws set forth in the FD&C Act and other federal laws, the FDA develops implementing regulations. This process is commonly known as "notice and comment rulemaking" and follows the requirements of the Administrative Procedure Act. Federal government regulations, including those promulgated by the FDA, have the force and effect of law. The FDA's regulations are codified in Title 21 Code of Federal Regulations (21 CFR). The CFR represents all current regulations and the volumes are published annually in paper format. They are also available on the internet at the Government Publishing Office's website (www.ecfr.gov). Medical device and electronic product regulations are in Parts 800-1299.

Conversely, FDA guidance is not law. Guidance documents describe the Agency's current thinking on regulatory issues and are not legally binding on the public or the FDA. Guidance documents facilitate the regulatory process by providing more information on the interpretation of medical device regulations and laws. The Agency develops guidance documents to articulate the FDA's recommendations for meeting safety and effectiveness requirements, including recommendations for various medical device types. Guidance documents that include new policies are issued in a draft. Once a draft guidance is issued, the FDA requests and considers comments before the Agency issues a final guidance. Guidance documents provide consistency and predictability, which help manufacturers gather appropriate non-clinical and clinical data, plan, and submit an organized and appropriate marketing submission.

More information on FDA laws, regulations, and guidances are available at http://www.fda.gov.

19.4 The Regulatory Environment for Bringing a Medical Device to Market

Anyone who wishes to commercially distribute medical devices or electronic products that emit radiation in the United States must comply with applicable U.S. laws and regulations, even if the product is authorized for marketing in another country. At present, the FDA does not recognize regulatory marketing approvals from other countries. However, it will evaluate all safety and effectiveness information submitted to it in the marketing submission. This can include foreign clinical data, provided that any such investigations that first enroll a subject on or after February 21, 2019 are conducted in accordance with good clinical practice.

Comprehensive information about the Center for Devices and Radiological Health and the regulation of medical devices can be found at https://www.fda.gov/Medical-Devices.

The device regulatory topics discussed in this chapter are also accessible from the FDA website "Device Advice: Comprehensive Regulatory Assistance": https://www.fda.gov/medical-devices/device-advice-comprehensive-regulatory-assistance.

Furthermore, the CDRH makes available to the public several medical device databases: https://www.fda.gov/medical-devices/device-advice-comprehensive-regulatory-assistance/medical-device-databases.

19.5 Regulatory Considerations for Medical Devices

19.5.1 Definition of Medical Devices

Section 201(h) of the FD&C Act defines the term "device" as "an instrument, apparatus, implement, machine, contrivance, implant, in vitro reagent, or other similar or related article, including a component part, or accessory which is (1) recognized in the official National Formulary, or the United States Pharmacopoeia, or any supplement to them, (2) intended for use in the diagnosis of disease or other conditions, or in the cure, mitigation, treatment, or prevention of disease, in man or other animals, or (3) intended to affect the structure or any function of the body of man or other animals, and which does not achieve any of its primary intended purposes through chemical action within or on the body of man or other animals and which is not dependent upon being metabolized for the achievement of any of its primary intended purposes. The term 'device' does not include software functions excluded pursuant to Section 520(o)."

The definition above is used to determine whether a product is a medical device and helps provide distinction between a medical device and other FDA-regulated products, such as drugs or biologics. In such cases, another FDA Center regulates the product.

19.5.2 Classification of Medical Devices

The FD&C Act established three categories of devices (class I, class II, and class III) and requires the FDA to classify devices intended for human use into one of these categories. The tiered classification is based on risk, including the associated level of regulatory control necessary to mitigate that risk (i.e., general controls, special controls, and/or premarket approval):

- Class I: General Controls
- Class II: General and Special Controls
- Class III: General Controls and Premarket Approval

The appropriate level of regulatory control necessary to provide a reasonable assurance of safety and effectiveness of a device type increases according to its classification, with class III devices requiring the most regulatory oversight. A more extensive discussion of these various controls are provided in later sections, while an overview of the tiered classification approach and examples are included in Table 19.1.

Table 19.1 Overview of the tiered classification approach

	Necessary regulatory control				
Class	General	Special	Premarket approval	Risk	Examples
I	X			Low-moderate	General surgical instruments, surgeon's gloves, medical adhesive tape
II	X	X		Moderate-high	Surgical suture, fetal heart monitor, feeding tube, hemodialysis system
III	X		X	High	Breast implant, dermal filler, percutaneous heart valves, implanted neurostimulators, and pacemakers

The publicly available product classification database provides a means for conducting a search of existing device types (http://www.accessdata.fda.gov/scripts/cdrh/cfdocs/cfPCD/classification.cfm). It includes the medical device name, three-letter product code unique to the CDRH, and identifies the device class as well as the type of premarket submission required, if any, for that device type. Other information also includes the device definition, regulation description, any applicable recognized consensus standard, and other important regulatory information about the generic device type. A screen shot of an example device type from such a search is included below in Fig. 19.1.

Class I devices are generally recognized as low-moderate risk devices. Reasonable assurance of safety and effectiveness is generally provided through the application of general controls for class I devices. Alternatively, class I devices are also those for which insufficient information exists to determine that general controls provide reasonable assurance of safety and effectiveness and to establish special controls that provide such assurance, but the devices are not for a use that is life supporting or sustaining or of substantial importance in preventing impairment of human life, and do not pose an unreasonable risk of illness or injury. The majority of class I devices do not require FDA premarket review prior to marketing. Certain class I devices can be reserved under Section 510(l)(1) of the FD&C Act and subject to premarket review via premarket notification under Section 510(k) of the FD&C Act.

Device	Syringe, Piston
Regulation Description	Piston syringe
Regulation Medical Specialty	General Hospital
Review Panel	General Hospital
Product Code	FMF
Premarket Review	Gastrorenal, ObGyn, General Hospital, and Urology Devices (OHT3) Drug Delivery and General Hospital Devices, and Human Factors (DHT3C)
Submission Type	510(k)
Regulation Number	880.5860
Device Class	2
Total Product Life Cycle (TPLC)	TPLC Product Code Report
GMP Exempt?	No
Summary Malfunction Reporting	Eligible
Recognized Consensus Standards	

* 1-79 ISO 26825 First edition 2008-08-15
Anaesthetic and respiratory equipment - User-applied labels for syringes containing drugs used during anaesthesia - Colours, design and performance
* 5-108 ISO 80369-6 First Edition 2016-03-15
Small bore connectors for liquids and gases in healthcare applications - Part 6: Connectors for neuraxial applications.

Figure 19.1 Screenshot of FDA's product code classification database.

Class II devices are generally moderate to high-risk devices for which general controls alone are not adequate to provide a reasonable assurance of safety and effectiveness and for which there is sufficient information to establish special controls to provide such assurance. Therefore, class II devices are or will eventually be subject to special controls in order to provide such assurance. Class II devices are typically subject to premarket review and clearance by the FDA through the 510(k) premarket notification process.

The highest risk devices (class III) are those for which insufficient information exists to determine that general and special controls are sufficient to provide reasonable assurance of safety and effectiveness for this class of devices. By definition, those that are life-sustaining/supporting or present a potential unreasonable risk of illness or injury, or are of substantial importance in preventing impairment of human health. Class III devices require a premarket approval (PMA) application prior to being legally marketed.

The flowchart in Fig. 19.2 summarizes the regulatory factors the CDRH considers consistent with 21 CFR Part 860 in determining whether to place a device into class I, II, or III during the classification or reclassification process.

Regulatory Considerations for Medical Devices | 245

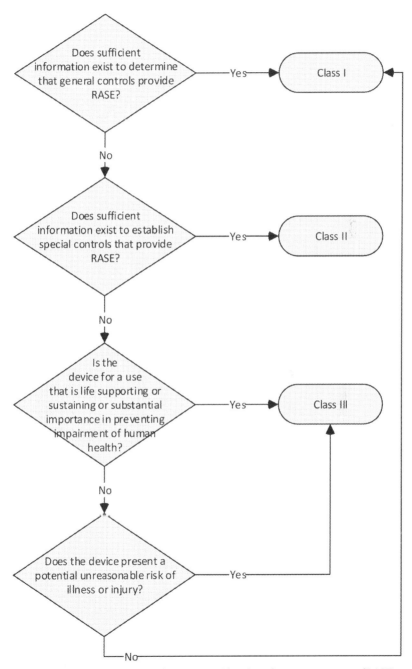

Figure 19.2 Flowchart explaining FDA's classification process (RASE, reasonable assurance of safety and effectiveness).

If the preceding information and resources are not sufficient to assess whether a product is a device, or its associated classification, the CDRH has additional resources available to the public. This includes the CDRH's Device Determination Officer, who can provide informal information, as well as formal mechanisms for obtaining information from the CDRH regarding device classification in accordance with Section 513(g) of the FD&C Act. The FDA's response to a 513(g) request for information will provide information regarding device classification and/or applicable regulatory requirements. Additional information regarding this resource is available: https://www.fda.gov/regulatory-information/search-fda-guidance-documents/fda-and-industry-procedures-section-513g-requests-information-under-federal-food-drug-and-cosmetic.

19.5.2.1 General controls

General controls are regulatory requirements authorized by the FD&C Act, under Sections 501, 502, 510, 516, 518, 519, and 520 of the FD&C Act. The FD&C Act requires these baseline controls for marketing and distribution, labeling, and monitoring safety and effectiveness once a device is on the market. Unless exemption is specifically provided in a classification regulation, all devices must comply with the following general controls, among others:

- Adulteration and misbranding
- Establishment registration and device listing
- Premarket notification (510k), unless exempt
- Labeling requirements
- Quality System Regulation/current Good Manufacturing Practices (cGMPs)
- Medical Device Reporting (MDR)

19.5.2.2 Special controls

Class II devices must meet applicable requirements designated as special controls. Special controls are device-specific and can include the following:

- Performance standards
- Postmarket surveillance

- Patient registries
- Development and dissemination of guidelines, including those for the submission of clinical data in a 510(k) submission
- Other appropriate actions as the FDA deems necessary to provide reasonable assurance of safety and effectiveness

All available medical device guidance documents are searchable at the following site: https://www.fda.gov/medical-devices/device-advice-comprehensive-regulatory-assistance/guidance-documents-medical-devices-and-radiation-emitting-products

19.5.2.3 Premarket approval

The highest risk devices (class III) require a PMA and

- are life sustaining and/or life supporting; or
- are of substantial importance in preventing impairment of human health; or
- present potential unreasonable risk of illness or injury.

Submission of a PMA requires each sponsor to submit valid scientific evidence to establish the reasonable assurance of that device's safety and effectiveness.

More information regarding regulatory controls used to provide reasonable assurance of safety and effectiveness for devices is available: http://www.fda.gov/MedicalDevices/DeviceRegulationandGuidance/Overview/GeneralandSpecialControls/default.htm

19.5.3 Convenience Kits

Certain convenience kits that meet the criteria in the Convenience Kit Interim Regulatory Guidance (https://www.fda.gov/regulatory-information/search-fda-guidance-documents/convenience-kits-interim-regulatory-guidance) are under enforcement discretion and do not require a 510(k). If a 510(k) is not required, documentation to support each of these determinations should be maintained in the kit assembler's/manufacturer's files in accordance with the Quality System Regulation (21 CFR Part 820) and should be available for FDA review if needed.

For a kit that does not contain a device for which a PMA is required and is not under the convenience kit interim guidance enforcement discretion policy, any 510(k) submission for a kit subject to 510(k) requirements should identify all devices provided in the kit and document the marketing status of each device in the kit by including the Kit Certification for 510(k)s (https://www.fda.gov/regulatory-information/search-fda-guidance-documents/kit-certification-510ks).

19.5.4 Labeling

Labeling includes all labels and other written, printed, or graphic matter (1) upon any article or any of its containers or wrappers, or (2) accompanying such article. This also includes descriptive and informational literature that accompany the device. The general labeling requirements for medical devices are specified in 21 CFR Part 801. These regulations specify the minimum labeling requirements for all devices. There are additional requirements needed for specific categories of devices such as labeling for in vitro diagnostic devices under 21 CFR Part 809, devices operating under an approved Investigational Device Exemption (IDE) application, radiation-emitting electronic products, and natural rubber (latex).

19.5.5 Adulteration and Misbranding

Medical devices are subject to the adulteration and misbranding provisions of the FD&C Act. There are numerous provisions related to both adulteration and misbranding under Sections 501 and 502 of the FD&C Act, respectively. For example, under Section 501(a)(1) of the FD&C Act, a device is considered to be adulterated if it includes any filthy, putrid, or decomposed substance, or if it is prepared, packed, or held under unsanitary conditions. Two examples of misbranding under Section 502(a)(1) and 502(f)(1) include provisions on device labeling that is false or misleading in any particular and that device labeling must bear adequate directions for use.

More information about adulteration and misbranding is available at: https://www.fda.gov/medical-devices/regulatory-controls/general-controls-medical-devices.

19.5.6 Establishment Registration and Medical Device Listing

Device establishments (both domestic and foreign) and initial importers of medical devices must register their establishments with the FDA and are required to pay a fee, register annually, and list their devices pursuant to 21 CFR 807 Subparts A-D. The initial registration and/or device listing information must be submitted within 30 days of an establishment beginning an activity or placing a device in commercial distribution. Afterwards, the fee-based registration information must be submitted every year between October 1 and December 31, even if no changes have occurred. Device listing must also be reviewed at the same time. However, device listing updates may be submitted at any time without additional fees. The registration process is done electronically through the FDA's FURLS system.

Foreign manufacturers must designate a U.S. Agent. Most establishments that are required to register are also required to list the devices that are made in those facilities, as well as activities that are performed on those devices. Registration and Listing is not a marketing authorization, certification, or license to market a device in the U.S. If a device requires 510(k) clearance or PMA approval before being legally marketed in the U.S., the manufacturer must wait until the FDA has cleared or approved the device. However, in other jurisdictions in the world, the term "Registration" may be equivalent to marketing authorization to sell and distribute a device.

More information on the required fees, clarification regarding who needs to register, how to register, payment process, U.S. agent, and other important reminders about registration and listing can be found at: https://www.fda.gov/medical-devices/how-study-and-market-your-device/device-registration-and-listing.

19.5.7 Quality System Regulation/Good Manufacturing Practices

Manufacturers of finished devices must comply with the Quality System Regulation under 21 CFR Part 820. The regulation includes requirements related to the methods used in the facilities and controls used for designing, purchasing, manufacturing, packaging, labeling, storing, installing, and servicing of medical devices in order to conform to cGMPs. The FDA inspects manufacturing facilities to assure compliance with the quality system requirements. Certain types of devices are exempt by regulation from Good Manufacturing Practices (GMPs). However, exemption from GMP requirements does not exempt manufacturers from maintaining complaint files or from general requirements concerning records. IDE clinical studies are exempt from GMP, except for design controls pursuant to 21 CFR 820.30 that require procedures to control the design of the device to ensure that the specified design requirements are met.

The International Organization for Standardization (ISO) developed and published ISO 13485. It is the international standard relating to Quality Management Systems for organizations involved in the manufacture of Medical Devices. The FDA is in the process of harmonizing the Quality System Regulation with ISO 13485. The revisions will supplant the existing requirements with the specifications of ISO 13485. The revisions are intended to reduce compliance and recordkeeping burdens on device manufacturers by harmonizing domestic and international requirements. The revisions will also modernize the regulation. For more information on the FDA Quality System Regulation, refer to Quality Systems at: https://www.fda.gov/medical-devices/postmarket-requirements-devices/quality-system-qs-regulationmedical-device-good-manufacturing-practices.

19.5.8 Medical Device Reporting

Reporting of adverse events is a mandatory post-marketing requirement for medical devices. Firms who have received complaints of device malfunctions, serious injuries, or deaths associated with medical devices are required to notify the FDA of the incident. User facilities (e.g., hospitals, nursing homes) are also required to report suspected medical device related deaths to both the FDA and the manufacturers. User facilities report medical device-related serious injuries only to the manufacturer unless the manufacturer is unknown. In such instances, the serious injury is reported to the FDA.

The MDR regulation is in 21 CFR Part 803. This regulation allows the FDA and manufacturers to identify and monitor significant adverse events involving medical devices. The objective of the regulation is to detect and correct problems in a timely manner. Under the MDR program, incidents in which a device may have caused or contributed to a death or serious injury must be reported to the FDA. Further, certain malfunctions must also be reported. The regulation specified that reports be filed on the FDA's Medwatch Form 3500A or an electronic equivalent. Manufacturers and importers are required to submit MDRs to the FDA in an electronic format that the FDA can process, review, and archive.

For more detailed information on postmarket requirements, i.e., reporting advssserse events, postmarket surveillance studies and post-approval studies, see: https://www.fda.gov/medical-devices/device-advice-comprehensive-regulatory-assistance/postmarket-requirements-devices.

How to report a medical device problem: https://www.fda.gov/medical-devices/medical-device-safety/medical-device-reporting-mdr-how-report-medical-device-problems.

19.5.9 Unique Device Identification

The FDA established the unique device identification system to identify devices sold in the United States from manufacturing through distribution and use. When fully implemented, the label of most devices will include a unique device identifier (UDI) in human- and machine-readable form, which will ultimately improve patient safety, modernize device postmarket surveillance, and facilitate medical device innovation. As part of the system, the device labelers are required to submit information to the FDA-administered Global Unique Device Identification Database (GUDID).

Information on the GUDID and UDI requirements are provided on the UDI homepage, including links to helpful education modules, guidances, and other UDI-related materials: https://www.fda.gov/medical-devices/device-advice-comprehensive-regulatory-assistance/unique-device-identification-system-udi-system.

19.5.10 User Fees

Under the user fee system, medical device companies pay fees to the FDA when they register their device establishments and list their devices with the Agency, whenever they submit an application or a notification to market a medical device in the U.S., and for certain other types of submissions. These fees help the FDA to fulfill its mission of protecting the public health and accelerating innovation in the industry.

Device user fees were first established in 2002 by the Medical Device User Fee and Modernization Act (MDUFMA). User fees were renewed in 2007, with the Medical Device User Fee Amendments (MDUFA) to the FDA Amendments Act (MDUFA II), 2012 with the MDUFA to the FDA Safety and Innovation Act (MDUFA III), and 2017 with the MDUFA to the FDA Reauthorization Act of 2017 (MDUFA IV). MDUFA IV will be in place from October 1, 2017 until September 30, 2022.

Information regarding MDUFA user fees, including information on how small businesses may be eligible for reduced/waived fees, is available: https://www.fda.gov/industry/fda-user-fee-programs.

19.6 Premarket Submissions for Medical Devices

There are multiple premarket submission types for medical devices. Ultimately, the submission type is dictated by the stage of the product in the development process (e.g., seeking the FDA's feedback on a clinical trial, requesting 510(k) clearance) as well as the risk associated with the device type and its intended use. This section will discuss a number of submission types, noting the two most common pathways in bringing a medical device to market are the Premarket Notification 510(k) clearance (21 CFR Part 807 Subpart E) and the PMA process (21 CFR Part 814). For purposes of seeking to distribute a device in interstate commerce in the U.S., the first step is to confirm that the product is a medical device according to the definition discussed in the section above. The next step is to determine which regulatory pathway is required.

19.6.1 eCopy Program for Medical Device Submissions

An electronic copy (eCopy) is an electronic version of your medical device submission stored on a CD, DVD, or a flash drive. A submission with an eCopy that does not meet the technical standards outlined in the eCopy guidance will be placed on eCopy hold until a valid eCopy is received.

The following resources will assist in understanding the eCopy program and provide direction for the successful creation and submission of an eCopy: https://www.fda.gov/medical-devices/how-study-and-market-your-device/ecopy-program-medical-device-submissions

19.6.2 Premarket Notification (510(k))

Section 510(k) of the FD&C Act requires each person who is required to register under this section and who proposes to begin the introduction or delivery for introduction into interstate commerce for commercial distribution of a device intended for human use shall, at least 90 days before making such introduction or delivery, submit a 510(k) to the FDA. The 510(k) regulatory process requires manufacturers to demonstrate "substantial equivalence." The FDA's goal is to complete its review of a 510(k) submission within 90 days.

19.6.2.1 Predicate device and substantial equivalence

The manufacturer must demonstrate that the proposed device is substantial equivalence (SE) to a legally marketed device that is not subject to PMA. The legally marketed device to which substantial equivalence is compared is also known as the "predicate" device. The predicate device is a device that was legally marketed prior to May 28, 1976, or a device which has been reclassified from class III to class II or I, a device which has been found to be substantially equivalent through the 510(k) premarket notification process, or a device granted marketing authorization through the De Novo classification process. A predicate device serves as a comparator to which substantial equivalence is to be established. Any legally marketed 510(k)-cleared device may be used as a predicate, provided that the predicate and subject devices meet the applicable criteria for establishing substantial equivalence. Demonstration of SE does not mean the new and predicate devices must be identical. Substantial equivalence is established with respect to intended use, design, energy used or delivered, materials, chemical composition, performance, safety, effectiveness, labeling, biocompatibility, standards, and other technological characteristics, as applicable.

A device is SE if it

- has the same intended use as the predicate; **and**
- has the same technological characteristics as the predicate; **or**
- has the same intended use as the predicate; **and**
- has different technological characteristics; but
  - does not raise different questions of safety and effectiveness; **and**
  - the information submitted demonstrates that the device is as safe and effective as the legally marketed device.

A device subject to 510(k) requirements may not be legally marketed in the U.S. until the submitter receives a letter declaring the device SE and "clears" the device for commercial distribution. However, if the aforementioned criteria are not met, the device may be found to be "not substantially equivalent (NSE)," necessitating submission of a De Novo request or PMA. Details regarding the 510(k) process and the preparation of a 510(k) submission, including its various types (i.e., Traditional, Special, Abbreviated, Safety and Performance Based Pathway), may be found at: https://www.fda.gov/medical-devices/premarket-submissions/premarket-notification-510k.

An additional useful resource is the guidance, The 510(k) Program: Evaluating Substantial Equivalence in Premarket Notifications [510(k)]-Guidance for Industry and Food and Drug Administration Staff: https://www.fda.gov/regulatory-information/search-fda-guidance-documents/510k-program-evaluating-substantial-equivalence-premarket-notifications-510k.

The FDA has online resources for identifying predicate device(s) and other regulatory information about a legally marketed device, such as product code, regulation number, and description.

More information can be found on the FDA's Medical Device Databases, such as the 510(k) Premarket Notifications database: http://www.accessdata.fda.gov/scripts/cdrh/cfdocs/cfpmn/pmn.cfm.

19.6.3 De novo Classification Request

In accordance with Section 513(f)(1) of the FD&C Act, devices that were not in commercial distribution before May 28, 1976 (the date of enactment of the Medical Device Amendments of 1976), generally referred to as "postamendments" devices, are classified automatically by statute into class III without any FDA rulemaking process. These devices remain in class III and require a PMA unless the device is classified or reclassified into class I or II. The Food and Drug Administration Modernization Act of 1997 (FDAMA) added the De Novo classification pathway under Section 513(f)(2) of the FD&C Act, establishing an alternate pathway to classify new devices into class I or II that had automatically been placed in class III after receiving an NSE determination in response to a 510(k) submission. Subsequently, in 2012, the FD&C Act was amended once again by the Food and Drug Administration Safety and Innovation Act (FDASIA), to provide a second option for De Novo classification. In this second pathway, a sponsor who determines that there is no legally marketed predicate may request a risk-based classification of their device and submit a Direct De Novo request without first receiving an NSE determination via a submitted 510(k).

The FDA goal under MDUFA IV is to generally make a final decision for a De Novo request in 150 days. If a De Novo request is granted, the new device can be legally marketed, provided that it is in compliance with applicable regulatory controls. The decision also establishes a new classification regulation in 21 CFR. The new device can serve as a predicate device for future 510(k) submissions.

The FDA posts a copy of the granting order on its website to show that we classified the device and granted marketing authorization. It generates and publicly discloses a transparency summary explaining the decision. It then publishes an order in the Federal Register announcing the new classification regulation and, for class II devices, the special controls.

19.6.4 Premarket Approval Application

For high-risk products, the pathway to market is typically through a PMA (21 CFR Part 814). These devices are class III, including those that support or sustain human life, or which is of substantial importance in preventing impairment of human health, or present an unreasonable risk of illness or injury, among other criteria. The PMA process is more involved and includes submission of clinical data to support the reasonable assurance of safety and effectiveness of the device. The time for the FDA to review the PMA and make a decision is generally 180 days if the review does not require the input of an advisory committee. Before approving or not approving a PMA, the appropriate FDA advisory committee may review the PMA at a public panel meeting and provide the FDA with the committee's recommendation on whether the FDA should approve the submission. For example, feedback from an advisory committee is typically sought for first-of-a-kind devices. The timeframe for review of a PMA and FDA decision is generally 320 days if input from an advisory committee is required.

The review of a premarket approval is a four-step review process consisting of

- administrative and limited review by FDA staff to determine completeness (acceptance and filing reviews);
- in-depth scientific, regulatory, and Quality System review by appropriate FDA personnel (substantive review);
- review and recommendation by the appropriate advisory committee, if requested by the Agency (panel review); and
- final deliberations, documentation, and notification of the FDA's decision.

Details regarding the preparation and review process of a PMA application are provided: https://www.fda.gov/medical-devices/premarket-submissions/premarket-approval-pma.

19.6.5 Accessory Requests

The appropriate classification of a legally-marketed accessory or a new accessory type can be requested under Section 513(f)(6) of the FD&C Act through the Accessory Request process. In accordance with Section 513(f)(6)(C) of the FD&C Act, manufacturers may submit an Accessory Request for appropriate classification of an accessory that is included in a PMA, PMA supplement, or 510(k), if the accessory has not been classified distinctly from another device. This New Accessory Request is appropriate for an accessory of a type that has not been previously classified under the FD&C Act, cleared for marketing under a 510(k) submission, or approved in a PMA. This request should be submitted together with the parent device submission. The FDA must grant or deny the New Accessory Request concurrently with the decision on the premarket submission with which the request was submitted.

Additionally, in accordance with Section 513(f)(6)(D) of the FD&C Act, manufacturers may submit an Accessory Request for appropriate classification of an accessory that has been granted marketing authorization as part of a submission for another device with which the accessory involved is intended to be used, through a premarket submission or a De Novo request for such other device (referred to as an "Existing Accessory Type"). The FDA must grant or deny the Accessory Request for an Existing Accessory Type within 85 days of receiving the request.

When the FDA grants a New Accessory Request or an Existing Accessory Request, the decision consists of an order establishing a new classification for such accessory for the specified intended use or uses of such accessory and for any accessory with the same intended use or uses as such accessory. Effective on the date of the granting order, the requester may immediately begin marketing the device subject to the general controls and any identified special controls.

For more information, see the following FDA guidance: https://www.fda.gov/regulatory-information/search-fda-guidance-documents/medical-device-accessories-describing-accessories-and-classification-pathways.

19.6.6 Investigational Device Exemption

An IDE allows an investigational device to be used in a clinical study to determine safety and effectiveness, typically in support of a future marketing submission. Studies conducted in the U.S. must be conducted consistent with Good Clinical Practices (GCP). When data from clinical investigations conducted outside the U.S. are submitted to the FDA, the requirements of 21 CFR 812.28 may apply. GCP refers to the regulations and requirements that must be complied with while conducting a clinical study. These regulations apply to manufacturers, sponsors, clinical investigators, institutional review boards (IRBs), and the medical device itself. The primary regulations address the following areas: Investigational Device Exemption (21 CFR Part 812); informed consent (21 CFR Part 50); IRBs (21 CFR Part 56); clinical investigators' financial disclosure (21 CFR Part 54); Good Laboratory Practice (GLP) for Nonclinical Laboratory Studies (21 CFR Part 58); and design controls (21 CFR 820.30).

Before a significant risk (SR) IDE study can begin, the FDA and the IRB must review and approve the study. An investigation may not begin until 30 days after the FDA receives the application, unless the FDA notifies the sponsor that the investigation may not begin. A non-significant risk (NSR) IDE study only requires IRB approval before the study can begin but must still comply with the abbreviated requirements of the IDE regulation under 21 CFR 812.2(b).

Details are available at: https://www.fda.gov/medical-devices/how-study-and-market-your-device/investigational-device-exemption-ide

19.6.7 Q-Submissions: Requesting Feedback and Meeting with FDA

There are mechanisms available to submitters through which they can request feedback from or a meeting with the FDA regarding potential or planned medical device premarket submissions. The FDA implemented the broader Q-Submission Program, which includes Pre-Submissions and additional opportunities to engage with the FDA. A Pre-Submission includes a formal written request from a submitter for feedback from the FDA that is provided in the form of a formal written response or, if the submitter chooses, formal written feedback followed by a meeting in which any additional feedback or clarifications are documented in meeting minutes. Such a Pre-Submission meeting can be in-person or by teleconference as the submitter prefers.

There are other mechanisms to engage with the FDA on medical device premarket submissions. A Submission Issue Request is a request for FDA feedback on a proposed approach to address issues conveyed in different hold letters, including those for premarket submissions. The Q-Submission Program is also used to track Study Risk Determinations, Informational Meetings, PMA Day 100 Meetings, and other Q-Submission types. To assist sponsors in the appropriate use of the Q-Submission Program, including Pre-Submissions, refer to the following guidance: https://www.fda.gov/regulatory-information/search-fda-guidance-documents/requests-feedback-and-meetings-medical-device-submissions-q-submission-program

19.6.8 Promoting Innovation: Breakthrough Devices

The Breakthrough Devices Program is a voluntary program for certain medical devices and device-led combination products that provide for more effective treatment or diagnosis of life-threatening or irreversibly debilitating diseases or conditions. This program is intended to help patients have more timely access to these medical devices by expediting their development, assessment, and review, while preserving the statutory standards for marketing submission decisions, consistent with the Agency's mission to protect and promote public health.

The Breakthrough Devices Program collectively supersedes the Expedited Access Program, Innovation Pathway, and the Priority Review Program. The Breakthrough Devices Program comprised two phases. The first is the Designation Request phase, in which an interested sponsor of a device requests that the FDA grant that device Breakthrough Device designation. The designation criteria provide for devices

- that provide for more effective treatment or diagnosis of life-threatening or irreversibly debilitating human disease or conditions; and

- that represent breakthrough technologies; for which no approved or cleared alternatives exist; that offer significant advantages over existing approved or cleared alternatives, including the potential, compared to existing approved alternatives, to reduce or eliminate the need for hospitalization, improve patient quality of life, facilitate patients' ability to manage their own care (such as through self-directed personal assistance), or establish long-term clinical efficiencies; or the availability of which is in the best interest of patients.

The second phase of the Breakthrough Devices Program encompasses actions to expedite development of the device and the prioritized review of subsequent regulatory submissions (e.g., Pre-Submissions, marketing submissions). The Breakthrough Devices Program has several principles, including interactive and timely communication, pre/postmarket balance of data collection, efficient and flexible clinical study design, review team support, senior management engagement, priority review, and manufacturing considerations for PMAs.

The request for Breakthrough Designation and, if designated, subsequent Breakthrough Interactions, should be submitted to the FDA as Q-Submissions (see Section 2.2.7—Q-Submissions). More details and resources regarding the Breakthrough Devices Program can be located here: https://www.fda.gov/regulatory-information/search-fda-guidance-documents/breakthrough-devices-program.

19.7 Summary

The laws in the 1938 FD&C Act, RCHSA, Medical Device Amendments, Title 21 Code of Federal Regulations and other statutes, regulations, and policies remain the underpinnings for the current U.S. medical devices regulatory framework to protect and promote the public health. The CDRH is the authority that regulates medical devices in the United States. If a product is labeled, promoted, or intended to be used in a manner that meets the definition of a device defined in Section 201(h) of the FD&C Act, it is regulated as a medical device; and is subject to certain premarket and postmarket regulatory controls. Foreign companies that manufacture medical devices and/or radiation emitting products that are imported into the United States must comply with applicable U.S. regulations before, during, and after importation into the U.S. or its territories. It is important to recognize that the same safety and effectiveness standards apply to all medical devices intended for marketing and distribution in the U.S. or its territories irrespective of the origin of manufacture.

Additional Resources

1. Device Advice is on-line comprehensive regulatory assistance. Available at: https://www.fda.gov/medical-devices/device-advice-comprehensive-regulatory-assistance.

2. CDRH Learn is a web-based tutorial consisting of a series of training modules. The different modules describe many aspects of medical device and radiological health regulation, covering both premarket and postmarket issues. Some modules are also available in Spanish and Chinese Mandarin. Available at: https://www.fda.gov/training-and-continuing-education/cdrh-learn.

3. FDA Medical Devices Website. Access to more information concerning other medical device topics, including current program initiatives, device clearances and approvals, recalls and alerts, public databases, and other information. Available at: https://www.fda.gov/Medical-Devices.

Chapter 20

Regulation of Combination Products in the United States

John Barlow Weiner and Thinh X. Nguyen

Office of Combination Products, Food and Drug Administration, WO32, Hub/Mail Room #5129, 10903 New Hampshire Avenue, Silver Spring, Maryland 20993, USA

John.Weiner@fda.hhs.gov, THUYNTT@its.jnj.com

Disclaimer

This chapter represents the views of the authors in their personal capacities. It does not represent an official statement by the US Food and Drug Administration.

    This chapter addresses the regulation of combination products in the United States. As the discussion below indicates, combination products present particular challenges. For example, issues may arise with respect to premarket data requirements and post-market regulatory requirements, as well as associated practical considerations such as information technology needs and collaboration with other regulated entities.

Medical Regulatory Affairs: An International Handbook for Medical Devices and Healthcare Products (Third Edition)
Edited by Jack Wong and Raymond K. Y. Tong
Copyright © 2022 Jenny Stanford Publishing Pte. Ltd.
ISBN 978-981-4877-86-2 (Hardcover), 978-1-003-20769-6 (eBook)
www.jennystanford.com

The following topics are addressed: What products are considered "combination products"; the standards for determining whether a product is a combination product and for determining which component of the US Food and Drug Administration (FDA) has primary responsibility for the regulation of a combination product; how to obtain a formal product classification or assignment determination from the FDA; premarket review and post-market regulatory considerations for combination products; the role of the FDA's Office of Combination Products (OCP); international convergence and coordination activities with foreign counterparts; and FDA resources for obtaining additional information.

20.1 What Products Are Considered Combination Products

In the United States, the term "combination product" is a product comprising any combination of a drug and a device; a device and a biological product; a biological product and a drug; or a drug, a device, and a biological product (21 CFR 3.2; see 21 USC 353(g)). Types of combination products include physically or chemically combined products (such as drug-eluting stents or syringes marketed prefilled with a drug or biological product); co-packaged products (such as an ampule of a drug packaged together with a delivery device, first aid kits that include bandages and drugs to treat wounds and other injuries, or surgical kits that include devices and drugs that might be used for a particular type of surgical procedure); and certain separately marketed but "cross-labeled" products (such as a laser and a light-activated drug that are marketed separately from one another but labeled specifically for use with one another). See 21 CFR 3.2(e).

20.2 The Standards for Determining If a Product Is a Combination Product

Because a combination product comprises two or more different types of medical products (constituent parts), a determination of whether a product is a combination product begins with determining the classification of the articles of which it is comprised. If the product includes at least two, distinguishable medical products that are classified differently from one another, it is a combination product. Generally, it may be fairly obvious whether a product is a combination product. A prefilled syringe, for example, clearly consists of a device (the syringe) and the drug or biological product contained in the syringe. In some cases, classification may be more difficult to determine. Regardless of how apparent the classifications may be, the answer depends on an assessment of which of the statutory definitions—that for biological product, device, or drug—best applies to each article of which a product is comprised.[1]

---

[1] Section 201(g) of the Federal Food, Drug, and Cosmetic (FD&C) Act (21 USC 321(g)) provides that the term "drug" means:

(A) articles recognized in the official United States Pharmacopoeia, official Homoeopathic Pharmacopoeia of the United States, or official National Formulary, or any supplement to any of them; and (B) articles intended for use in the diagnosis, cure, mitigation, treatment, or prevention of disease in man or other animals; and (C) articles (other than food) intended to affect the structure or any function of the body of man or other animals; and (D) articles intended for use as a component of any articles specified in clause (A), (B), or (C)... .

Section 201(h) of the FD&C Act (21 USC 321(h)) provides that the term "device" means:

... an instrument, apparatus, implement, machine, contrivance, implant, in vitro reagent, or other similar or related article, including any component, part, or accessory, which is—

(1) recognized in the official National Formulary, or the United States Pharmacopeia, or any supplement to them,

(2) intended for use in the diagnosis of disease or other conditions, or in the cure, mitigation, treatment, or prevention of disease, in man or other animals, or

(3) intended to affect the structure or any function of the body of man or other animals, and

which does not achieve its primary intended purposes through chemical action within or on the body of man or other animals and which is not dependent upon being metabolized for the achievement of its primary intended purposes.

Section 351(i) of the Public Health Services Act (42 USC 262(i)) provides that:

The term "biological product" means a virus, therapeutic serum, toxin, antitoxin, vaccine, blood, blood component or derivative, allergenic product, protein (except any chemically synthesized polypeptide), or analogous product, or arsphenamine or derivative of arsphenamine (or any other trivalent organic arsenic compound), applicable to the prevention, treatment, or cure of a disease or condition of human beings.

20.3 The Standards for Determining Which FDA Component Has Primary Responsibility for Regulating a Combination Product

The FDA has three components, called "Centers," that regulate medical products for humans: the Center for Biologics Evaluation and Research (CBER); the Center for Devices and Radiological Health (CDRH); and the Center for Drugs Evaluation and Research (CDER). Section 503 of the FD&C Act requires that the determination as to which of these Centers will have primary responsibility for the regulation of a combination product be based upon the "primary mode of action" (PMOA) of the combination product. See 21 USC 353(g). There are three modes of action for purposes of this analysis, that of a biological product, a device, or a drug (21 USC 353(g); 21 USC 3.2(k)). If the PMOA is provided by a biological product constituent part, the combination product is typically assigned to CBER. If the PMOA is provided by a device constituent part, the combination product is typically assigned to CDRH, and if the PMOA is provided by the drug, the combination product is typically assigned to CDER. See 21 USC 353(g)(1); 21 CFR 3.4(a).

PMOA is defined as the mode of action that provides the greatest contribution to the overall therapeutic effect of the combination product (21 CFR 3.2(m)). As for classification, the PMOA may be obvious in some cases. For example, the drug or biological product provides the PMOA for a prefilled syringe because the syringe serves merely to deliver the drug or biological product that then treats the disease or condition the combination product is intended to address. In other cases, PMOA may be more difficult to determine, for example, if more than one constituent part of the combination product directly contributes to the treatment. In such cases, data on the relative contribution of the different constituent parts may be needed to make the determination.

Further, in some cases, it may not be possible to determine PMOA directly. For example, a product may consist of two articles that address different diseases or conditions, so that there is no overall therapeutic effect to which either contributes more than the other. In such circumstances, PMOA is determined indirectly through the use of an algorithm (21 CFR 3.4(b)). The first step of the algorithm asks whether one of the Centers already regulates a combination product that raises similar questions of safety and effectiveness to those raised by this combination product. If there is one such Center, the combination product will be assigned to that Center. If there is not such a Center, the second step of the algorithm applies. Under the second step, the combination product is assigned based on which Center has the most experience with the most significant safety and effectiveness issues presented by the combination product.

20.4 Requests for Designation

Section 563 of the FD&C Act provides that a request for designation (RFD) may be submitted to obtain a formal, binding determination from the FDA as to the classification or Center assignment for a product. 21 USC 360bbb-2. The FDA is required to respond to RFDs not later than 60 days after they are filed. See 21 USC 360bbb-2(b). If the agency does not respond within that time frame, the classification or assignment recommended by the RFD submitter applies. See 21 USC 360bbb-2(c). A classification or assignment made through the RFD process can only be changed either with the consent of the RFD submitter or for public health reasons based on scientific evidence. See 21 USC 360bbb-2(b),(c); see 21 CFR 3.9.

RFDs are submitted to the OCP, which is charged both with determining whether products should be classified as biological products, devices, drugs, or combination products, and with assigning them to the appropriate Center. The FDA permits and encourages sponsors to seek guidance from the OCP before submitting an RFD, to confirm whether one is needed and to help ensure that the RFD will be complete for filing. Among other requirements, an RFD must describe the product, its components and ingredients; state the product's intended use; provide the submitter's understanding of how the product works; and provide a justification for the product's PMOA if the product is a combination product. See 21 CFR 3.7. For further guidance, see How to Write a Request for Designation (RFD), Guidance for

Industry (http://www.fda.gov/RegulatoryInformation/Guidances/ucm126053.htm). The OCP is also available to provide preliminary feedback on product classification and assignment, including to assist in determining whether an RFD is needed or how to ensure that an RFD provides sufficient information.

20.5 Premarket Review Considerations

There are no special premarket standards or requirements specifically for combination products. The basic concern for them, as for all medical products, is to ensure that they are safe and effective for their intended uses. While the lead Center for a combination product is determined based on the PMOA of the product as discussed above, other Centers may participate in the premarket review of combination products, bringing to bear their expertise and informing the review process. See, e.g., 21 USC 360bbb-2(g)(1), (4)(C).

A principal consideration for combination products is to ensure that product development and testing addresses all relevant considerations relating to each constituent part and their interactions. Accordingly, for example, in assessing whether a syringe is an appropriate delivery device for a particular drug, relevant considerations would include whether the syringe interacts with the drug, whether it will deliver the correct drug dosage, and whether the integrity of the syringe material can be maintained in accordance with the combination product's shelf-life. Similarly, in the case of a drug-eluting stent for example, considerations for the drug constituent part would include (in addition to its effectiveness to achieve its intended therapeutic purpose) such factors as whether the formulation of the drug is appropriate in light of the need to control the elution rate and resist flaking from the stent.

Strong, clear business arrangements between regulated entities can be important to obtaining marketing authorization. Access to third-party proprietary data submitted by one manufacturer, for example, can reduce data development demands and expedite product review and approval, including with respect to post-market changes. A related issue with regard to cross-labeled combination products in particular, is whether it makes better sense to seek a single marketing authorization for the combination product or to seek separate marketing authorizations for each of the separately marketed constituent parts of the combination product. Separate applications may facilitate further development of these constituent parts for independent uses not involving the other constituent part. However, reliance upon separate applications may also pose challenges, for example, with respect to coordination of post-market modifications to either constituent part, particularly if the applications are held by different sponsors. The FDA has the authority to require two applications in appropriate cases for combination products (21 CFR 3.5(c)).

20.6 Post-Market Regulatory Considerations

In the context of two rulemakings, for current good manufacturing practices (CGMPs) and for post-marketing safety reporting (PMSR) for combination products, the FDA has stated that combination products are subject to the legal requirements applicable to their constituent parts, and that combination products comprise a distinct category of medical products that can be regulated in light of the distinct regulatory considerations they raise. See Current Good Manufacturing Practice Requirements for Combination Products (final rule), 78 FR 4307 (https://www.gpo.gov/fdsys/pkg/FR-2015-01-27/pdf/2015-01410.pdf); Post-marketing Safety Reporting for Combination Products (proposed rule), 74 FR 50,744 (http://edocket.access.gpo.gov/2009/pdf/E9-23519.pdf).

The FDA has taken a similar approach in both these rulemakings. Specifically, for each topic the agency reviewed the regulations applicable to biological products, drugs, and devices, and assessed in which ways those sets of regulations overlap and in which ways they may be distinct. Based on this review, the FDA developed approaches it states are intended to enable regulated entities to comply with the multiple sets of regulations applicable to their combination product without unnecessary redundancy of requirements or burden. In both cases, the FDA has proposed an approach under which some regulations applicable to constituent parts of the combination product must be implemented in their entirety while only specified provisions of other such regulations need to be met. These two rulemakings offer some insight as to the FDA's thinking regarding what requirements apply to combination products and how to clarify and streamline appropriate measures to comply with them.

As in the premarket context, various staff from Centers other than the lead Center and from other FDA components may participate in post-market regulatory activities for combination products. Such coordination may be appropriate, for example, to ensure staff with appropriate expertise participate in site inspections and the review of post-market changes to products.

20.7 Role of Office of Combination Products

The OCP is a statutorily mandated office whose responsibilities include ensuring the timely, effective premarket regulation of combination products and their consistent and appropriate post-market regulation. See 21 USC 353(g)(4). The OCP undertakes a wide range of activities in accordance with this mandate, including responding to RFDs as discussed above; facilitating the consultation process between Centers on combination product regulatory issues; offering input on product-specific regulatory issues; convening working groups and coordinating initiatives to develop policies; participating in agency working groups and initiatives that may affect regulation of combination products; and resolving disputes within the FDA and between the FDA and regulated entities. In addition, the OCP acts as a resource to combination product developers and manufacturers, to answer regulatory questions and facilitate interactions with the agency, including Center review staff, field staff, and enforcement and inspectional personnel.

20.8 International Harmonization and Coordination Activities with Foreign Counterparts

The FDA works with foreign counterpart agencies on regulatory issues relating to combination products and has expressed interest in working with foreign counterparts to promote convergence and coordination of activities with respect to combination products. In addition, the FDA continues to provide technical assistance to foreign counterparts throughout the world, and the FDA continues to participate in regional and international standard-setting activities relating to combination products.

20.9 FDA Resources for Obtaining Additional Information

For further information, the OCP's webpage on the FDA website is an excellent resource: http://www.fda.gov/CombinationProducts/default.htm. In addition, as noted above, the OCP itself is an important resource to address questions and seek assistance for engaging with the FDA on regulatory issues for combination products and medical product classification issues. Contact information for the OCP is available on its webpage.

Chapter 21

European Union Medical Device Regulatory System

Arkan Zwick[a] and Gert Bos[b]

[a]*CROMA PHARMA GmbH, Regulatory Affairs Department, Industriezeile 6, 2100 Leobendorf, Austria*
[b]*Qserve Group, Utrechtseweg 310, Building B42, 6812 AR Arnhem, The Netherlands*

arkan.zwick@croma.at

21.1 Glossary of Terms

AIMD	Active Implantable Medical Devices Directive 90/385/EEC
BBP	Benzyl Butyl Phthalate
CE	European Conformity
CEN	European Committee for Standardization
CENELEC	European Committee for Electrotechnical Standardization
CMR	Carcinogenic, Mutagenic and Toxic to Reproduction
COCIR	European Radiological, Electromedical and Healthcare IT Industry

Medical Regulatory Affairs: An International Handbook for Medical Devices and Healthcare Products (Third Edition)
Edited by Jack Wong and Raymond K. Y. Tong
Copyright © 2022 Jenny Stanford Publishing Pte. Ltd.
ISBN 978-981-4877-86-2 (Hardcover), 978-1-003-20769-6 (eBook)
www.jennystanford.com

DBP	Dibutyl Phthalate
DEHP	Di-(2-ethylhexyl) Phthalate
EC	European Community
EEA	European Economic Area
EEC	European Economic Community
EFTA	European Free Trade Association
EMIG	European Medical Devices Industry Group
ETSI	European Telecommunications Standards Institute
EU	European Union
FSCA	Field Safety Corrective Action
IMDRF	International Medical Device Regulators' Forum (IMDRF)
IAF	International Accreditation Forum
IEC	International Electrotechnical Commission
ISO	International Standards Organization
IVD IVDR	In-Vitro Diagnostics Medical Devices Directive 98/79/EC In-Vitro Diagnostics Medical Devices Regulation EU 2017/746
MD	Medical Device
MDD MDR	Medical Device Directive 93/42/EEC Medical Devices Regulations EU 2017/745
MDCG	Medical Device Coordination Group
MEDDEV	Medical Devices Guidance Document
MEDTECHEUROPE	Alliance between EUCOMED and EDMA
NANDO	New Approach Notified and Designated Organizations information system: database operated by the European Commission that catalogs all the notified bodies and the directives that the organizations technically are qualified to review.
NB	Notified Body
NB-MED	Co-ordination of Notified Bodies Medical Devices on Council Directives

NBOG	Notified Body Operations Group
PMS	Post-Market Surveillance
STED	Summary Technical Documentation for Demonstrating Conformity to the Essential Principles of Safety and Performance of Medical Devices
TEAM-NB	European Association for Medical Devices of Notified Bodies
UDI	Unique Device Identifier

21.2 European Union: Medical Device Market and Structure

The European Union (EU) as of 2020 comprises 27 countries which represent 447.7 million inhabitants (source Eurostat 2020) (Table 21.1). The United Kingdom (UK) has left the EU on January 31, 2020 (so called Brexit) in accordance with the withdrawal agreement with a transition period until end of 2020. The medical device market is estimated to be of 120 billion Euros total sales amount making up 27% of the world market for medical technologies. It is the second largest medical technology market in the world after the United States (US). The medical device sector is consisting of around 32,000 companies, the majority of them (95%) are small and medium sized enterprises (SMEs). Based on manufacturer prices the largest markets in the EU are Germany (27%), France (14%), the United Kingdom (11%), Italy (10%) followed by Spain (6%), the Netherlands, Sweden, Belgium and Austria (all between 5% and 2% market share) (Source MedTechEurope 2020, the European Medical Technology Industry).

Those countries constitute the EU where the free circulation of goods, services and humans is established. Medical Devices bearing a marking indicating the European Conformity Assessment to regulations the—CE mark—are considered to be cleared in the European Union market. CE marking indicates that a product has been assessed by the manufacturer and deemed to meet EU safety, health and environmental protection requirements hence to be in compliance with the EU conformity assessment regulations.

It is required for products manufactured anywhere in the world that are then marketed in the EU.

Table 21.1 Member states of the European Union (EU)

Date of entry into the EU	EU member states
1958	Belgium, France Germany, Italy, Luxembourg, the Netherlands
1973	Denmark, Ireland, the United Kingdom (Brexit on January 31, 2020)
1981	Greece
1986	Portugal, Spain
1995	Austria, Finland, Sweden
2004	Czech Republic, Cyprus, Estonia Hungary, Latvia, Lithuania, Malta, Poland, Slovak Republic, Slovenia
2007	Bulgaria, Romania
2013	Croatia
2020	Withdrawal of the United Kingdom

The European Economic Area (EEA) was established on January 1, 1994, following an agreement between the member states of the European Free Trade Association (EFTA) and the European Community: Iceland, Liechtenstein and Norway are allowed to participate in the EU's Internal Market without a conventional EU membership. In exchange, they are obliged to adopt all EU legislation related to the single market, for Medical Devices for instance. One EFTA member, Switzerland, has not joined the EEA. CE marking of Medical Devices is, however, recognized in Switzerland by means of a Mutual recognition Agreement (MRA) between EU and Switzerland, in which every law specifically needs to be confirmed into the MRA structure.

CE marking is also in place in Turkey, following the Ankara Agreement between EU and Turkey. Mutual acceptance between Turkey and EFTA is arranged in an agreement between these parties based on CE marking. The EU has also signed MRAs to promote trade in goods between the European Union and third countries and facilitate market access outside Europe with Australia and New Zealand (Table 21.2).

European Union | 277

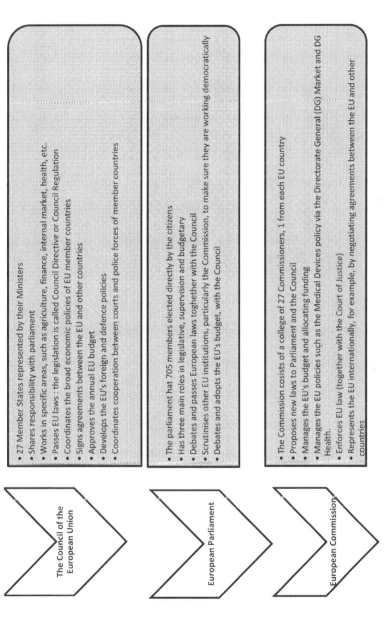

Figure 21.1 Structure of the EU institutions.

Table 21.2 Associated states with EU accepting conformity assessment results

Association	States
EFTA/EEA	Iceland, Lichtenstein, Norway
Mutual Recognition Agreements (MRA)	Switzerland, Turkey, Australia, New Zealand

21.3 EU Regulations on Medical Devices

The EU has with 2017 launched new Regulations on Medical devices 2017/745 (MDR) and In vitro Diagnostics Regulation 2017/745 (IVDR) with a transition period of 4 respectively 5 years and Date of Application of May 2021 for MDR and May 2022 for IVDR. These new Regulations will replace the existing Directives. The Directive is addressed to the EU Member States and to be transposed into national law in each country whereas the Regulation is directly applicable in each EU member state without national transposition necessary. Detailed technical requirements and characteristics are further provided by voluntary product standards and guidance document developed by national and international expert groups such as the international standards organization (ISO www.iso.org, e.g. EN ISO 10993 series on biocompatibility, ISO 14155 on clinical investigation of medical devices or ISO 14971 on risk management). The international reference standard for medical device quality systems requirements is ISO 13485 is fully accepted in the EU by competent authorities and notified bodies.

21.4 Harmonized Standards, Presumption of Conformity and State of the Art

A standard stands for an agreed understanding of key words and terminology and methods, which, when applied by the manufacture, ensure a common global level of safety of medical care to patient.

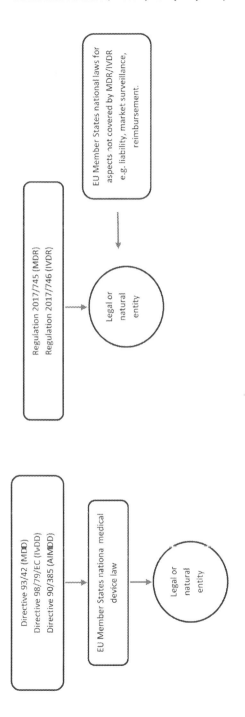

The formal definition of a standard is: standards are documented agreements containing technical specifications or other precise criteria to be used consistently as rules, guidelines or definitions of characteristics, to ensure that materials, products, process are fit for their purpose. Standards can

- provide reference criteria that a product, process or service must meet;
- provide information that enhances safety, reliability and performance of products, processes and services;
- assure consumers about reliability or other characteristics of goods or services provided in the marketplace.

Harmonized standards are European standards prepared in accordance with the General Guidelines agreed between the Commission and the European standards harmonization, and follow a mandate issued by the Commission after consultation with the Member States. Harmonized standards are published in the official journal of the European Union in the framework of the implementation of Council Directive 93/42 or Regulation 2017/745 as new standards are added to the list or revisions are to be applied. In case of revisions, the communication also gives clarity on the transition period until which the new version has to be complied with.

Conformity with harmonized standards as published in the official journal, confers a presumption of conformity with the Essential Requirements of the medical devices directive 93/42/EEC. Doing so harmonized standards become binding with the same legal quality as EU or national medical device laws. In addition to harmonized standards, regulators in the EU request economic operators and especially manufacturers to comply with latest standard versions available from standardization organizations (e.g. www.iso.org) in the context of the state of the art (SOTA) requirements as stipulated by ISO 13485 requirements and national medical device laws. The SOTA benchmark requires the industry to actively and regularly screen latest developments of standards and document gaps and necessary actions to comply with the latest version within due time. Non-compliance to harmonized standard or SOTA standards leads to non-conformity and possibly loss of CE marking for the concerned device.

For getting information about standards under review or creation and for acquisition of the published standard, the manufacturer has to apply and buy them from the standard organizations at the European level (Table 21.3) or at national level.

Table 21.3 List of committees for standardization

Committee name	Scope
CEN	CEN is the European Committee for Standardization. CEN contributes voluntary technical standards which promote free trade, the safety of workers and consumers, environmental protection, exploitation of research and development programs, and public procurement.
CENELEC	CENELEC is the European Committee for Electrotechnical Standardization and is responsible for standardization in the electrotechnical engineering field.
ETSI	The European Telecommunications Standards Institute (ETSI) produces globally applicable standards for Information and Communications Technologies (ICT), including fixed, mobile, radio, converged, broadcast and internet technologies.
IEC	IEC is the International Electrotechnical Commission. IEC is World's leading organization that prepares and publishes International Standards for all electrical, electronic and related technologies.
ISO	ISO is International Standards Organization: it is an international standard-setting body composed of representatives from various national standards bodies.

21.5 European Associations

The European associations which are active in Europe and contribute to shape the regulatory environment and offer platforms for regulators and industry to communicate with each other and work on harmonization. A non-exhaustive list of some organization is listed in Table 21.4.

Table 21.4 List of European associations

Association name	Competences/terms of reference	Participants
Medical Device Coordination Group (MDCG)	MDCG provides advice to the Commission and assists the Commission and the Member States in ensuring a harmonized implementation of medical devices Regulations (EU) 2017/745 and 2017/746.	Organization and Member States
Notified Body Operations Group (NBOG)	The Notified Body Operations Group's aim is to contribute to improvement of the overall performance of notified bodies in the medical devices sector by primarily identifying and promulgating examples of best practice to be adopted by both notified bodies and those harmonization responsible for their designation and control. They review the "recommendations" issued by the NB-MED (group where all the EU notified bodies participate) and act as a "Mirror Group" following GHTF work relating to notified bodies.	Competent Authorities/designating Authorities (experts), COM services (On focal points, a notified body representative can be invited to participate)
TEAM-NB	Team-NB stands for The European Association for Medical Devices of Notified Bodies, a non-profit association of notified bodies under the medical device regulations The Association aims • to actively support the transition to the new medical devices and in vitro diagnostic regulations 2017/745 and 2017/746 and the new regulatory framework; • to promote high technical and ethical standards in the functioning of notified bodies; and • to protect the legal and commercial interests of notified bodies in their vital role in the functioning of the three medical device directives.	Medical device notified bodies

Association name	Competences/terms of reference	Participants
MedTech Europe **Medical Devices Technology Industry Association** MedTech Europe from diagnosis to cure	MedTech Europe represents the medical technology industry in Europe. MedTech Europe members include both national and pan-European trade and product associations as well as medical technology manufacturers, representing around 22,500 designers, manufacturers and suppliers of medical technology used in the diagnosis, prevention, treatment and amelioration of disease and disability.	Industry representatives
EDMA **European Diagnostic Manufacturers Association**	EDMA is the trade association that represents the In Vitro Diagnostic (IVD) industry active in Europe. EDMA membership brings together 22 National Associations in European countries and 43 major companies engaged in the research, development, manufacture or distribution of IVD products. Through its affiliated National Associations, EDMA represents in total more than 500 companies across Europe. Since its establishment in 1979, EDMA acts in co-operation with other European and international trade associations representing medical devices, pharmaceuticals and biotechnology in general, as well as with scientific societies and patients organizations to make a real difference in health and life quality.	Industry representatives
COCIR	European Radiological, Electromedical and Healthcare IT Industry non-profit trade association founded in 1959, representing the medical technology industry in Europe. COCIR's aim is to represent the interests and activities of its members and act as a communication channel between its members and the European institutions and other regulatory bodies. It also cooperates with other organizations on issues of common interest.	Industry representatives

(Continued)

Table 21.4 (*Continued*)

Association name	Competences/terms of reference	Participants
	COCIR seeks to promote the development of harmonized international standards and regulatory control which respect the quality and effectiveness of the medical devices and healthcare IT systems without compromising the safety of patients and users and promote the free worldwide trade of these products. COCIR encourages the use of technology in delivering cost-efficient and state of the art healthcare systems.	
EUROM VI-Medical Technology	European small and medium enterprises involved in medical devices for: –surgery (including dental), anesthesia, respiration and inhalation, operating theatre, gas supply, sterilization, disinfection, internal and external orthopedics, opto-medical and ophthalmology area, rehabilitation and handicaps fields, infusion and transfusion. Its objectives are to represent European Medical Technology Industry to promote cooperation between members but also with other European organizations; to encourage worldwide trade by being involved in harmonization of legislation, standardization, mutual recognition and certification procedures, to be a partner on works with UE Commission and Standardization Bodies; to defend European Industry views on international activities.	Industry representatives

21.6 Overview of New Medical Devices Regulations

The EU has revised its laws governing medical devices and in vitro diagnostics to align with the developments of the sector. The priority was to ensure a robust, transparent and sustainable regulatory framework and maintain a high level of safety, while supporting innovation. Two new regulations on medical devices and in vitro diagnostic medical devices entered into force in May 2017 and are progressively replacing the existing directives after a transition period. The existing directives on medical devices and in vitro diagnostics are currently replaced by new regulations with their date of application (DOA) (Table 21.5).

Table 21.5 List of applicable regulations to medical devices

Repealed directives	Scope	New EU Regulations	Date of application (DOA)	Examples
AIMDD 90/385/EEC	Active Implantable Medical Devices	**Regulation 2017/745/ EU (MDR)**	26.05.2021	Cardiac pacemaker, defibrillator Sutures, syringes, implants, software, ophthalmic or orthopedic devices, devices without medical purpose
MDD 93/42/EEC	Medical Devices			
IVD 98/79/EC	In Vitro Diagnostics Medical Devices	**Regulation 2017/746/ EU (IVDR)**	26.05.2022	Cancer diagnostics, self-tests, immunoassays, glucose monitors

Other products which do not fall under the scope of Medical Devices are regulated by different Directives and Regulations such as

- Biocides Directive 98/8/EC;
- Directive for Medicinal Products 2001/83/EC, applicable to medicinal products/drugs;
- Regulation 1223/2009 establishing the safety and efficacy requirements for cosmetic products in the EU applicable to cosmetic products.

21.7 Guidelines MEDDEV/NB-MED

The interpretation of the medical device laws is supported by the elaboration of guidelines entitled "MEDDEV" and issued within the Medical Device Experts Group: they are elaborated through a process of consultation with Competent Authorities and Commission representatives, notified bodies, industry and other interested parties in the medical devices sector. They remain voluntary; however, it is anticipated that the guidelines will be followed within the Member States and, therefore, ensure uniform application of relevant Directive provisions. Guidelines are subject of a regular updating process. The MEDDEVs are primarily applicable for devices under the Directive (MDD 93/42) but are also applicable for MDR 2017/745 devices as long as no new guidance exist. MEDDEVs will consecutively be amended and replaced by MDCG Guidance documents under the new Regulation (MDR 2017/745) as described below. In addition, the European Association for Medical Devices Notified Bodies (TEAM NB) has issued several guidance documents in order to facilitate the implementation of the medical devices regulations (NB-MED recommendations) (Table 21.6). Please note NB-MED is changing its name to NBCG (Notified Body Coordination Group).

Table 21.6 Examples of MEDDEV and NB-MED guidance (the exhaustive list in their latest revision is available on line on the MEDDEV and NBMED website)

Scope	Title	MEDDEV/NBMED reference
MDD Scope, filed of application, Definitions	Borderline products, drug-delivery products and medical devices incorporating, as integral part, an ancillary medicinal substance or an ancillary human blood derivative	MEDDEV 2.1/3 rev 3
MDD Classification of devices	Guidelines to the classification of medical devices	MEDDEV 2.4/1 rev.9
	Manual Borderline Classification Medical Devices	Version 1.22 (05-2019)
Technical Dossier, life cycle	Technical documentation	NB-MED/2.5.1/Rec5
	Reporting of design changes and of changes of the quality system	NB-MED/2.5.2/Rec2

Scope	Title	MEDDEV/NBMED reference
Clinical investigation, clinical evaluation	Clinical evaluation: Guide for manufacturers and notified bodies	MEDDEV 2.7.1 Rev4
	Guidelines for Competent Authorities for making a validation/assessment of a clinical investigation application under directives 90/385/EEC and 93/42/EC	MEDDEV 2.7/2 rev. 2
		MEDDEV 2.7/3 rev.3
		MEDDEV 2.7/4
	Clinical investigations: Serous adverse event reporting	
	Guidelines on Clinical investigations: a guide for manufacturers and notified bodies	
Market surveillance and vigilance	Medical Devices Vigilance System	MEDDEV 2.12/1 rev.8
	Clinical Evaluation–Post Market Clinical Follow-up	MEDDEV 2.12/2 rev.2
	Post-Marketing Surveillance (PMS) post market/production	MEDDEV 2.12/Rec1
	Manufacturer Incident Report (MIR) Form	Version 2020

21.8 Guidelines MDCG and Common Specifications

Under the new Regulations, the European Commission provides a range of guidance documents to assist stakeholders in implementing the medical devices Regulations. Legally non-binding guidance documents, adopted by the medical device coordination group (MDCG) these documents pursue the objective of ensuring uniform application of the relevant provisions of the regulations within the EU. The scope of these guidance documents provide details in line with the new requirements of the MDR and IVDR with regards to Unique Device Identifier (UDI), the new EUDAMED database, requirements for notified bodies and transition period as well as clinical evaluation and investigation and the new person responsible for regulatory compliance (PRRC).

Table 21.7 Examples of MDCG Guidelines and Common Specifications (the exhaustive list in their latest revision is available online on the European Commission website)

Scope	Title	Reference
UDI	Guidance on UDI for systems and procedure pack	MDCG 2018-3
		MDCG 2018-1
	Guidance on basic UDI-DI and changes to UDI-DI	MDCG 2018-6
	Clarifications of UDI related responsibilities in relation to article 16 MDR	
EUDAMED	Timelines for registration of device data elements in EUDAMED	MDCG 2019-4
		MDCG 2019-5
	Registration of legacy devices in EUDAMED use of the EUDAMED actor registration module and of the Single Registration Number (SRN) in the Member States	MDCG 2020-15
Notified bodies and transition	Significant changes regarding the transitional provision regarding to devices covered by MDD or AIMDD certificates	MDCG 2020-3
		MDCG 2019-13
		MDCG 2019-10
	Guidance on sampling of devices for the assessment of the technical documentation	MDCG 2020-2
	Transitional provisions concerning validity of certificates issued in accordance to the directives	
	Class I transitional provisions under Article 120 (3 and 4)—(MDR)	
Clinical investigation and evaluation	Clinical evaluation assessment report template	MDCG 2020-13
		MDCG 2020-6
	Guidance on sufficient clinical evidence for legacy devices	MDCG 2020-5
		MDCG 2019-9
	Guidance on clinical evaluation—Equivalence	
	Summary of safety and clinical performance (SSCP)	
Person Responsible for Regulatory compliance	Guidance on article 15 of the medical device regulation (MDR) and in vitro diagnostic device regulation (IVDR) on a "person responsible for regulatory compliance" (PRRC)	MDCG 2019-7
Reprocessing of single-use devices.	Commission Implementing Regulation (EU) 2020/1207 of 19 August 2020 laying down rules for the application of Regulation (EU) 2017/745 of the European Parliament and of the Council as regards common specifications for the reprocessing of single-use devices	Regulation (EU) 2020/1207

The MDR also foresees a new set of requirements in the form of common specifications (CS) (Table 21.7), which means a set of technical and/or clinical requirements, other than a standard, that provides a means of complying with the legal obligations applicable to a device. According to where no harmonized standards exist or where relevant harmonized standards are not sufficient, or where there is a need to address public health concerns, the European Commission, after having consulted the MDCG, may, by means of implementing acts, adopt CS in respect of the general safety and performance requirements set out in MDR Annex I, the technical documentation set out in MDR Annexes II and III, the clinical evaluation and post-market clinical follow-up set out in MDR Annex XIV or the requirements regarding clinical investigation set out in MDR Annex XV. It is also foreseen that specific product groups such as the MDR Annex XVI devices without medical purpose are subject to CS. As of October 2020 the Commission has published one set of CS via Implementing Regulation (EU) 2020/1207 of 19 August 2020. This CS is laying down rules for the reprocessing of single-use devices.

21.9 Definitions

21.9.1 Medical Device

A "medical device" according to MDR EU 2017/745 Art. 2 means any instrument, apparatus, appliance, software, implant, reagent, material or other article intended by the manufacturer to be used, alone or in combination, for human beings for one or more of the following specific medical purposes:

- Diagnosis, prevention, monitoring, prediction, prognosis, treatment or alleviation of disease
- Diagnosis, monitoring, treatment, alleviation of, or compensation for, an injury or disability
- Investigation, replacement or modification of the anatomy or of a physiological or pathological process or state
- Providing information by means of in vitro examination of specimens derived from the human body, including organ, blood and tissue donations

and which does not achieve its principal intended action by pharmacological, immunological or metabolic means, in or on the human body, but which may be assisted in its function by such means.

The following products shall also be deemed to be medical devices:

- Devices for the control or support of conception
- Products specifically intended for the cleaning, disinfection or sterilization of devices as referred to in Article 1(4) MDR and of those referred to in the first paragraph of this point

21.9.2 Devices without Medical Purpose

According to Art.1 (2) MDR the Regulation shall also apply, as from the date of application of common specifications adopted pursuant to Article 9 MDR, to the groups of products without an intended medical purpose that are listed in MDR Annex XVI, so called non-medical purpose devices. These group of products enters into the scope of the MDR in line with the common specifications to be adopted by the end of 2020. The MDR is regulating these devices having no intended medical purpose such as non-medical contact lenses, product modifying the anatomy, substances for dermal implantation, equipment to modify adipose tissue and others listed in MDR Annex XVI.

21.9.3 CE Mark

The CE Mark is the proof from the manufacturer that the product is in conformity with the Regulations.

CE Marking attests that the products are in conformity with the General Safety and Performance Requirements (GSPR) of the MDR and IVDR and that the products were subjected to the procedure of conformity assessment envisioned in the directives.

CE Marking is affixed before marketing the product and releasing it to the EU market for distribution or use.

CE Marking allows freedom of movement of the medical device in the territory of the European Union (EU) and EFTA/EEA.

It engages the responsibility of the legal manufacturer, on all aspects relative to the products safety and efficacy as the legal (CE) manufacturer being the EU market authorization holder centrally responsible for the safety and efficacy of the product and all related activities such as design, manufacturing, sterilization and labeling.

The manufacturer must maintain the technical documentation of the device and supply it to the regulatory authorities in case of quality audits or inspections at any time.

Devices other than custom made devices and those intended for clinical investigations, that are put on the market or brought into service in Europe must be CE Marking according to the provisions of the MDR Annex V.

The manufacturer chooses a procedure for CE Marking, whether or not to utilize a notified body, in particular according to the class of the medical device.

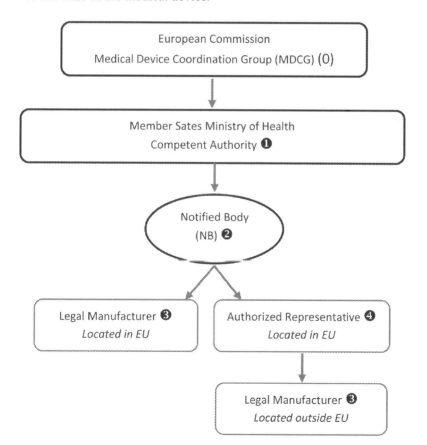

21.9.4 European Commission and MDCG (0)

The European Commission helps to shape the EU's overall strategy, proposes new EU laws and policies such as the MDR and IVDR regulations, monitors their implementation and manages the EU budget. Within the European Commission, there are several Directorate Generals (DGs) that are in charge for the different policies. The DG internal market and DG health are responsible for the implementation and supervision of the EU MDR and IVDR regulations as the medical devices and IVD sector are essential to the provision of healthcare to citizens and is an important player in both the European and global economy.

The Medical Device Coordination Group (MDCG) provides advice to the Commission and assists the Commission and the Member States in ensuring a harmonized implementation of medical devices Regulations (EU) 2017/745 and 2017/746. The MDCG consists of Member States and Organizations involved in MD and IVD policy and consists of the following working groups involved in the implementation of the regulations and advocacy and guidance (e.g. MDCG guidance documents):

- Notified Bodies Oversight (NBO) Working Group
- Standards Working Group
- Clinical Investigation and Evaluation (CIE) Working Group
- Post-Market Surveillance and Vigilance (PMSV) Working Group
- Market Surveillance Working Group
- Borderline and Classification (B&C) Working Group
- New Technologies Working Group
- EUDAMED Working Group
- Unique Device Identification (UDI) Working Group
- International Matters Working Group
- In vitro Diagnostic Medical Devices (IVD) Working Group
- Nomenclature Working Group
- Annex XVI MDR devices Working Group

21.9.5 Competent Authority ❶

The Competent Authority is a government agency or other entity in an EU Member Sate that exercises a legal right to control the use or sale of medical devices within its jurisdiction respectively territory. It has the ability to take enforcement action to ensure that medical products marketed within its jurisdiction comply with legal requirements.

Its roles and responsibilities are as follows:

- Accreditation and inspection of the notified bodies
- Supervises the market
- Centralizes and evaluates the vigilance data until EUDAMED is operational
- Takes suitable enforcement measures in case of violation of the laws.

Examples of Competent Authority:

- UK: MHRA (Medicines and Healthcare products Regulatory Agency)
- France: ANSM (Agence Nationale de Sécurité du Médicament et des produits de Santé)
- Ireland: IMB (Irish Medicines Board)
- Germany: BfArM (Bundesinstitut für Arzneimittel und Medizinprodukte)
- Spain: AEMPS (Spanish Agency of Medicines and Medical Devices)

21.9.6 Notified Body—Conformity Assessment Body ❷

A notified body or Conformity Assessment Body (as per MDR terminology) is a certification organization authorized by the relevant Member State's Competent Authority to perform conformity assessment tasks specified in the Regulation. As per Art. 38 MDR Conformity assessment bodies shall submit an application for designation to the authority responsible for notified bodies (ARN).

The assessment of the application is further described in Art 39ff MDR and involves the ARN, European Commission, the MDCG and a joint assessment team including an onsite audit of the applicant. After assessment of the application, the audit and corrective measure completion, the ARN shall submit its final assessment report and the NB designation to the Commission, the MDCG and the joint assessment team. After a final positive opinion, the ARN is responsible for the designation decision.

The notified body is responsible to evaluate the conformity of the product with the General Safety and Performance Requirements (GSPRS) and issues the CE mark certificate and quality management certifications. The notified body is an organization indicated and supervised by the proper authority of the Member State and must meet the criteria of Annex VII of MDR 2017/745. Notified bodies have to undergo designation until May 2021, date of Application (DOA) of the MDR in order to be able to continue their activities under the new Regulation.

The manufacturer calls upon the notified body of his choice: among the 56 (out of the original 87) registered notified bodies designated under the medical devices directive 93/42/EEC only 44 applications have been submitted for designation under the MDR. The number of accredited notified bodies has decreased in the past year and will continue to do as quality requirements for the NB designation are increasing in the EU.

Nando Information System

Notification is an act whereby a Member State informs the Commission and the other Member States that a body, which fulfils the relevant requirements, has been designated to carry out conformity assessment according to a directive. Notification of notified bodies and their withdrawal are the responsibility of the notifying Member State (Table 21.8). The lists of notified bodies can be searched on the NANDO web site. The lists include the Regulation under which the NB is designated (e.g. MDR EU 2017/745 or IVDR EU 2017/746), the identification number of each notified body as well as the tasks for which it has been notified, and are subject to regular update.

Table 21.8 Examples of notified body

Notified body, country		Identification number
GMED, France	GMED	0459
British Standards Institution, the United Kingdom	BSI	0086
National Standards Authority of Ireland, Ireland	NSAI	0050
TÜV SÜD Product Service GmbH, Germany	TÜV	0123

21.9.7 The Manufacturer ❸

The Manufacturer or legal manufacturer means the natural or legal person with responsibility for the design, manufacturing, packaging and labeling of a device before it is placed on the market under his own trademark, regardless of whether these operations are carried out by that person himself or on his behalf by a third party, as defined in each relevant section of the Regulations.

The manufacturer must ensure that it is manufactured to meet or exceed the required standards of safety and performance. This includes the three phases (design/development/testing, manufacturing, packaging and labeling) that lead to a product being ready for the market. The Legal Manufacturer is responsible of all the operations necessary to design, manufacture, label and package the product throughout its lifecycle from design to distribution to the final customer. The Legal Manufacturer is responsible to choose the notified body and affixing the CE Mark on the product once it is obtained. Depending on the organization, the Legal Manufacturer can be the Design Centre but not necessarily.

The manufacturer has an obligation to ensure that a product intended to be placed on the Community market is designed and manufactured, and its conformity assessed, to the GSPRS in accordance with the provisions of the applicable Regulations.

The manufacturer may use finished products, ready-made parts or components, or may subcontract these tasks. However, he must always retain the overall control and have the necessary compctence to take the responsibility and liability for the product according to Art. 10 MDR.

The Manufacturers shall also, in a manner that is proportionate to the risk class, type of device and the size of the enterprise, have measures in place to provide sufficient financial coverage in respect of their potential liability under EU law and national law.

The Manufacturer has to establish a person responsible for regulatory compliance (PRRC) in his organization according to Art 15 MDR.

21.9.8 Authorized Representative ❹

The authorized representative means any natural or legal person established within the Union who has received and accepted a written mandate from a manufacturer, located outside the Union, to act on the manufacturer's behalf in relation to specified tasks with regard to the latter's obligations under this Regulation. According to Art. 11 MDR where the manufacturer of a device is not established in an EU Member State, the device may only be placed on the Union market if the manufacturer designates a sole authorized representative. This representative must fulfill at least the tasks as described in the Regulation such as the following:

- Verification that the EU declaration of conformity and technical documentation have been drawn up and, where applicable, that an appropriate conformity assessment procedure has been carried out by the manufacturer
- Keeping available a copy of the technical documentation, the EU declaration of conformity and, if applicable, a copy of the relevant certificate, including any amendments and supplements, at the disposal of competent authorities
- Complying with the registration obligations and verify that the manufacturer has complied with the registration obligations in EUDAMED
- Responding to a request from a competent authority, provide that competent authority with all the information and documentation necessary to demonstrate the conformity of a device, in an official Union language determined by the Member State concerned
- Forwarding to the manufacturer any request by a competent authority of the Member State in which the authorized

representative has its registered place of business for samples, or access to a device and verify that the competent authority receives the samples or is given access to the device
- Cooperate with the competent authorities on any preventive or corrective action taken to eliminate or, if that is not possible, mitigate the risks posed by devices

The authorized representative is explicitly designated by the manufacturer, and he may be addressed by the authorities of the Member States instead of the manufacturer with regard to the latter's obligations under the Regulations.

21.10 Classification

21.10.1 Medical Devices

The EU Classification of Medical Devices is based on the risk of the device and follows the GHTF classification. The classification is made based on the rules laid down in Annex VIII of the MDR EU 2017/745. The classification is based on a risk assessment of the product, when used as intended. The "risk" is composed of the duration of use and the level of invasiveness, as defined by the intended purpose stated by the manufacturer. The "Intended purpose" means the use for which the device is intended according to the data supplied by the manufacturer on the labeling, in the instructions and/or in promotional materials.

Medical devices are divided into four classes named **Class I (s/r), Class IIa, Class IIb, and Class III** according to the level of risk the device as based on the following criteria (Table 21.9):

- Duration of use (transient, short-term, long-term)
- Invasiveness (invasive devices, body orifice, surgically invasive device)
- Active/non-active
- Implantable
- Specific hazards (e.g. contact with the Central Nervous System, animal tissues, absorbable material, ionizing radiation, Medical device with ancillary pharmaceutical substance)

Table 21.9 Examples of medical devices per class

Class	Risk	Examples
Class I	Moderate Risk	Surgical instruments Non-invasive tubing to evacuate body liquids Examination gloves Hospital beds Dental curing lights
Class Is	Moderate Risk (Sterile)	Sterile body liquid collection devices Sterile absorbent pads
Class Im	Moderate Risk (with a measuring function)	Syringe without needle (graduated barrel) Device for measuring body temperature Non-active device for measuring intra-ocular pressure
Class IIa	Moderate—Average Risk	Syringe with needle Tubing intended for use with an infusion pump Non-medicated impregnated gauze dressings Short term corrective contact lenses
Class IIb	Average—Elevated Risk	Hemodialysers Long term corrective contact lenses Urinary catheters intended for long term use Insulin pens Intraocular lenses (IOL)
Class III	Elevated Risk	Neurological catheters Prosthetic heart valves Aneurysm clips Pre-filled syringe for vascular access device flushing Spinal needle Absorbable sutures

21.10.2 Active Implantable Medical Devices

The active implantable medical devices are at maximum risk according to MDR EU 2017/745: by definition they are any active medical device which is intended to be totally or partially introduced, surgically or medically, into the human body or by medical intervention into a natural orifice, and which is intended to remain after the procedure.

Following are the examples of active implantable medical devices:

- Implantable cardiac pacemakers
- Implantable defibrillators
- Leads, electrodes, adaptors for above examples
- Implantable nerve stimulators
- Bladder stimulators
- Sphincter stimulators
- Diaphragm stimulators
- Implantable active drug administration device

By default, all accessories to AIMDs are covered under AIMD themselves. They cannot be classified separately under the MDD.

21.10.3 In Vitro and Diagnostics Medical Devices

The classification of in vitro and diagnostics medical devices is based on the risk of usage in line with IVDR EU 2017/746 (Table 21.10). The new rule-based risk classification system is more flexible than the list-based system from the IVDD that it replaces. Instead of naming specific IVD devices or medical conditions, the risk classification of a device is determined by its intended purpose and takes into consideration not only the risk to the individual but also the risk to public health. To classify their device, manufacturers should consult the rules listed Annex VIII of the IVD Regulation. If more than one rule applies, the rule resulting in the highest classification should be followed. In line with the international principles of classification, the four classes are:

Table 21.10 In vitro diagnostic medical devices risk class

Class	Risk	Conformity assessment
A	Low individual risk and low public health risk	Self-certified by their manufacturers (if non-sterile)
B	Moderate individual risk and/or low public health risk	Conformity assessment by a notified body
C	High individual risk and/or moderate public health risk	
D	High individual risk and high public health risk	

21.11 Conformity Assessment Procedures

Figure 21.2 shows the different steps a manufacturer has to go through from the device classification through the notified body selection to the final CE mark certificate and Declaration of Conformity issuance.

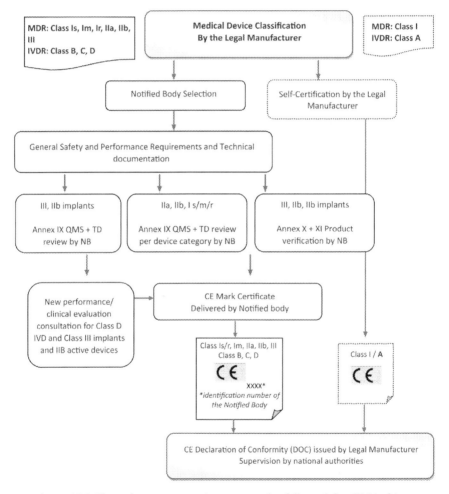

Figure 21.2 Flow chart representing steps to be followed for CE Marking under MDR/IVDR.

The Conformity Assessment Procedures correspond to modules described in the Regulations, which the manufacturer has to follow in order to affix the CE mark on the device. The procedures may be used single or in association and are aimed to cover the Design and the production quality of the device.

The possible procedures are defined in the annexes of each Regulations (Table 21.11):

- Annex IX: Conformity Assessment based on quality management system and assessment of the technical documentation
    - Roles and responsibilities of the Manufacturer:
        - Maintain a complete system of quality assurance including the design, manufacture and control
        - Provide appropriate information about the product, and documentation about the Quality System
        - Maintain the technical dossiers
    - Roles and responsibilities of the notified body: Quality System Verification
        - Inspect the technical dossier for class I and II products during audits
        - Inspect the buildings and the processes and the quality system
        - Examine the design of the product for class III devices and issue design examination certificate for class III devices.
- Annex X + XI: Conformity Assessment based on CE Type Examination (Certificate by the notified body that a sample representative of production satisfied the Essential Requirements)
    - Roles and responsibilities of the Manufacturer:
        - Provide the documentation about the design, manufacture, and performance of the product and a sample
    - Roles and responsibilities of the notified body:
        - Examine and evaluate the documentation
        - Verify that the sample is manufactured in conformance with the documentation
        - Perform, or subcontract suitable inspections and necessary tests
- Annex XIII: Procedure for custom made devices

Table 21.11 Overview of process per Regulation Annex

Process	MDR EU 2017/745	IVDR EU 2017/746
Quality management system and technical documentation	IX	IX
Type examination	X	X
Product conformity verification	XI	—
Production quality assurance	—	XI
Custom made devices	XIII	—

Table 21.12 Combination of procedures for CE marking of medical devices as per MDR EU 2017/745

Class I	Art 52 (7) MDR: Self-certification (no intervention of a notified body) base on technical documentation.
Class Is (sterile device) Class Im (measuring function) Class Ir (reusable)	Annex IX or Annex XI NB involvement is limited to sterility, metrology, cleaning/reuse
Class IIa	Annex IX + TD assessment of at least one representative per device group Or Annex XI
Class IIb	Annex IX + TD assessment of at least one representative per device group Or Annex X + XI
Class IIb implant	Annex IX + TD assessment of each device. Except for sutures, staples, dental fillings, dental braces, tooth crowns, screws, wedges, plates, wires, pins, clips and connectors, Or Annex X + XI
Class III	Annex IX + TD assessment of each device Or Annex X + XI
Class III implant and IIb active devices intended to administer or remover medicinal products	Annex IX + Clinical Evaluation consultation procedure according to Art 54 MDR Except for renewals, modification or common specification compliance

Custom made devices	Annex XIII
Devices with ancillary drug substances	Annex IX Involvement of a competent authority for medicinal substance according to Directive 2001/83/EC
Devices from tissues or cells of Human origin	Annex IX Involvement of a competent authority for human tissues and cells according to Directive 2004/23/EC

Table 21.13 Combination of procedures for CE marking of in vitro diagnostic medical devices as per IVDR 2017/746

Class D	Annex IX + TD assessment of each device Or Annex X + Annex XI For companion diagnostics, the notified body shall consult a competent authority designated by the Member States in accordance with Directive 2001/83/EC For first certification for a type of device, the notified body shall consult the relevant experts referred to in Article 106 of Regulation (EU) 2017/745
Class C	Annex IX + TD assessment of at least one representative per device group Or Annex X + XI For companion diagnostics, the notified body shall consult a competent authority designated by the Member States in accordance with Directive 2001/83/EC
Class B	Annex IX + TD assessment of at least one representative per device group
Class A	Self-certification (no intervention of a notified body) base on technical documentation NB involvement in case of sterility

21.12 General Safety and Performance Requirements

The General Safety and Performance Requirements (GSPR) lay down the necessary elements for protecting the public interest with regard to safety and performance of products placed on the market: they are mandatory to be complied with. Only products complying with the GSPR may be placed on the market and put into service in the European Union (EU).

General Safety and Performance Requirements must be applied as a function of the hazards inherent to a given product: they correspond to the technical requirements with which a device must comply in terms of safety and performance in order to be CE marked: they are to be met to demonstrate the safety and functionality of a device and are defined in the Annex I of each applicable Regulation (MDR EU 2017/745 and IVDR EU 2017/746).

The Essential Requirements checklist is divided into the following chapters:

1. General Requirements including
   - Risk management system
   - Storage and Transport
   - Performance of the device
2. Requirements regarding Design and Manufacture
   - Biocompatibility, leachable, extractable
   - CMR and/or endocrine-disrupting substances, phthalates
   - Infection and microbial contamination
   - Materials of biological origin
   - Interaction of the device with the environment
   - Measuring function
   - Protection against radiation
   - Programmable systems
   - Active devices and active implantable devices
   - Mechanical and thermal risks
   - Protection against supplying energy
3. Requirements regarding labeling
   - Labels and Instruction for use

21.13 Labeling

The labeling of Medical Devices in very important for the product identification and tracaebility. The Manufacturer must meet the requirements from the Annex I of the Essential Requirements of each applicable Directive:

- Annex I section 23 "Requirements regarding Information supplied with the device" of the MDR EU 2017/745
- Annex I section 20 "Requirements regarding Information supplied with the device" of the IDVR EU 2017/746

National regulations in the European Community usually requires the information referred to in the device label and Instructions for Use to be in their national language(s) or in another Community language when a device reaches the final user, regardless of whether it is for professional or other use. The use of symbols that conform to harmonized standards (e.g. ISO 15223-1) ensures consistency throughout the European Community and means that there is no need to translate certain information.

21.14 Technical Documentation

The Technical Documentation must allow assessment of the conformity of the product with the requirements of the Directive. Its content is described in each Annex of the Directive, depending on the Conformity Assessment route chosen by the manufacturer to demonstrate conformity of the device with the provisions of the Directive.

Depending on the medical device class, a notified body is involved for the quality system review or Technical Documentation review. The Technical Documentation can be reviewed either during an audit on site by a notified body or as part of the CE certification, depending on the risk classification of the product.

Table 21.14 provides some guidance about how to design the layout of a Technical Documentation for CE Marking: it details the level of information which should be covered in a Technical File devices.

Table 21.14 Content of technical documentation

Modules	MDR EU 2017/745	IVDR EU 2017/746
1. Device Description and specifications including variants and accessories	X	x
1.1. Device description including name, intended purpose and intended users, UDI, principles of operations, rationale for qualification as device, risk class, novel features, accessories, variants, key functional elements and compositions, raw materials, technical specifications	X	x
1.2. Reference to previous generations of the device	X	x
2. Information to be supplied by the manufacturer with completed set of labels and the instructions for use in the languages accepted in the Member States where the device is envisaged to be sold	X	x
3. Design and manufacturing information including design stages, specifications, including the manufacturing processes and their validation, their adjuvants, the continuous monitoring and the final product testing and identification of all sites, including suppliers and sub-contractors, where design and manufacturing activities are performed	X	x
4. General safety and performance requirements (GSPR) that apply to the products, methods to demonstrate conformity, harmonized standard or CS applied, precise identity of the controlled documents offering evidence of conformity with each harmonized standard, CS or other method applied to demonstrate conformity with the general	X	x
5. Risk management file	X	x
6. Product Verification and Validation	X	x
6.1. Pre-clinical and clinical data including results of tests, such as engineering, laboratory, simulated use and animal tests, and evaluation of published literature applicable to the device, taking into account its intended purpose, or to similar devices, regarding the pre-clinical safety of the device and its conformity with the specifications;		

Biocompatibility, electrical safety, software verification and validations tests, stability and transport validation studies; Clinical evaluation plan (CEP) and reports (CER), Post market clinical follow up plans (PMCFP) and Reports (PMCFR)	X	X Information on analytical clinical performance
6.2. Additional information in specific cases such as documentation of medicinal substances used with the device, tissues or cells of human or animal origin, substances that are absorbed or disbursed in the human body, CMR or endocrine disrupting substances, validation of process and packaging for sterile devices, accuracy of measuring for devices with measuring function, combination to other devices	X	
7. Post-market surveillance plan (PMSP)	X	x
8. Post-market surveillance reports (PMSR) and periodic safety update reports (PSUR)	X	x

21.15 Quality Management System

A quality system implemented on the basis of the ISO 13485 standard gives a presumption of conformity with the respective conformity assessment procedures with regard to the provisions in the modules that these standards cover, and provided that the quality system enables the manufacturer to demonstrate that the products fulfill the Essential Requirements of the directive in question. This means that the manufacturer must specifically address regulatory needs when implementing and applying a quality system, in particular the following:

- The quality objectives, quality planning, quality manual and control of documents must fully take on board the objective of delivering products that conform to the Essential Requirements.
- The manufacturer must identify and document the Essential Requirements that are relevant for the product and the harmonized standards to be used or other technical solutions that will ensure fulfillment of the Essential Requirements.
- The identified standards or other technical solutions must be used as design input, and as verification that design output ensures that the Essential Requirements will be met.

- The measures taken by the organization to control production must ensure that the products conform to the identified safety requirements.
- The organization in its measurement and control of the production process and finished products must identify and use methods which are identified in standards or other appropriate methods to ensure that the Essential Requirements are met.
- Quality records, such as inspection reports and test data, calibration data, qualification reports of the personnel concerned, must be suitable to ensure the fulfillment of the applicable Essential Requirements.
- The manufacturer has the responsibility to implement and continuously operate the quality system in such a way that regulatory needs are respected. The notified body must ensure in its assessment, approval and continued surveillance, that this is the case.

The objective of ISO 13485 is to set out requirements for a Quality Management System that is capable of consistently providing safe and effective medical devices and consistently meeting customer requirements (including regulatory requirements).

The Quality Management System has to be evaluated by a third party (notified body or general registrar) through periodic surveillance inspections (audit). The quality system certificates which are issued by a notified body have a limited validity and must be renewed every 3 years. (only exception of 5 years is Germany who is not part of the global IAF organization) IAF is International Accreditation Forum: This is the global organization where member states discuss and review ISO accreditation rules, which they then use to supervise notified bodies issuing ISO certificates. They for instance provide a document which identifies how many audit days are needed for what size of organization, issue guidance on how many days for surveillance versus renewal versus initial audits, etc.

21.16 Risk Management

The ISO 13485 recognizes that risk management is a key quality system element that pervades the entire life cycle of a medical device.

The ISO 14971 sets for guidance related to risk management.

The manufacturer shall establish, document and maintain throughout the life-cycle an ongoing process for identifying hazards associated with a medical device, estimating and evaluating the associated risks, controlling these risks, and monitoring the effectiveness of the controls. This process shall include the following elements:

- Risk analysis
- Risk evaluation
- Risk control
- Production and post-production information

The risk/benefit analysis is the basis to be conducted to ensure a high level of protection of health and safety: its purpose is to assess the acceptable risks when weighted against benefits to the patient. The risk reduction is then the elimination or mitigation as far as possible of the risks for a safe design and construction. It addresses protection measures to be taken and information to be provided the user and patient for each residual risk.

21.17 Clinical Evaluation

The Clinical Evaluation corresponds to the medical assessment of a Medical Device: the aim is to demonstrate that the intended purpose and claim of the device are achieved and to confirm that the device is safe and performs as expected. In this aim, the conformity of the device characteristics and performances as defined in the Essential Requirements, needs to be assessed, based on the evaluation of the side-effects and the acceptability of the benefit/risk ratio.

Clinical data is data which is relevant to the various aspects of the clinical safety and performance of the device. This may include data from prospective and retrospective clinical investigations of the device concerned as well as market experience of the same or similar devices and medical procedures and information from the scientific literature.

The Clinical Data Evaluation is the process by which clinical data from all selected sources (literature, results of clinical investigations and other) is assessed, analyzed and deemed appropriate and adequate to establish conformity of the device with the pertinent Essential Requirements of the Directive as they relate to safety and performance, and to demonstrate that the device performs as intended by the manufacturer. The outcome of this process is a report which includes a conclusion on the acceptability of risks and side effects when weighed against the intended benefits of the device.

The clinical data evaluation report should be made based on the risk of the product and based on the assessment of the combination of several of the following critical evaluation:

- Relevant scientific literature currently available relating to the safety, performance, design characteristics and intended purpose of the device where the data adequately demonstrate compliance with the relevant Essential Requirements
- Results of existing clinical investigations issued from clinical studies
- Data from market experience of same/similar devices already regularly placed on the market
- Additional requirements of applicable standards and guidance documents
- Performance claims
- Risk analysis report

The conclusion should compile all risk/benefit assessment from each part: the rationale should be included to justify the acceptability of each remaining risk when weighted against the intended benefits from use of the device. Statements concerning the field of use of the device and its indications and contraindications, effects and side effects, should be consistent with the device labeling, including the instructions for use.

Where demonstration of conformity with Essential Requirements based on clinical data is not deemed appropriate, adequate justification has to be given based on risk management output and under consideration of the device/body interaction, the clinical performances intended and the claims of the device.

The Clinical Evaluation should be reviewed, signed and dated by a dully qualified medical practitioner or other expert (Medical expert: Medical Director or Associate).

21.18 CE Mark Certificate and Declaration of Conformity

The CE marking symbolizes the conformity of the product with the applicable European Community requirements imposed on the manufacturer.

The CE marking affixed to products is a declaration by the person responsible that the product conforms to all applicable European Community provisions, and that the appropriate conformity assessment procedures have been completed.

A CE Mark Certificate indicating the procedure of evaluation applied and the range of the products is established by the notified body. It is valid for a maximum of 5 years and can be renewed. During its validity, it can be amended to allow for changing products and line-extensions.

If the selected procedure utilizes a notified body, then the product labeling has to include on all levels of labeling/packaging the following symbol:

$C \epsilon$XXXX, where XXXX is the identification number of the notified body

In the case of self certification (Class I, non-sterile, without a measuring function), the labeling of the product comprises the CE Symbol only:

$$C \epsilon$$

The Declaration of Conformity is the final step in the relevant conformity assessment procedure for all Medical Devices, which fall under the scope of one of each of the Directives and is incumbent to the Manufacturer: this is whereby the Manufacturer ensures and declares that the products concerned meet the provisions of the Directives. It is a mandatory document which must imperatively be established and signed before any product release for sale and which must be kept at regulators disposal. A CE Declaration of conformity is established by the manufacturer whatever the class of the device. It ends any conformity assessment procedure.

21.19   Post-Market Surveillance

The Post-market Surveillance (PMS) is an information gathering process to ensure that any problems or risks associated with the use of a manufacturer's medical device, once freely marketed, are identified, reported to National Competent Authorities and Corrective actions are taken to mitigate the problem or risk.

The post-market surveillance plan includes post-market clinical follow-up, reporting of serious incidents to Competent Authorities (Vigilance System), application of Field Safety Corrective Actions (FSCA, previously named Recall). A Field Safety Corrective Action is an action taken by the Legal Manufacturer to reduce a risk of death or serious deterioration in the state of health associated with the use of a medical device that is already placed on the market. Such actions should be notified via a Field Safety Notice, which is the communication of a Field Safety Corrective Action to users.

The post-market surveillance system is based on the information received from the field in the post-production phase, which is the part of the life-cycle of the product after the design has been completed and the medical device has been manufactured. The post-market surveillance system shall consist of the review and assessment for effectiveness and or further action of the following:

- Corrective and preventive actions associated with the Complaints, Medical Device Vigilance Reports and Field Safety Corrective Action activity
- The results of customer surveys
- Feedback from manufacturing plant surveillance activities
- Output reports from post-market clinical evaluation studies
- Feedback from users via sales representatives
- Reports and or interactions with regulatory authorities
- Literature reviews and media analysis where applicable
- Output from expert user/focus groups if applicable to the product platform

The post-market surveillance activities are aimed at obtaining some of the following types of knowledge or feedback:

- Detection of manufacturing problems and detection of needs for product quality improvement
- Confirmation of risk analysis
- Knowledge of performance of the device in different user populations and on ways in which the device is misused
- Feedback on indications for use, on Instructions for Use, training needed for users, use with other devices, customer satisfaction and on continuing market viability

21.20 From MDD to MDR

A new Medical Device Regulation EU 2017/745 (MDR) has been published in the EU Official Journal on May 5th 2017.

The new regulation includes 123 Articles and 17 Annexes unifying in a single text the old Medical Device Directive (MDD 93/42/EEC) and Active Implantable Medical Device (AIMD) provisions.

Within the framework known from the MDD 93/42/EEC based on general requirements for safety and performance, use of harmonized standards, risk based classification and third party conformity assessment based on risk, the new MDR introduces new requirements in many areas in the European Union (EU).

- Notified bodies: a new designation procedure and quality system requirements for notified bodies is introduced including joint assessment and increased supervision and unannounced audits and renewals of their designation.
- The 14 Essential Requirements are replaced with 23 General Safety and Performance Requirements with more details.
- Classifications are revised to capture a host of new devices into Class III and introduce special requirements for Class I reusable surgical instruments. Certain class III and IIb devices are subject to a scrutiny procedure involving the EU commission for market approval beside the notified body and clinical evaluation consultations.
- There are increased requirements for Risk Management, Postmarked Monitoring and Clinical Evaluation.

- New organizational function such as a Qualified Person (QP) is introduced in manufacturer's organization.

21.21 Transition and MDR Impact

The MDR was published and came into force in May 2017 and will be fully applicable after a transition period of four years on May 26th 2021 date of application (DOA). During this transition period notified bodies are obliged to re-designate under the new rules and market authorization holders (CE manufacturers) are obliged to prepare their quality system and products to the new requirements of the MDR. With the MDR DOA notified bodies will only be allowed to issue new certifications according to the new regulation MDR EU 2017/745. Old MDD 93/42/EEC certificates can be maintained valid until a maximum date of May 25th 2024. MDD products must be cleared from the supply chain and cannot be made available on the market or put into service after 26 May 2025 ("sell off" date) as described in Art 120 MDR. The following date must be taken into consideration as for devices:

Table 21.15 MDR EU 2017/745—Transitional provisions

MDR Date of application (DOA)	May 26, 2021
Maximum validity of MDD certifications	May 26, 2024
"Sell off" date of MDD devices	May 26, 2025

The following section provides a summary of the most important changes of the MDR compared to the MDD.

21.22 General Safety and Performance Requirements

The old MDD Essential Requirements (ER) are replaced by the new MDR Annex I with more detailed set of GSPR. There are new requirements relating to software, devices incorporating medicines or biological origin materials, and risks associated with lay use of devices. The requirements for sterile devices and for biological safety are both considerably expanded, with specific requirements related to specified carcinogens, endocrine disruptors and phthalates. The old Essential Requirement 13 on labeling is expanded to a complete Chapter III in the new text. The new Regulations retain reference to harmonized standards, but also add in relevant parts of the European pharmacopoeia and introduce Common Specifications additionally to the known harmonized standards.

21.23 Stronger Notified Body Oversight

One of the central pillars of the MDR is the strengthening of the oversight—and capacity—of notified bodies. The new regulations bring a much more rigorous designation process, with multiple parties from national member states and the European Commission via joint assessment teams involved in the designation process to ensure consistency and transparency. All notified bodies are required to re-apply for designation, with a lengthy review process. The expectation is that the end of this process a part of the existing notified bodies will either restricting their scope or discontinue their device activities due to the higher burden.

21.24 Economic Operators

The MDD 93/42/EEC essentially provided requirements for manufacturers and their authorized representatives. The new Regulations introduce the concept of economic operator. That includes manufacturers, importers, distributors and market representatives and notified bodies and the regulations place specific requirements on each of these. Most notably:

- Distributors and Importers are required to maintain detailed records and to actively participate in post-market monitoring including supporting adverse event reporting and recall processes.
- European Representatives are to be held jointly and severally liable for safety of medical devices. Expect EC reps to insist on insurance cover and fees to be adjusted

accordingly. And speaking of insurance, the preamble to the new Regulations specifically requires manufacturers to take out product liability insurance.

21.25 Clinical Evidence, Postmarked Follow-Up and Risk Reviews

The strengthening of clinical requirements is one of the pillars of the reform. The text of the MDR retains the explicit requirements for clinical evaluation of all devices irrespective of class and is specific that clinical evaluation shall include consideration of all information sources including pre-clinical data, risk management, literature and direct clinical experience. Class III devices are required to have direct clinical trial data except in the specific circumstance of design changes from a previous device—where a justification can be provided based on review of testing, or by comparison with another manufacturer's devices as a predicate—but only if the other manufacturer will provide complete and perpetual access to the technical file information of the predicate.

Clinical Evaluation reports (CER) are expected to be prepared based on a clinical evaluation plan (CEP) by a team of experts including biostatisticians, medical writers, product specialists and a clinician—who must sign off the report. Experts are required to be appropriately trained or experienced with degree level qualifications or lengthy direct experience.

Clinical Evaluations must be reviewed periodically (Annually for Class III devices) and the reviews used to recalibrate risk management files. All class III devices and any device which relied on predicate comparisons must be subject to active post-market clinical follow up and again the information gained must feed into risk management review.

A summary of safety and clinical performance (SSPC) needs to be drafted and published in order to be publicly available.

21.26 PSUR/PMCF Reports

Manufacturers will need to prepare Post Market Surveillance (PMS), Periodic Safety Update Reports (PSUR) and Post Market

Clinical Follow Up (PMCF) plans and reports. For higher risk devices these will need to be reviewed by the notified body, along with updates to the Clinical Evidence Report.

Eudamed

All of this expanded regulation and oversight is to be managed by a pan-European Database (EUDAMED) which will hold all of the various reports, records of assessments of manufacturers, and of notified bodies, as well as details of all economic operators and of all devices in the form of Unique Device Identification (UDI) supplied in Europe. The Regulation had called for the Commission to complete a plan to build the database within 2018 and to have it operational by 2022. While full operation of EUDAMED is delayed, the Commission has decided to deploy the actor registration module as of 1st December 2020 (see MDCG 2020–15). The members of the MDCG strongly encourage the use of the actor registration module by all relevant actors on their territories, including the use of the single registration number by actors as stipulated in the MDR (e.g. indicating the SRN on certificates).

EUDAMED will consist of the following modules according to its functional specifications:

- Actor registration for manufacturers, importers, authorized representatives
- Device registrations—UDI
- Certificates and notified bodies
- Clinical investigations
- Vigilance
- Market surveillance

21.27 UK Leaving the EU—BREXIT

After the UK has exitid the EU and from January 1, 2021, the Medicines and Healthcare products Regulatory Agency (UK MHRA) will take on the responsibilities for the UK medical devices market that are in the past undertaken through the EU system. Here is a summary of key requirements for placing a device on the Great Britain market as impacted by BREXIT. The impact of

BREXIT relates to registration requirements of devices in a UK specific database, the nomination of a UK responsible person and a new route to market UK CA mark.

The transition time is as follows:

- CE marking will continue to be used and recognized by UK MHRA until 30 June 2023.
- Certificates issued by European Economic Area (EEA) based notified bodies will continue to be valid for the Great Britain market until 30 June 2023.
- A new route to market and product marking will be available for manufacturers wishing to place a device on the Great Britain market from 1 January 2021 the UKCA mark that will become compulsory as of July 1st 2023.
- From January 1, 2021, all medical devices and in vitro diagnostic medical devices (IVDs) placed on the UK market will need to be registered with the MHRA. There will be a grace period for registering:
    - 4 months for Class IIIs and Class IIb implantables, and all active implantable medical devices
    - 8 months for other Class IIb and all Class IIa devices
    - 12 months for Class I devices

The requirements for a UK Responsible person are the following:

Manufacturers that are based outside the UK and wish to place a device on the UK market, will need to establish a UK Responsible Person who will take responsibility for the product in the UK. The UK Responsible Person will act on behalf of the outside-UK manufacturer to carry out specified tasks in relation to the manufacturer's obligations. This includes registering with the MHRA before the manufacturer's devices can be placed on the UK market. In summary, the UK Responsible Person must:

- Ensure that the declaration of conformity and technical documentation have been drawn up and, where applicable, that an appropriate conformity assessment procedure has been carried out by the manufacturer.
- Keep available a copy of the technical documentation, a copy of the declaration of conformity and, if applicable, a

copy of the relevant certificate, including any amendments and supplements for inspection by the MHRA.
- In response to a request from the MHRA, provide the MHRA with all the information and documentation necessary to demonstrate the conformity of a device.
- Forward to the manufacturer any request by the MHRA for samples, or access to a device, and ensure that the MHRA receives the samples or has been given access to the device.
- Cooperate with the MHRA on any preventive or corrective action taken to eliminate or, if that is not possible, mitigate the risks posed by devices.
- Immediately inform the manufacturer about complaints and reports from healthcare professionals, patients and users about suspected incidents related to a device for which they have been designated.
- Terminate the legal relationship with the manufacturer if the manufacturer acts contrary to its obligations under these Regulations and inform the MHRA and, if applicable, the relevant notified body of that termination.

UKCA mark

The UKCA (UK Conformity Assessed) mark is a new UK product marking that will be used for certain goods, including medical devices, being placed on the Great Britain market after the transition period. The UKCA mark will not be recognized in the EU, EEA or Northern Ireland markets, and products currently requiring a CE marking will still need a CE mark for sale in these markets. Manufacturers will be able to use the UKCA mark from January 1, 2021.

From July 1, 2023, to place a device on the Great Britain market, manufacturers will need to meet the requirements for placing a UKCA mark on their device.

References

1. Council Directive 93/42/EEC of 14 June 1993 concerning medical devices.

2. Directive 98/79/EC of the European Parliament and of the Council of 27 October 1998 on in vitro diagnostic medical devices.
3. Council Directive 90/385/EEC of 20 June 1990 on the approximation of the laws of the Member States relating to active implantable medical devices.
4. REGULATION (EU) 2017/745 OF THE EUROPEAN PARLIAMENT AND OF THE COUNCIL of 5 April 2017 on medical devices, amending Directive 2001/83/EC, Regulation (EC) No 178/2002 and Regulation (EC) No 1223/2009 and repealing Council Directives 90/385/EEC and 93/42/EEC.
5. REGULATION (EU) 2017/746 OF THE EUROPEAN PARLIAMENT AND OF THE COUNCIL of 5 April 2017 on in vitro diagnostic medical devices and repealing Directive 98/79/EC and Commission Decision 2010/227/EU.
6. COMMISSION NOTICE The "Blue Guide" on the implementation of EU products rules 2016.

Chapter 22

Regulation of Combination Products in the European Union

Gert Bos

Qserve Group,
Utrechtseweg 310, Building B42, 6812 AR Arnhem, The Netherlands

Gert.Bos@Qservegroup.com

22.1 Introduction: Legal Basis

In Europe, there are two regulatory routes for drug–device combination products, either as medical devices incorporating ancillary medicinal substances or as medicinal products utilizing a delivery device.

Following are the regulations covering these two options:

- Medical Devices Directive 93/42/EEC (MDD)/Active Implantable Medical Devices 90/385/EC (AIMDD) which is being transitioned into Medical Device Regulation EU 2017/745 (MDR)
- Medicinal Products Directive 2001/83/EC (MPD)

As a general rule, a combination product is regulated either by the MDD/AIMDD/MDR for the conformity assessment procedure relevant to medical devices or by the MPD for the marketing authorization procedure applicable to medicinal products. The procedures of both Directives do not apply cumulatively; however, some cross-references are made within one regime to specific requirements of the other regime, most notably also the additional change to the pharma legislation made in the last update of the medical device regulations (MDR Article 117).

The two procedures have different requirements, and the correct demarcation is crucial for development and regulatory approval of a combination product in Europe. The appropriate regulatory procedure depends upon the principal mode of action of the combination product.

- Drug-delivery products presented as an integral combination with a medicinal product are regulated as **medicinal products**.

    Example: pre-filled syringes

    See Section 22.2.

- Drug-delivery products presented separately from the medicinal product are regulated as **medical devices**.

    Example: drug delivery pump

    See Section 22.3.

- Medical devices incorporating, as an integral part, an ancillary medicinal substance are regulated as **medical devices but with additional requirements for the ancillary medicinal substance**.

    Example: drug-eluting stent

    See Sections 22.4, 22.5 and 22.6.

Guidance on which regulations are applicable, particularly in cases where the principal mode of action is not clear, can be found in the MEDDEV 2.1/3: EC Guidance document on "Borderline products, drug-delivery products and medical devices incorporating, as an integral part, an ancillary medicinal substance or an ancillary human blood derivative". See the European Commission website [1]. The information provided in the following sections is based, to a large extent, on the text of the MEDDEV 2.1/3 document, as no new guidance has been provided since the publication of the EU MDR in 2017. The reader is directed to this document for more detailed information and examples of drug–device combination products.

22.1.1 Definitions

22.1.1.1 Medical device

Article 1(2) (a) MDD (changes in the MDR Article 2.1 in bold) defines a medical device as

> Any instrument, apparatus, appliance, software, **implant, reagent**, material or other article, **intended by the manufacturer to be** used, alone or in combination, for human beings for **one of** the **following specific medical** purposes:
>
> - diagnosis, prevention, monitoring, **prediction, prognosis**, treatment or alleviation of disease,
> - diagnosis, monitoring, treatment, alleviation of or compensation for an injury or **disability**,
> - investigation, replacement or modification of the anatomy or of a physiological **or pathological** process **or state**,
> - **providing information by means of in vitro examination of specimens derived from the human body, including organ, blood and tissue donations**,
>
> and which does not achieve its principal intended action by pharmacological, immunological or metabolic means, **in or on the human body**, but which may be assisted in its function by such means.

The medical device function is usually achieved by physical means (including mechanical action, physical barrier, replacement of or support to organs or body functions). The MDR has provided minor tweaks only, and no key shift in products moving between the two legislations.

22.1.1.2 Medicinal product

Article 1(2) MPD defines a medicinal product as follows:

> (a) Any substance or combination of substances presented as having properties for treating or preventing disease in human beings; or

(b) Any substance or combination of substances which may be used in or administered to human beings either with a view to restoring, correcting or modifying physiological functions by exerting a pharmacological, immunological or metabolic action, or to making a medical diagnosis.

This definition comprises two limbs, one relating to presentation and the other to function. A product constitutes a medicinal product if it is covered by one or other or both of those limbs.

The definition of a medicinal product is applied on a case by case basis and determination takes into account current European Case Law. The Medicinal Products Directive (MPD), Article 2(2), is clear that, in cases of doubt, taking into account all of a product's characteristics, if a product meets both the definition of a medical device and a medicinal product, then the MPD will apply.

22.1.1.3 Combination products: Principal mode of action

The principal intended action of a combination product is determined based on the mechanism of action of the device and medicinal substance aspects.

The manufacturer's labelling and claims are also considered, but they should be in line with, and not contradict, current scientific data. The product manufacturer should be able to justify scientifically their rationale for the design and classification of the combination product.

Useful definitions from MEDDEV 2.1/3

"**Pharmacological means**" is understood as an interaction between the molecules of the substance in question and a cellular constituent, usually referred to as a receptor, which either results in a direct response, or which blocks the response to another agent. Although not a completely reliable criterion, the presence of a dose-response correlation is indicative of a pharmacological effect.

"**Immunological means**" is understood as an action in or on the body by stimulation and/or mobilization of cells and/or products involved in a specific immune reaction.

"**Metabolic means**" is understood as an action which involves an alteration, including stopping, starting or changing the speed of the normal chemical processes participating in, and available for, normal body function. Medical devices may be assisted in their function by pharmacological, immunological or metabolic means, but as soon as these means are not ancillary with respect to the principal intended action of a product, the product no longer fulfils the definition of a medical device and the product will be regarded as a medicinal product.

Examples

Bone cements and gentamicin-loaded polymethylmethacrylate (PMMA) beads are useful examples of combination products for illustration of the classification rules:

- Plain bone cement without antibiotics is a medical device since it achieves its principal intended action (the fixation of prosthesis) by physical means.
- Bone cements containing antibiotics, where the principal in-tended action remains fixation of prosthesis, are also medical devices. In this case, the action of the antibiotic, which is to reduce the possibility of infection being introduced during surgery, is clearly ancillary.
- If, however, the principal intended action of the combination product is to deliver the antibiotic (for example, gentamicin-loaded beads for insertion into and withdrawal from an infected bone cavity), this will be regarded as a medicinal product with the beads acting as a delivery device for local action of the gentamicin.

22.1.1.4 Borderline products: MEDDEV 2.1/3

The EC guidance document MEDDEV 2.1/3 is an extremely useful document which provides detailed information on the classification criteria between medical devices and medicinal products. In cases where it is not clear which regulatory regime applies, numerous examples and explanatory notes are provided to aid in the decision-making process. In time, it will be updated to be a formal match to the new MDR legislation; at that stage, it will be published on the website of the European Commission.

22.1.1.5 Borderline products: Manual of decisions

There have been a number of cases of disagreement about classification over the years, between manufacturers and notified bodies and also between different notified bodies and competent authorities (the EU national regulatory bodies for medicines).

A Medical Device Expert Group (MDEG) on Borderline and Classification issues has been set up to discuss products where there is a divergence of opinion across Europe. The Expert Group meets regularly to help reach agreement on classification issues. The decisions of the MDEG on Borderline and Classification are published in the Manual of Decisions available on the European Commission website [2]. However, the manual is designed to aid case-by-case application of the legislation by the competent authorities and is not a legal document itself.

The runner-up of the MDEG under the MDR is the Medical Device Coordination Group (MDCG); it is anticipated that in time this group will modify the manual to fit the MDR; at that stage, it will be published on the website of the European Commission.

22.2 Combination Products Regulated as Medicinal Products

This classification refers to devices that are intended to administer a medicinal product in the case where the device and the medicinal product form a single integral product, which is intended exclusively for use in the given combination and which is not reusable.

According to the MDD/MDR, this single product is governed by the MPD but the relevant essential requirements of Annex I to the MDD/MDR apply as far as the safety and performance-related device features are concerned.

22.2.1 Examples of Combination Products Regulated as Medicinal Products

- Pre-filled syringes
- Aerosols containing a medicinal product
- Patches for transdermal drug delivery

- Implants containing medicinal products in a polymer matrix whose primary purpose is to release the medicinal product
- Intrauterine contraceptives whose primary purpose is to release progestogens
- Temporary root canal fillers incorporating medicinal products, whose primary purpose is to deliver the medicinal product

22.3 Combination Products Regulated as Medical Devices

This category refers to devices that are intended to administer a medicinal product, but are not combined with a specific medicinal product. They may be re-useable and may be used with different medicinal products. In this case, that device is governed by the MDD or the AIMDD.

22.3.1 Examples of Combination Products Regulated as Drug-Delivery Devices

- Drug delivery pump
- Implantable infusion pump
- Iontophoresis device
- Nebulizer
- Syringe, jet injector
- Spacer devices for use with metered dose inhalers
- Port systems

22.4 Combination Products Regulated as Devices Incorporating, as an Integral Part, an Ancillary Medicinal Substance

Under the old regime of the MDD/AIMD, these products relate to a device that incorporates, as an integral part, a medicinal substance which, if used separately, may be considered to be a medicinal product and which is liable to act upon the body with action that is ancillary to that of the device.

In this case, the principal mode of action is attributable to the device element and the medicinal substance has a lesser, secondary effect.

The device is assessed and authorized in accordance with the MDD or the AIMDD, with additional requirements in line with MPD applied to the medicinal substance component.

According to the directives, the substance incorporated in the device must meet the three following conditions:

- The substance, if used separately, may be considered to be a medicinal product.
- The substance is liable to act upon the human body.
- The action of this substance is ancillary to that of the device.

In addition, a medical device incorporates a medicinal substance *as an integral part*, within the meaning of the directives only if the device and the substance are physically or chemically combined at the time of administration (i.e., use, implantation, application, etc.) to the patient.

22.4.1 Examples of Devices Incorporating an Ancillary Medicinal Substance

- Drug eluting stents
- Catheters coated with heparin or an antibiotic agent
- Bone cements containing antibiotic
- Root canal fillers which incorporate medicinal substances with ancillary action
- Bone void filler intended for the repair of bone defects where the primary action of the device is a physical means or matrix, which provides a volume and a scaffold for osteoconduction and where an additional medicinal substance is incorporated to assist and complement the action of the matrix by enhancing the growth of bone cells
- Condoms coated with spermicides

22.4.2 Examples of Drug Substances Incorporated into Devices

- Antibiotics, e.g., gentamicin, vancomycin

- Other anti-microbials, e.g., silver, chlorhexidine, rifampin
- Anti-proliferatives, e.g., paclitaxel, sirolimus
- Heparin
- Dexamethasone
- Nonoxinol
- Benzocaine

Please note that under the MDR the wording of "liable to act on the human body" have been removed; that means that for whatever purpose a medicinal product is added to the device, it is assumed that the drug will have an effect on the patient, and that it is added for that purpose!

22.4.3 Assessment of the Medicinal Substance Aspects of a Device Incorporating an Ancillary Medicinal Substance

For devices incorporating, as an integral part, an ancillary medicinal substance, under the MDD/AIMD system, the notified body should first of all verify the usefulness of the substance as part of the medical device themselves. They must then seek a scientific opinion from one of the EU competent authorities for medicines on the quality and safety of the substance including the clinical benefit/risk profile of the incorporation of the substance into the device.

The aspect of "usefulness" is interpreted as the rationale for using the medicinal substance in relation to the specific intended purpose of the device. It refers to the suitability of the medicinal substance to achieve its intended action, and whether the potential inherent risks (aspect of "safety") due to the medicinal substance are justified in relation to the benefit to be obtained within the intended purpose of the device.

Under the MDR, a similar process is required, in which the drug agency will be verifying *the quality, safety and usefulness of the substance by analogy with the methods specified in Annex I to Directive 2001/83/EC.* This means that for usefulness from now on the pharmaceutical interpretation will be utilized, which is more geared towards the effect(-iveness) of the added pharmaceutical component.

22.5 The Consultation Process

The notified body and the manufacturer may choose the EU competent authority with whom they consult. However, the European Medicines Agency (EMA) must be consulted for all medical devices incorporating ancillary human blood derivatives, e.g., human albumin or medicinal products manufactured using biotechnological processes (i.e., those falling within the scope of Annex I to Regulation (EC) No. 726/2004 [1]).

In accordance with MEDDEV 2.1/3, Section C, the notified body should ensure that data supplied by the manufacturer in relation to the device and its intended use includes a specific segment regarding the medicinal substance being incorporated with ancillary purpose. Detailed submission guidance is available on the websites of EMA, MHRA and other competent authorities. It typically expects the relevant portions of a CTD to be filled out.

22.6 Information to Be Provided on the Ancillary Medicinal Substance

22.6.1 General

Information addressing the safety, quality and usefulness of the medicinal substance should be prepared by the manufacturer, submitted to the notified body, and then forwarded by the notified body to the competent authority (MHRA in UK). In addition, a report on the usefulness of the medicinal substance should be prepared by the notified body and included with the application form.

Because of the wide range of medical devices which incorporate, as an integral part, an ancillary medicinal substance, a flexible approach to the data requirements is necessary. However, the information should be based in principle and to the relevant extent on Annex 1 to Directive 2001/83/EC (MPD).

22.6.1 Quality

Relevant information should be provided on both
- the drug substance itself and
- the drug substance as incorporated into the medical device.

With regard to the drug substance, evidence should be provided that the drug substance is manufactured to a high, reproducible quality and is suitably controlled by an appropriate specification.

Where a European Pharmacopoeia (PhEur) monograph exists for a substance, the specification should be that of the PhEur as a minimum.

With regard to the combination product, information relevant to the medicinal substance should be provided, i.e., on the quantitative composition, details of manufacture, control of critical excipients, and control of intermediate and finished products. Validation data should be provided where appropriate, in particular for analytical methods. The data submitted should demonstrate that the amount of medicinal substance incorporated is based on safety and efficacy considerations, that it can be suitably controlled within specified limits, that it is evenly distributed within/across the device as necessary and maintains satisfactory performance over the shelf-life of the product.

22.6.2 Safety and Usefulness (Clinical Benefit/Risk)

There should be a clear rationale and safety consideration for including the ancillary medicinal substance at the proposed dose.

Where well-known medicinal substances for established purposes are involved, original data on all aspects of safety and usefulness may not be required and many of the headings of Annex 1 to Directive 2001/83/EC will be addressed by reference to literature sources, including standard textbooks, experience and other information generally available. However, all headings should be addressed, with either relevant data or justification for absence of data, based on the manufacturer's risk assessment.

For new active substances and for known medicinal substances for a non-established purpose, comprehensive data is required to address the requirements of Annex 1 to Directive 2001/83/EC. The evaluation of such active substances would be performed in accordance with the principles of evaluation of new active substances.

22.6.3 Guidance

MEDDEV 2.1/3, Section C, provides guidance on the consultation procedure for devices incorporating ancillary medicinal substances or ancillary human blood derivatives [1]. In addition, the EMA and national competent authorities publish guidance on procedures and documentation requirements for the consultation.

There are also a number of useful European guidelines relating to the quality, safety and efficacy of medicinal substances as used in medicinal products which can be found on the EMA website [3] and in these consultation guidance documents.

It is not intended that the guidelines should be strictly adhered to for ancillary medicinal substances used in devices, however, as for medicinal product evaluation, justification for the use of different approaches should be provided.

22.7 Combination Products Regulated as Drugs Incorporating, as an Integral Part, an Ancillary Medicinal Substance

Under the Medical Devices Regulation (MDR), Article 117, manufacturers placing drug-device combination products onto the market as an integral device and marketing them as a "medicinal product" are required to seek a Notified Body Opinion on the safety and performance of the device component, in line with the requirements for medical devices in the MDR. In the procedure, the notified body confirms if the device is compliant with the relevant General Safety and Performance Requirements (GSPR) via their opinion report. The manufacturer then includes this report in the Market Authorization Application.

Only a very limited amount of opinions have been initiated to date, and hopefully based on the first experiences some further guidance will be provided from the EU commission, drug agencies and notified bodies.

22.8 Other Combination Products

This chapter has focused on the combination of medical devices with chemical medicinal substances. There are also other combination products in the early developmental stage, e.g., gene therapy, cell therapy, tissue engineered products combined with medical devices such as scaffolds, implants, stents and extracorporeal circuits.

These combinations are regulated as advanced therapy medicinal products (ATMP) and are assessed centrally by the Committee for Advanced Therapies (CAT) at EMA.

In this case, consultation with a notified body is required on the device aspects in a similar manner to consultation with a competent authority on the medicinal substance aspects of devices incorporating an ancillary medicinal substance.

More information on ATMP products can be found on the EMA [3] and MHRA websites [4].

Other combinations are feasible as well. Think for example on drug delivery systems and access ports that come with a stuffed animal attached, that after the procedure can be removed and given to the young patient to cuddle. The stuffed animal would be CE marked under the toys directive, so that set of essential requirements will apply to the toy component.

References

1. MEDDEV 2.1/3: "EC Guidance document on 'Borderline products, drug-delivery products and medical devices incorporating, as an integral part, an ancillary medicinal substance or an ancillary human blood derivative," retrieved (October 2020) from https://ec.europa.eu/docsroom/documents/10328/attachments/1/translations.
2. Medical Devices Expert Group on Borderline and Classification: Manual of Decisions, retrieved (October 2020) from https://ec.europa.eu/health/sites/health/files/md_topics-interest/docs/md_borderline_manual_05_2019_en.pdf.
3. European Medicines Agency website: Ancillary medicinal substances: Regulatory and procedural guidance, retrieved (October 2020) from https://www.ema.europa.eu/en/human-regulatory/overview/advanced-therapy-medicinal-products-overview.
4. MHRA website: Borderline with Medicines, retrieved (October 2020) from https://www.gov.uk/government/publications/borderlines-between-medical-devices-and-medicinal-products Aboutadvancedtherapymedicinalproducts/index.

Chapter 23

Medical Device Regulatory Affairs in Latin America

Carolina Cera[a] and Gladys Servia[b]

Regulatory Affairs Professionals—Latin America Expertise
carolinacera.regulatory@gmail.com, ra.serviagladys@gmail.com

23.1 Introduction

The Latin America region (Fig. 23.1) is a group of territories and countries known for speaking romance languages or Latin languages and is considered the largest area in America. It covers 42 regions from the southern border of North America—Mexico—passing through Central America and Caribbean and ending at the southern tip of South America. The four subregions, North America, Central America, Caribbean, and South America house 525.2 million people with nominal GDP (gross domestic product) of USD 4.775 trillion in 2014 [1].

Portuguese and Spanish are the main languages spoken among other languages such as French, Dutch, and native/indigenous languages. Portuguese is spoken only in Brazil (Brazilian

---

Medical Regulatory Affairs: An International Handbook for Medical Devices and Healthcare Products (Third Edition)
Edited by Jack Wong and Raymond K. Y. Tong
Copyright © 2022 Jenny Stanford Publishing Pte. Ltd.
ISBN 978-981-4877-86-2 (Hardcover), 978-1-003-20769-6 (eBook)
www.jennystanford.com

Portuguese) and Spanish is the official language in most of the rest of the countries in the Latin America region as well as Cuba, Puerto Rico (English is co-official and it is commonwealth of the United States of America) and Dominican Republic.

Figure 23.1 Latin America region (*Source*: *Wikipedia*).

Since the late 1990s, medical device manufacturers have been discovering a wealth of opportunity in Latin America [2], which can be attributed to Brazil, Mexico, Colombia, Chile, and Argentina, countries which together make up the third largest economy in the world [3]. With the demand growth comes the need for better controls and assurance of safe and effective products for the population.

23.2 Latin America Market Analysis

23.2.1 Medical Device Market in Latin America

The rise of the healthcare in some countries has been opening door for health care multinationals suppliers. This opening allowed an increase of 18% in the CAGR (compound Annual Growth Rate) from 2004 to 2009, it means that in 2009 the top six Latin American countries recorded $7.6 billion in imports [4].

As the economy in Latin America blooms, healthcare economy is one of many fields that are positively affected, leading to increased consumer demands as well as health care programs and incentives. Brazil and Mexico are the main manufacturers of medical devices in Latin America. The Brazilian Association of Importers of Medical Equipment, Products and Supplies (ABIMED) considers that the positioning of Latin America among the best manufacturers of medical devices may be achieved by

(i) Promoting the harmonization of regulations and the value of technology in Latin America;
(ii) sharing regional experiences to reduce counterfeiting and smuggling of medical supplies;
(iii) exchange of information to promote differentiation between devices and drugs.

23.2.2 Trade Blocs

Imports have been playing a very important role in the Latin America economies and helping in the resurgence of the area. The international medical device industry has been set in the region for a long time, bringing an increase of active medical device companies in Latin America's territory. In Brazil, for example, there are more than 100 large companies whose principal ownership lies outside the country. This is a very promising scenario not only for Brazil but also for other countries located in the Latin American territory, mainly for those that have trade agreements.

Trade agreements have been facilitating market access providing benefits to their members by reducing barriers and facilitating the entry of new technology into the region. MERCOSUR (Common Southern Market formed by Argentina, Brazil, Paraguay, Uruguay and Venezuela and the Pacific Alliance (formed by Chile, Colombia, Mexico and Peru) are the largest trade bloc in the Latin America region, representing about 93% of the GDP in the region. Nonetheless small trade blocs such as NAFTA (North American Free Trade Agreement connecting United States, Mexico and Canada) also contribute to the region growth.

MERCOSUR also provides medical devices guidance to the state's political party, it means, the technical regulation MERCOSUR 40/00, which adopts elements from the US, Canadian and EU regulatory systems, has a high influence in the medical regulation in the region where it is effective.

23.3 Overview of Medical Device Regulation in Latin America

23.3.1 Evolution of the Medical Device Regulation in Latin America

The emergent Latin American countries have been offering unique opportunities for medical device commercialization due to their rapidly growing market. However, not only the market but also the medical devices regulations are evolving.

Countries that have never had any medical device regulation are becoming regulated, e.g., El Salvador, and countries that already have regulations are becoming more stringent regarding requirements and enforcement such as Brazil and Mexico. Table 23.1 shows countries where medical devices are not regulated.

Table 23.1 List of countries with no medical device regulation

Anguilla	Chile	Paraguay
Antigua & Barbuda	Dominica	Saint Kitts and Nevis
Aruba	Dominican Republic	Saint Lucia Island
Bahamas	Grenada	Saint Vincent and Grenadines
Barbados	Guyana	Suriname
Belize	Haiti	Trinidad and Tobago
British Virgin Islands	Montserrat	Turks and Caicos
Cayman Islands	Netherland Antilles	

Chile and Paraguay are in the early stages of developing a medical device system; however, nothing has been enforced so far; Puerto Rico and Virgin Islands are under the US (FDA) jurisdiction.

Table 23.2 shows that countries that not only regulate medical devices commercialization but importation of goods as well.

Table 23.2 List of countries with medical device regulation

Argentina	Ecuador	Nicaragua
Brazil	El Salvador	Panama
Bolivia	Guatemala	Uruguay
Colombia	Honduras	Venezuela
Costa Rica	Jamaica	
Cuba	Mexico	

French Guyana, Guadeloupe, and Martinique are part of France and are under the French regulatory authority; in other words, EU laws are applicable in these countries.

Besides, MERCOSUR has a strong influence in the state's political party medical device regulation, other countries has been adopting regulatory structure from other successfully implemented medical device system, such as FDA (US), Health Canada (Canada) and European Union Directives.

23.3.2 Regulatory Environment in Latin America

23.3.2.1 Finding local registration holder

Latin America region has many attractions; however, to pursue the introduction of new technology in the region, a few considerations should be taken. Partnership is one of the most important steps to take before any pre market action is made.

Regulatory bodies from Argentina, Brazil, Colombia and Mexico, for example, require an in-country representative acting on behalf of the international company in terms of product registration and distribution in the region. The registration license will be issued to the local representative, which in other words is the license holder.

Only in Colombia, the medical device registration license ownership can be granted to either qualified importer or Colombian legal representative. In the first case, the importer holds the ownership license while the legal representative only manages the registration process and the legal manufacturers retain the ownership of the license.

This partnership between local distributor and the legal manufacturer is sealed through a **Letter of Authorization**, which has legal value only when legalized by the country's embassy and is one of the mandatory requirements for product registration in most of the regulated countries.

23.3.2.2 Harmonization of medical device

Intentions to propose a harmonized system for medical device regulation are still under review in the Latin America region, and currently most of the countries have adopted elements from the US, Canadian and EU regulatory systems. As the medical device market has been increasing in the region, the regulations in the area have been evolving too.

Since 1994, the Pan American Health Organization has been working with the countries of Latin America and the Caribbean in establishing and strengthening the regulation of medical devices, with the technical cooperation of the Office of Medical Devices in Canada. The Food and Drug Administration (FDA) of the United States and the Emergency Care Research Institute (ECRI) Collaborating Center PAHO/WHO, have joined in this activity.

Countries have been increasing their efforts to develop and strengthen regulatory programs, nationally and collectively through agreements and regional forums in collaboration with the Pan American Health Organization (PAHO), founded in 1902 [4], is the world's oldest international public health agency.

The fact that most of the countries are mainly importers of medical devices and therefore technologically dependent on its supply has made governments to give priority to strengthening their capacity to regulate medical devices. There are a number of factors that have determined the need for agencies or programs to regulate medical devices, which are mostly based on solid regulatory systems such as the FDA in the United States.

Although Argentina and Brazil share the same MERCOSUR directive for medical devices, each country has its own particularity for additional requirements in terms of medical device registration. This is due the fact that besides the harmonized regulation, both countries have many other secondary laws and regulations that act as additional requirements for product registration.

23.3.2.3 Challenges in the region

In the early 1990s, few countries in Latin America and the Caribbean had regulation programs for medical devices, such us Argentina, Brazil, Colombia, Mexico, Peru, and Panama, but with a limited capacity of action.

With the technical cooperation of the Pan American Health Organization (PAHO), Regional Office of the World Health Organization for the Americas region and the support of the regulatory agencies in Canada and the United States of America, it was possible for those and other countries in the region to develop and strengthen the capacity to regulate medical devices.

In parallel, manufacturers of medical devices discovered an important opportunity to develop this industry in the Latin American market. It grew very fast since the medical device regulations were still evolving.

Like other areas of the world, Latin America will eventually move to harmonized regulations. Until then, to commercialize a medical device in a specific country, first the companies should to obtain the license approval in that target country.

The growth of this industry made countries that have never had any medical device regulation to become regulated, and regulated countries are becoming more stringent in their enforcement. These trends will continue. Therefore, it is crucial to have good communication channels with all ministries of health. Nowadays, ministries allow the companies to be part of the creation of new regulations, or modifications on the current ones. This generates the opportunity to negotiate, to let ministries learn more from the industry, and to consider special situations that the regulations do not and they can apply exceptions to a general rule.

Staying abreast of changing regulations also will help companies to avoid roadblocks in their attempts to commercialize medical device in Latin America.

23.4 Argentina and Brazil Medical Device System

In 23.4 Argentina and Brazil, imported or domestic devices are required to be registered with the local regulatory agency before they are marketed in the territory. Registration of devices can be done by group or families of products, depending on the intended of use and characteristics of the device. Medical Device registration process in Argentina and Brazil is very similar as both follow the MERCOSUR directive and this similarity is shown in Fig. 23.2.

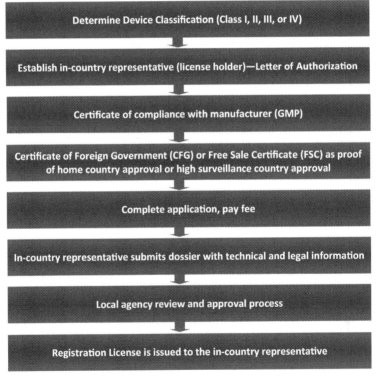

Figure 23.2 High-level registration process in Argentina and Brazil.

Currently, submission should be done in paper form since both countries do not accept electronic registration and the same should be written in local language, Spanish (Argentina) and Brazilian Portuguese (Brazil); English is acceptable only for large studies. The application does not have any value if it is not signed by the technical responsible, whose is the one responsible for all communications and follow-up with local regulatory agency.

23.4.1 Definition of Medical Device

A medical device is "a health product such as equipment, apparatus, material, item or system or application use medical, dental or laboratory, for the prevention, diagnosis, treatment, rehabilitation or that does not use contraception or pharmacological, immunological or metabolic to perform its primary function in humans, can in meanwhile be assisted in its function by such means."

23.4.2 Classification of Medical Devices

Premarket evaluation process is determined by risk classification which is very similar to the European Medical Devices Directive MDD93/42/EEC, including amendment 2007-47-EC and it requires four-layer risk classification (Class I, II, III and IV). Table 23.3 shows the classification levels compared with the European classification.

Table 23.3 Classification of medical devices

Risk	Level of risk	European classification equivalency
Class I	Low	Class I
Class II	Low-medium	Class IIa
Class III	Medium-high	Class IIb
Class IV	High	Class III

23.4.3 Argentina Medical Device System

23.4.3.1 Regulatory authority and medical device regulation

ANMAT (Administración Nacional de Medicamentos, Alimentos y Tecnología Médica) created in 1992 under the Ministry of Health is an independent entity in charge of controlling medical

devices distribution in Argentina. ANMAT major regulations are as follows:
- **Disposición 2318/02** identifies product classification and its related requirements for product registration.
- **Disposición 727/13** replaces the Provision 5627-2006, which specifies requirements for renewals and modifications, as well as simplification of Class I that should be used in order to obtain a complete list of requirements for new registration, renewals, and product modifications.
- **Disposición 2319/2002** which provides requirements for GMP (good manufacturing practices) for foreign manufacturers.
- **Disposición 2124/2011:** post-market surveillance program for medical devices.
- **Disposición 3266/2013:** MERCOSUR Technical Regulation of Good Manufacturing Practices for Medicinal Products and Products for Use In Vitro Diagnostics.
- **Disposición 3265/2013** Common procedures for inspections of manufacturers of medical products and products for in vitro diagnostic use in the state's political party parties.
- **Techno-Vigilance Program for Medical Devices**: Post-surveillance activities.

23.4.3.2 Registration process of medical devices

The local distributor has to obtain ANMAT GMP for product distribution in accordance to Disposición 3266-2013 and authorization permit according to Disposición 2319-2002. The first certification should consider the type and product classification desired to be registered and distributed. The registration process in Argentina is similar to the process shown in Fig. 23.2.

23.4.3.3 Documents and labeling requirements

Attachment III.C of the Disposición 2318-2002 combined to the requirements of the Disposición 0727 from February-4-2013 (replaces de Provision 5627-2006) should be consulted to guarantee if the required documentation to proof safety and efficacy of the medical devices is provided. The list of documentation varies according to product classification risk.

A few requirements, among others, required for product registration in Argentina are listed as follows:

- Free Sale Certificate (FSC) or Certificate To Foreign Government (CFG) issued by the competent health authority in the country of manufacture or commercialization
- Letter of Authorization from legal manufacturer
- Affidavit agreeing to report any recall and field safety corrective actions in any markets where the device is sold
- Device description and specification
- Detailed Manufacturing process for finished devices
- Description of safety and efficacy (non-clinical)
- Packaging instructions
- Labeling
- Risk Management (ISO 14971) reports
- Historical of commercialization, including countries were the product is marketed

Additionally, in order to provide documentary evidence to show that the medical device complies with the medical device regulation, the license holder also has to submit a draft of the labels to be affixed on the product as well as copy of the instructions of use. Disposición 2318-2002: Sections 2 and 3 Attachment III.B—Medical Devices Labels and Instructions of Use Information and Disposición ANAMAT No 593-1998 provides specific labeling requirements for medical device registration and commercialization.

Labeling should in the local language and follow the main requirements described in the regulation, such as

- name and address of the manufacturer and importer;
- technical responsible for the group;
- name and description of the device;
- specific storage conditions;
- product registration number;
- warning and precautions.

23.4.3.4 Official registration fee

ANMAT has assigned values for submission according two categories: National Product and Imported Products (Table 23.4).

Table 23.4 Fees for medical device registration

Submission	National product fee (USD)*	Imported product fee (USD)*
Class I	133	158
Class II	146	205
Class III	193	170
Class IV	273	382

*Real exchange rate USD 4.00.

23.4.3.5 Timeline

For all risk classes, the statutory time for review is 180 days after submission; however, the time lines have changed and are shown in Table 23.5.

Table 23.5 Time frame from submission to approval

Device Class	Risk	Timing from submission to approval
Class I	Low	8–10 months
Class II	Moderate	10–12 months
Class III	High	12–14 months
Class IV	High	14–16 months

Market approval is valid for a period of five years. The license holder can request renewals for the same period any time within three months before the expiration date to guarantee business continuity. License remains valid until the renewal license is granted.

23.4.3.6 Regulatory action for changes and device modifications

Modifications or changes in products can be under requirement, which should contain documentation, used in the original submission related to the change. Approval is required in most of the cases and the same timeline provided in the Item 1.4.3.4 should be considered.

23.4.3.7 Post-market surveillance

Post-Market studies are not required for regulatory approval of medical devices; however, medical devices manufacturers and importers are required to have well-documented post-market surveillance system in place as part of the GMP requirements. The same should document evidences of adverse events and recalls related to the medical device; for the latter the company has to send recall form to ANMAT.

23.4.4 Brazil Medical Device System

ANVISA (Agencia Nacional de Vigilancia Sanitaria) is an independent regulatory body of the Brazilian government established in 1999 but connected to the Ministry of Health.

The regulations in Brazil are constantly evolving and some of them are not clear; these, in some cases, make the registrations of medical devices complex and slow. Major regulations are:

- **RDC 185/01**: Technical regulation for Medical device Registration
- **RDC 16/13**: Brazilian GMP Requirements
- **RDC 27/11**: Electro Medical Device Compulsory Certification
- **RDC 35/14**: requirements for medical device constituted by plastic blood bags
- **RDC 40/15**: requirements for *Cadastro* of products Class I and II

23.4.4.1 Registration process of medical devices in Brazil

The registration process and requirements vary according to the product risk classification and also the product nature (e.g., plastic blood bags). As mentioned before, to obtain approval to import and market product into the Brazilian region, a few more steps should be taken, which are shown in Fig. 23.3.

According to the RDC No. 40/2015, for class I and II, it is necessary to prepare a technical dossier with IFU and labels, to have the legal documents, as stated in RDC 185/2001. These files are maintained by the local license holder, in case ANVISA decides to perform on-site inspections. It is a simple submission format (called *Cadastro*) with less initial information, (however, ANVISA requests to submit details of raw materials, i.e., concentration of each material (pigment: base polymer), metal international standard).

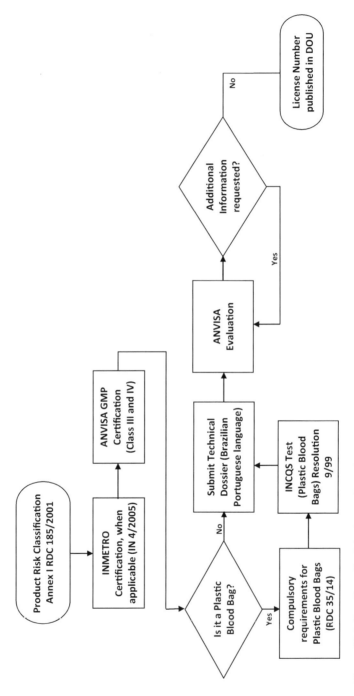

Figure 23.3 Brazil medical device registration process.

INMETRO and ANVISA GMP Certification are requirements for product registration; and the manufacturer has to follow the RDC 27/2011 and RDC 16/2013 standards, respectively. It is important to consider that both certification processes are extensive (require audits) and demand time, mainly the ANVISA GMP Certification, mandatory for Class III and IV registration, which can take up to 2 years to be released, whereas it is not require for for Class I and II products.

INMETRO certification is often required for electro-medical devices subject to IEC 606013rd edition, as well as some other medical devices, it is valid for 5 years, and annual audits and fees are required. The certification must be held by a Brazilian entity, which should be the same entity that is appointed as license holder.

Medical Device constituted by plastic blood bags should follow specific requirements for product registration—RDC 35/14 and Resolution 9/99—which are in accordance to the local good practices for blood bank centers.

23.4.4.2 Documents and labeling requirements

Depending on the product risk classification the documentation requirements vary from filling specific form with technical information about the product (Class I and II) to extensive technical documentation (Class III and IV).

Requirements for Class I, II, III and IV registration can be found in the RDC 185/01 and RDC 40/15. Among others, a few requirements (Class III and IV) are listed below, for product registration in Brazil:
- Copy of the payment bank receipt provided by ANVISA
- Free Sale Certificate (FSC) or Certificate To Foreign Government (CFG) issued by the competent health authority in the country of manufacture or commercialization
- ANVISA GMP Certification
- INMETRO Certification, when applicable—this certification is applicable for all product risk classification
- Letter of Authorization from legal manufacturer
- Device description and specification
- Detailed Manufacturing process for finished devices
- Device diagram and technical drawings

- Description of safety and efficacy (non-clinical)
- Packaging instructions
- Labeling
- Risk Management (ISO 14971) reports

Labeling requirements are presented in the main regulations described above and should be in the local language. A few requirements are described below:

- Name and address of the manufacturer and importer
- Technical responsible for the group
- Name and description of the device
- Specific Storage conditions
- Product registration number
- Warning and precautions

23.4.4.3 Official registration fee

On September 2015 the government issued new resolution increasing the existing fee, before specified in the RDC 22/2006. Fee payments vary based on the size of the company (annual sales). In Table 23.6, a comparison between original fee and revised fee for large sized companies (annual sales greater than USD 12.5 mi/year). Every beginning of year, the license holder has to provide annual sales report to ANVISA in order to adjust their category (small to high sized company) and define fees for future submission process in ANVISA.

Table 23.6 Comparison of original and revised fee by submission (large-sized companies)

Submission	Original fee (USD)*	Revised fee (USD)*
New product registration	3,000	5,809
Product renewal	1,800	5,809
GMP certification (International)	9,250	27,153
License modification	450	1,307

*Real exchange rate USD 4.00.

23.4.4.4 Timeline

For all risk classes, the statutory time for review is 90 days after submission; however, this timeline can vary from 2 months (Class I and II) to 4 years (Class III and IV).

The timeline for Class III and IV depends on (1) whether the manufacturer **already has** ANVISA GMP Certification for the class of product to be registered or (2) whether the manufacturer **does not have** ANVISA GMP Certification for the class of product under registration. The process to get GMP Certification usually takes 2–3 years to be released (audit and final report analysis). Table 23.7 presents timelines according to product classification.

Table 23.7 Risk classification and related timeline for approval in Brazil

Risk classification	Timeline for approval
Class I and Class II	2–6 months
Class III	Up to 4 years
Class III*	14–18 months
Class IV	Up to 4 years
Class IV*	18–30 months

*The manufacturer already has ANVISA GMP Certificate issued.

For Class I and II, there is no validity period for device registration; however, for Class III and IV devices, the market approval is valid for a period of five years and the license holder can request renewals for the same period, within 1 year to 6 months before the current license expires.

23.4.4.5 Regulatory action for changes and device modification

Any modification in the information provided previously for device registration may require modification and authorization from ANVISA. In some cases, a simple notification without any approval is allowed; however, most of the changes require approval. In this case, relevant documentation to the modification should be provided.

Modifications or changes in products can be under requirement, which should contain documentation, used in the original submission related to the change. Requirements can vary according to the type of the modification and main regulations should be consulted for detailed information.

23.4.4.6 Post-market surveillance

The medical device post-market is very significant in Brazil although there is no regulation in place or requirements of post-market studies as an approval condition but the process can be delayed if ANVISA requires additional information. The RDC 16/13 demands complete traceability of Class III and IV medical devices.

Medical device reporting is done through Technical Surveillance System and Sentinel Hospitals, where anybody can file reports into the Technical Surveillance System. This system is also used to receive and monitor worldwide recalls information through the ECRI Institute. Based on this information, ANVISA can contact the license holder and request further explanations about the recall or even any medical device report received through their system.

23.5 Colombia Medical Device System

23.5.1 Regulatory Authority and Medical Device Regulation

INVIMA (Instituto Nacional de Vigilancia de Medicamentos y Alimentos) is Colombia's national institute for food and drugs that regulates the medical device importation and commercialization.

Imported or domestic products should be registered in INVIMA before they can be marketed in the Colombian territory. The registration process can be done by group or families of products depending on the intended of use and characteristics and it needs to follow these major regulations:

- **Decreto 4725/05** regulates registration, permits for commercialization and techno surveillance requirements for medical devices. This regulation applies for national and foreign devices and is very close to the European Medical Device Directive—MDD 93/42/EEC.
- **Resolucion 4816/08:** post-market surveillance national program.
- **Decreto 3275/09:** requirements for usage, prescription, manufacturing, and commercialization of medical devices.

23.5.2 Definition of Medical Device

Medical Device is any instrument, apparatus, machine, software, biomedical equipment or other similar or related article, whether used alone or in combination, including its components, parts, accessories and software necessary for its proper application, proposed by the manufacturer for use in

- diagnosis, prevention, monitoring, treatment, or alleviation of disease;
- diagnosis, prevention, monitoring, treatment, alleviation or compensation for an injury or handicap;
- investigation, replacement, modification or support the anatomical structure or physiological process;
- diagnosis of pregnancy and conception control;
- care during pregnancy, birth or thereafter, including newborn care;
- products for disinfection and/or sterilization of medical devices.

Medical devices for human use shall not have the principal intended action through pharmacological, immunological, or metabolic pathways.

23.5.3 Classification of Medical Devices

Colombia's medical device classification scheme follows the European Union closely, which is a four-layer risk model (Class I, Class IIa, Class IIb, and Class III). Chapter II of the Decreto 4725/05 provides the necessary criteria and rules for product classification. Table 23.8 shows the risk and level of risk of medical devices classification in Colombia.

Table 23.8 Classification of medical devices in Colombia

Risk	Level of risk
Class I	Low
Class IIa	Medium
Class IIb	Medium-high
Class III	High

23.5.4 Registration of Medical Devices

23.5.4.1 Regulatory process

The registration process in Colombia is quite similar to the rest of the countries. Decreto 4725/05 provides all the steps and requirements for medical device registration in Colombia.

Figure 23.4 shows a high-level registration process of medical devices in Colombia.

Despite the fact that all the classes (I, IIa, IIb, and III) require approval from INVIMA for product importation and distribution into Colombian territory, the category called *Automatic Registration* is granted to products classes I and IIa; this means that once the legal and technical documentation is submitted, INVIMA will grant authorization to start commercialization. The agency will still review the application and may request additional information which must be provided within 30 working days. Colombian health authority can reject the application in case the company did not provide the required information.

23.5.4.2 Documents required

Depending on the product risk classification, the documentation requirements vary from filling specific form with technical information about the product (Class I and IIa) to extensive technical documentation (Class IIb and III). Documents required for high risk classification are listed below:

- Free Sale Certificate (FSC) or Certificate To Foreign Government (CFG) issued by the competent health authority in the country of manufacture or commercialization
- Letter of Authorization from legal manufacturer
- Device description and specification
- Detailed manufacturing process for finished devices
- Device diagram and technical drawings
- Clinical studies
- Packaging instructions
- Risk Management (ISO 14971) reports

Colombia Medical Device System | 355

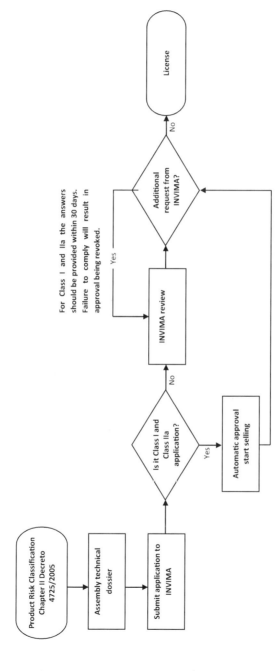

Figure 23.4 High-level medical device registration process in Colombia.

23.5.4.3 Official registration fee

Classification	INVIMA Fee (USD)
Class I	615
Class IIa	615
Class IIIb	702
Class III	702

23.5.4.4 Timeline

Table 23.9 shows the timelines to obtain approval after submission according to the risk classification.

Table 23.9 Timeline for approval after submission in Colombia

Risk	Timeline for approval
Class I	Instantly
Class IIa	Instantly
Class IIb	3–5 months
Class III	3–5 months

Market approval is valid for a period of 10 years and the license holder can request renewals for the same period of time three months before license expiration.

23.5.4.5 Regulatory action for changes and device modifications

Modification to the registration licenses is not allowed when the changes affects the functionality, safety and effectiveness of the medical device and also when the change implies any modification on the design, material chemical composition, manufacturing process or energy source. For others, modifications that involve changes in the technical content of the documentation submitted will require supporting information and the respective technical or legal document that supports the modification 30 days before the modification is effective.

23.5.4.6 Post-market surveillance

Colombia has implemented a complete post-market surveillance system (Resolución 4816/08) which guarantees the program is correctly implemented and followed by the medical device customers. Basically, it is founded on the reports of adverse events, recalls, and any event that can potentially cause death or serious injuries to the patients.

The program is divided into three levels: national, district and local. Each level has its own responsibilities, and for more information, the regulation, mentioned before, should be consulted.

Although INVIMA does not require post-market studies, the program authorizes the agency to order these studies at its discretion.

23.6 Mexico Medical Device System

23.6.1 Regulatory Authority and Medical Device Regulation

COFEPRIS (Comisión Federal para la Protección contra Riesgos Sanitarios) is the authority responsible for the control and regulation of medical devices products in Mexico in addition to drug food and beverages, tobacco products, other healthcare supplies (vaccines, blood and tissues, etc.), healthcare services, cosmetics and other consumer goods, pesticides, plant nutrients and toxic substances, national health emergencies, occupational health, and environmental risks.

Imported or domestic products are required to be registered before COFEPRIS before the product is marketed in the Mexican territory. Devices are regulated by

- **General Health Law:** articles 179, 180 and 184;
- **2010 Equivalence Agreement;**
- **NOM-137-SSA1-2012**: labeling requirements;
- **NOM-240-SSA1-2012:** post-market surveillance program.

It is important to notice that regulations are divided into mandatory standards (NOM's—Normas Oficiales Mexicanas) and voluntary standards (NMX's—Normas Mexicanas).

On August 26, 2010, COFEPRIS introduced a legislation to enable equivalence agreements with the United States, Canada, and Japan with the goal to expedite access for safe and effective medical device sold in Mexico since most of them are authorized in US, Canada, and Japan. This agreement leads to a more simplified procedure for device registration and it is applicable only for devices that are approved in countries with the equivalence agreement.

Excluding the process that are submitted under the equivalence agreement, COFEPRIS has authorized certain "third parties" to conduct reviews for the three classes of medical devices with the intention to reduce the review time up to half of it.

23.6.2 Definition of Medical Device

A medical device is a substance, mixture of substances, material, device or instrument (including computer program necessary for proper use or application); used alone or in combination with them in diagnosis, monitoring or prevention of diseases in humans; or aids in the treatment of the same and of disabilities, such as those employed in the replacement, correction, restoration or modification of the anatomy or human physiological processes. For the correct application of the criteria for the classification of medical devices based on their level of risk, these products are divided as follows:

- implantable medical device
- active medical device
- active medical device for diagnosis
- active therapeutic medical device
- invasive medical device
- surgically invasive medical device

23.6.3 Classification of Medical Devices

The risk classification is very similar to the European Medical Devices Directive MDD93/42/EEC but with some clear modifications. Table 23.10 shows the three-layer classification and its risks.

Table 23.10 Medical device risk classification in Mexico

Risk	Level of risk
Class I	Low
Class II	Medium
Class III	High

In August 2011, COFRERIS proposed a new device classification (Class 1A) with the intent to simplify regulatory process for low risk products for nearly 100 products that would be exempted for registration (e.g., tongue depressors, bandages, gauze). A comprehensive list is available on the COFEPRIS website [7]. Nonetheless, no regulation was issued for the registration of this class of product.

23.6.4 Registration of Medical Devices

23.6.4.1 Registration process

Figure 23.5 shows the registration procedure in Mexico.

23.6.4.2 Documents and labeling requirements

Documentation requirements vary according to the path chosen 1) standard process or 2) equivalence agreement. Corresponding regulations for each alternative should be consulted in order to have an accurate list of requirements.

Although there is a regulation specific for labeling requirements, that will not be in force until further notice.

23.6.4.3 Official registration fee

Table 23.11 shows registration fees related to the year of 2013.

Table 23.11 Registration fee for submissions in Mexico

Registration	Fee (USD)
Class I	701
Class II	1028
Class III	1309

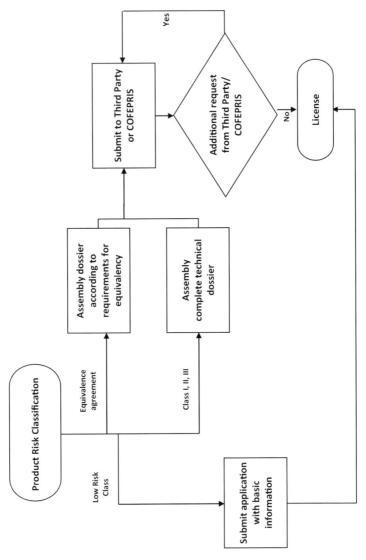

Figure 23.5 Medical device registration procedure in Mexico.

23.6.4.4 Timeline

Statutory timelines for review and approval are 30, 35, and 60 days for Class I, II, and III, respectively. However, the timeline is superior to 12 months for standard application. The review timeline varies according to the "type" of application. Table 23.12 shows a matrix comparing timelines for currently existing review processes.

Table 23.12 Matrix comparing type of submissions and its timeline

Risk class	Standard	Equivalence agreement	Third-party review
Class IA	3–4 months	—	—
Class I	10–14 months	6–8 months	6 months
Class II	10–16 months	6–8 months	6 months
Class III	10–16 months	6–8 months	6 months

Market approval is valid for a period of five years and the license holder can request renewals for the same period of time. Renew request should be done six months before the actual license expires to guarantee business continuity.

23.6.4.5 Regulatory action for changes and device modifications

Any modification in the information previously provided for device registration requires modification and authorization from COFEPRIS. In this case, relevant documentation to the modification should be provided. The modifications are split between Administrative and Technical. Requirements vary according to the type of modification. Ley General de Salud should be consulted in order to obtain detailed information regarding requirements.

23.6.5 Post-Market Surveillance

Post-market surveillance documentations are not required for regulatory approval of medical devices; however, medical devices manufacturers and importers are required to have well-documented system in place as part of the GMP requirements.

Medical device distributors are obligated to inform the secretary of health of adverse events and suspected adverse events under the techno vigilance activities.

References

1. *Latin America and Caribbean Developing*, http://data.worldbank.org/region/LAC, retrieved February 17, 2015.
2. Flood, Patricia M., *Latin American Medical Device Regulations*, Medical Device and Diagnostic Industry (MDDI), July 1, 2000, http://www.mddionline.com/article/latin-american-medical-device-regulations, retrieved February 17, 2016.
3. Raamt, Helgert V., *The Market for Medical Devices in Latin America: Leaving the Years of Unpredictability Far Behind, and Growing Steadily into an Opportunity to Build a Stable Business*, http://www.reuters.com/article/research-and-markets-idUSnBw245523a+100+BSW20150924, retrieved February 17, 2016.
4. Aslan, Fred, Huge markets for devices emerging in brazil, Latin America, Emerging MedTech Markets, *IN VIVO: The Business & Medicine Report*, October 2012.
5. PAHO website, http://www.paho.org/hq/index.php?lang=en, retrieved February 17, 2016.
6. ANMAT (Argentina), http://www.anmat.gov.ar, retrieved February, 2016.
7. ANVISA (Brazil), http://portal.anvisa.gov.br/wps/portal/anvisa/home, retrieved February 2016.
8. INVIMA (Colombia), https://www.invima.gov.co, retrieved February 2016.
9. COFPERIS (Mexico), http://www.cofepris.gob.mx/Paginas/Inicio.aspx, retrieved February 2016.

Chapter 24

Saudi Arabia: Medical Device Regulation System

Ali Aldalaan

Global Harmonization Working Party Chair,
Vice Executive President, Medical Devices Sector, Saudi FDA
4904 Northern Ring Branch Rd., Hittin Dist., Riyadh 13513, Kingdom of Saudi Arabia

amdalaan@sfda.gov.sa

Disclaimer

The regulatory information contained in this report is intended for market planning and is subject to change frequently. Translations of laws and regulations are unofficial.

24.1 Introduction

The Kingdom of Saudi Arabia (KSA) is a leading country in economy and an influential market among Middle East, North Africa, and The Gulf Cooperation Council. It is one of only a few fast-growing countries in the world with a relatively high per capita income of $23,139 (2019); therefore, the Saudi Arabian medical devices market is the biggest market in the Middle East, with growth rate of 10%. Saudi Arabia has a highly sophisticated healthcare facilities system associated with 478 hospitals occupying more than 54,000 beds, 2037 primary healthcare centers, 175 artificial kidney centers and approximately 217 private medical centers. However, the Saudi Arabia medical device market is expected to grow tremendously in the next few years due to the heavy investment in health care. The Saudi Arabian Government established a medical device regulation to comply with the revolution of medical devices technology and to safe guard the public health of Saudi Arabia.

In accordance with the Food and Drug Authority (SFDA) law issued by the royal decree (M/6) on the 13/02/2007, which gave the Food and Drug Authority the responsibility to regulate medical devices in Saudi Arabia. The council of ministers decision No. 181 issued on 18/06/2007 came to stress the importance of regulating medical devices which was issued specifically for medical devices to address the need of the rapid growth of Saudi Arabia medical devices market to ensure safety, quality and effectiveness of medical devices throughout their life cycle (pre-market, on-market and post-market).

24.2 Legislative Responsibilities

1. Providing technical support to the local manufacturers, authorized representatives and importers.
2. Set regulations, guidelines and standards for medical devices, IVDs, and radiation emitting electronic devices, which affect the human health.
3. Set quality management system requirements for all medical devices establishments and manufactures.
4. Set policies and procedures for medical devices testing.
5. Set requirements for medical devices marketing and distribution and all related issues.
6. Set requirements for medical devices advertising.

24.3 Executive Responsibilities

1. Testing of medical devices, IVDs, prescription eye glasses, contact lenses, to ascertain their quality, safety, effectiveness and compliance with regulations and standard approved by the SFDA.

2. Licensing of medical devices establishments.
3. Issuance of marketing authorization for medical devices after evaluating its technical documents.
4. Permitting the marketing of locally manufactured medical devices, IVDs, prescription eye glasses, contact lenses and their solutions when they have meet the marketing authorization requirements.
5. Establish a central reference laboratory for medical devices in SFDA headquarter, and specialized branched laboratories in the Kingdom's provinces.
6. Accreditation and licensing the private medical devices laboratories that relate to medical devices work field.
7. Establish a medical devices database, and exchange information with local, regional and international bodies.
8. Establish a main research center to conduct research and applied study relating to medical devices field.
9. Conduct researches, studies and provide consultancy services related to medical devices and cooperate with medical devices establishments, authorities, universities, scientific research centers and other institutions conducting similar activities.
10. Conduct training programs that enhance the competency of those working in the medical devices field.
11. Promote consumer awareness of medical devices and sector's responsibilities.
12. Represent the Kingdom of Saudi Arabia in regional and international authorities and organizations relating to the sector's activities.
13. Establish National Center for Medical Devices Reporting (NCMDR).
14. Establish Medical Devices National Registry (MDNR).

24.4 Surveillance Responsibilities

1. Monitor the implementation of regulations, guidelines, and procedures relating to the licensing of medical device manufacturers.

2. Inspect medical device establishments to assure their compliance with the SFDA requirements.
3. Monitor the market of medical devices, IVDs, prescription eyeglasses, contact lenses and their solutions to ensure the establishment has the establishment license and marketing authorization.
4. Follow up medical device recalls with hospitals, healthcare providers in the kingdom, manufacturers and imports and take appropriate corrective and/or preventive action.

24.5 Regulation Overview

The Saudi FDA issued medical device interim regulation on December 27, 2008 and amended by Saudi Food and Drug Authority Board of Directors decree number (4-16-1439) dated 27 December 2017; the interim regulation is replaced with the medical devices law issued by Royal Decree No. (M/54) dated 6/7/1442 AH. Thus, Medical devices may be placed on the market and/or put into service only if they comply with the applicable provisions of this medical devices law, as signified by the SFDA issuing the manufacturer with a written marketing authorization. The SFDA may exempt any medical device and shall announce the exempt medical devices on its website taking into consideration the public interest.

The main objectives of the medical device law are to protect and maintain public health within the KSA by the implementation of provisions ensuring a high level of safety and health protection of patients, users, and third parties with regard to the use of medical devices as it relates to their manufacture, supply, and use during their lifecycle and to mandate measures, and allocate responsibilities to ensure that medical devices placed on the market and/or put into service within the KSA comply with all relevant requirements and provisions of the medical devices law.

24.6 Registration Requirements

- Manufacturers established within the KSA, authorized representatives, importers and distributors of medical devices, healthcare providers importing medical devices, and any party who is involved in importing medical devices shall:
    A. register their establishments with the SFDA;
    B. list medical devices with the SFDA.
- When the manufacturer is located outside the KSA, he shall appoint an authorized representative to act on his behalf.
- The SFDA shall issue a national registry number for each establishment.
- The SFDA shall issue a listing number for medical devices.

24.7 Information to Be Provided to the SFDA

1. Each organization involved in the importation and/or the distribution of medical devices within the Kingdom of Saudi Arabia (KSA) require an establishment license, issued by the SFDA, before it undertakes such activities.
2. Such an organization is normally an 'importer' or 'distributor' but may be a local manufacturer distributing its own or another manufacturer's medical device, or an authorized representative of an overseas manufacturer involved in the importation or distribution of medical devices, or a retail pharmacy distributing medical devices.
3. Organizations involved in both importation and distribution will have both activities included on the establishment license.
4. Each organization intending to import and/or distribute medical devices in the KSA shall have obtained an Establishment National Registry Number from the SFDA, before it applies for an establishment license.

5. Each importer shall, in cooperation with the authorized representative(s) of one or more manufacturers shall decide which category(ies) or group(s) of medical devices it intend to import into the KSA and establish the contact details of the manufacturer(s) concerned.
6. Each distributor shall, in cooperation with either the local manufacturer(s) or with the authorized representative(s) of one or more manufacturers, as relevant, shall decide which category(ies) or group(s) of medical devices it distribute within the KSA and establish the contact details of the manufacturer(s) and authorized representative(s) concerned.
7. The device category is selected from: active implantable devices; anaesthetic and respiratory devices; dental devices; electro mechanical medical devices; hospital hardware; In Vitro Diagnostic medical devices; non-active implantable devices; ophthalmic and optical devices; reusable devices; single use devices; assistive products for persons with disability; diagnostic and therapeutic radiating devices; complementary therapy devices; biologically derived devices; healthcare facility products and adaptations; laboratory equipment.
8. The application for establishment licensing will be made electronically through the Medical Device Establishment Licensing (MDEL) which is found on the SFDA website and must be completed by each importer and distributor.
9. Once satisfied that the application meets the relevant requirements, the SFDA shall issue the applicant with an establishment license that is renewable annually.
10. Licensed organizations shall revise the information provided to the SFDA within 10 days of the change occurring when
    - there is a change to its contact information;
    - there is a change to the category or group of medical device it imports or distributes;
    - it adds another manufacturer to those it already represents.

24.8 Medical Device Marketing Authorization

To obtain marketing authorization, medical devices shall comply with the SFDA regulatory requirements, and additionally with national provisions.

- Marketing authorization is required for
  - All medical devices whatever their classification;
  - Contact lenses for cosmetic as well as for medical purposes;
  - Laser surgical equipment intended for cosmetic as well as medical purposes.
- A manufacturer or the authorized representative of the overseas manufacturer (if wishes) shall submit Medical Device Marketing Authorization (MDMA) application.
- Information is submitted to the SFDA using the electronic application "GHAD System—Marketing Authorization Services".
- Medical device shall comply with the "Essential Principles of Safety and Performance".
- Medical device manufacturer shall:
  - Prepare, hold and update the "Medical Device Technical Documentation" and/or "IVD Technical Documentation" that confirm to "Essential Principles of Safety and Performance"
  - Establish, document and maintain an effective quality management system (QMS) according to the ISO standard (SFDA.MD/GSO ISO 13485:2017) or any identical adopted standard for the same issue/version.
- Medical devices may be bundled/grouped within one application based on the criteria of each category below:
  1. Medical Devices:
     - Medical Devices Family
     - Medical Devices System
     - Medical Devices Procedure Pack
  2. IVD Medical Devices

  Note: TOTAL NUMBER of medical device that are grouped/bundled within a single application shall NOT EXCEED 50 items.

24.9 Medical Device Listing

To obtain an establishment license, the importer and the distributor have to indicate the categories of medical device they intend to supply to the KSA market but do not have to provide full details of the devices. Before particular medical devices within each category or group are placed on the KSA market for the first time these devices/must have been authorized by the SFDA and when these devices are imported or distributed within the KSA market, the importers and the distributors must provide listing information to the Medical Device National Registry (MDNR) for the devices concerned.

24.10 Registration Fees

24.10.1 Medical Device Marketing Authorization (MDMA)

Table 24.1 illustrates Medical Device Marketing Authorization.

Table 24.1 Medical Device Marketing Authorization

| Table for medical device marketing authorization processing fees ||||
Fee groups	The basis of the application for SFDA Marketing Authorization	Three years or less (SR)	More than three years (SR)	Lead time (working days)
FG (1)	All Class I/General IVD (other/ Exempt IVD (TGA)	15,000*	N/A	35
FG (2)	All Class II/IIa/Self-test IVD, Listable IVD	19,000	21,000	35
FG (3)	ALL Class IIb/CLASS III (CA/PAL)/Annex II List B (IVD)	21,000	23,000	35
FG (4)	All other Class III/Class IV/ AIMD/Annex II List A (IVD)/ Registrable IVD	23,000	25,000	35

*For Class I, the Medical Device Marketing Authorization issued by SFDA will be valid for three (3) years.

Note: For all other classes, the Medical Device Marketing Authorization issued by SFDA will be valid for the remaining validity of the original license or for three (3) years for license with undefined validity.

24.10.2 Medical Device Authorized Representative License

Table 24.2 illustrates the Medical Device Authorized Representative license (AR).

Table 24.2 Medical device authorized representative license (AR)

Medical Device Authorized Representative License (AR) Processing Fees	
The basis of application for SFDA authorized representative	Annual fee (SR)
Per manufacturer mandate	2,600

24.10.3 Medical Device Establishment Licensing

Table 24.3 illustrates the Medical Device Establishment license.

Table 24.3 Medical Device Establishment license

Medical device establishment license processing fees	
Establishment class	Annual fees (SR)
(A)	25,000
(B)	15,000
(C)	8,000
(D)	5,000

24.11 General Information and Documentary Evidence Need to Be Provided to the SFDA

- Where the applicant is a local manufacturer, an overseas manufacturer or the AR of the overseas manufacturer (if overseas manufacturer wishes) shall provide his company name and contact information, together with the contact details of the person responsible for the medical device MDMA application. Also, he shall provide his Establishment National Registry Number and Authorized Representative License Number

- The applicant shall provide information that will allow the medical device that is the subject of the application to be identified unambiguously. Where the MDMA application procedure covers more than one medical device type, the requirements in the above two paragraphs must be met. If they are not, the MDMA application will be rejected and the applicant must resubmit multiple applications.
- The applicant shall provide a documentary evidence that the medical device that is the subject of the MDMA application complies with the medical devices regulation, the applicant must provide evidence that the device complies with requirements specific to the KSA. These concern labeling, any a/c power supply, environmental factors, handling/transportation/storage, and advertising.

The applicant shall provide the full technical documentations that include but not limited to

- Device Description and Specification, Including Variants and Accessories
- Information to be Supplied by the Manufacturer
- Design and Manufacturing Information
- Essential Principles of Safety and Performance
- Benefit-risk Analysis and Risk Management
- Product Verification and Validation
- Vigilance and Post-market Surveillance

24.12 Labeling Requirement for Medical Device

- Labeling shall be complied with relevant requirements specified in:
    - "Essential Principles of Safety and Performance for Medical Devices" of "Guidance on Requirements for Medical Device Listing and Marketing Authorization (MDS-G5)".
    - "National Provisions and Requirements for Medical Devices".

- "Guidance on Labelling Requirements for Medical Devices (MDS-G10)".
- For home use medical devices, if the user is a lay person:
  - The label (including that on any display) and IFU shall be, wherever feasible, in both Arabic and English languages. Where this is not feasible, the language shall be Arabic.
  - The IFU shall be provided in a paper format.
  - The advertising and marketing material shall be provided in both languages (Arabic and English).
- For home use medical devices, label and/or IFU shall:
  - contain contacts information for the center(s) within the KSA providing technical assistance for users.
  - be supplemented by machine-readable forms, such as radio-frequency identification (RFID) or bar codes.
  - be written in a simple language to indicate the safe use of the medical device and all of its functions, and shall contain all necessary information and precautions
- Label and IFU shall contain an indication of any special storage and/or handling conditions that apply (e.g., when the device needs to be stored, operated, transported or used in temperatures specified by the manufacturer).
- Label shall contain an unambiguous indication of the date until when the medical device may be used safely (e.g., on medical devices supplied sterile), where this is relevant.
- Manufacturer's instructions for handling, storage, transportation, installation, maintenance (including service manuals), and disposal of the medical devices shall be made available to the SFDA or any relevant party in English and, when requested, in Arabic.
- Manufacturer shall ensure that the medical device will maintain its specified performance and will perform as intended when subject to the environmental and/or conditions of use that may be encountered within the KSA. The IFU and label shall provide information on any measures taken to accommodate those conditions, such as local temperature and humidity conditions for operating, transportation and storage.

24.13 Clinical Investigation

Means a systematic and planned process to continuously generate, collect, analyse and assess the clinical data pertaining to a device in order to verify the safety and performance, including clinical benefits, of the device when used as intended by the manufacturer.

1. To plan, continuously conduct and document a clinical evaluation, manufacturers shall:
   (a) establish and update a clinical evaluation plan, which shall include at least:
   - an identification of the essential principals of safety and performance that require support from relevant clinical data;
   - a specification of the intended purpose of the device;
   - a clear specification of intended target groups with clear indications and contra-indications;
   - a detailed description of intended clinical benefits to patients with relevant and specified clinical outcome parameters;
   - a specification of methods to be used for examination of qualitative and quantitative aspects of clinical safety with clear reference to the determination of residual risks and side-effects;
   - an indicative list and specification of parameters to be used to determine, based on the state of the art in medicine, the acceptability of the benefit-risk ratio for the various indications and for the intended purpose or purposes of the device;
   - an indication how benefit-risk issues relating to specific components such as use of pharmaceutical, non- viable animal or human tissues, are to be addressed; and
   - a clinical development plan indicating progression from exploratory investigations, such as first-in-man studies, feasibility and pilot studies, to confirmatory investigations.
   (b) identify available clinical data relevant to the device and its intended purpose and any gaps in clinical evidence through a systematic scientific literature review;

(c) appraise all relevant clinical data by evaluating their suitability for establishing the safety and performance of the device;
(d) generate, through properly designed clinical investigations in accordance with the clinical development plan, any new or additional clinical data necessary to address outstanding issues; and
(e) analyze all relevant clinical data in order to reach conclusions about the safety and clinical performance of the device including its clinical benefits.

2. The clinical evaluation shall be thorough and objective, and take into account both favorable and unfavorable data. Its depth and extent shall be proportionate and appropriate to the nature, classification, intended purpose and risks of the device in question, as well as to the manufacturer's claims in respect of the device.
3. A clinical evaluation may be based on clinical data relating to a device for which equivalence to the device in question can be demonstrated. The following technical, biological and clinical characteristics shall be taken into consideration for the demonstration of equivalence:

- technical: the device is of similar design; is used under similar conditions of use; has similar specifications and properties including physicochemical properties such as intensity of energy, tensile strength, viscosity, surface characteristics, wavelength and software algorithms; uses similar deployment methods, where relevant; has similar principles of operation and critical performance requirements;
- biological: the device uses the same materials or substances in contact with the same human tissues or body fluids for a similar kind and duration of contact and similar release characteristics of substances, including degradation products and leachable;
- clinical: the device is used for the same clinical condition or purpose, including similar severity and stage of disease, at the same site in the body, in a similar population, including as regards age, anatomy and physiology; has the same kind of user; has similar relevant critical performance in view of the expected clinical effect for a specific intended purpose.

The characteristics listed in the first paragraph shall be similar to the extent that there would be no clinically significant difference in the safety and clinical performance of the device. Considerations of equivalence shall be based on proper scientific justification. It shall be clearly demonstrated that manufacturers have sufficient levels of access to the data relating to devices with which they are claiming equivalence in order to justify their claims of equivalence.

4. The results of the clinical evaluation and the clinical evidence on which it is based shall be documented in a clinical evaluation report which shall support the assessment of the conformity of the device.

The clinical evidence together with non-clinical data generated from non-clinical testing methods and other relevant documentation shall allow the manufacturer to demonstrate conformity with the essential principals of safety and performance and shall be part of the technical documentation for the device in question. Both favorable and unfavorable data considered in the clinical evaluation shall be included in the technical documentation.

24.14 Post-Market Surveillance Requirement

24.14.1 General

SFDA takes all-appropriate measures to ensure medical devices authorized on the KSA market are comply with the requirements of the Interim Regulation and subject to post-marketing surveillance. Post-marketing surveillance comprises two activities, namely proactive and reactive activities.

1. Proactive activities including: post-marketing clinical evaluation studies, auditing the clinical investigation studies, assess the compliance of healthcare facilities with the requirements of the safe use of medical devices within healthcare facilities as well as market control activities, risk communication with healthcare providers, public and manufacturers and conducting annual monitoring plan for medical devices based on Risk assessment approach.

2. Reactive activities that include medical device adverse event management, Field Safety Corrective Actions' handling of which a medical device vigilance system is an integral part.

SFDA issued the following regulatory requirements related to proactive and reactive activities:

1. (Reporting and investigation of Incidents and Adverse Events of Medical Devices)
2. (Medical Device Field Safety Corrective Actions).
3. Post-market Clinical Follow up Studies.
4. Quality, safety and effectiveness for Medical Devices at Healthcare Facilities.

24.14.2 Proactive Surveillance and Risk Assessment Activities

Manufacturers should be obliged to have an appropriate post-market surveillance plan to investigate and assess the residual risks while the device is placed on the market. This investigation aims to collect and accumulate real-world data that address the patients' safety and the device effectiveness at the post-market phase through systematic and appropriate post-market clinical follow-up studies. Healthcare providers shall ensure the safe use of relevant medical devices to enhance the performance, effectiveness of medical devices in order to protect staff, patients, and public.

SFDA issue safety communications notices with advices and recommendations on the safe use of medical devices and to avoid any potential/associated risk may impose with the use of medical devices. Safety communication is directed to healthcare providers and/or the public. SFDA continue monitoring the safety, efficacy and performance of the medical devices placed in the kingdom through analyzing the associated risk with the identified safety signals and taking the appropriate actions towards it. Safety signals can be identified – but not limited to – the following circumstances:

- A trigger that results from any previous clinical investigation, including adverse events, or from post-market surveillance activities;

- risks identified from the literature or other data sources for similar marketed devices;
- Risks that relate to the variations on the local behavior, and/or environmental parameters.
- SFDA may also request PMCF studies in situation that raise post-market questions, during the post market phase, with the purpose of:
- Understanding the nature, severity, or frequency of suspected problems reported in adverse event reports or in the published literature.
- Obtaining more information on the device performance associated with real-world clinical practice.
- Addressing long term or infrequent safety and effectiveness issues for implantable and other devices for which the premarket testing provided limited information, and
- Defining the association between problems and devices when unexpected or unexplained serious adverse events occur after a device is marketed

24.14.3 Reactive Surveillance/Vigilance Activities

SFDA has identified the channels for reporting incidents and complaints related to medical devices by manufacturers, health care users, and other parties. Reporting MD incidents and complaints shall be through National Centre for Medical Device Reporting (NCMDR) or Saudi Vigilance site. The National Center for Medical Device Reporting (NCMDR) as a database management system contained information on medical device field safety corrective actions and incidents reports. Moreover, considered the main channel between SFDA and the concerned parties for reporting and exchange information related to MD incidents and adverse event and field safety corrective actions (FSCA). SFDA ensures the appropriate and efficient vigilance procedures by manufacturers and other concerned parties included health care provides. Authorized Representative and Importer and distributor shall have a tracking system to record the data and information of all imported and distributed medical devices within Saudi Arabia.

24.14.3.1 Medical devices' management for incidents and adverse events

Manufacturers, Authorized representatives, Importers, Distributors shall report to the SFDA any relevant adverse events, incidents, complaints, of which it becomes aware, and provide SFDA with all documents and information related to the incident and the concerned medical devices. The Reporting period to SFDA is defined as below scale:

- Reporting within two working days from the date of receiving the report and being informed of it when the accident or problem poses a serious threat to public health.
- Reporting within (10) working days from the date of receiving the notification and being informed of it, when the accident leads to death or serious injury.
- Reporting within (30) days from the date of receiving the notification and being aware of it for all accidents, problems and complaints of medical devices and products not associated with high risks or injuries from their use.
- In the event that the authority communicates with manufacturers and their legal representatives, suppliers and distributors regarding an accident report, problem or complaint related to a medical device

Manufacturer shall complete the investigation of incident and provide a final investigational report with recommended corrective or preventive action to ensure the safety, efficiency, and quality of medical devices within defined periods as below:

- Initial report: The report containing the preliminary information about the device and the incident or complaint.

  Follow-up report. It contains additional information, developments in the investigation, procedures, and information about the accident or complaint. And it must be provided in the event that the investigation of the report took more than 30 days, with clarification of the necessary justifications.
- Final report: It is the last report to be submitted on the incident and that it fulfills all the information and details of the accident, the procedures taken and the final recommendations, and is subject to the evaluation by SFDA

and determines the type of corrective or preventive action with the manufacturer in accordance with what guarantees the safety, safety and efficiency of the performance of the medical device.

24.14.3.2 Medical device field safety corrective actions

- Manufacturer/Authorized Representative or importer, shall Reporting FSCA to SFDA when the KSA is affected (imported and/or placed on the market and/or put into service in KSA) within two days.
- Manufacturer/Authorized Representative are required to submit a corrective action plan to SFDA for any FSCA affecting the Kingdom.
- Manufacturer/Authorized Representative shall implement or follow-up FSCA according to the approved plan.
- In case the KSA market is affected by FSCA, and after confirming implementing the corrective actions for all affected medical devices in the KSA and collecting all implementation proofs, then manufacturer/Authorized representative are required to provide "Confirmation Statement for Completing the Field Safety Corrective Actions (FSCA)"
- In case the KSA market is not affected by FSCA (none of the affected medical devices included in the Field Safety Notice (FSN) were imported and/or placed on the market and/or put into service in KSA), then manufacturer/Authorized representative are required provide "Statement Confirming KSA is Not Affected by FSCA".

24.14.4 Confidentiality of Information

All parties involved in market surveillance are required to preserve the confidentiality of both commercially sensitive information and of personal data in respect of providing information to parties other than those directly involved in a particular post-marketing surveillance activity.

The obligation described in the previous paragraph does not apply to reporting information to the NCMDR;

- information provided to support the manufacturer's field safety corrective action activities;
- the publication of Field Safety Notices;
- the publication of notices to support the SFDA's safeguard responsibilities;
- informing the relevant IMDRF Founding Member RA where the device concerned has been authorized to be marketed in the KSA through the provisions of the Interim Regulation;
- Providing information on relevant incident/adverse events, to other National Regulatory Authorities.

Appendix: Definitions

KSA: the Kingdom of Saudi Arabia
SFDA: Saudi Food and Drug Authority
The Board: the SFDA board of directors
Party: any natural or legal person
Medical device: any instrument, apparatus, implement, machine, appliance, implant, in vitro reagent or calibrator, software, material or other similar or related article:

A. Intended by the manufacturer to be used, alone or in combination, for human beings for one or more of the specific purpose(s) of
   - diagnosis, prevention, monitoring, treatment or alleviation of disease,
   - diagnosis, monitoring, treatment, alleviation of or compensation for an injury or handicap,
   - investigation, replacement, modification, or support of the anatomy or of a physiological process,
   - supporting or sustaining life,
   - control of conception,
   - disinfection of medical devices,
   - providing information for medical or diagnostic purposes by means of in vitro examination of specimens derived from the human body, and

B. Which does not achieve its primary intended action in or on the human body by pharmacological, immunological or metabolic means, but which may be assisted in its intended function by such means

Accessory: a product intended specifically by its manufacturer to be used together with a medical device to enable that medical device to achieve its intended purpose

Advertising of medical devices: any form of information, canvassing activity or inducement intended to promote the supply or use of medical devices

Applicant: any party established within the KSA required providing information for establishment licensing purposes

Authorized representative: any natural or legal person established within the KSA who has received a written mandate from the manufacturer to act on his behalf for specified tasks, including the obligation to represent the manufacturer in its dealings with the SFDA

CAB: a conformity assessment body (third party), established within the KSA, independent of both the manufacturer and user of the medical device that is subject to assessment

Distributor: any natural or legal person in the supply chain who, on his own behalf, furthers the availability of a medical device to the end user

Establishment any place of business within the KSA that is involved in the manufacture and/or placing on the market and/or distribution of medical devices; or acting on behalf of the manufacturer

Fully refurbished medical device: a used device that has been returned to a state which would allow it to be subject to the same conformity assessment procedures as applied to the original device

Global Harmonization Task Force (GHTF): countries working to achieve harmonization in medical device regulation among themselves; these countries are Australia, Canada, Japan, the USA, and the EU/EFTA

Importer: any natural or legal person in the supply chain who is the first to make a medical device, manufactured in another jurisdiction, available in the KSA

In vitro medical device: a medical device, whether used alone or in combination, intended by the manufacturer for the in vitro examination of specimens derived from the human body solely or principally to provide information for diagnostic, monitoring or compatibility purposes; this includes reagents, calibrators, control materials, specimen receptacles, software and related instruments or apparatus or other articles

Labeling: written, printed or graphic matter

A. Affixed to a medical device or any of its containers or wrappers
B. Information accompanying a medical device, related to identification, technical description
C. Information accompanying a medical device, related to its use, but excluding shipping documents

Manufacturer: any natural or legal person with responsibility for the design and manufacture of a medical device with the intention of making it available for use, under his name, whether or not such a medical device is designed and/or manufactured by that person himself or on his behalf by another person

Medical Devices National Registry (MDNR): the database of registered establishments and the medical devices they manufacture, import, or distribute

National Center for Medical Device Reporting (NCMDR): an organization managing a database of information on safety and performance related aspects of medical devices and capable of taking appropriate action on any confirmed problems

Placing on the market: the first making available in return for payment or free of charge of a medical device, with a view to distribution and/or use within the KSA, regardless of whether it is new or fully refurbished

Putting into service: the stage at which a device has been made available to the final user as being ready for use for the first time in the KSA for its intended purpose

Registrant: any party established within the KSA required to provide information for establishment registration or medical device listing purposes.

References

1. SFDA, Medical Devices Interim Regulation, https://sfda.gov.sa/sites/default/files/2020-09/MD-InterimRegulation-en_0.pdf.
2. SFDA, MDS-IR1 Implementing Rule on SFDA's Requirements for: Quality Management System Auditing Organization and, Conformity Assessment Bodies Conducting Medical Devices Technical Review, https://www.sfda.gov.sa/sites/default/files/2020-09/MDS-IR1e.pdf.
3. SFDA, MDS-IR2 Implementing Rule on Establishment Registration, https://www.sfda.gov.sa/sites/default/files/2020-09/%28MDS%E2%80%93IR2%29en.pdf.
4. SFDA, MDS-IR3 Implementing Rule on Medical Devices Listing, https://old.sfda.gov.sa/ar/medicaldevices/regulations/DocLib/(MDS%E2%80%93IR3)en.pdf.
5. SFDA, MDS-IR4 Implementing Rule on Establishment Licensing, https://www.sfda.gov.sa/sites/default/files/2020-09/%28MDS%E2%80%93IR4%29en.pdf.
6. SFDA, MDS-IR5 Implementing Rule on Licensing of Authorized Representatives, https://www.sfda.gov.sa/sites/default/files/2020-09/%28MDS%E2%80%93IR5%29en.pdf.
7. SFDA, MDS-IR6 Implementing Rule on Marketing Authorization, https://www.sfda.gov.sa/sites/default/files/2020-09/%28MDS%E2%80%93IR6%29en.pdf.
8. SFDA, MDS-IR7 Implementing Rule on Post-Marketing Surveillance, https://www.sfda.gov.sa/sites/default/files/2020-09/%28MDS%E2%80%93IR7%29en.pdf.
9. SFDA, MDS-IR8 Implementing Rule on Safeguard Procedures, https://old.sfda.gov.sa/ar/medicaldevices/regulations/DocLib/(MDS%E2%80%93IR8)en.pdf.

Part 4

Medical Device Regulatory System in Asia-Pacific Region

Chapter 25

Australian Medical Device Regulations: An Overview

Petahn McKenna

*ASEAN Regulatory Affairs Manager,
Johnson & Johnson Medical Singapore*

PMCKENN1@its.jnj.com

25.1 Medical Device Market in Australia

One of the wealthiest healthcare markets in Asia-Pacific, Australia's spending on health is at par with European markets such as Finland, Norway, and the United Kingdom. Similar to many other developed countries in the world, increased life expectancy, income, and demands for a higher quality of life are driving an increase in health expenditure.

Healthcare provision in Australia is made up of public and private funding: Public expenditure is funded by the Commonwealth (central) Government under a health care system called Medicare: Australia's universal, tax-financed insurance scheme covering medical, pharmaceutical, and public hospital services. Private

---

*Medical Regulatory Affairs: An International Handbook for Medical Devices
and Healthcare Products* (Third Edition)
Edited by Jack Wong and Raymond K. Y. Tong
Copyright © 2022 Jenny Stanford Publishing Pte. Ltd.
ISBN 978-981-4877-86-2 (Hardcover), 978-1-003-20769-6 (eBook)
www.jennystanford.com

expenditure is largely funded through insurance. However, Medicare also pays 75% of the schedule fee for services and procedures provided in private hospitals.

Australia's medical device manufacturing industry is also growing: There were 655 medical devices companies in Australia in 2007, 30 more than the previous year. In 2009, Australia had domestic medical device sales of AUD 6 billion and earned export revenue of AUD 1.7 billion, employing at least 17,500 people.

Table 25.1 Australian health demographics

Population (June, 2011) [1]	22,620,600	
Population Growth Rate (over previous year) [1]	1.4%	
Life Expectancy (years) [2]	82	
Economics		
GDP Per capita (USD) [3]	39,975	
GDP Growth Rate [2]	2.5%	
Healthcare statistics		
Total expenditure on health as percent of gross domestic product (2009) [3]	8.5% (2009)	8.4% (2009)
Per capita total expenditure on health (PPP int. $) [3]	3,382 (2009)	2,980 (2005)

25.2 Medical Device Regulations

25.2.1 Overview

In Australia, medical devices are regulated by the Therapeutic Goods Administration (TGA). The Australian medical device regulatory requirements, which were implemented in 2002, are largely based on the European Council Medical Device Directive (MDD) 93/42/EEC (1). Since July 2010, In vitro diagnostic devices (IVDs) have also been regulated as a subset of medical devices.

The TGA regulates medical devices via the following platforms:
(i) pre-market assessment
(ii) post-market monitoring and enforcement of standards
(iii) licensing of Australian manufacturers and verifying overseas manufacturers' compliance with the same standards as their Australian counterparts

The following text will concentrate on pre-market assessment and the requirements that must be fulfilled in order to place a medical device on the Australian market. Much of the information within this section has been extracted from the Australian Regulatory Guidelines for Medical Devices (ARGMD), accessible via the TGA website. Wherever any more detail or clarification is required, readers should always refer to the TGA website or contact the TGA directly.

For details of the Australian post-market vigilance and monitoring requirements, please refer to Section 22 of the ARGMD.

25.2.2 Regulating Authority

The body responsible for regulating medical devices in Australia is the Office of Devices Authorisation, within the Therapeutic Goods Administration (TGA). The TGA is a division of the Australian Government Department of Health and Ageing.

25.2.3 Legislation and Guidance

All regulatory decisions applicable to medical devices are made using legislation as set out in the Therapeutic Goods Act 1989, Therapeutic Goods Regulations 1990, and the Therapeutic Goods (Medical Devices) Regulations 2002.

The TGA, in collaboration with the medical devices industry sector, has developed a consolidated reference document detailing the Australian regulatory requirements for medical devices. This document is called the Australian Regulatory Guidelines for Medical Devices.

25.3 Definition of Medical Device

The Therapeutic Goods Act 1989 defines a medical device as
> any instrument, apparatus, appliance, material or other article (whether used alone or in combination, and including the software necessary for its proper application) intended by the person under whose name it is to be supplied, to be used for human beings for the purpose of one or more of the following:
> (i) Diagnosis, prevention, monitoring, treatment or alleviation of disease;

(ii) Diagnosis, monitoring, treatment, alleviation of or compensation for an injury or handicap;
(iii) Investigation, replacement or modification of the anatomy or of a physiological process; control of conception;

and that does not achieve its principal intended action in or on the human body by pharmacological, immunological or metabolic means, but which may be assisted in its function by such means*; or

(i) an accessory to such an instrument, apparatus, appliance, material or other article.

*There are therapeutic goods that have both a medicine and a medical device component, and it is the combination of the two components that deliver the desired therapeutic effect. In deciding how these products are regulated, the TGA considers

(i) the primary intended purpose
(ii) the mode of action of the product

as they relate to the definition of a medicine and a medical device.

Further guidance is provided in Section 14 of the ARGMD: Medical Devices Incorporating a Medicine.

The regulations define a medical device to be an IVD if it satisfies the following criteria:

(i) It is a reagent, calibrator, control material, kit, specimen receptacle, software, instrument, apparatus, equipment or system, whether used alone or in combination with another diagnostic product for in vitro use; and
(ii) It is intended by the manufacturer to be used in vitro for the examination of specimens derived from the human body, solely or principally for
   (a) giving information about a physiological or pathological state or a congenital abnormality; or
   (b) determining safety and compatibility with a potential recipient; or
   (c) monitoring therapeutic measures.

25.4 Classification of Medical Devices

Medical devices are classified on the basis of risk, through consideration of the manufacturer's intended use, degree of invasiveness in the human body, and duration of use. The classification follows the EU system except that the Australian system includes a separate class for active implantable medical devices. The classification levels are shown in Table 25.2.

Table 25.2 Classification of medical devices

Classification	Level of risk
Class I	Low
Class I—supplied sterile Class I—incorporating a measuring function Class IIa	Low–medium
Class IIb	Medium–high
Class III	High risk
Active implantable medical devices (AIMD)	High risk

Note: Differences between the Australian and European classification systems are detailed in Section 8 of the ARGMD.

The TGA has adopted a risk-based approach to regulation of medical devices, through consideration of the risk to the patient, versus potential benefit of the device. The level of TGA regulatory control increases with the level of risk of the medical device.

25.4.1 Classification of IVD Medical Devices

The Australian medical device regulatory framework has a separate classification system for IVD medical devices (IVDs). The manufacturer is responsible for determining the class of an IVD (as described in Table 25.3) using a set of classification rules with regard to

(i) the manufacturer's intended use of the device; and
(ii) the level of risk to the patient and the public (taking into account the likelihood of harm and the severity of that harm).

Table 25.3 Classification of IVD medical devices

Classification	Level of risk
Class 1 IVD	No public health risk or low personal risk
Class 2 IVD	Low public health risk or moderate personal risk
Class 3 IVD	Moderate public health risk or high personal risk
Class 4 IVD	High public health risk

Note: The same classification rules apply to both commercial IVDs and in-house IVDs.

25.5 Inclusion of Medical Devices on the ARTG

The majority of medical devices must be included in the Australian Register of Therapeutic Goods (ARTG) before being made available for supply in Australia. Exceptions to this requirement are devices that are supplied through one of the four mechanisms for supplying medical devices in Australia not included in the ARTG:

(i) clinical trials in Australia
(ii) authorised prescribers
(iii) Special Access Scheme
(iv) personal importation
(v) custom-made medical devices

For a medical device to be included in the ARTG, the TGA must be satisfied that available evidence demonstrates that the device is safe and effective and that an appropriate system is in place for monitoring the ongoing performance and safety of the device. As such, a sponsor can only apply to include a medical device in the ARTG if

(i) the device complies with the Essential Principles
(ii) appropriate conformity assessment procedures have been applied to the device

Applications for inclusion of a medical device in the ARTG are submitted through the TGA eBusiness (eBS) services, accessed via the TGA website, http://www.tga.gov.au.

There are different levels of access for registered users and the general public. Only an Australian sponsor can apply to include a medical device in the ARTG.

Following the inclusion of a medical device in the ARTG, the TGA issues a Certificate of Inclusion, which can be downloaded by the Sponsor from the eBS.

Class 1–3 in-house IVD medical devices are required to be entered on an in-house IVD database that is separate from the ARTG—For more information, please refer to Part 6A, Schedule 3, of the Regulations.

25.5.1 Process for Supplying a Medical Device in Australia

The following processes must be followed in order to include a medical device on the ARTG:

(i) process for all Class I non-sterile and non-measuring devices
(ii) process if the medical device is manufactured in Australia (other than Class 1)
(iii) process if the medical device is manufactured overseas (other than Class 1)
(iv) process if the device contains a medicine or materials of animal, microbial recombinant, or human origin

This chapter only outlines the latter two processes as these will be of most interest to the majority of readers. Please refer to the ARGMD on the TGA website for a detailed description of the up-to-date processes.

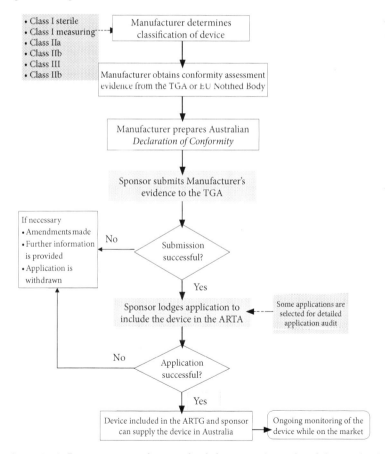

Figure 25.1 Process to supply a medical device in Australia if the medical device is manufactured overseas.

The following processes must be followed in order to include an IVD medical device on the ARTG:
(i) process for including Class 1 IVD medical devices (other than export only) in the ARTG
(ii) process for including IVD medical devices (other than Class 1) in the ARTG

For most Class I and II devices, successful electronic submission of the medical device application form is followed by inclusion of the device in the ARTG.

Applications for Class III, AIMD, and certain Class IIb devices (refer to Section 11, ARGMD) are subjected to a mandatory Level 2 application audit prior to inclusion in the ARTG whereby all supporting documentation will be submitted to the TGA for evaluation.

In addition, any application for a medical device of any Class may be randomly selected for a Level 1 Application Audit. In such cases, the sponsor is required to provide the requested documentation to the TGA within 20 working days.

Since the Australian regulatory system is closely aligned with the European system, evaluation requirements for all devices (other than those containing a medicine or material of animal, human or microbial origin) are significantly reduced for products which have already been approved for CE Marking by a European Notified Body. Products which have not been CE Marked must go though full Conformity Assessment evaluation by the TGA.

Additionally, an abbreviated approval route through the Aust-EC Mutual Recognition Agreement (MRA) is available for devices other than those containing a medicine or material of animal, human or microbial origin, which are substantially manufactured within the European Community (and have a EC legal manufacturer). Devices with MRA certificates issued by a European Notified Body may be approved by the TGA within 1 week.

Full TGA Conformity Assessment evaluation is required for devices that include human blood or plasma derivative, derivatives of animal or microbial origin, or incorporate a medicine. TGA approval for such products may involve a TGA audit of the manufacturing facility.

Once issued, TGA Conformity Assessment certificates are valid for 5 years.

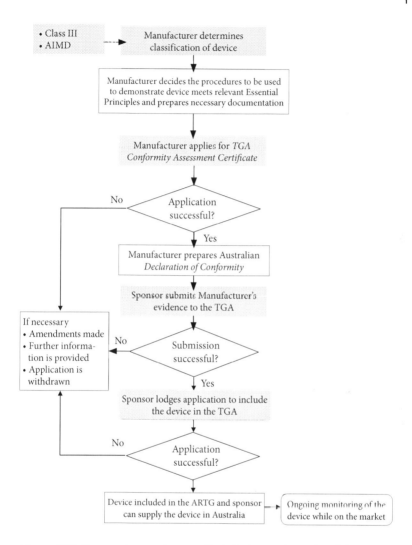

Figure 25.2 Process to supply a medical device in Australia if the device contains a medicine or materials of animal, microbial recombinant, or human origin.

25.5.2 Process for Including Class 1 IVD Medical Devices (Other Than Export Only) in the ARTG

The flowchart in Fig. 25.3 summarises the steps for including in the ARTG a Class 1 IVD that is to be supplied in Australia:

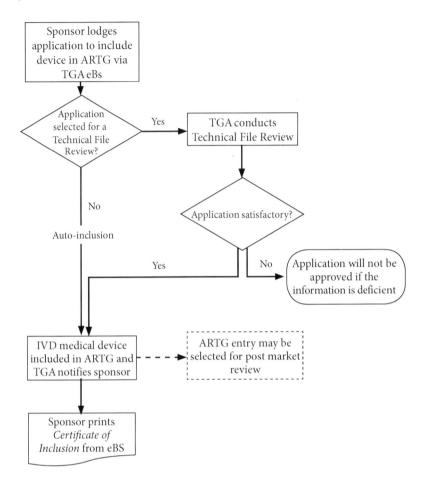

Figure 25.3 Steps for including in the ARTG a Class 1 IVD that is to be supplied in Australia.

25.5.3 Process for Including IVD Medical Devices (Other Than Class 1) in the ARTG

The flowchart in Fig. 25.4 summarises the steps for including in the ARTG an IVD that is to be supplied in Australia, other than a Class 1 IVD:

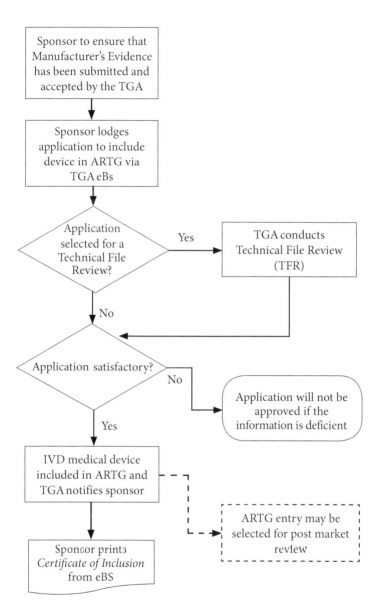

Figure 25.4 Steps for including in the ARTG an IVD that is to be supplied in Australia, other than a Class 1 IVD.

25.6 Same Kind of Medical Device (SKMD)

In the case of Class I, Class I sterile, Class I measuring, Class IIa, and Class IIb medical devices, and Class 1, Class 2, Class 3 IVDs and Class 4 immunohaematology reagents (IHRs), one medical device is considered to be of the "same kind" as another medical device, if both devices

(i) have the same manufacturer
(ii) have the same sponsor and
(iii) are the same classification
(iv) have the same GMDN code

Medical devices which are identical in respect to these criteria can be grouped under one ARTG inclusion. There is no record kept in the ARTG of the product family name, model numbers, or catalogue numbers for these classes of device.

For Class III and Class AIMD medical devices, and Class 4 IVDs (that are not IHRs) a further requirement is added to the definition of same kind of medical device—they must have the same Unique Product Identifier (UPI), described in the following section.

25.7 Unique Product Identifier

As specified in Regulation 1.6, of the Therapeutic Goods (Medical Devices) Regulations 2002, the unique product identifier (UPI) is the combination of words, numbers, symbols, or letters assigned by the manufacturer to uniquely identify the device and any of its variants. The UPI is distinct from the catalogue identifier assigned to the device.

The TGA recognizes that different manufacturers identify their product lines in different ways such as

(i) using family names to identify a range of similar devices
(ii) uniquely identifying each device with a model number
(iii) adopting a combination of both of these approaches

Figure 25.5 explains UPI showing the example of a family of prosthetic heart valves, as provided within the ARGMD. In this example, the family name does not uniquely identify all of the device models in the product range. Therefore, the term "Globus prosthetic heart valves" is not considered a UPI, because it does not distinguish between the different intended purposes of each model in the product range—atrial-versus mitral-valve replacement.

However, the model names

(i) Globus atrial prosthetic heart valve
(ii) Globus mitral prosthetic heart valve

are considered UPIs because the model/catalogue numbers are only variations of the diameter of the device that do not change its intended purpose.

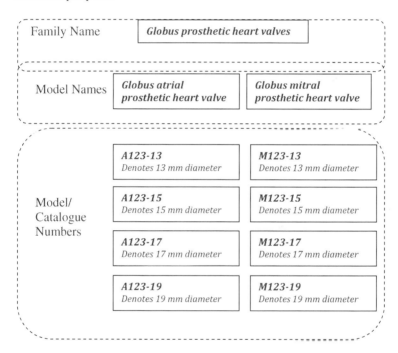

Figure 25.5 Example of a UPI provided within the ARGMD.

25.8 In vitro Diagnostic UPIs

The UPI for an IVD uniquely identifies an individual IVD, or a combination of IVDs which together constitute an IVD closed system (a combination of reagents, calibrators and quality control materials that share a common intended purpose; and are to be used only in combination with each other as components of a single assay).

25.9 Renewal

In Australia, the requirement for sponsors to renew or resubmit their regulatory approvals on a periodic basis is restricted to those supported by a TGA Conformity Assessment certificate, which requires renewal every 5 years.

Once a medical device is included in the ARTG, the sponsor is obligated to ensure the medical device remains in compliance with the Essential Principles and that changes are submitted to the TGA for evaluation as required.

Sponsors are required to pay an annual fee for each ARTG inclusion, equivalent to the original fee for inclusion on the ARTG.

25.10 Documentation Requirements

25.10.1 Conformity Assessment Applications

Manufacturers who apply for a TGA Conformity Assessment Certificate are required to prepare technical documentation to demonstrate that the medical device complies with the Essential Principles. This will vary on a case by case basis, depending on the, type of device, risk associated with its manufacture and use, and period that it has been on the market. However, the minimum technical documentation will always include the elements described in Fig. 25.6.

Documentation Required
Conformity Assessment
Copy of all current conformity assessment evidence for the medical device and/or manufacturer
Clinical evidence
Risk management records (ISO 14971)
Essential Principles compliance summary (e.g., Essential Principle checklist or similar)
Evidence to support compliance with any standards or test methods utilised for compliance (for example, test reports or assessment reports, labels and Instructions for Use)

Microbial, or recombinant origin; or medicinal substances
Evidence to support the quality and safety of animal derived material, in accordance with the TGA approach to minimising the risk of exposure to Transmissible Spongiform Encephalopathies (TSEs) through medicines and medical devices, available on the TGA website.
Drug Master File and GMP Clearance for medicine manufacturer.

Figure 25.6 Minimum documentation required, applications for Conformity Assessment Certificate.

The TGA refers sponsors to the GHTF Summary Technical Documentation for Demonstrating Conformity to the Essential Principles of Safety and Performance of Medical Devices (STED), for guidance on the technical documentation that should be assembled and submitted to demonstrate conformity to the Essential Principles.

Devices containing materials of human blood or plasma derivatives, animal, microbial or recombinant origin or medicinal substances are regulated more stringently by the TGA. As such, the TGA reserves the right to conduct an on-site audit of manufacturing facilities as part of the application process if they are not completely satisfied with the supporting documentation.

At a minimum, the TGA requires evidence to support the quality and safety of animal derived material, in accordance with the TGA approach to minimising the risk of exposure to Transmissible Spongiform Encephalopathies (TSEs) through medicines and medical devices, available on the TGA website.

25.11 Application Audits

The TGA will write to the sponsor requesting the information that is required to conduct the application audit. The TGA may ask for any documentation relating to the device and/or manufacturer (Fig. 25.7).

Level 1
- Original or correctly notarised copy of the manufacturer's Australian Declaration of Conformity
- Copy of the latest and current conformity assessment evidence for the medical device and/or manufacturer
- Information about the device, including copies of the
  label
  instructions for use
  advertising material such as brochures, web pages, advertisements

Level 2
- All the documentation listed above for a Level 1 audit
- Risk management report
- Clinical evaluation report
- Efficacy and performance data for medical devices that disinfect including sterilisation of other medical devices

Figure 25.7 Minimum documentation requirements for each level of an application audit.

25.12 Access to Unapproved Medical Devices

There are four mechanisms for accessing unapproved medical devices in Australia which do not require inclusion in the ARTG prior to supply:

(i) clinical trials in Australia
(ii) authorised prescribers
(iii) the Special Access Scheme
(iv) personal importation

Importantly, these schemes cannot be used to facilitate the commercial supply of therapeutic goods.

Further information on these mechanisms is available on a web page titled "Accessing unapproved products", on the TGA website, http://www.tga.gov.au/hp/access.htm.

References

1. Australian Demographic Statistics (June 2011). *Australian Bureau of Statistics.* Retrieved (18 May 2012) from http://www.abs.gov.au/.
2. AusMedtech (28 Nov 2011). Retrieved (22 February 2012) from http://www.ausbiotech.org/ausmedtech/.
3. Department of the Treasury. *Australia to 2050: Future Challenges, January 2010 (Intergenerational Report 2010).*
4. TGA. Australian Regulatory Guidelines for Medical Devices (ARGMD) (May 2011). *Therapeutic Goods Administration* Version 1.1. Retrieved (25 January 2012) from www.tga.gov.au.
5. TGA. Accessing unapproved products. *Therapeutic Goods Administration.* Retrieved (25 January 2012) from http://www.tga.gov.au/hp/access.htm.
6. World Health Organisation. Global Health Observatory (GHO)(2011). *Country Statistics* 2011. Retrieved (18 January 2012) from http://www.who.int/gho/countries/en/.
7. Year Book Australia (2008). *Australian Bureau of Statistics.* Retrieved (22 January 2012) from http://www.abs.gov.au/ausstats/abs@.nsf/0/A50BD9743BF2733ACA2573D2001078D8?opendocument.

Chapter 26

China: Medical Device Regulatory System

Jack Wong

Founder of Asia Regulatory Professionals Association

Jack.wong@arpaedu.com

Disclaimer

The regulatory information contained in this chapter is intended for market planning and is subject to change frequently. Translations of laws and regulations are unofficial.

26.1 Introduction

China's overall registration framework is under revision and amendment, and certain regulation outlined below may have been revised recently and was updated to September 30, 2020. Certain explanation of the regulation may also be updated accordingly. Please refer to the most updated regulation from the National Medical Products Administration official Chinese website or consult your local regulatory affairs staff/consultant for up-to-date explanation and practice.

Medical Regulatory Affairs: An International Handbook for Medical Devices and Healthcare Products (Third Edition)
Edited by Jack Wong and Raymond K. Y. Tong
Copyright © 2022 Jenny Stanford Publishing Pte. Ltd.
ISBN 978-981-4877-86-2 (Hardcover), 978-1-003-20769-6 (eBook)
www.jennystanford.com

26.2 Market Overview

China's medical device market is forecasted to have a value of US$ 89.8 billion in 2019 and estimated US$ 109.3 billion in 2020 with strong growth in two-digit annual growth rate. China has the world's largest population. In 2020, it was officially estimated at 1.4 billion.

Data showed that the total investment on healthcare was increased through 2009–2019 significantly (Fig. 26.1). In 2019, the total investment was RMB 3,015,055.96 billion (USD 430,722 billion).

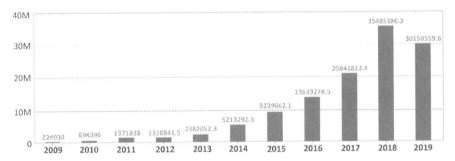

Figure 26.1 Total investment on healthcare through 2009–2019 (RMB 100 million).

The prospects for medical device spending are huge. The government has committed heavily in the construction of thousands of hospitals, healthcare centers and clinics, and this will inevitably lead to spending on capital goods, most notably medical devices.

26.3 Overview of Regulatory Environment and What Laws/Regulations Govern the Medical Devices

China's regulatory system for medical devices is still in its rapid development phase, which unavoidably leads to a certain degree of difficulty to catch up on the latest regulatory requirement. Language can also be a major problem for foreign manufacturers as most of regulation and standard are in Chinese. Product registration submission files are also required to be in Chinese as well.

The medical device regulatory system is made up of administrative documents and technical documents. Currently, the effective regulatory framework in China is the second generation.

There are three levels of legislation in administrative documents, all of which are considered mandatory and apply to both local and foreign manufacturers:

- Regulations for the Supervision and Administration of Medical Devices, Decree No. 650 of the State Council and were effective as of June 1, 2014, was updated to Decree 680 since May 4, 2017, on some of the items.
- NMPA Orders, which provided the details for lifecycle management and implementing the Regulations.
- Normative documents of the central or local DAs

Regulations for the Supervision and Administration of Medical Device basic content include:

- Chapter I: General Provisions
- Scope: This Regulation was formulated in order to ensure the safety and efficacy of medical devices and protect the health and safety of human body. The research, development, manufacturing, trading and usage of medical devices, and their supervision and administration within the territory of the People's Republic of China shall comply with this Regulation.
    - Competent authority: NMPA and local DAs
    - Classification: Class I, II and III
- Chapter II: Registration and Filing of Medical Devices
    - Filing of Class I medical devices
    - Registration of Class II and Class III medical devices
    - Application documents
    - Acceptance, review and approval
    - Registration variation
    - Registration renewal

○ Requirements of clinical evaluation
○ Clinical trial notification for clinical trial of high-risk medical devices
- Chapter III: Manufacture of Medical Devices
- Chapter IV: Distribution and Use of Medical Devices
- Chapter V: Handling of Adverse Events and Recall of Medical Devices
- Chapter VI: Supervision and Inspection of Medical Devices
- Chapter VII: Legal Responsibilities
- Chapter VIII: Supplementary Provisions

There are two main types of public technical documents:

- Guidance, all of which are considered recommendatory
- Standards, some of which are considered mandatory, while some of which are considered as recommendatory.

26.3.1 Guidance

The current valid regulations are the second generation. NMPA has issued quite a lot of regulations and guidance documents for application and reviewing, most of which are about how to prepare registration application documents for a certain kind of medical devices under different pathway, also includes how to conduct a clinical trial and how to write user manuals, etc. It is very useful for applicants to get access to these guidance documents, which can be available on the NMPA website www.nmpa.gov.cn, and there is English-friendly website for international visitors, http://english.nmpa.gov.cn.

Technical guideline for medical device etc. can also be available on the website of the Center for Medical Device Evaluation (www.cmde.org.cn).

26.3.2 Standards

China has National Standard (GB xxxxx or GB/T xxxxx) and Industry Standard (YY xxxx or YY/T xxxx) for medical device, including Mandatory standard (GB xxxxx or YY xxxx) and Recommended standard (GB/T xxxxx or YY/T xxxx). A medical device must follow China Mandatory standard specification. More than 35% of all IEC and ISO standards have now been adopted by China, but many are not direct transpositions and contain China-specific requirements. Table 26.1 lists some of the main medical device related international standards and their Chinese equivalents. The standards ISO 14155 and ISO 13485 have also been adopted.

Table 26.1 Primary medical device related international standards and their Chinese equivalents

International standard	Chinese equivalent
ISO 13485: 2016	YY/T 0287–2017
ISO 14971: 2007	YY/T 0316–2016
IEC 60601-1: 2012:2012	GB 9706.1–2020
IEC 60601-1-2: 2004	YY 0505–2012
ISO 10993-1:2009	GB/T 16886.1–2011

National medical devices standards can be found at National Standardization Technical Committee's website (www.sac.gov.cn).

Industry standards can be found at the websites of on the NMPA website www.nmpa.gov.cn and the Center for Medical Device Evaluation (www.cmde.org.cn).

26.4 Regulatory Body

The NMPA is directly under the State Administration for Market Regulation, and is in charge of the safety, standards management, regulate the registration, quality management, post-market risk management, organize and guide the supervision and inspection, engage in international exchange and cooperation in the regulation of drugs (including traditional Chinese medicines (TCMs) and ethno-medicines, the same below), medical devices and cosmetics.

The detailed organization chart of NMPA can be found at http://english.nmpa.gov.cn/NMPAorganizations.html.

NMPA contact details
No. 1 Beiluyuan Zhanlan Road, Xicheng District, Beijing, China
Postcode: 100037
Tel: 8610-6831-1166

26.5 Regulatory Overview

26.5.1 Definition of Medical Device

Medical devices refer to the instrument, apparatus, appliance, in vitro diagnostic reagents and calibrators, materials, and other similar or related articles including the needed computer software that are directly or indirectly used on human body; their efficacy are mainly obtained through physical and related means rather than pharmacological, immunological or metabolic means that can only play a supporting roles despite their possible involvement; the purposes of medical devices are as follows:

1. Diagnosis prevention, monitoring, treatment or mitigation of the diseases
2. Diagnosis, monitoring, treatment, mitigation or functional compensation of the injuries
3. Testing, replacement, modification or support of the physical structure or physiological processes;
4. Life support or maintenance
5. Control of conception
6. Examination of human samples to provide information for medical or diagnostic purposes

The institutional users of medical devices refer to institutions that use medical devices to provide others with medical and other technical services, including medical institutions with the practice license of medical institution, family planning technical service institutions with the practice license of family planning technical service institution, and blood bank, plasma collection stations, rehabilitation and assistive device adaptation institutions that do not require the practice license of medical institution by the law.

26.5.2 Classification of Medical Devices

Classification for medical device in China is not the same as those used in the European Union or USA, and the Class III classification is much broader than many manufacturers may be used to.

There are many reasons for these classification differences in China. The two key reasons are

- historical reason, i.e., some devices are classified as different classification or even as drug even before medical device regulation came into being and hence the device may follow the old classification;
- social reason, i.e., the NMPA may consider some device to have higher risk in China, e.g., after some adverse events.

For further information on the classification of devices, readers are referred to "Rules for Classification of Medical Devices (Decree No. 15 of SFDA)" and "Classification Catalogue of Medical Devices." Also, the NMPA publishes the announcements of classification definition or adjustment for some devices per needs. It has recently considered adjusting the classification rules and catalogue of medical devices.

For a newly developed medical device not yet listed in the classification catalogue, the applicant may directly apply for a class III medical device registration, or may determine the classification of the product according to the classification rules and apply to The Center for Medical Device Standardization Administration of NMPA for a class confirmation, then apply for product registration.

When a class III medical device registration is directly applied for, NMPA will determine the class according to its risk degree after review.

The process of pre-market medical device administration depends on the device's classification. For Class I medical device, filing process is conducted. NMPA announced a list for devices identified as class I, for those not listed in the list, sponsor should apply for classification from Center for Medical Device Standardization Administration of NMPA before going through the filing process.

For Class II or III medical device, registration process is conducted.

The handling of administration matters relating to medical devices is also determined by the device's classification. For locally produced devices,

- Class I devices are administered by a city level regulatory authority;

- Class II devices are administered by a provincial/municipal regulatory authority;
- Class III devices are administered by center DA which is called NMPA.

All imported medical devices are administered by NMPA.

26.5.3 Filing Process for Class I Medical Device

Each Class I medical device, which is in accordance with catalogue of Class I medical devices of China, should conduct the filing process before it can be placed on the market in China. For locally produced devices, the filing entity should provide the application documents to city level regulatory authority for filing.

For imported Class I medical devices, the filing entity should be submitted through its representative office established in China or an enterprise legal person in China designated by it as its agent, submit the documents to NMPA for filing.

An imported medical device being applied for filing should be one which has already got permission for distribution in the country (region) where the filing entity is registered, or the manufacture is carried out. If the product is not managed as medical device in the country (region) where the filing entity is registered or the manufacture is carried out, the filing entity should provide relevant proof documents, including the permission for distribution of sold product in the country (region) where the filing entity is registered, or the manufacture is carried out.

For every imported medical device, before filing, the registrant should establish the Product Technical Requirements as the first step. The Product Technical Requirements should be written in accordance with the Guidance for Writing Product Technical Requirements of Medical Devices (CFDA Announcement No. 9, 2014). Then, the registrant should make sure the provided specification and method meets China related standard and guideline, and also comply with product design framework through conducting self-testing or consign other institution or company testing to ensure that the product passes the test as per the Product Technical Requirements and make a clinical evaluation where no clinical trial is needed. Once the NMPA Administrative Acceptance Service Center has all the required information, the filing process has been completed.

NMPA will publish the filing information on the website. The documents needed to submit for filing include

- filing Information Form for Class I medical device;
- risk analysis of the product;
- product Technical Requirement;
- self-testing or consignment testing report of the product;
- clinical evaluation;
- instructions for use and sample of label of the product;
- manufacture information;
- approval documents;
- declaration of conformity.

NMPA Announcement No. 26, 2014 provides more detailed information about the filing process. For a filed medical device, if any information that recorded in the filing information list or the Product Technical requirement change, the filing entity should apply a filing alteration process to NMPA.

26.5.4 Registration Process for Class II or III Medical Device

In order to standardize the administration of medical device registration and filing, "Provisions for Medical Device Registration" was promulgated by Decree No. 4 of CFDA and effective as of October 1, 2014. For the registration of In-Vitro Diagnostic Reagent, "The Provisions for In-Vitro Diagnostic Reagent Registration" (Decree No. 5 of CFDA) should be followed.

26.5.4.1 Medical device registration certificate

Each Class II or III medical device or medical device family should have a Medical Devices Registration Certificate before it can be placed on the market in China. The certificate is owned by registrant and must be renewed every 5 years (the old regulation required the expiry date after a four-year term). The precise requirements for product registration vary from device type and degree of complexity but can include sample testing, clinical evaluation/investigation, evaluation and site inspection.

For imported medical devices, the applicant should, through its representative office established in China or an enterprise legal person in China designated by it as its legal agent, submit the application documents to NMPA for Registration.

An imported medical device being applied for registration should be one which has already got permission for distribution in the country (region) where the applicant is registered or the manufacture is carried out. If the product is not managed as medical device in the country (region) where the applicant is registered or the manufacture is carried out, the applicant should provide relevant proof documents, including the permission for distribution of sold product in the country (region) where the applicant is registered, or the manufacture is carried out.

For every imported Class II or III medical device, before registration, the applicant should write the Product Technical Requirements as the first step. Then, the applicant should arrange product Registration Testing by the national Testing Centre to ensure that the product passes the test as per the Product Technical Requirements and make a clinical evaluation, where a clinical trial is needed or not needed. Nowadays, the submission is under e-submission pathway. Once NMPA has all the required information, the application is passed to the technical review stage at Center for Medical Devices Evaluation (CMDE).

CMDE is responsible for reviewing the application documents and examining the evaluation of the applicant for the safety and effectiveness of the medical device. If necessary, CMDE may consult external expert panel in the form of a meeting or just letters. After reviewing the application documents, CMDE may issue a notice for supplementary documents to ask some questions or additional information to the appliance. The appliance has 1 year as maximal time to supplement the documents through e-submission.

Both the notice from CMDE and the supplement from the appliance just have one Chance. During the review process, if necessary, CMDE may notify the Center for food and Drug Inspection (CFDI) of NMPA to carry out the inspection according to GMP regulations.

When the supplementary documents are submitted, CMDE will continue the review process. Then CMDE provide the review comments to the Department of Medical Devices Registration at NMPA, then to the Director of the Medical Devices Registration Department, and then to the Director of NMPA for final approval.

Finally, the result of the application and certificate of approval is sent back to the Application Receiving Office for collection by the applicant. If the application documents and supplementary documents are not meet safety and effectiveness requirements, NMPA will make a disapproval decision.

A simplified flow chart depicting the process for product filing or registration is shown in Fig. 26.2.

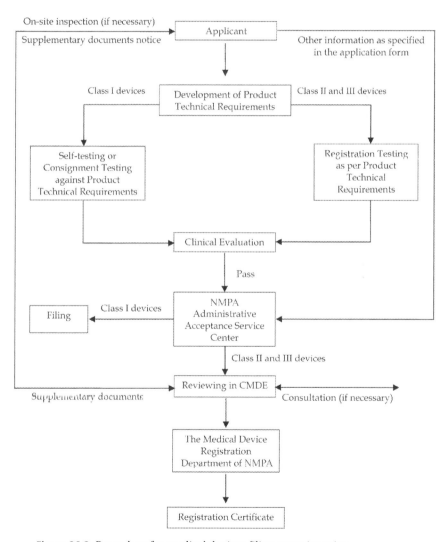

Figure 26.2 Procedure for medical devices filing or registration.

26.5.4.2 Documents needed for registration application

According to CFDA Announcement No. 43, 2014, the documents needed to submit for Class II or III medical device registration application include
- application form;
- approval documents;
- list of basic requirements for safety and effectiveness of medical devices;
- summary documents:
- summary;
- product description;
- model;
- packaging;
- applied range and contraindications;
- congeneric or previous products for reference (if any);
- other contents need explanations;
- research documents:
- product performance;
- biocompatibility evaluation;
- biosafety evaluation;
- sterilization and disinfection process;
- effective period and packaging;
- animal experiment;
- software application;
- others;
- manufacture information:
- passive product/active product production process;
- production site;
- test report:
- test report;
- pre-evaluation opinion;
- instructions for use and sample of label of the product:
- instruction for use;
- pattern of label for minimum sales package;
- declaration of conformity.

For the application documents of In-Vitro Diagnostic Reagent, CFDA Announcement No. 44, 2014 should be followed.

26.5.4.3 Product technical requirements and registration testing

The applicant should prepare the Product Technical Requirements for the medical device to be registered. Product technical requirements mainly include the performance indicators and testing method of the finished medical device. Performance indicators refer to functional and safety indicators and other quality control indicators of finished medical devices that can be measured objectively. Generally, performance indicators are not including biocompatibility tests. The performance indicators should follow China National Standard and Industry Standard as well as requirement from other specific guideline.

In the absence of China standard, the applicant can transform an ISO/IEC standard to Product Technical Requirement. The Product Technical Requirement should be prepared in accordance with Guidance for Writing Product Technical Requirements of Medical Devices (CFDA Announcement No. 9, 2014). The medical devices marketed in China should comply with the Product Technical Requirements which are approved by NMPA.

When applying for class II and class III medical device registration, registration testing should be carried out in accordance with the Product Technical Requirements by the medical device testing institution (Testing Center), which must have the relevant qualification of the applying medical device testing. And the medical device testing institutions will carry out pre-evaluation of the Product Technical Requirement.

The products being tested should represent the safety and effectiveness for other products in the same registration unit. Clinical trials or registration application can be carried out only after the products passed the registration testing. There is 1-year validation duration for testing report to initiate local clinical trial since the report is being issued.

26.5.4.4 Biocompatibility evaluation

The applicant should submit the evaluation documents in the research documents, generally following the GB/T 16886 idt. ISO 10993 series standards, on the biocompatibility of materials in finished products directly and indirectly contacting patient and user.

Biocompatibility documents should include

- basis and method for biocompatibility evaluation;
- description of materials and the nature of contacting human body;
- reason and demonstration for execution or exemption of biological test;
- evaluation on existing data or test result.

When biological test is involved in medical device biocompatibility evaluation, the biological test report could be submitted as research documents by applicants. All biological tests should be entrusted with the biological laboratory with medical devices testing qualification, within whose qualified test field. The tests should be performed in accordance with the relevant standards. If the test report is issued by a foreign biological laboratory, a quality assurance document about that which is in accordance with GLP laboratory requirements should be submitted.

26.5.4.5 Clinical evaluation and clinical trials

There are three clinical evaluation paths for Class II or III medical devices:

- Clinical Evaluation of Products Listed in "The Catalogue of Medical Devices Exempted from Clinical Trial"
- Analysis and Evaluation through Data Obtained in Clinical Trials or Clinical Use of Medical Devices of the Same Kind approved in China
- Clinical evaluation/clinical trials
    - For a medical device for which the clinical trial is conducted within the territory of China, the clinical trial shall be conducted by a qualified clinical trial institution following requirements of GCP for medical devices. When applying for registration, the registration applicant shall submit the clinical trial protocol and the clinical trial report.
    - For an import medical device the clinical trial of which is conducted overseas, the clinical trial shall comply with related technical requirements in relevant Chinese laws and regulations and registration guidance, The

applicant shall also submit related supportive information concerning whether the clinical performance and/or safety of the product have ethnic differences.
- If the applicants use multi-center clinical trial data, synchronously carried out in our country and overseas, as registration application materials, allocation basis for trial size in China should be clearly indicated so as to further evaluate conformity to relevant requirements for registration in our country.
- For high risk medical device, clinical evaluation pathways above or local clinical trial will be needed. Please be noted that a clinical trial notification (CTN) prior local clinical trial will be needed.

As an additional requirement, the applicant of imported medical devices should submit the clinical evaluation data upon which foreign competent authority has approved the product to be sold in the market.

Clinical evaluation data should be submitted in accordance with "Guidance for Clinical Evaluation for Medical Devices" (CFDA Announcement No. 14, 2015).

26.5.4.5.1 Clinical Evaluation through proving the accordance with the catalogue of medical devices that are exempted from clinical trial

For the clinical evaluation of products listed in "The Catalogue of Medical Devices Exempted from Clinical Trial" (hereinafter referred to as "the Catalogue"), the registration applicant shall submit comparative information about the products under application and the contents in the Catalogue and also a description of comparison of the products under application and medical devices in the Catalogue having been approved for registration within the territory of China. Requirements about clinical evaluation information that shall be submitted are detailed as follows:

(I) Comparative information about the products under application and the contents in the Catalogue shall be submitted;

(II) A description of comparison of the products under application and medical devices in the Catalogue having been approved for registration within the territory of China shall be submitted. The description shall include a Comparative Table of the Product under Application and Medical Devices in the Catalogue Having Been Approved for Registration within the Territory of China (see Table 26.2) and corresponding supportive information.

Table 26.2 Comparative table of a product under application and medical devices in the catalogue having been approved for registration within the territory of China

Comparison item	Product in the catalogue	Product under application	Difference	Summary of Supportive Information
Basic principle (working principle/ action mechanism)				
Structural composition				
Manufacturing material of product or manufacturing material of human contact part				
Performance requirements				
Sterilization/ disinfection method				
Scope of application				
Application method				
...				

Note: More comparison items may be included, as appropriate.

The above information submitted shall prove the product under application is equivalent to the products in the Catalogue; otherwise, corresponding work shall be conducted following other requirements in this guidance.

26.5.4.5.2 Clinical Evaluation through the analysis and assessment made on the basis of the data obtained from clinical trial or application of a medical device of the same kind

A medical device of the same kind refers to a product having been approved for registration within the territory of China that is basically equivalent to the product under application in terms of basic principle, structural composition, manufacturing material (manufacturing material in contacting with the human body for active products), manufacture technology, performance requirements, safety evaluation, standards-compliance and intended use.

A product under application and a medical device of the same kind may be considered basically equivalent if the difference between will not exert adverse influence on safety and efficacy of the product.

For analysis and evaluation through data obtained in clinical trials or clinical use of medical devices of the same kind to prove a medical device is safe and effective, the registration applicant shall first compare the product under application to one or more medical devices of the same kind to prove they are basically equivalent.

The comparison items shall include but not be limited to the items listed in below table; the compared contents shall include the qualitative and quantitative data, the verification and validation results. The similarities and differences between the two shall be expounded. Whether the differences between the product under application and products of the same kind impact the safety and efficacy of the product under application can be verified and/or validated through nonclinical study data, clinical literature data, clinical experience data and clinical trials on differences conducted within the territory of China. Collection, analysis and evaluation of related data shall comply with related requirements. Clinical trials shall comply with related GCP requirements.

The registration applicant shall provide comparative information in a tabular form (Tables 26.3 and 26.4). In case of any item inapplicable, the inapplicability shall be justified.

Also, refer to the evaluation process illustrated in Fig. 26.3.

Table 26.3 Comparison items of a product under application and medical devices of the same kind (passive medical devices)

Passive medical devices	Comparison item
	1. Basic principle
	2. Structural composition
	3. Production technology
	4. Manufacturing material (such as material mark, animal-derived material, allograft material, ingredient, pharmaceutical ingredient, bioactive substance, valid standard and other information)
	5. Performance requirement
	6. Safety evaluation (such as biocompatibility, biosafety, etc.)
	7. Valid national/industry standard for product
	8. Scope of application
	(1) Applicable people
	(2) Applicable site
	(3) Contact mode with human body
	(4) Indications for use
	(5) Applicable illness phase and degree
	(6) Operation environment
	9. Application method
	10. Contraindications
	11. Precautions and warnings
	12. Delivery status
	13. Sterilization/disinfection mode
	14. Packaging
	15. Labeling
	16. Instruction for use

Table 26.4 Format of comparative table of a product under application and medical devices of the same kind

Comparison item	Medical device of the same kind	Product under application	Difference	Summary of supportive information
Basic principle				
Structural composition				
...				
...				
...				

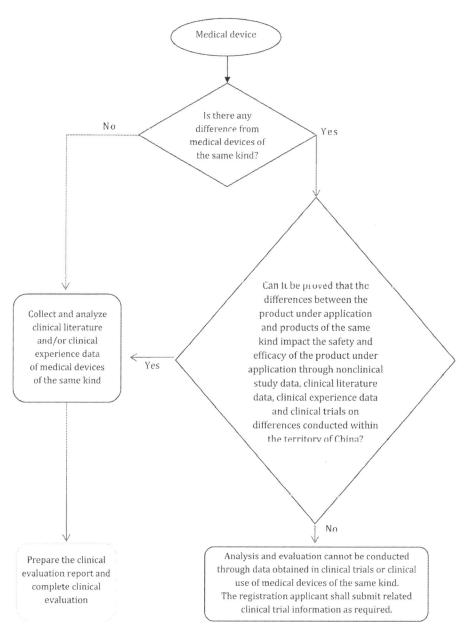

Figure 26.3 Clinical evaluation through the analysis and assessment made on the basis of the data obtained from clinical trial or application of a medical device of the same kind.

26.5.4.5.3 Clinical evaluation through a clinical trial

The clinical trial of medical devices should be conducted within a qualified clinical trial institution according to the requirements of the Good Clinical Practice (GCP) for medical devices, and shall be filed with the local FDA of the province where the applicant, or its agent, for clinical trial is located.

If clinical trial is required, the clinical evaluation materials to be submitted should include clinical trial protocol and clinical trial report.

For a medical device registration, an overseas clinical trial may be accepted by NMPA:

Overseas clinical trial materials submitted by the applicants should include at least: clinical trial protocols, ethic opinions and clinical trial reports. Clinical trial reports should contain analysis of the complete clinical trial data and conclusions.

According to the clinical evaluation pathways selected by the applicants, overseas clinical trial data might be provided either as clinical trial materials or as verification materials used to prove differences with the same variety of devices will not have negative influences on the safety and effectiveness of the products. Generation process of clinical trial data for the latter includes: data generated from overseas clinical trials regarding differences with the same variety of devices; and overseas clinical data already possessed by the applicants covering clinical trials regarding differences with the same variety of devices.

Overseas clinical trial data that meets relevant requirements for registration in our country and which is scientific, complete and sufficient will be accepted. For overseas clinical trial data that meets the basic requirements set out in this Guideline, however, supplementary clinical trials can be carried out within or outside our country when some documents are required to be supplemented according to the relevant technical requirements for registration of our country; if the supplementary trial data and the previous overseas trial data comply with relevant technical requirements for registration in China, these data shall be accepted.

26.5.4.5.4 Notification of a clinical trial for high-risk medical device

If a Class III medical device poses high risks to human body, the local clinical trial of the device should be subject to NMPA for notification before initiation. The catalogue of class III medical devices subject to clinical trial notification is compiled, adjusted, and published by NMPA.

The examination and notification of a clinical trial is a process that NMPA conducts a comprehensive analysis to the degree of risk the device may entail, clinical trial protocol, comparative analysis report on clinical benefit and risk of the device to be put on clinical trial.

If a clinical trial notification is required, the applicant should submit the application documents to NMPA in accordance with related requirements. The application documents will be reviewed by CMDE of NMPA. If the application of clinical trial is accepted, the Letter for Clinical Trial of Medical Device will be issued. If the application is disapproved, the reasons will be stated in written form.

Clinical trials of medical devices should be conducted within 3 years after CTN letter is issued. If it is not conducted in due time, the original letter should be invalid automatically. If the clinical trial still needs to be conducted, a new application should be submitted.

26.5.4.6 Enforcing GMPs

When conducting review process for an imported class II or class III medical device, if the CMDE of NMPA deems it is necessary to conduct a quality management system inspection, it will notify the Center for food and Drug Inspection (CFDI) of NMPA to carry out the inspection according to GMP regulations. The time for quality management system inspection will not be calculated in the overall evaluation time.

NMPA officially implemented the new medical device GMP regulations as a trial in 2009. In order to meet the new regulations and provisions for manufacture supervision, NMPA revised the GMP regulations and published it on December 29, 2014. The revised GMPs are divided into 13 chapters covering institutions and personnel, plant and facilities, equipment, document management, design and development, procurement, production management, quality control, and other sections equivalent to international GMP requirements. Also, NMPA published several appendixes of GMP regulations to specify the requirements for some certain types of medical devices one after another.

The most detailed sections of the new regulations cover production management. The production management guidelines require manufacturers to devise, implement and document production processes at all steps under their control and formulate guidelines for each device they manufacture. Each product must be marked and monitored during the entire production process to identify and prevent improper use of the device prior to or after official release.

To enforce the new GMPs, NMPA is using a risk-based approach similar to that used by its US counterpart. This means that high-risk devices such as Class III and sterile Class II devices are being prioritized for NMPA review during the initial inspection and review process. After NMPA has expanded its inspection teams and processed the backlog of high-risk inspections required during the first phase of this endeavor, other manufacturers can expect to be visited.

26.5.4.7 Timeframes

Figure 26.4 lists the timeframes involved in each activity for obtaining registration of a product or approval for a clinical trial. Please note this is just an example, and timeline will be varies from case by case.

This is just a rough timeline reference; the timeline actually varies from product to product, and case by case. The review cycle time will be longer if a panel meeting will be required. The review time set in regulations for registration alteration is the same as registration at the first time. However, the real time spent may be less, because that both the preparation of documents and CMDE review mainly aimed at the change part, and the variation application will not go to NMPA registration department.

26.5.4.8 Registration alteration

For a registered class II or class III medical device, if the content of the medical device registration certificate and its attachment(s) has an alteration, the registrant should apply to the original registration department for registration alteration and submit application materials according to relevant requirements.

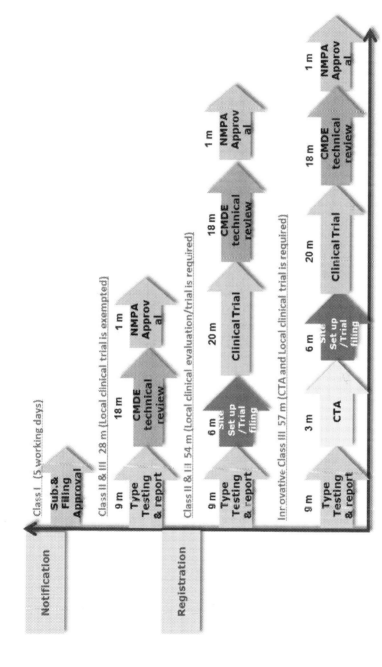

Figure 26.4 NMPA review system and process.

If the alteration is for product name, model, specifications, structure, and components, scope of application, product technical requirement, etc., and for an import medical device, production address, the registrant should apply to the original registration department for change of permission items. If the alteration is for name, residence of the applicant and the agent, the registrant should apply to the original registration department for change of registry items.

The medical device registration alteration document should be used in conjunction with the original medical device registration certificate and its period of validity should be the same with that of the original registration certificate. After obtaining the registration alteration document, the registrant should alter the product technical requirements, instructions and labels by himself.

26.5.4.9 Registration renewal

If it is required to extend the period of validity of the medical device registration certificate every 5 years, the registrant should submit application for registration renewal to the medical products regulatory department at least 6 months before the expiration of period of validity and submit application documents according to corresponding requirements.

26.5.4.10 The registration fee

Table 26.5 lists the fee for registration, alteration, renewal or the approval of clinical trial.

Table 26.5 Fee for registration, alteration, renewal or the approval of clinical trial

Category		Fee (×RMB 10K)	
		Domestic	Importing
Class II	New application	Provincial standard	21.09
	Variation	Provincial standard	4.20
	Renewal (every 5 years)	Provincial standard	4.08
Class III	New application	15.36	30.88
	Variation	5.04	5.04
	Renewal (every 5 years)	4.08	4.08
	CTN	4.32	4.32

26.5.4.11 Innovative medical devices

NMPA encourages industry innovation. In No. 13 normative document issued in 2014, NMPA states that applicants can apply for special approval procedures for innovative medical devices. The conditions for this procedure are

- through technology innovation, the applicant has invention patent on the core technology of the device by law, or own the patent or rights to use the patent in China by law; or its application for invention patent on the core technology of the device is published by patent administration department under State Council;
- the main working principle/mechanism is domestic initiative in China, product performance or safety has significant improvement compared to other similar products, and the technology used is at international advanced level and has significant clinical application values; and
- the applicant has finished preliminary research and has finalized product, research process is real and well controlled, and research data has excellent integrity and traceability.

After NMPA receives application on special approval procedures on innovation medical devices, innovative medical device review office will conduct expert review process and give comments within 40 working days after acceptance. Once identified as an innovative medical device, it will have priority in reviewing and approval process.

26.5.4.12 Drug/device combination products

For a drug/device combination product, if the main function part is drug, then it will register as a drug, and vise verse. If the product has not been approved in China, applicant should seek for product attribute definition from CFDA Administrative Acceptance Service Center before submitting registration application documents.

For first imported combined product, the drug should have been registered in China or approved to launch in its country of origin. For such products, CMDE may evaluate and review together with Center for Drug Evaluation of MMPA.

26.6 Monitoring Adverse Events

In order to strengthen the monitoring and re-evaluation of adverse events of medical devices, timely and effectively control their post-marketing risks, and ensure human health and life safety, NMPA released <Provisions for Medical Device Adverse Event Monitoring and Re-evaluation> effective on January 1, 2019, in accordance with the Regulations on Supervision and Administration of Medical Devices. This is to provide the requirement for monitoring, re-evaluation and supervision of medical device adverse events in the territory of the People's Republic of China.

26.6.1 AE Reporting

If the MAH discovers or learns of a suspicious medical device adverse event, it shall promptly investigate the cause; any event results in death shall be reported within 7 days; any event results in serious injury or may results in serious injury or death shall be reported within 20 days. If a medical device distributor or institutional user discovers or learns of a suspicious medical device adverse event, it shall promptly inform the MAH. Of the events, those that result in deaths shall be reported within 7 days, and those result in serious injury, or may results in serious injury or death shall be reported within 20 days, through the National Medical Device Adverse Event Monitoring Information System. Foreign MAHs of imported medical devices and MAHs of domestically produced medical devices sold abroad shall take the initiative to collect adverse of their medical devices that occur outside China. The agents designated by foreign MAHs and MAHs of the domestic medical devices shall report those abroad-occurred adverse events which result in or may result in serious injury or death within 30 days from the date of discovery or knowledge. After reporting a medical device adverse event or having learned of a relevant event from the National Medical Device Adverse Event Monitoring Information System, the MAH shall conduct follow-up investigation, analysis and evaluation as required; the results of the evaluation shall be reported to the local provincial monitoring agency in a timely manner (those result in death, within 30 days; those result in serious injury, or may results in serious injury or death, within 45 days). A supplementary report shall be provided if there are new findings or new understanding about the event.

26.6.2 Periodic Risk Assessment Report

The MAHs shall conduct continuous research on the safety of their medical devices on the market, summarize and analyze the adverse event reports, monitoring data and risk information at home and abroad, evaluate the risks and benefits of their products, record adopted risks control measures, and form periodic post-marketing risk assessment reports.

The MAHs shall complete the periodic risk assessment reports (PRAR) of their products' safety profile in the previous year within 60 days after every full year (counted from the date of the first approval of registration or filing of the products). PRARs of medical devices registered at the State Drug Administration shall be submitted to the national monitoring agency; PRARs of devices registered at provincial drug regulatory authorities shall be submitted to the local provincial monitoring agencies. PRARs of Class I medical devices shall be kept by the MAHs for future reference.

For medical devices that have renewed registration, the PRARs of the present registration cycle shall be completed by the time of the next application for renewal of registration, and shall be kept by the MAHs for future reference.

26.7 Managing Recalls

The Measures for Administration of Medical Device Recall was effective on May 1, 2017, to strengthen the supervision and management of medical devices, control the defective medical devices, eliminate safety risk of medical devices, ensure the safety and effectiveness of medical devices, as well as human health and life safety.

The expression 'medical device recall' described herein means the actions that a medical device manufacturer has taken, such as warning, inspection, repair, re-labelling, IFU revision and improvement, software update, replacement, withdrawal or destroy, for a certain category, model or batch of a commercial available product with defects in accordance with the relevant specified procedures.

According to the severity of medical device defects, recalls are divided into following:

1. Class I Recall: The use of the medical device is likely to cause or has caused a severe health hazard;
2. Class II Recall: The use of the medical device could cause or has caused a transient or irreversible health hazard;
3. Class III Recall: The use of the medical device is less likely to cause hazards, but it is still necessary to recall the device.

Medical device manufacturers should, based on specific circumstances, determine the level of recall and design recall plan scientifically and organize the implementation according to the recall level and the sales and use of the affected medical device.

Once deciding to recall any medical device, the manufacturer shall notify the relevant operators, use units or users within 1 day in the case of a Class I recall, or 3 days in the case of a Class II recall, or 7 days in the case of a Class III recall.

The recall notice shall cover the following:

Basic information such as the name, model and specification and batch of the medical device;

1. (2) The reasons for the recall;
2. (3) The requirements for the recall: for example, an immediate suspension of the sale and use of the product, a recall notice shall be forwarded to the relevant operators or use units;
3. (4) The disposal methods of the medical device recalled.

Once deciding to recall any medical device, the medical device manufacturer shall immediately submit the Reporting Form for Medical Device Recall Event to the supervision and administration department of the province, autonomous region or municipality directly under the central government where it is located and the regulators that have approved the registration or conducted the filing of the product, and submit the investigation and evaluation report and the recall plan to the same authorities within five (5) working days for the record.

The food and drug supervision and administration department of the province, autonomous region or municipality directly under the Central Government where the medical device manufacturer is located shall report the recall situation to NMPA within 1 working day after receiving the reporting form.

The investigation and evaluation report shall include the following:

1. (1) The details of the medical device recalled, including basic information such as name, model, specification and batch;
2. (2) The reasons for the recall;
3. (3) Results of the investigation and evaluation;
4. (4) Classification of the recall.

The recall plan shall include the following:

1. (1) The production and sales of the medical device and the quantity intended to be recalled;
2. (2) The specific contents on recall measures, including the organization, scope and time limits for implementation, etc.;
3. (3) Disclosing channel and scope of recall information;
4. (4) The expected effects of the recall;
5. (5) The disposal measures for the medical device recalled.

The medical device manufacturer shall evaluate the recall effect within 10 working days after the completion of the recall and submit a summary assessment report to the supervision and administration department of the province, autonomous region or municipality directly under the Central Government where the medical device manufacturer is located.

Supervision and administration department of the province, autonomous region or municipality directly under the Central Government where the medical device manufacturer is located shall review the report within 10 working days from the date of receiving the summary assessment report and evaluate the recall effect; the above authority shall request in writing the manufacturer to recall again if it is believed that the recall has not eliminated product defects or controlled risks on the product effectively. Medical device manufacturers should conduct the recall again in accordance with the requirements from supervision and administration departments.

Where the supervision and administration department, through the investigation and evaluation, considers that the medical device manufacturer should have recalled the defective product but failed to do so voluntarily, the manufacturer shall be ordered to recall the said device.

The mandatory recall decision may be made by the supervision and administration department of the province, autonomous region or municipality directly under the Central Government where the medical device manufacturer is located, or by the medical products supervision and administration department that has approved the registration or filing of the medical device. The medical products supervision and administration department that has made the decision shall make public the information concerning the mandatory recall on its website.

Chapter 27

Hong Kong: Medical Device Regulatory System

Jack Wong and Linda Chan

Asia Regulatory Professionals Association

Jack.wong@arpaedu.com

The regulatory information contained in this report is intended for market planning and is subject to change frequently. Translations of laws and regulations are unofficial.

27.1 Market Overview

27.1.1 Market Environment

(i) Hong Kong is a prosperous economy and acts as a hub for trade throughout Asia.
(ii) The population is about 8 million and very receptive to the use of advanced medical products. All medical and healthcare products, subject to the China and Hong Kong Closer

---

Medical Regulatory Affairs: An International Handbook for Medical Devices and Healthcare Products (Third Edition)
Edited by Jack Wong and Raymond K. Y. Tong
Copyright © 2022 Jenny Stanford Publishing Pte. Ltd.
ISBN 978-981-4877-86-2 (Hardcover), 978-1-003-20769-6 (eBook)
www.jennystanford.com

Economic Partnership Arrangement's (CEPA) rules of origin, are able to enjoy duty-free access to the China mainland. The medical device market in Hong Kong is summarized in Fig. 27.1.

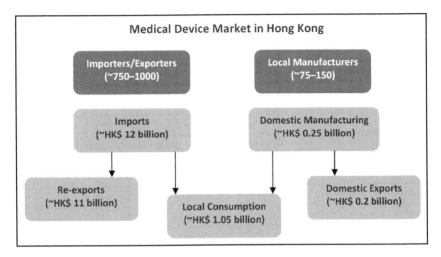

Figure 27.1 Medical device market in Hong Kong. *Source*: 2006 data (Regulatory Impact Assessment Report (2008). US$1 = HK$7.8.

27.1.2 Overview of Regulatory Environment and What Laws/Regulations Govern Medical Devices

Medical device regulation is developing in Hong Kong. The new regulatory system will largely be based on the recommendation of the Global Harmonization Task Force (GHTF). The GHTF is a voluntary consortium with representatives from the trade and regulatory authorities from the USA, Canada, Australia, Japan, and the European Union formed in 1992 to harmonize the standards and principles of regulating medical devices.

Currently, the importation or sale of medical devices in Hong Kong except those containing pharmaceutical products or emitting ionizing radiation does not require registration.

An "irradiating apparatus" is defined by the Radiation Health Unit as one intended to produce or emit ionizing radiation, or capable of producing or emitting ionizing radiation at a dose rate exceeding 5 microsievert per hour at a distance of 5 cm from any accessible point of the surface of the apparatus. The

Radiation Health Unit's Irradiating Apparatus Licence Section processes applications for and handles the renewal of licenses required to import, sell, manufacture, produce, deal in/with, possess, and/or use irradiating medical equipment.

The Consumer Goods Safety Ordinance (Cap.456) provides protection against the supply, manufacture, or import of unsafe products, including some medical devices that can be regarded as consumer goods, unless otherwise specified in the schedule. The Electrical Products (Safety) Regulation (Cap.406G) provides protection against the supply of unsafe electrical products, including medical devices designed for household use except those products specified otherwise [1].

The statutory regulation of certain health care professionals, whereby practitioners are required to ensure the safe and appropriate treatment for patients, also provides incidental control on the use of medical devices.

The Undesirable Medical Advertisements Ordinance (Cap.231) prohibits advertisements related to the curative or preventive effects of products on diseases listed in the ordinance [2].

To raise public awareness on the safe use of medical devices and enable traders to familiarize themselves with future mandatory requirements, the Medical Device Control Office (MDCO) was established in July 2004 (*Note*: The Medical Device Control Office has been renamed as the Medical Device Division with effect from October 1, 2019.). Its mission is to set up a risk-based and cost-effective regulation on the supply and use of medical devices with reference to harmonized standards and procedures recommended by the GHTF. The government launched a Medical Device Administrative Control System (MDACS) on November 26, 2004. The first phase of MDACS starts with the voluntary listing of Class IV (high risk) medical devices. The second phase of MDACS, which includes voluntary listing of Class II and Class III (medium risk) medical devices, was launched on November 14, 2005.

27.1.3 Regulatory Body

The regulatory body in Hong Kong is the Medical Device Division (MDD), formerly known as the Medical Device Control Office (MDCO) which was established in July 2004.

Contact information for medical devices
Medical Device Division
Department of Health
Room 604, 6/F, 14 Taikoo Wan Road, Taikoo Shing, Hong Kong
Email: mdd@dh.gov.hk
Website: https://www.mdd.gov.hk/en/about-us/contact-us/index.html

Contact information for irradiating apparatus
Electrical & Mechanical Services Department (EMSD),
Health Sector Division,
16/F Multi-Centre Block A,
Pamela Younde Nethersole Eastern Hospital,
Chai Wan,
Hong Kong
Medical Electronics Projects and Procurement Subdivision 1,
Fax: 852-2553-7887
Mr. Albert K. F. Poon
Tel: (852) 2505 0276
Website: http://www.emsd.gov.hk/en/home/index.html

27.2 Regulatory Overview

27.2.1 The Definition of Medical Device (It follows GHTF)

Medical devices range from sophisticated equipment such as cardiac pacemakers used by health care professionals to simple products such as bandages and thermometers bought over the counter by the public.

In essence, a medical device refers to any instrument, apparatus, appliance, material, or other article, excluding drugs, used for human beings for diagnosis, prevention, treatment, monitoring of diseases or injuries; or for rehabilitation purposes; or for the purposes of investigation, replacement or modification of body structure or function. In addition, it includes devices used for examination of human specimens.

An accessory to a medical device is subject to the same regulations that apply to the medical device itself. However, devices designed for the treatment or diagnosis of diseases and injuries in animals are outside the scope of the proposed regulatory framework.

An "irradiating apparatus" is defined by the Radiation Health Unit as one intended to produce or emit ionizing radiation, or capable of producing or emitting ionizing radiation at a dose rate exceeding 5 microsievert per hour at a distance of 5 cm from any accessible point of the surface of the apparatus. The Radiation Health Unit's Irradiating Apparatus Licence Section processes applications for and handles the renewal of licenses required to import, sell, manufacture, produce, deal in/with, possess, and/or use irradiating medical equipment.

27.2.2 Classification of Medical Devices (Class I, II, III, IV)

The Principle

Medical devices are classified according to the risk level associated with their intended use. In general, the risk level depends on the design of a medical device as well as its intended use. The actual classification of each device also depends on one or several of the following factors, such as the duration of device in contact with the body, the degree of invasiveness, whether the device delivers medicines or energy to the patient, whether they are intended to have a biological effect on the patient and local versus systematic effects (e.g., conventional versus absorbable sutures). These factors may, alone or in combination, affect device classification. For details, please refer to the rules of classification in the MDD's guidance document GN-01, *Overview of the Medical Device Administrative Control System* [3].

(*Note*: The "Classification Rules for Medical Devices" given in Appendix 1 of this guidance document may be updated from time to time. You may wish to refer to the most updated version included in the Technical Reference [TR-003] Classification Rules for Medical Devices published by Department of Health.)

Adopting from the Principles of Medical Devices Classification proposed by the GHTF, medical devices are grouped into four classes according to the risk level associated with their intended use. Class IV medical devices bear the highest risk, whereas Class I devices bear the lowest risk (Table 27.1).

Table 27.1 Classification of medical devices

Classification	Risk level	Examples of medical devices
I	Low	Tongue depressor, bandage, dressing
II	Medium-low	Suction pump, gastroscope, transdermal stimulator
III	Medium-high	Lung ventilator, orthopedic implant, X-ray machine, medical laser
IV	High	Prosthetic heart valve, implantable cardiac pacemaker, heparin-coated catheter

Similar to Europe, it is the manufacturer's responsibility to assign the classification of their medical devices.

27.2.3 The Role of Distributors or Local Subsidiaries

The Local Responsible Person (LRP) is the one who applies for the inclusion of a medical device into The List of Medical Devices under the MDACS. They have the following roles related to the device in the List:

(i) effective communication with the manufacturer, importers, users, the public and Department of Health
(ii) keeping of distribution records
(iii) arrangement of maintenance and services
(iv) tracking of specific medical devices
(v) management of alerts and recalls

The reporting of adverse incidents follows the MDD guidance document [GN-03], *Guidance Notes for Adverse Incident Reporting by Local Responsible Persons*.

It is the manufacturer's responsibility to assign the classification of their medical devices. The LRP should understand the classification rationale. More than one LRP for the same product is allowed in Hong Kong, but the manufacturer should have a proper recall system to identify which LRP should recall which batch of the same product. The LRP's detailed roles can be found in MDD's Code of Practice Document—[COP-01], *Code of Practice for Local Responsible Persons*.

27.2.4 Product Registration or Conformity Assessment Route and Time Required

The MDACS is basically divided into three parts: pre-market approval, post-market surveillance, and recognition of conformity assessment bodies.

Pre-market approval is in the form of voluntary listing of medical devices, importers, and local manufacturers complying with specific requirements. They will be listed in one of the following lists:

- The List of Devices, including Class IV, Class III, and Class II
- The List of Importers
- The List of Local Manufacturers

A summary of requirements is presented in Table 27.2.

Table 27.2 Summary of the requirements

	Class I	Class II	Class III	Class IV
Product registration	Not required	Required	Required	Required
Registration of local manufacturers	Required	Required	Required	Required
Registration of overseas manufacturer or its local representative	Not required	Required	Required	Required
Registration of importer	Required	Required	Required	Required
Registration of retailer	Not required	Not required	Not required	Not required

Post-market surveillance is the review of the experience of using those devices that have been placed on the market. The surveillance system would monitor and co-ordinate the management of adverse incidents, safety alerts, and recalls related to the use of medical devices.

Recognition of Conformity Assessment Bodies (CAB) is the acceptance of a list of agencies from which manufacturers could ask them to assess their medical devices for compliance with the requirements related to safety and performance. BSI Product Service is the first CAB approved in Hong Kong on April 13, 2007. TUV SUD and SGS were also approved as CABs afterwards.

If there is no marketing approval of the device in any of the GHTF founding members, namely Australia, Canada, the European Union (EU), Japan, and the USA, the device listing application will not be processed unless a MDACS conformity assessment certificate issued by one of the CABs recognized by the MDCO could be provided.

The applicant can indicate whether it is an application for a single medical device, a medical device family, a medical device series, or a medical device system.

An approval or conditional approval for listing a device will be valid for five years. The LRP must submit an application for the continuation of the listing to the MDCO at least 3 months before the expiry of this five-year validity period.

27.2.4.1 Suggested registration routes/steps

(i) The manufacturer understands the essential principles and gets Quality System Certification, e.g., ISO 13485.
(ii) The manufacturer appoints an LRP.
(iii) The LRP submits product registration to the MDCO or the CAB.
(iv) The MDCO or the CAB conducts a review.
(v) The documentation receives acknowledgement and the application number is given 2 weeks after submission.
(vi) The MDCO takes about 12 weeks for review after getting all required documentation.

27.2.4.2 Technical material requirement

For Class II to IV, please follow [GN-02], *Guidance Notes for Listing Class II, III & IV Medical Devices* (July 2011 edition).

The following documents in English are required for submission:

 (i) copy of ISO 13485 certificate of manufacturer
 (ii) copy of quality management system of LRP if available
 (iii) additional information in format similar to MDS-01 for a medical device family, medical device series or a medical device system if applicable
 (iv) documented procedures for managing product recalls and field safety notices, reportable adverse incidents in Hon Kong, distribution records and tracking of specific medical devices
 (v) a summary of all recalls, suspensions, reportable adverse incidents, banning of the device in other countries or post-market surveillance studies
 (vi) risk Analysis report or summary
 (vii) test reports and certificates for any type tests
 (viii) biological safety data, biocompatibility report and certificate of analysis of the materials/substances for devices containing biological materials or medicinal substances and/or materials that will come into contact with body tissues and/or fluids
 (ix) clinical evaluation report
 (x) approval document in GHTF founding member countries if available; if the medical device is approved for marketing in EU, a copy of the EC Declaration of Conformity shall also be submitted together with a copy of the EC certificate
 (xi) essential requirements checklist in accordance with the EU Medical Device Directives and Essential Principles Declaration of Conformity OR Essential Principles Conformity Checklist MD-CCL (available in MDCO website)

The technical documentation must be submitted in English.

27.2.4.3 The labeling requirement of medical device

Please follow the technical reference document in MDCO website: [TR-005], *Additional Medical Device Labelling Requirements*.

Note that the contact details of the manufacturers and the LRP shall be in both English and Chinese wherever applicable. The instructions for use as user manuals provided shall be in both Chinese and English. If any one of the two languages is not available, it shall be specified as per the requirements of Special Listing Information. Other labeling information such as maintenance manuals provided shall preferably be in both English and Chinese or shall be at least in one of these two languages.

The product labeling requires English and Chinese for self-use.

27.2.4.4 Post-marketing surveillance requirement

Please follow MDD's guidance document [GN-03], *Guidance Notes for Adverse Incident Reporting by Local Responsible Persons*.

Upon the issuance of alerts, modification notices, and recalls by the manufacturer or overseas authorities, the LRP shall inform the MDD about the related details and actions will be taken in Hong Kong as soon as possible, and not later than 10 calendar days after their issuance. The LRP shall follow up the actions and shall submit progress reports to the MDD as requested until the case is concluded. It is preferred that prior arrangements be made such that within 4 hours of the issuance of an alert, recall, or modification notice by the manufacturer, the same be also e-mailed direct to the MDD. At present, there are no product recall guidelines for medical devices.

27.2.4.5 Manufacturing-related regulation

Do manufacturers need registration/authorization? If yes, what is the overview of the authorization process?

Voluntary listing is recommended. Please follow [GN-08], *Guidance Notes for Listing of Local Manufacturers*.

27.2.4.6 Clinical trial related regulation

Do medical device clinical trials regulated? If yes, what is the overview of the authorizations process?

No at the moment, but consultation document was released before [4]. The process is very similar to pharmaceuticals.

27.2.4.7 Is there a procedure for mutual recognition of foreign marketing approval or international standards?

Yes, Hong Kong recognizes ISO standards. Product registration approvals in GHTF countries will simplify the registration process.

27.3 Commercial Aspects

Any price control of medical device?
No at this moment.

Are parallel imports allowed?

As the MDACS is only a voluntary system, there is no regulatory requirement on parallel imports. However, only those products supplied through (or with consent of) the LRP could use the assigned listing number, as it is impractical, if not impossible, to verify whether a parallel-imported product is of the same quality and performance as those listed products supplied through the LRP. In this way, the LRP could effectively control and recall the product when required.

Any advertisement regulation of medical device?

The Undesirable Medical Advertisements Ordinance (Cap.231) prohibits advertisements related to the curative or preventive effects of products on diseases listed in the ordinance [5].

27.4 Upcoming Events

The MDD proposed a new medical device regulation to the Legislative Council after a number of consultations. The earliest rollout of the new regulation will be in 2019 with 2–3 years' grace period.

References

1. http://www.justice.gov.hk/home.htm.
2. Consultation Document on Regulation of Medical Devices in Hong Kong (page 6), July 2003, http://www.info.gov.hk/archive/consult/2003/meddevice_e.pdf.

3. http://www.mdco.gov.hk/english/mdacs/mdacs_gn/files/gn_01.pdf.
4. http://www.mdco.gov.hk/english/mdacs/mdacs_gn/files/dgn_eng.pdf.
5. http://www.justice.gov.hk/home.htm.

Chapter 28

India: Medical Device Regulatory System

Kulwant S. Saini

Johnson & Johnson Medical, India,
A division of Johnson & Johnson Ltd,
9/43, Kirti Nagar, New Delhi 110015, India

ksaini@its.jnj.com, kulwantsaini@hotmail.com

Disclaimer

The regulatory information contained in this report is intended for market planning and is subject to change frequently. Translations of laws and regulations are unofficial.

28.1 Market Overview

28.1.1 Market Environment

- India, a parliamentary democracy and an emerging market in Asia Pacific, is the most populous (1.24 billion) democracy in the world.

- India's next five-year plan is poised to increase the healthcare expenditure to 3% of GDP and to 5% by 2020. Nearly 60 million middle-class households get added every 10 years, and this increased affordability and access would drive the 75% of the growth in healthcare. Healthcare remains in the top three categories in terms of consumption growth. India has the potential to show the fastest growth over the next 30–50 years and slated to be third largest economy by 2030.
- The total healthcare expenditure per capita has grown from US$ 28 in 2000 to US$ 43 in 2010 at a compound annual growth rate of 4.4% pa. The medical device industry is estimated to be approximately US$ 3.0 billion and the medical equipment industry is around half a billion. The medical device industry is growing at a rate of over 15%. India has been recognized for its quality pharmaceuticals and is largely self-sufficient, but for medical devices it is mostly dependent on imports (65–75%). There are a lot of local manufacturers that are also exporting their products worldwide, but barring a few simpler medical devices, most of the advanced medical devices and implants are being imported into India.
- Healthcare spending trends
  - Self-pay will continue to be major part of healthcare spend (currently at 58% going down to ~50% by 2017).
  - Increased penetration of insurance (no. of insured [private] to grow at 15% over the next seven years).

28.1.2 Overview of Regulatory Environment and What Laws/Regulations Govern Medical Devices

Currently, there are no separate medical device (MD) regulations in India, and there is a serious mix up of drug and device regulations. The MD regulations are evolving and the direction is that the MD regulations would shape up in line with Global Harmonization Task Force (GHTF) guidelines.

The Drugs and Cosmetics Act 1940 (Act) and the Drugs and Cosmetics Rules 1945 are currently applicable to drugs as well as medical devices. In addition to the aforementioned act, following other acts and rules also are applicable to regulate the drugs.

- Pharmacy Act, 1948
- Drugs and Magic Remedies (Objectionable Advertisements) Act, 1954
- Narcotic Drugs and Psychotropic Substances Act 1985
- Medicinal and Toilet Preparations (Excise Duties Act, 1955
- Drugs (Price Control) Order, 1995 (under the Essential Commodities Act)

Following are the other laws that have a bearing on the manufacture, distribution and sale of drugs and cosmetics:

- Industries (Development and Regulation) Act, 1951
- Trade and Merchandise Marks Act, 1958
- Indian Patents and Design Act, 1970
- Factories Act, 1948.
- Legal Metrology Act 2009 (in force from 1.4.2011),
- Draft Legal Metrology Packaged Commodities Rules 2011

The **Drugs and Cosmetics Act 1940** regulates the import, manufacture, distribution and sale of drugs and cosmetics (and medical devices). It is a comprehensive act for the requirements applicable for the drugs and cosmetics but not yet for the medical devices (except for Schedule M III, provisions applicable to hypodermic needles and syringes) which have been mentioned in some sections (drug definition, labelling, standards for MDs) under drugs.

When we talk of MD regulations in India, we talk of two different periods, one before October 2005 and the other after October 2005, when 10 categories of medical devices were notified by the Government of India (MOH) through Gazette notification SO No. 1468(E) and GSR 627(E).

Before 2005, the following medical devices were regulated as drugs as these fall under either the definition of the drug or were part of the monograph of a Pharmacopoeia or were of a biological origin (Schedules C and C1):

- Blood grouping sera
- Ligature, surgical sutures (and staplers)
- Intra uterine devices
- Condoms
- Tubal rings
- Surgical dressings

- Umbilical tapes
- Blood/blood component bags
- Disposable hypodermic syringes and needles, disposable perfusion sets (GSR 365(E), dated March 17, 1989)
- In vitro diagnostic devices for HIV, HbsAg and HCV

The first major notification on medical devices was released in October 2005 post court intervention in one of the case (filed by the Maharashtra FDA), involving a Mumbai hospital using unapproved drug-eluting stents (classification issue, drug or device) which were not approved in country of the origin. The court directed the Drugs Controller General of India (DCGI) to regulate these critical medical devices. The following list of categories of sterile products, intended for internal or external purpose, was notified:

- Cardiac stents
- Drug-eluting stents
- Catheters
- Intra ocular lenses
- IV cannulae
- Bone cements
- Heart valves
- Scalp vein sets
- Orthopaedic implants
- Internal prosthetic replacements

Later, in August 2006, a clarification was issued that the devices which are imported non-sterile and sterilized by the hospitals before use would also fall under the ambit of the October 2005 notification (SO No 1468(E)).

Each brand under the aforementioned categories of notified medical device needs to be registered with the regulators before these can be imported into India.

Post October 2005 notification, other lists of products and accessories were put on the Central Drugs Standard Control Organization (CDSCO) website (September 2007) but were never implemented or notified through a gazette notification, as the act states under the definition of drug. Hence, those were still being freely imported. On 5 March 2012, the DCGI sent a circular to State Drug Controllers and CDSCO branch offices categorizing 11 products from the aforementioned list into various notified categories of the main gazette notification of products notified in October 2005. This correlation of additional products to notified categories eliminates the need for a tedious gazette notification process. Table 28.1 shows a list of these products:

Table 28.1 Additional medical devices classified under categories notified in October 2005

Name of devices	Class of notified devices
Spinal needle	Disposable hypodermic needle
Insulin syringes	Disposable hypodermic syringes
Three-way stop cock as an accessory of IV cannula/catheter/perfusion set	Disposable perfusion set
Introducer sheath	IV cannula
Cochlear implant	Internal prosthetic replacement
Close wound drainage set	Catheter
AV fistula needle	Disposable hypodermic needle
Extension line as an accessory of infusion set	Disposable perfusion set
ANGO kit/PTCA/cath lab kit	IV cannula/disposable perfusion set
Measure volume set	Disposable perfusion set
Flow regulator as an accessory of infusion set	Disposable perfusion set

Another requirement for drug-eluting stents was put in September 2007, which stated that before approval/registration, a manufacturer needed to perform a clinical trial in India on 100 patients for six months (products already in use and approved by USFDA or CE marked) or 12 months (new products and not yet approved elsewhere). Through this clarification, peripheral stents were also brought under the list of regulated products.

Radiation-emitting devices need to be approved by the Bhabha Atomic Research Centre for the radiation safety.

Operationally, the regulatory mechanism in India consists of a two-tier system, one in the centre, the DCGI, and one each in the states (total 28) and in the union territories (total 7). These two regulators operate independently, and the DCGI reports to the central government and the state regulators (the Food and Drug

Administration or the Drugs Control Department) report to their ministry of health. Both the central regulator and the state regulator are responsible for the implementation of the Drugs and Cosmetics Act provisions. Their major responsibilities are listed in the following subsections.

Under the Drug and Cosmetics Act, the regulation of the manufacture, sale and distribution of drugs is primarily the concern of the state authorities, while the central authorities are responsible for approval of new drugs, clinical trials in the country, laying down the standards for drugs, control over the quality of imported drugs, coordination of the activities of state drug control organizations and providing expert advice with a view to bringing about uniformity in the enforcement of the Drugs and Cosmetics Act.

28.1.3 Functions Undertaken by DCGI and Central Government

28.1.3.1 Statutory functions

- laying down standards of drugs, cosmetics, diagnosis and devices
- laying down regulatory measures, amendments to acts and rules
- regulating market authorization of new drugs
- regulating clinical research in India
- approving licenses to manufacture certain categories of drugs as the Central License Approving Authority, i.e. for blood banks, large-volume parenterals and vaccines and sera.
- regulating the standards of imported drugs
- carrying out tasks relating to the Drugs Technical Advisory Board (DTAB) and Drugs Consultative Committee (DCC)
- overseeing testing of drugs by central drugs laboratories
- overseeing publication of Indian Pharmacopoeia

28.1.3.2 Other functions

- coordinating the activities of the State Drugs Control Organization to achieve uniform administration of the act; policy guidance
- providing guidance on technical matters
- participating in the WHO GMP certification scheme
- monitoring adverse drug reaction (ADR)

- conducting training programs for regulatory officials and Government Analysts
- looking after the distribution of quotas of narcotic drugs for use in medicinal formulations
- screening drug formulations available in the Indian market
- evaluating/screening applications for granting no-objection certificates for the export of unapproved/banned drugs

28.1.4 Functions Undertaken by the FDA and State Governments

28.1.4.1 Statutory functions

- licensing of drug manufacturing and sales establishments
- licensing of drug testing laboratories
- approval of drug formulations for manufacture
- monitoring of quality of Drugs and Cosmetics, manufactured by respective State units and those marketed in the State
- investigation and prosecution in respect of contravention of legal provisions
- administrative actions
- pre- and post-licensing inspection

Importers and manufacturers face difficulties in compliance and registration of their products as there had been no separate regulations written and implemented for MDs. For the registration of MDs, the drug registration/approval process has been followed since 2006 although a brief list of contents of the technical dossier was published in 2006, which was modified later and published (last quarter of 2010) on the website (http://cdsco.nic.in/).

To streamline the work for designing proper MD regulations, a core group of industry members and regulators was constituted. Various industry agencies (Federation of Indian Chambers of Commerce and Industry [FICCI], Confederation of Indian Industry [CII], American Chamber of Commerce in India [AMCHAM], Advanced Medical Technology Association [ADVAMED]) were also involved into the process through their MD regulatory subgroups. The efforts of this core group during 2009 and 2010 culminated into the compilation of a base document (Schedule M III) which was based on the European Medical Device Directive and the GHTF guidelines. This comprehensive document was submitted to the

DCGI office at the end of 2010 and the highlights of this document are as follows:
1. includes the GHTF definition of a medical device
2. based on the risk-based (GHTF) classification of medical devices
3. includes the QMS certification (ISO 13485) and Conformity Assessment Procedures (MDD and GHTF) involving notified bodies and the regulators (for Class C and D products)
4. post marketing surveillance (PMS)

Schedule M III document need to be wetted by the legal ministry and has to be linked with various sections of Drugs and Cosmetics Act in such a way that it becomes a standalone guideline for MDs within the act. Currently there is no proper definition of the MD in the Drugs and Cosmetics Act, which is the main hindrance in the inclusion of the MD guidelines/regulations in the act. A Drugs and Cosmetics (Amendment) Bill was introduced in the Parliament in 2007. This bill has been modified in the last four years to amend the act with respect to the provisions for the drugs as well medical devices. It includes a proper definition for the MD and separate provisions for MDs in the act. It even recommends the change of the name of the Act to "Drugs, Cosmetics and Medical Devices Act" so as to provide proper recognition to MDs. This bill, which is now pending approval by the Parliament, would enable the government to make rules for MDs and inclusion of the Schedule M III into the act.

As the passing of the aforementioned bill by the Parliament is a time-consuming exercise, as it is busy with other priorities, in the meantime the DCGI office has come out with guidelines to streamline the process of registration, manufacture, PMS and clinical trials. These guidelines have been published on the CDSCO website. A dialogue is going on between the DCGI and industry members to further refine these interim guidelines.

28.1.5 Guidance Documents

- Guidance Document on Common Submission Format for **Registration of Medical Devices** in India (4 August 2010)
- Guidance Document on Common Submission Format for **Import License in Form 10** of Medical Devices in India (25 August 2010)
- Requirements for **Conducting Clinical Trial(s)** of Medical Devices in India (4 August 2010)

- Guidance document on application for grant of **License in Form-28** for manufacture of Medical Devices in India under Central Licensing Authority Scheme (12 August 2010)
- Guidance on clinical trial inspection (1 November 2010)
- Guidance Document on Common Submission Format for Registration of Notified Diagnostics Kits in India (5 January 2011)
- Guidance Document on Common Submission Format for Import License of Notified diagnostics kits in India (5 January 2011)
- Guidance Document on Common Submission Format for Import License of Non-Notified diagnostic kits in India (5 January 2011)

28.1.6 Indian Pharmacopoeial Commission

The Indian Pharmacopoeial Commission (IPC) is an autonomous institution under the Ministry of Health and Family Welfare, Government of India, dedicated to laying down standards for drugs, pharmaceuticals and healthcare devices/technologies besides providing reference substances and training.

28.1.7 Detail of Key Regulator(s)

The CDSCO headquarters are located in New Delhi, and it functions under the Directorate General of Health Services (see Figs. 28.1–28.3 for details).

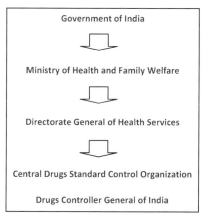

Figure 28.1 The Central Drugs Standard Control Organization: The central agency.

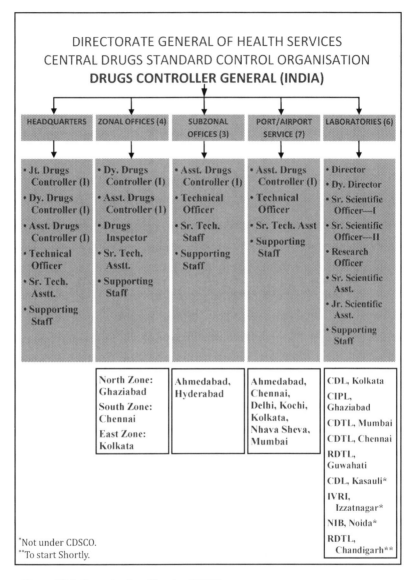

Figure 28.2 Organization Chart—CDSCO.

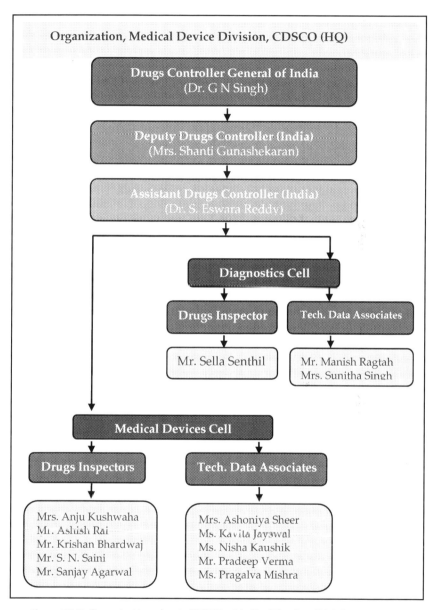

Figure 28.3 Organization chart: CDSCO—Medical Devices Division.

Contact Information
Drugs Controller General of India
Dr. G. N. SINGH
FDA Bhawan, Kotla Road, New Delhi 110002
Phones: +91 11 23236965 (D)
Fax: +91 11 23236973.
Email: dci@nb.nic.in
CDSCO website: http://cdsco.nic.in/
Medical Device Division website: http://cdsco.nic.in/Medical_div/medical_device_division.htm

28.2 Regulatory Overview

28.2.1 Definition of Medical Device

Medical devices are partially regulated in India and some (as listed Section 28.1.2) of the critical devices have been notified as drugs and regulated as drugs. Rest of the devices are still being freely imported into India.

Currently, the MD has been defined under the definition of the drug in Section 3(b)(iv) of the Drugs and Cosmetics Act. This definition is not close to GHTF definition and requires a gazette notification for any addition of MD to the list of notified devices:

> 3 (b) (iv)
> ... such devices intended for internal or external use in the diagnosis, treatment, mitigation or prevention of disease or disorder in human beings or animals, as may be specified from time to time by the Central Government by notification in the Official Gazette, after consultation with the board.

In view of the partial and incomplete MD regulations, currently, there is no definition of the accessory in the Drugs and Cosmetics act.

The proposed definition of the MD in the Drugs and Cosmetics Amendment bill is based on the GHTF definition of the MD. This bill still awaits the approval by the Parliament before these new provisions can be made part of the Act. Similarly, the Schedule M III (document compiled for base regulations for MDs) now available with the regulator also has a GHTF definition of MD. The contents of the bill and the base regulations would become part of the act in due course after review and approval by the Parliament and other ministries involved in the process. This complete definition would eliminate the need for the involvement and review by the advisory board in future.

28.2.2 Classification of Medical Devices

In absence of the base medical device regulations, currently, there is no system for the risk-based classification of MDs. Medical devices have been notified from time to time as per the recommendation of the Advisory Board although the last notification contained mostly the critical and implantable MDs.

The proposed base MD regulations in the comprehensive Schedule M III document are in line with the GHTF and the European MD Directives. A comprehensive risk-based classification on the lines of GHTF (Class A, B, C, D) requirements has been included in this document. This considers various factors such as the duration of contact, degree of invasiveness, biological origin and combination with drugs. Once implemented, this document also proposes a phased implementation of regulations starting with Class D devices.

28.2.3 Role of Distributors or Local Subsidiaries (LRP)

A notified MD can be sold in India by a local manufacturer (having a manufacturing license) or the foreign manufacturer or importer (licensee), after registration of the manufacturing site and obtaining an import license for the products manufactured at that particular site, either by the manufacturer himself or by an affiliate of the manufacturer having an office in India or by an agent/importer/distributor. This person is responsible for the compliance to the provisions (also the GMPs) of the Drugs and Cosmetics Act in India on behalf of the manufacturer (if situated outside India):

- An authorization from the manufacturer to be its agent in India in the form of a Power of Attorney.
- Need to have a valid wholesale license for sale and distribution of products in any state of India, issued by the state's FDA/DCA.

- Make sure the premises where the imported products are to be stocked are equipped with proper storage accommodation and equipment (temperature/humidity).
- The importer is a link between the regulators and the source company for compliance:
    - Customer complaints management in coordination with manufacturer and adverse incident reporting; also responsible for the reporting of the MDR to the regulator in India
    - Management of the field actions as communicated by the manufacturer
- The manufacturer shall at all times observe the undertaking given by them or on their behalf in Form 9.
- The licensee shall allow any inspector authorized by the licensing authority to enter with or without notice any premises where the imported products are stocked, to inspect and to take samples, if required, for testing.
- If the licensing authority directs, the licensee shall not sell or offer for sale any batch of the products.
- For traceability, the licensee shall maintain a record of all sales showing particulars of the products and of the person to whom products have been sold.

28.2.4 Product Registration or Conformity Assessment Route and Time Required

The pre-market product and site registration process currently followed in India is based on the drug regulations. Figure 28.4 illustrates the process of the registration of the site and product (mandatory for Notified MDs). The evaluation process as defined for MDs in 2006 for expedited review is based on the prior approval (USFDA, CE Mark or Conformity Assessment Procedure, approval in any of the GHTF countries) of the products. A new product not approved previously in any of the countries may be asked to go for the clinical trials for the generation of evidence for any of the new indication and then would go for an additional complete evaluation (for Form 45 new product approval). Almost all of the MDs approved by DCGI office since year 2006 have gone through the "Me Too" route, which involves two steps:

- Application in Form 40 and DCGI approval in Form 41 (Registration Certificate)
- Application in Form 8 and DCGI approval in Form 10 (Import License)

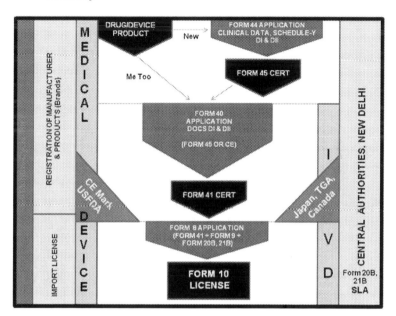

Figure 28.4 Manufacturing site and product registration: process flow chart.

This evaluation for the indigenous manufacturers involves DCGI as well as the state FDAs. For foreign manufacturers, it involves only the DCGI office. The process also involves a site inspection by the joint team of CDSCO and FDA for indigenous manufacturers. If required (as per conditions of the license), a site inspection may be performed for a foreign manufacturer, although such inspection has not been performed since 2006. Generally, the Conformity Assessments Procedures performed by the NBs are acceptable for this purpose.

For in vitro diagnostics (IVDs) (non-critical), only the second step is required and no site registration is needed. A registration certificate (RC) for a particular site is generally valid for three years and then re-registration is required. The import license is valid till the registration certificate is valid. During the currency of the

Registration Certificate, additional products from the same site can be endorsed on the RC. The import license in this case would be valid for the remaining period of the RC. The manufacturing license issued to an Indian manufacturer is valid for five years.

The dossier format (various forms), contents and submission process is defined in the guidance documents listed in Section 28.2.8 and also available on the CDSCO site and the Drugs and Cosmetics Act. In case of "Me Too" products, the DCGI office evaluation process/documentation review takes about 6–9 months to issue the RC and the import license. In case of a new product/new indication, performing a clinical trial and the issue of Form 45 may need additional cycle time. In such cases, the technical dossier may also be additionally referred to an expert committee for review/comments.

28.2.5 Quality System Regulation

Currently as MDs are notified as drugs, it is expected that the local manufacturer will comply with Schedule M GMPs (Drugs and Cosmetics Act). Source companies situated outside India, generally, comply with ISO 13485, or USFDA QSRs or the QMS requirements in the country of origin. These are also acceptable, currently, as part of the submission for the registration certificate to be issued by the Indian regulators. The Plant Master file required in India is generally replaced by the Quality Manual of the source company for the submission.

Future proposed regulations in the Schedule M III draft would be based on ISO 13485 requirements as per GHTF.

28.2.6 Product Registration and Quality System Regulation for Combined Device–Drug Product

As mentioned above, there are no base MD regulations yet in place in India. MDs are notified and regulated as drugs. Hence, there is no definition of combination product and no specific requirements laid down for the combination products.

Proposed Schedule M III has provisions similar to GHTF and would require such combination products to be evaluated by drug as well as device section of the CDSCO depending on the status of approval of drug in India. Combination product definition is proposed to be similar to GHTF and the product as such would be treated as MD.

28.2.7 Registration Fee

Table 28.2 shows the fees that are applicable for the various activities/submissions/functions carried out by the CDSCO, state FDA and test laboratories.

Table 28.2 Fees applicable for various activities/submissions/functions carried out by CDSCO, state FDA and test laboratories

Type of license	Form name as per D&C Act	Type of registration/ activity	Fee
Registration and re-registration	Form 40	Mfg. site registration	US$ 1,500
		Product registration (per product)	US$ 1,000
Import license	Forms 8 and 9	With one product	INR 1,000
		Subsequent additional Product	INR 100
Foreign manufacturer site inspection with respect to site registration	As per conditions of license	GMP inspection (if required)	US$ 5,000
Clinical trials	Form 122A		INR 50,000
Manufacturing licenses	Form 27 (Sterile)	Site fees	INR 6,000
		Product fees (per product)	INR 1,250
	Form 24 (non-sterile)	Site fees	INR 3,750
		Product fees (per product)	INR 750
Renewal of plant layout			INR 5,000
Test license	Form 12	1st product	INR 100
		Subsequent product	INR 50
Government lab testing fees	Rule 7, 48 and 168-F	Product testing	Refer to Schedule B, B-1 of Act

Abbreviation: INR, Indian rupee.

28.2.8 Technical Material Requirement

The dossier submissions for the site and medical device product registration follow the guidance document available on the CDSCO website under the MD section (Guidance Document on Common Submission Format for Registration of Medical Devices in India (4 August 2010), Guidance Document on Common Submission Format for Import License in Form 10 of Medical Devices in India (25 August 2010)). These are also listed in Section 28.1.5 for MDs and IVDs.

28.2.9 The Labelling Requirement of Medical Device

The labelling requirements for drugs (and MDs) are defined under Rule 96 (Manner of Labelling) of the Act. Additional labelling requirements for the Non-Sterile Surgical Ligatures and Sutures are provided under Rule 102. For special products, the labelling requirements are provided under Rule 109-A, wherein it is stated that the labelling of MDs shall conform to "Indian Standard Specifications" laid down from time to time by the "Bureau of Indian Standards" in addition to other requirements.

Generally, for imported medical devices, ISO or GHTF guidelines are considered sufficient, but they are occasionally superseded by Indian regulations.

According to the Drugs and Cosmetics Act, it is a mandatory requirement that the product should have 60% residual shelf-life period as on the date of import.

Additionally, as per "Legal Metrology Act, 2009" (Declaration on Pre-packaged Commodities), all products sold in retail must carry the following information prior to commercial access:

(a) name and address of the importer
(b) generic or common name of the product
(c) net quantity in terms of standard units of weights and measures
(d) month and year of packing in which the product is manufactured or packed or imported
(e) maximum retail sale price (inclusive of all taxes) at which the product in packaged form may be sold to the ultimate consumer

(f) the approval number (import license number) for the notified devices category

(g) customer contact number

This labelling is generally done (if not labelled by the manufacturer) in warehouses in India by the importers before the product can be sold in the market.

Recent requirements by the Ministry of Health specify the inclusion of the GS1 bar code and serialization on products supplied to the MOH. The Director General of Health Services also specified the products to be exported from India for the purpose of tracking products and to identify counterfeits.

28.2.10 Post-Marketing Surveillance Requirement

The Drugs and Cosmetics Act specifies the reporting requirements for the Complaints and Adverse Reactions for drugs under Schedule M (GMPs) and additionally, under the Schedule Y for the new drug clinical trials. There are currently no specific requirements for MDs.

The proposed Schedule M III has a separate section on PMS and is based on the GHTF guidelines.

28.2.11 Manufacturing-Related Regulation

The manufacturing site and the product (if notified) need to be registered in India with the regulators. Licenses to manufacture medical devices in India can usually be obtained from the state FDA/DCA, after joint inspection from State FDA and CDSCO. However, for recently notified devices (see the list of 10 categories of MDs, Section 28.1.2), DCGI approval is needed. The required information for a device manufacturing license includes the implementation of Good Manufacturing Practices (GMP), as well as some other requirements. Licensing usually involves an inspection by the CDSCO and the state FDA. The license is valid for five years and is renewable.

GMPs are provided under Schedule M of the act. As with the product registration process, much of its requirements were prepared with drugs in mind. Its provisions include the adequacy of the manufacturing site, production area, quality control process, record keeping, testing and adverse event and recall handling system. Schedule M also calls for the maintenance of a Site Master File, as well as a self-inspection to be conducted before the official inspection. For the companies having CE-marked products, the QMS as per ISO 13485 is considered. Schedule L-1 lays down the "Good Laboratory Practices and Requirements of Premises and Equipment".

Future proposed regulations (Schedule M III) are based on the Conformity Assessment Procedures (GHTF) involving notified bodies and the CDSCO (Class C and D MDs).

Finally, if a product such as a drug-eluting stent contains a patented or proprietary drug, a good deal of information on the drug being manufactured must be submitted with a manufacturing license application.

After the DCGI office formulated guidelines for the import and manufacture of medical devices in the country, it also specified that all importers, stockists and retailers of medical devices will have to obtain sales licenses from the state FDAs within three months.

28.2.12 Clinical Trial-Related Regulation

The requirements for the new drugs and clinical Trials are provided under Part X-A of the act (Import or Manufacture of New Drug for Clinical Trials or Marketing). Rule 122 and its various subparts specify the process of Application for Permission to Import New Drug (and device) and to obtain permission for conducting the clinical trials. The details of requirements for the clinical trials are provided in Schedule Y of the act. These requirements are written basically for drugs/pharmaceuticals.

Following are the clinical trial draft guidelines specific to MDs, available on the CDSCO website under MDs section:

- Requirements for Conducting Clinical Trial(s) of Medical Devices in India (4 August 2010)
- Guidance on clinical trial inspection (1 November 2010)

Following additional clarification was issued on 5 September 2007 pertaining to the clinical trial requirements for drug-eluting stents:

- In case of manufacturers whose stents are already in use abroad and they wish to introduce the same in India, a six-month clinical trial has to be carried out on 100 patients.

- In case of stents which are new in nature and have not been used anywhere, a 12-month clinical trial has to be carried out on 100 patients.
- Global clinical trials: Whenever a firm applies only for a global clinical trial to be carried out in the country, this can be permitted as per the present norms followed in case of drugs. However, if the same company wishes to also market the product in India, it has to comply with the aforementioned norms.
- It is clarified that "All Peripheral Stents" are covered under the provisions of the Drugs and Cosmetics Rules and, hence, need to be registered for import and licenses approved by Central License Approving Authority (CLAA) for indigenous manufacture.

28.2.13 Is There a Procedure for Mutual Recognition of Foreign Marketing Approval or International Standards?

There are no mutual recognitions currently in place with any countries, but India recognizes ISO standards. Product registration approvals by the USFDA or in GHTF countries and the CE mark are recognized, and these have formed the basis for the site and product registrations since 2006. The proposed Schedule M III regulations are based on the GHTF guidelines and European regulations.

28.3 Commercial Aspects

28.3.1 Any Price Control of Medical Device

MDs are notified and regulated as drugs in India. For drugs, there are price control requirements (Drugs (Prices Control) Order 1995) in India. The National Pharmaceutical Pricing Authority (Department of Pharmaceuticals, under the Ministry of Chemicals and Fertilizers) is an organization responsible for the price controls (to fix/revise) and availability of medicines in India. In the last three years, there have been some attempts by the authority to bring MDs under the price control, due to high visibility of the prices of stents and orthopaedic implants. There have been many interactions and representation on this subject from the industry to dissuade the agency from implementing the manufacturing cost-based price controls on MDs.

In the mean time, the Ministry of Health and Family Welfare has put certain ceiling limits for reimbursement of stents (drug-eluting stents, bare-metal stents and non-coronary stents) for its Central Government Health Scheme beneficiaries. These are based on the type of approval the product has in India (approved by DCGI) or in the country of origin (USFDA approved or CE marked). There has been strong representation from importers as well as local manufacturers on this *ad-hoc* approach to pricing.

28.3.2 Are Parallel Imports Allowed?

For importing the notified MDs, one needs to have an import license from the DCGI. This is issued on the basis of the Registration Certificate for a manufacturing site of a foreign company. These licenses can be used by the authorized agent to whom these have been issued. An importer may authorize another importer to use its Registration Certificate and allow the issue of another import license to the latter by the DCGI. Both of these need to maintain the distribution records as per the requirements of the act.

28.3.3 Any Advertisement Regulation of Medical Device?

MDs are regulated as drugs in India. Rule 106 A states,

> No drug (or MD) may purport or claim to prevent or cure or may convey to the intending user thereof any idea that it may prevent or cure, one or more of the diseases or ailments specified in Schedule J.

The Drugs and Magic Remedies (Objectionable Advertisement) Rules, 1955: This Act is meant to control the advertisements related to drugs (or MD).

28.4 Upcoming Regulation Changes

Since the notification (October 2005) of 10 categories of MDs, the registration and product approval process followed by the DCGI office has been the same as followed for drugs. This has led to a mix-up of the drug and device regulations. The industry and the regulators in India have realized this, and some attempts have been made by them collectively to resolve this in the near future. Some of the things are listed below:

- Currently there is no proper definition of a medical device in the act. A complete definition of the MD (in line with GHTF) has been proposed in the Drugs and Cosmetics amendment Bill. This bill was reviewed by Parliament Standing Committee in 2009–2010, and the bill after modification was re-submitted to the Parliament again in October 2010. This bill is now pending the Parliament approval. The Parliament's approval of this bill will lead to the inclusion of the definition of a medical device and related regulations/guidelines in the Drugs and Cosmetics Act. Following are the features of the bill:
    o The bill proposes a proper definition of a medical device and a clinical trial and also proposes separate chapter II A in the Act for the provisions related to MDs.
    o The preamble changes in the act have been proposed, and the proposed name would be "Drugs, Cosmetics and Medical Devices Act".
    o The bill proposes the Powers of Central Government to make rules regarding classification, pre-market evaluation, standards, conformity assessment procedures, labelling and registration of MDs, etc.
- A core industry–regulator group has compiled (during 2008–2010) a comprehensive document Schedule M III containing MD regulations. These can form part of the Drugs and Cosmetics Act once the MD definition becomes part of the act through the passage of the aforementioned bill by Parliament. Schedule MII is based on the following:
    o the GHTF guidelines and the European Medical Device Directive
    o risk-based classification of MDs as per GHTF guidelines
    o the conformity assessment procedures involving the notified bodies
    o the adverse event reporting in line with GHTF guidelines
    o requirements for the clinical trials

- As the aforementioned bill approval by Parliament and finalization of the Schedule M III is a time-consuming process, the DCGI office has come out with MD-specific interim guidelines. These guidelines cover the registration of the site/product and import license submission requirements (common submission dossier) and approval process for MDs and in vitro devices, clinical trials, etc. These are available on the CDSCO website under the MDs section.

28.5 Related Agencies/Departments and Ministries

- Ministry of Health (MoH)
  - Directorate General of Health Services (DGHS)
  - Central Drugs Standards Control Organization (CDSCO)
- Ministry of Chemicals and Fertilizers
  - Department of Pharmaceuticals (DoP)
  - National Pharmaceutical Pricing Authority (NPPA)
- Department of Science and Technology (DST)
- Federation of Indian Chambers of Commerce and industry (FICCI)
- Confederation of Indian Industries (CII)

Chapter 29

Indonesia: Medical Device Regulatory System

Mita Rosalina

PT. Johnson & Johnson Indonesia
mrosalin@its.jnj.com

Disclaimer

The regulatory information contained in this chapter is intended for market planning and is subject to change frequently. Translations of laws and regulations are unofficial.

29.1 Introduction

Indonesia, with a population of more than 220 million, is a medical device market that has a large potential. However, foreign manufacturers of medical devices have always been faced with numerous obstacles and challenges in entering this market. This is despite the fact that Indonesia relies mainly on imported medical devices due to the lack of any established local manufacturer.[4]

Medical Regulatory Affairs: An International Handbook for Medical Devices and Healthcare Products (Third Edition)
Edited by Jack Wong and Raymond K. Y. Tong
Copyright © 2022 Jenny Stanford Publishing Pte. Ltd.
ISBN 978-981-4877-86-2 (Hardcover), 978-1-003-20769-6 (eBook)
www.jennystanford.com

Penetrating this huge market of medical devices is difficult due to the wide and extensive geographical boundaries in the country. Also, highly bureaucratic medical device import and trading regulations have made it more difficult for foreign manufacturers to enter the market. Indonesia also imposes tariffs of up to 30% on imported medical devices depending on the type, use, and value of these devices.[4]

Indonesia has a centralized healthcare system, headed by the Ministry of Health (Kementerian Kesehatan R.I.). The Ministry of Health purchases most of the medical disposables and hospital equipment for the 900 public hospitals around the country. Due to the highly rigid nature, there is limited budget for most of the hospitals in the country. Recent changes, however, have been encouraging in de-centralizing healthcare services in the country. This will spur the growth of the medical devices market in the country as individual regions collect and plan the healthcare budget individually themselves.[4]

There is a market disparity in the standard of healthcare between rural and urban areas. The capital city, Jakarta, enjoys relatively good levels of primary care as well as a range of modern private specialist facilities, while healthcare coverage in remote regions tends to be insufficient.

The majority of the Indonesian medical device market is supplied by imports, which dipped by 1.5% over the previous year to US$ 348.4 million in 2009. Imports have grown at a CAGR of 35.2% in 2005–2009.[5] During the 24th ASEAN Consultative Committee on Standards and Quality Meeting on August 3–4, 2004, the formation of a Product Working Group on Medical Device (ACCSQ-MDPWG) was proposed to implement specific measures on medical device under the road map for healthcare integration. This is to be in line with the ASEAN leaders' decision on the establishment of the ASEAN Economic Community (AEC) by 2020 and fast-track integration of the 11 priority sectors, including the healthcare sector. Therefore, as one of ASEAN member country, Indonesia is committed to the implementation of ASEAN Harmonization in 2014.

To welcome it, at this time, Indonesia has some regulations relating to medical devices:

- Ministerial Regulation of Health Republic Indonesia No. 1189/MENKES/Per/VIII/2010. *Production of Medical Devices and Household Device*. Issued on August 23rd, 2010.
- Ministerial Regulation of Health Republic Indonesia No. 1190/MENKES/Per/VIII/2010. *Marketing Authorization Medical Devices and Household Device*. Issued on August 23rd, 2010.
- Ministerial Regulation of Health Republic Indonesia No. 1191/MENKES/Per/VIII/2010. *Distribution Medical Devices and Household Device*. Issued on August 23rd, 2010.

29.2   Regulating Authority

Medical devices in Indonesia are regulated by the Ministry of Health of the Republic Indonesia; Directorate of General Pharmaceutical Service and Medical Device; and Directorate of Medical Device Production and Distribution Development with the following address of office:

Kementrian Kesehatan RI
Jl. H.R. Rasuna Said Blok X-5
Kav No. 4–9; Jakarta 12950
Indonesia
Web site: www.binfar.depkes.go.id

29.3   Definition of Medical Device

A medical device means any instrument, apparatus, implement, machine, appliance, implant, in vitro reagent, software, material, or other similar or related article

(a) intended by the manufacturer to be used, alone or in combination, for human beings for one or more of the specific purpose of
- diagnosis, prevention, monitoring, treatment, or alleviation of disease
- diagnosis, monitoring, treatment, or alleviation of or compensation for an injury
- investigation, replacement, modification, or support of the anatomy or of a physiological process

- supporting or sustaining life
- control of conception
- disinfection of medical device
- providing information for medical or diagnostic purposes by means of in vitro examination of specimens derived from the human body

(b) that does not achieve its primary intended action in or on the human body by pharmacological, immunological or metabolic means, but may be assisted in its intended function by such means.

29.4 Classification of Medical Devices

Medical devices are classified into three classes based on risk[2]:

- **Class I**: In case of malfunction or misuse, these low-risk devices would not cause serious harm. The evaluation of these health instruments shall focus on the quality and the product.
- **Class IIa**: In case of failure or misuse, these medical devices can result in a significant impact on the health of patients, but not a serious accident. Before distribution, such health instruments will have to comply with a sufficient requirement for evaluation, but they do not require a clinical test.
- **Class IIb**: In case of failure or misuse, these medical devices can have significant effect on the health of patients, but not a serious accident. Before distribution, these health instruments shall have to comply with a complete requirement for evaluation, including a risk analysis and security facts for the evaluation, but they do not require a clinical test.
- **Class III**: The failure or misuse of these medical devices can result in serious implications for patients or nurses/operators. Before distribution, these health instruments shall have to comply with a complete requirement for evaluation, including a risk analysis and security facts for the evaluation, and they also require a clinical test.

Currently Indonesia does not have any guidelines on combination products. Therefore, it is strongly recommended that prior to the registration process, the product specification, intended use, and product claims be discussed with the Ministry of Health staff to ensure the classification of the product. Attached classification of product from legal manufacturer on application is recommended as reference.

29.5   Registration of Medical Devices

All medical devices (for all classes) with each code should be registered with the Ministry of Health before they can be freely imported, distributed, and sold in the market. Unit package that needs to be registered depends on unit sales.

Safety, quality, and efficacy of the product will be considered during evaluation to grant marketing authorization.

Abridged evaluation is not recognized by Indonesia authority. A full assessment is required for all classes of product. Approval status from others countries serves only as reference for the Ministry of Health staff in the reviewing process.

29.5.1   Process

The application must have distribution license "IPAK" from the Ministry of Health and the company that as applicant officially appointed by the source company or the legal manufacturer as the sole agent. This letter of appointment from source/legal manufacturer explains that the representative company in Indonesia has authorization to register, import, and distribute the product in Indonesia. The minimum appointment granted by the source company is two years, and this document needs to be legalized by the Indonesian embassy. It should be noted that the company that has the right to import the product is the same company that is the license holder for the products.

Currently, application is done by submitting the hard copy of registration documents and follow-ups directly to the Ministry of Health. English and Indonesian are the acceptable languages for the documents.

All classes of medical devices follow the registration process as shown in Fig. 29.1.

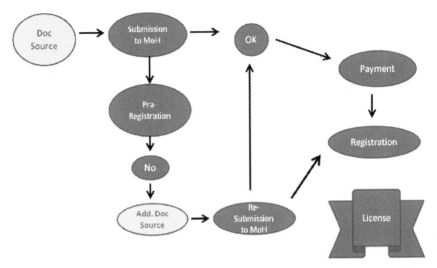

Figure 29.1 Medical device registration process followed by the Ministry of Health.

29.5.2 Documents Required

Please refer to Appendix 1.

29.5.3 Official Registration Fee

Type of medical device	New registration (mio rp)	Re-registration/variation/renewal (mio rp)
Class III	5	1
Class II	3	1
Class I	1.5	0.5
Explanation letter: 0.5 Mio Rp		

29.5.4 Time Line

The time frame needed to get approval is as follows:
- Class I: 30 working days after pre-registration
- Class II : 60 working days after pre-registration
- Class III: 90 working days after pre-registration

In general, it takes around six to nine months from the date of submission to get approval. This applies for new registration, re-registration, variation, and renewal.

29.5.5 Validity of Product License

The product license issued by the Ministry of Health will be effective for five years or in accordance with the term of validity of the letter of appointment. The certificate of license needs to be renewed three months before the expiry date.

29.5.6 Indonesian Labeling Requirement

Refer to Ministerial Regulation of Health Republic Indonesia No. 1190/MENKES/Per/VIII/2010: Marketing Authorization Medical Devices and Household Device; article 26.

Following is the minimum information that needs to be stated on the labeling:
- trade name/brand name/product name
- name and address of legal and physical manufacturer
- name and address of Importer
- product component, formula (household device)
- active ingredient with percentage (household device)
- intended use and direction for use (must be in Indonesian)
- warning, precaution, and adverse effect (must be in Indonesian)
- expiry date
- batch number/production code/serial number, registration number, and net quantity of the product

29.5.7 Regulatory Action for Changes and Device Modifications

This is not specified in the regulations. Changes can be registered via a data amendment process, variation, or re-registration, and the requirements are judged on a case-by-case basis.

Any changes that have an impact on the following should be considered:
(1) size
(2) packaging specification
(3) LABELING
(4) tax payer number (NPWP)

29.6 Post-Market Surveillance System

There can be severe consequences of faulty or substandard medical devices for public health and safety. Regulatory bodies all over the world are thus striving to achieve an optimal system of medical device controls that accommodates a device in all its permutations and combinations, balancing quality and safety with faster access to the market.

Regulatory systems need to be in place to manage the potential residual risks posed to the intended user. Because of this, the Indonesia authority is developing a Post-Market surveillance process for medical devices. Post-market surveillance mechanisms are required to monitor the medical devices already in the market. This is essential for such devices. In the event of any complications, the authorities are able to direct swift corrective actions.

In Indonesia, post-market surveillance is still voluntary. However, it has been recommended by the authority for cases that have adverse health consequences for the patient or cause death. In such cases, it is mandatory for the company to report to the authority in 24 hours with details of the corrective action.

References

1. Ministerial Regulation of Health Republic Indonesia No. 1189/MENKES/Per/VIII/2010. *Production of Medical Devices and Household Device.* Jakarta, August 23, 2010.
2. Ministerial Regulation of Health Republic Indonesia No. 1190/MENKES/Per/VIII/2010. *Marketing Authorization Medical Devices and Household Device.* Jakarta, August 23, 2010.
3. Ministerial Regulation of Health Republic Indonesia No. 1191/MENKES/Per/VIII/2010. *Distribution Medical Devices and Household Device.* Jakarta, August 23, 2010.
4. Wee, A. (2006). *Medical Device Distribution in Indonesia-Partnering Established Companies for Optimum Penetration.* Retrieved March 7, 2012, from http://www.frost.com/prod/servlet/market-insight-top.pag?docid=72640483.
5. Espicom (2012). *The Medical Device Market: Indonesia.* Retrieved March 7, 2012, from http://www.espicom.com/prodcat2.nsf/Product_ID_Lookup/00000554?OpenDocument.

Appendix 1

Indonesia CSDT Format

Attachment A: Administrative Documents

Sections	Documents required	Details	Class I	Class II a	Class II b	Class III
Administrative Documents	1. Copy of Local Company Licence as Medical Devices Distributor	– No need to do addendum process to add new principle in distribution license as long its products category already available in the list – If product category has not yet in the list, addendum process will be needed – JJMI will provide this himself.	*	*	*	*
	2. Power of Attorney	POA (Power of Attorney) from the legal manufacturer to local company as sole agent or sole distributor in Indonesia—Need ID embassy legalization from the country of legal manufacturer	*	*	*	*
	3. Certificate of Free Sale	No need ID embassy legalization Endorsed by regulatory body in country of physical manufacturer or country of legal manufacturer, including the following information: product name, product code, description and the manufacturing site	*	*	*	*

(Continued)

Attachment A (Continued)

Sections	Documents required	Details	Class I	Class II a	Class II b	Class III
4. Executive Summary	A. Overview	*Brief description of device *Intended uses and indications for use *Any novel features	*	*	*	*
	B. Commercial Marketing History	*Date of first introduction and use *List of countries where it is marketed	*	*	*	*
	C. Intended Use and Indications	*Instructions for Use (IFU)	*	*	*	*
	D. List of regulatory approval or marketing clearance obtained and status of any pending request for market clearance	*Registration status in reference agencies *Copies of certificates or approval letters from the reference agencies (approval certificates from the 5 benchmark countries) *Declaration on labeling, packaging and instructions for use (IFU)	*	*	*	*
	E. Important safety/ performance related information	*Summary of adverse events and recalls *Description of the following (if the medical device contains these items): 1. Drugs (description about safety and efficacy drugs usage, doses, adverse effect, pharmacology, etc.) If applicable 2. Animal or human cells, tissues and/or derivatives thereof, rendered non-viable (e.g. porcine heart valves, catgut sutures, etc.); If applicable	*			*

Sections	Documents required	Details	Class I	Class II a	Class II b	Class III
		3. Cells, tissues and/or derivatives of microbial or recombinant origin (e.g. dermal fillers based on hyaluronic acid derived from bacterial fermentation processes); If applicable				
		4. Irradiating components, ionizing (e.g. x-ray) or non-ionizing (e.g. lasers, ultrasound, etc.). If applicable				
5. Essential Principles	Essential principles conformity checklist	*Essential Principles of Conformity Checklist for the device. This checklist should contain all of the essential principles described in the Guidance on Essential Principles for Safety and Performance of Medical Devices	*	*	*	*
6. Evidence conformity with the standard	Evidence conformity with the standard	*Certificate of Analysis (COA) of products. COA should be matched with product specification. The template should mention about parameter, limitation and nominal result of test. COA should mention name of product that we register, batch/lot no. *ISO 13485 from Physical manufacture and legal manufacture *IEC 60-601 for electrical safety – if applicable *Declaration of conformity	*	*	*	*

Attachment B: Information of Device

Sections	Documents required	Details	Class I	Class II a	Class II b	Class III
Device Description	1. Description ABCD	Description about: – Direction for use – Indication – Brochure – Raw material specification – Expiry date (only for sterile products) – JJMI will provide this himself.	*	*	*	*
	2. Device Description and Features	*Description and principles of operation	*	*	*	*
		*Risk Class and applicable classification rules	(If Applicable)	(If Applicable)	(If Applicable)	(If Applicable)
		*Description of accessories and other devices to be used with the medical device				
		*Description of key function elements of the medical device				
		*Labeled pictorial representation (product drawing, picture, photo)				
		*Explanation of any novel features				
	3. Intended use, 4. Indications, 5. Instructions of Use, 6. Contraindications, 7. Warnings, 8. Precautions and 9. Potential Adverse Effects	*Information typically found in the Instructions for Use (IFU); for electromedic--> Manual book should be provided. These information should be provided in Indonesia language as well.	*	*	*	*

Sections	Documents required	Details	Class I	Class II a	Class II b	Class III
	10. Alternative Therapy	*Description of any alternative practices or procedures for diagnosing, treating, curing or mitigating the disease or condition for which the device is intended	*	*	*	*
	11. Materials	*List of materials of the medical device making direct contact (e.g. with the mucous membrane) or indirect contact (e.g. during extracorporeal circulation of body fluids) with a human body.		*	*	*
		*Include information regarding chemical material, biological and physical characterization of the medical devices		*	*	*
		*For medical devices intended to emit ionizing radiation, information on radiation source (e.g. radioisotopes) and the material used for shielding of unintended, stray or scattered radiation–if applicable		*	*	*
	12. Other Descriptive Information-manufacturer information	*Description of any other important characteristics of the medical device that is not addressed in the preceding sections. *Finished product specification test (protocol and result) –example PPQ protocol and result	*	*	*	*

(*Continued*)

Attachment B (Continued)

Sections	Documents required	Details	Class I	Class II a	Class II b	Class III
	13. Manufacturing Process	Must-cover method of manufacturing process, condition, environment, facilities and controls used for process, packaging, labeling and storage devices. This process can be represent by flow chart of manufacturing process. Include a summary of methods and sterilization processes (for sterile products)	*	*	*	*
		*Identify all sites (name and full address) for design and manufacturing activities (including contract manufacturers and contract sterilizers)				
		1. Identify the activities conducted at different manufacturing sites				
		2. Provide Quality Management System (QMS) certificates for all design and manufacturing sites (including contract manufacturers and contract sterilizers)				
	13a. Sterilization validation (for sterile products)	*Information on the sterilization validation method used, sterility assurance level (SAL) attained; standards applied, sterilization protocol, summary of validation results	*	*	*	*

Sections	Documents required	Details	Class I	Class II a	Class II b	Class III
		*Initial sterilization validation including bioburden testing, pyrogen testing, testing for sterilant residues (if applicable) and packaging validation. If initial sterilization validation is not performed, adequate justification must be provided	*	*	*	*
		*Evidence of the ongoing revalidation of the process (e.g. evidence of revalidation of the packaging and sterilization processes)		*	*	*
		*Post-sterilization functional test on the medical device	*	*	*	*
		*If the sterilant is toxic or produces toxic residuals, test data and methods that demonstrates the post-process sterilant and/or residuals are within acceptable limits	*	*	*	*
		*Data demonstrating that the relevant performances and characteristics of the medical device are maintained throughout the claimed shelf life which the "expiry" date reflects, including:	*	*	*	*
	13b. Stability Studies (for devices with a shelf life)	1. Prospective studies using accelerated ageing, validated with real time degradation correlation; or	*	*	*	*

(Continued)

Attachment B (*Continued*)

Sections	Documents required	Details	Class I	Class II a	Class II b	Class III
		2. Retrospective studies using real time experience, involving e.g. testing of stored samples; review of the complaint history or published literature etc.; or	*	*	*	*
		3. A combination of (1) and (2).	*(If applicable)	*(If applicable)	*(If applicable)	*(If applicable)
		*Summary of stability studies conducted, stability protocols used, results of studies. This summary should include conclusions with respect to storage conditions and shelf life, and, if applicable, in-use storage conditions and shelf life.	*	*	*	*
13c. Projected Useful Life (for devices without expiry dates)		* For devices that do not have expiry dates (e.g. infusion pump, digital thermometer), the projected useful life of the medical device must be provided. Manufacturers may refer to TS/ISO 14969 (Medical devices – Quality management systems – Guidance on the application of ISO 13485:2003) for information on how to determine the projected useful life	*	*	*	*

Attachment C: Specification and Quality Assurance Device

Sections	Documents required	Details	Class I	Class II a	Class II b	Class III
Specification and Quality Assurance Device	1. Specification	Finished product specification: functional characteristics and technical performance specifications for the device including, as relevant, accuracy, sensitivity, specificity of measuring and diagnostic devices, reliability and other factors; and other specifications including chemical, physical, electrical, mechanical, biological, software, sterility, stability, storage and transport, and packaging to the extent necessary to demonstrate conformity with the relevant Essential Principles.	*	*	*	*
	2. Other Relevant Specifications	List of features, dimensions and performance attributes of the medical device, its variants and accessories that would typically appear in the product specification made available to the end user, e.g. in brochures and catalogues		*	*	*
	3. Summary or reference or design verification and design validation	This section should summarize or reference or contain design verification and design validation data to the extent appropriate to the complexity and risk class of the device: Such documentation should typically include: declarations/certificates of conformity to the "recognized" standards listed as applied by the manufacturer; and/or		*	*	*

(*Continued*)

Attachment C (Continued)

Sections	Documents required	Details	Class I	Class II a	Class II b	Class III
		– summaries or reports of tests and evaluations based on other standards, manufacturer methods and tests, or alternative ways of demonstrating compliance. The data summaries or tests reports and evaluations would typically cover, as appropriate to the complexity and risk class of the device: a listing of and conclusions drawn from published reports that concern the safety and performance of aspects of the device with reference to the Essential Principles; • engineering tests; • laboratory tests; • biocompatibility tests; • animal tests; • simulated use; • software validation				
Clinical Study	4. Pre-clinical Studies	*Biocompatibility Testing (Protocol and result) *Physical Testing (Protocol and result) *Animal Testing (Protocol and result) *Software validation Studies (Protocol and result)			* (If Applicable)	* (If Applicable)
	5. Devices containing Biological material (If applicable)	* results of studies in relation to the risks associated with transmissible agents. Statement letter for disease and virus free	* (If Applicable)	* (If Applicable)	* (If Applicable)	* (If Applicable)

Sections	Documents required	Details	Class I	Class II a	Class II b	Class III
	6. Software verification and validation studies (if applicable)	*Manufacturer should give design validation and software development. The manufacturer and/or device sponsor must provide evidence that validates the software design and development process. This information should include the results of all verification, validation and testing performed in-house and in a user's environment prior to final release, for all of the different hardware configurations identified in the labeling, as well as representative data generated from both testing environments.			*	*
	7. Clinical Evidence	*Clinical evaluation report of the device			*	*
		*Full reports of all studies referenced in clinical evaluation report			*	*
		Note: The documented evidence submitted should include the objectives, methodology and results presented in context, clearly and meaningfully. The conclusions on the outcome of the clinical studies should be preceded by a discussion in context with the published literature.
Clinical evidence of effectiveness may comprise device-related investigations conducted domestically or other countries. It may be derived from relevant publications in a peer-reviewed scientific literature				

(Continued)

Attachment C (Continued)

Sections	Documents required	Details	Class I	Class II a	Class II b	Class III
		Use of Existing Bibliography Copies is required of all literature studies, or existing bibliography, that the manufacturer is using to support safety and effectiveness. These will be a subset of the bibliography of references. General bibliographic references should be device-specific as supplied in chronological order. Care should be taken to ensure that the references are timely and relevant to the current application				
		Clinical evidence can be obtained from publications relating scientific literature which has been discussed with the experts.				
Risk Management	8. Risk Management Documents	*Risk Management report for the design of the device and its manufacturing process		*	*	*
		*Accompanying documents referenced in the report, including the risk management plan, results of risk assessment and risk control				
Raw material specification	raw material specification	Specification of raw material or components of product.		*	*	*
Packaging Material Specification (for diagnostic product)	Packaging Material Specification (for diagnostic product)	Packaging Material Specification (for diagnostic product),		*	*	*
Analysis Test Report	Analysis Test Report	Analysis Test Report and or clinical study for reagent/in vitro diagnostic product		*	*	*

Attachment D: Labeling and Intended of Use

Sections	Documents required	Details	Class I	Class II a	Class II b	Class III
Device Labeling *Labels are required for each and every product codes which are to be registered.	1. Device Labeling	*Copies (in original color) of:				
		1. Labels on the device and its packaging (primary and secondary levels)	*	*	*	*
		2. Instructions for use/IFU (including operating manual and user manual) –> Should be Indonesian language		*	*	*
		3. Patient information leaflet (where applicable)		*	*	*
		4. Product picture with address of manufacture for Electromedic device	*	*	*	*
		*Promotional material (including brochures and catalogues) *Note*: These following information must be available: – Name and address physical manufacturer medical devices and/or Household Medical Supply – IFU need to be translated in Indonesian language, at least with below information: – Indication and Direction of use must be in the Indonesian language – Adverse effect-warning signs must be in the Indonesian language – Importer-name and address as well as marketing authorization number				
Lot Numbering system		Lot numbering system or explanation letter regarding the lot system	*	*	*	*

Attachment E : Post-Market Evaluation

Sections	Documents required	Details	Class I	Class II a	Class II b	Class III
Post-Market Evaluation	1. SOP about handling product complaint, recall and stop shipment	This SOP should come from the legal manufacturer/source and local as applicant			*	*

Chapter 30

Japan: Medical Device Regulatory System

Atsushi Tamura and Keizo Matsukawa

Pharmaceuticals & Medical Devices Agency,
Shin-Kasumigaseki Bldg., 3-3-2 Kasumigaski, Chiyoda-Ku, Tokyo, Japan
tamura-atsushi@pmda.go.jp

Disclaimer

The regulatory information contained in this chapter is intended for market planning and is subject to change frequently. Translations of laws and regulations are unofficial.

30.1 Introduction

Japan is an island nation in East Asia, located in the Pacific Ocean with around 127 million people as of January 1, 2020, according to the Ministry of Internal Affairs and Communications.[1]

Regarding international harmonization of medical device, Japan has been a founding member of the Global Harmonization Task Force (GTHF) and currently is a member of the International Medical Device Regulators Forum (IMDRF).

Medical Regulatory Affairs: An International Handbook for Medical Devices and Healthcare Products (Third Edition)
Edited by Jack Wong and Raymond K. Y. Tong
Copyright © 2022 Jenny Stanford Publishing Pte. Ltd.
ISBN 978-981-4877-86-2 (Hardcover), 978-1-003-20769-6 (eBook)
www.jennystanford.com

Production, import and export values of medical devices are 1950 (1990), 1621 (1649) and 668 (619) billion yen, respectively in FY2018 (FY2017), respectively.[2]

30.2 Regulatory Agency in Japan

30.2.1 Ministry of Health Labour and Welfare/MHLW

The Ministry of Health, Labour and Welfare (MHLW) designs and implements social systems and structures that are closely related to people's lives throughout their lifetime and entire life cycle. The MHLW covers a wide range of areas from medical care to employment and childcare support and each department of the ministry serves "for people, for life, and for the future" on a daily basis.[3]

The mission of the Ministry is to build a society that ensures lifelong security of each and every person—from birth through post-retirement. It will also support the basis of economic growth of Japan. In order to look after people and their lives and support this country into the future, the Ministry is working to promote various measures.

The Pharmaceutical Safety and Environmental Health Bureau

The Pharmaceutical Safety and Environmental Health Bureau is working to ensure quality, efficacy, and safety of pharmaceuticals, etc., and to prevent the occurrence/expansion of health risks through approval reviews and safety measures of pharmaceuticals/medical devices manufactured and sold in Japan and by development of pharmacy/pharmacist systems. The Bureau is also working to make people's lives safer and more hygienic through formulation of food standards, supervision/guidance, improvement of environmental health, and the provision of safe tap water, etc.

30.2.2 Pharmaceuticals and Medical Devices Agency

The Pharmaceuticals and Medical Devices Agency (PMDA) is Japanese regulatory agency, working together with the MHLW.

PMDA's obligation is to protect the public health by assuring safety, efficacy and quality of pharmaceuticals and medical devices. PMDA conducts scientific reviews of marketing authorization application of pharmaceuticals and medical devices, monitoring of their post-marketing safety. PMDA is also responsible for providing relief compensation for sufferers from adverse drug reaction and infections by pharmaceuticals or biological products.[4]

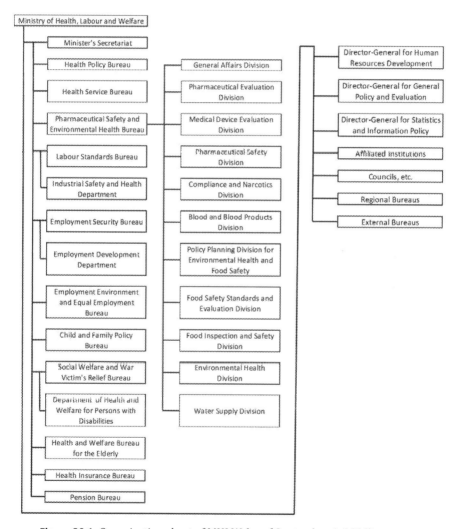

Figure 30.1 Organization chart of MHLW (as of September 1, 2020).

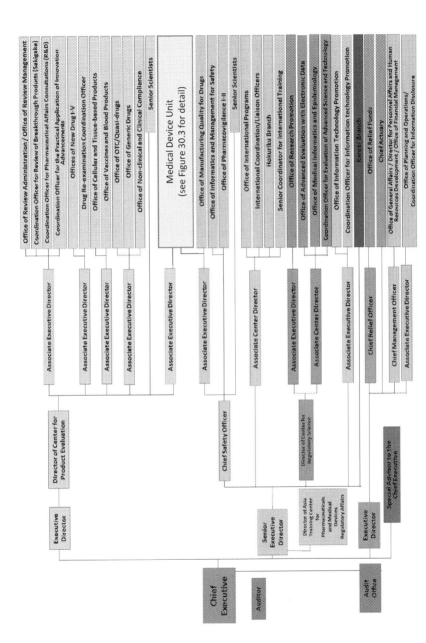

Figure 30.2 Organization Chart of PMDA (As of April 1, 2021).

30.2.3 PMDA Medical Device Unit

Establish a department specializing in medical devices, while strengthened collaboration and coordination among each division of the medical device field and established a system to efficiently carry out tasks in order to take on more highly specialized tasks based on the characteristics of medical devices

Office of Medical Devices I
Office of Medical Devices II
Office of Medical in Vitro Diagnostics
Office of Software as a Medical Device
Office of Manufacturing Quality and Vigilance for Medical Devices

- Division of vigilance of medical devices
- Division of manufacturing quality of medical devices
- Division of drugs for In Vitro Diagnostics

Office of Standards and Compliance for Medical Devices

- Division of standards for medical Devices
- Division of non-clinical and clinical Compliance of medical devices
- Division of governing registered certification bodies

30.2.4 Shared Responsibility of MHLW and PMDA on Medical Device Regulation

MHLW has ultimate responsibilities in policies and administrative measures, for example, the final judgment on approval, product withdrawal from market, and so on. On the other hand, PMDA plays a role as a technical arm of the MHLW, through an actual review, examination, data analysis, etc., to assist MHLW's measures, for instance, approval Review of medical devices, Quality Management System/Good Laboratory Practice/Good Clinical Practice inspection, and collection and analysis of Adverse Event Reports.

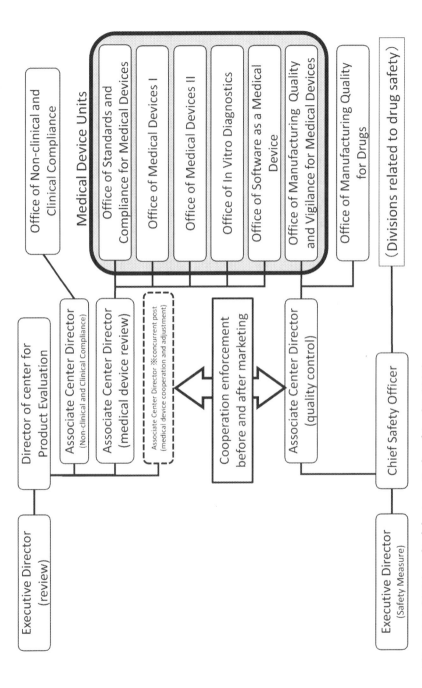

Figure 30.3 PMDA medical device units (as of April 1, 2021).

30.3 Legislation of Medical Devices

Medical device administration in Japan is maintained in accordance with a number of laws concerned (Table 30.1). Among them, the Act on Securing Quality, Efficacy and Safety of Products Including Pharmaceuticals and Medical Devices which is called as the Pharmaceuticals and Medical Devices Act in short and it is abbreviated as PMD Act, is the supreme law. Some related cabinet ordinance, ministerial ordinance and ministerial notification are issued for enforcement of the PMD Act. Furthermore, notifications by the director general of bureau and/or notifications by director of division/office are also published as guidance documents for smooth enforcement of the PMD Act.

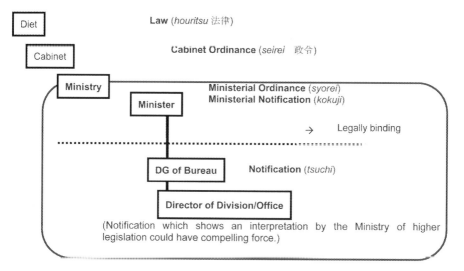

Figure 30.4 Hierarchy of legislation.

Medical devices are defined under the PMD Act, as follows:

The term "medical device" as used in this Act refers to appliances or instruments, etc., which are intended for use in the diagnosis, treatment or prevention of disease in humans or animals, or intended to affect the structure or functioning of the bodies of humans or animals (excluding regenerative medicine products), and which are specified by Cabinet Order.

Table 30.1 Key medical device legislation

Law (houritsu)	Act on Securing Quality, Efficacy and Safety of Products Including Pharmaceuticals and Medical Devices (PMD Act, 1960)[5]
Cabinet Ordinance (seirei)	Order for Enforcement of the Act on Securing Quality, Efficacy and Safety of Products Including Pharmaceuticals and Medical Devices (Cabinet Ordinance on PMD Act, 1961)[6] Cabinet Ordinance on Pharmaceutical Affairs and Food Sanitation Council, 2000
Ministerial Ordinance (shorei) [7]	Regulation for Enforcement of the Act on Securing Quality, Efficacy and Safety of Products Including Pharmaceuticals and Medical Devices (Ministerial Ordinance on PMD Act, 1961)[8] Ministerial Ordinance on Standards for Manufacturing Control and Quality Control of Medical Devices and In Vitro Diagnostic Reagents 2004 Ministerial Ordinance on Good Clinical Practice for Medical Devices Ministerial Ordinance on the Standards of the System for Manufacturing Management or Quality Control of Medical Devices and In Vitro Diagnostic Reagents 2014 Ministerial Ordinance on Good Clinical Practice for Medical Devices 2005 The Ministerial Ordinance on Good Vigilance Practice for drugs, quasi-drugs, cosmetics, and medical devices 2004 Good Quality Practice (GQP), 2004, etc.
Ministerial Notification (kokuji)	Essential Principles[9] Certification standards for Designated Controlled Medical Devices and Designated Specially Controlled Medical Devices Classification of medical devices[10,11] List of orphan designation, etc.
Notification (tsuchi) by Director General of Bureau or Director of Division/Office[12]	Information on application procedures Guidelines for clinical evaluation, etc.

All medical devices have a generic name which is called the Japanese Medical Device Nomenclature (JMDN). In JMDN, not only single product but also some grouping products like surgical kit, ablation system and so on is defined.

There is no clear definition in this moment regarding a product comprising two or more regulated components, so called a combination product.

In order to bring medical devices in to Japanese market, under the PMD Act, a legal procedure is required for products, companies and manufacturing sites, respectively (Table 30.2).

Table 30.2 Prerequisites to bring MDs into the Japanese market

Product	Minister's Approval (*shonin*) (Art. 23-2-5) or Third Party Certification (*ninsho*) (Art. 23-2-23) or Marketing Notification (*todokede*) (Art. 23-2-12)
Company (MAH)	License for Marketing Authorization Holder (Art. 23-2)
Manufacturing Site/ Manufacturer	Registration of Manufacturer (Art. 23-2-3) Registration of Foreign Manufacturer (Art. 23-2-4)

Medical devices have at least to be indicated the name and address of market authorization holder, the brand name, the generic name, the serial number or manufacture sign, and the device category on device itself, wrappers or containers clearly in Japanese. Since there are the other labelling requirements depending upon the type of the devices, usage of the specific materials and so on, it is necessary to check carefully the PMD Act and related regulations.

30.3.1 Classification of Medical Devices

The classification identifies the level of regulatory control that is necessary to assure the safety and effectiveness of a medical device. In Japan, medical devices are classified mainly into four classes, on basis of risk-based system following the GHTF rule (Table 30.3).[4] Japanese classification rules basically follow the GHTF rules, but there are some deviations in some medical devices.

Table 30.3 Classification of medical devices

GHTF rule	Category	Examples	Risk description to the patient and/or user
Class I	General Medical Devices	Surgical Knife, Cotton wool	Low
Class II	Controlled Medical Devices	CT, Pulmonary catheter	Low-Medium
Class III	Specially Controlled Medical Devices	Dialyzer Artificial Knee	Medium-High
Class IV	Specially Controlled Medical Devices	Artificial Heart Stent graft	High

30.3.2 Type of Product's Registration

30.3.2.1 Notification of marketing (Art. 23-2-12)

General Medical Devices in PMD Act are those devices that have a relatively low risk to the patient and/or user. A marketing authorization holder (MAH) who intends to market a General Medical Devices is not required to obtain the Minister's approval, and is allowed to launch a medical device onto the Japanese market by submitting the marketing notification for the medical device to PMDA.

30.3.2.2 Certification (Art. 23-2-23) (third-party certification)

Designated Controlled Medical Devices and Designated Specially Controlled Medical Devices are those devices that are certified by verifying compliance with certification standards. A MAH who intends to market Designated Controlled Medical Devices and Designated Specially Controlled Medical Devices must receive certification for each of such items. To register and market a designated controlled medical device, the MAH needs to file Pre-Market certification application with a Registered Certification Body (third-party certification body) and obtain their certification. Application dossiers for pre-market certification have to be written in Japanese and the technical data and supporting information have to be submitted following the summary technical documentation (STED) format.[13]

A person who intends to market specially-controlled medical devices, controlled medical devices or in vitro diagnostic reagents (IVDs) with specified standards designated by the MHLW (hereinafter referred to as "designated specially-controlled medical devices"), or a person who is engaged in manufacturing, etc., of designated specially-controlled medical devices exported to Japan in foreign countries (hereinafter referred to as "foreign manufacturer of designated specially-controlled medical devices") and intends to have a marketing authorization holder appointed pursuant to the provisions of Article 23-3, paragraph (1) to market designated specially-controlled medical devices must receive certification for each of such items by a person who is registered for marketing by the MHLW (hereinafter referred to as a "registered certification body"), pursuant to the provisions of Order of the MHLW.

30.3.2.3 Approval (Art. 23-2-5)

Specially Controlled Medical Devices that do not comply with the certification standards must be approved by the MHLW. When a MAH intends to launch a "specially controlled medical device" onto the Japanese market, the Minister's approval to market the medical device is required. The Minister's approval is granted based on the scientific review by PMDA.

Controlled Medical Devices other than Designated Controlled Medical Devices are also subject to Pre-market Approval. In the case that no applicable certification standard has been established or that the product is deemed as a new medical device, the MAH is required to submit an application to PMDA to obtain the Minister's approval for the product.

Application dossiers for pre-market approval have also to be written in Japanese and the technical data and supporting information have to be submitted following the summary technical documentation (STED) format.[13]

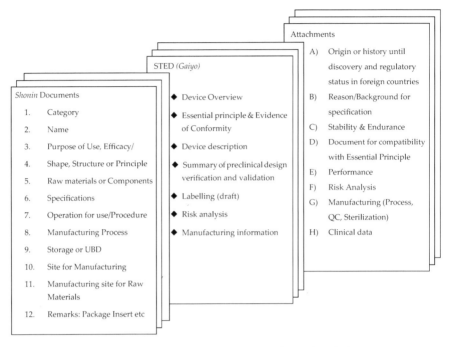

Figure 30.5 Image of application documents set for high-risk device.

30.3.3 Marketing Licenses (Art. 23-2)

In accordance with the criteria for medical devices or IVDs set forth in the left hand columns of Table 30.4, no person other than the one who has obtained a license from the MHLW specified in the right hand columns of the same table, respectively, may engage in the business of marketing medical devices or IVDs. A person intending to manufacture medical devices or IVDs in a foreign country that are exported to Japan (hereinafter referred to as a "foreign manufacturer of medical devices") may obtain registration from the MHLW for each manufacturing facility. The license prescribed in the preceding paragraph expires when a period specified by Cabinet Order of not less than three years passes, unless the license is renewed within each specified period.

Table 30.4 Overview of classification and pre-market regulation for medical devices

Category		Technical standards for certification	Type of regulation	Reviewed by	QMS
General MDs		NA	Registration (to PMDA)	Self-declaration by MAH	Some exemption
Controlled MDs	Designated Controlled Medical Devices	YES	Third Party Certification	Registered Certification Body	Applied
	Others than above	NO	Minister's Approval	PMDA and MHLW	
Specially Controlled MDs	Designated Specially Controlled Medical Devices	YES	Third Party Certification	Registered Certification Body	
	Others than above	NA	Minister's Approval	PMDA and MHLW	

30.3.4 Marketing Authorization Holder (Art. 23-2)

The marketing authorization holder (MAH) is a person who is responsible for the distribution of medical device products at the same time as shipping and marketing products, regardless of whether or not they manufacture medical devices. The MAH must actively collect, analyze, and evaluate not only quality (manufacturing) but also safety (information) and take necessary measures.

In addition, "Notification", "certification" and "approval" of medical devices are also responsible for the MAH. The MAH plays the most important role through the total product life cycle of medical device because the MAH has a responsibility on not only pre-market document control for medical devices and post market safety measurements but also quality control of the device. To market, lease and/or grant the products (which are manufactured or imported), it is necessary to have the appropriate class of MAH license (Table 30.5). The MAH has to have the MAH general manager, a quality assurance manager and a post-marketing safety manager.

Table 30.5 Types of license for MAH (Art. 23-3)

Criteria for medical devices or IVDs	Criteria for license
Specially-controlled medical devices	First-class marketing license for medical devices
Controlled medical devices	Second-class marketing license for medical devices
General medical devices	Third-class marketing license for medical devices
In vitro diagnostic reagents	Marketing license for in vitro diagnostics

30.3.5 Accreditation of Manufacturing Businesses (Art. 23-2-3)/Accreditation of Foreign Manufactures of Medical Devices (Art. 23-2-4)

A person who intends to be engaged in the business of manufacturing medical devices or IVDs must obtain registration from the MHLW for each manufacturing facility (of the manufacturing processes for Medical devices or IVDs, limited to those for designing, assembling, sterilization and others specified by Order of the MHLW, pursuant to the provisions of Order of the MHLW.

A person intending to manufacture medical devices or iIVDs in a foreign country that are exported to Japan (hereinafter referred to as a "foreign manufacturer of medical devices") may obtain registration from the MHLW for each manufacturing facility.[14–16]

The registration prescribed before expires when a period specified by Cabinet Order of not less than three years passes, unless the registration is renewed within each specified period. A manufacturer is a person who manufactures medical devices on a contract basis for MAH. The license as a manufacturer only extends to manufacturing contracted products, not to applying approval/certification application for such products nor selling such products to distributers, in contrast of the rights permitted to MAH.

Table 30.6 Scope of registration of manufacturers

Process		General Medical Device	Other than General Medical Device	Stand-alone software	Recording media for a stand-alone software
Design	Facilities where there is a person responsible for design and development.		Y	Y	Y
Main manufacturing process	Facilities that are responsible for the realization of products	Y	Y		
Sterilization	Facilities for sterilization	Y	Y		
Final shipping warehouse in Japan	The facility where the product is stored at the time of the shipment decision to the market	Y	Y		Y

30.3.6 Registered Certification Bodies

The Third Party Certification System launched in 2005. This certification systems have been introduced to deliver relatively low-risk medical devices and IVDs to patients faster. Manufacturer of Designated Controlled Medical Devices and Designated Specially Controlled Medical Devices and IVDs may obtain more rapid marketing clearance. There are 13 registered certification bodies (RCBs) registered pursuant to the provisions of Article 23-6, paragraph 1 of the PMD act as of September 1, 2020. Half of the RCBs are Japanese companies, and the other half are foreign-affiliated companies.

As for the foreign-affiliated companies, the Japanese branch is registered by the MHLW.

Either Yes or No indicates whether the RCB can certify Medical Devices and IVDs or not. All RCBs can certify medical devices, but only 7 RCBs can certify IVDs as of April 1, 2020.

To certify a product, both "product review" and "QMS inspection" must be passed. If the product fails to pass either product review or QMS inspection, certification cannot be given. In principle, product review and QMS inspection conducted by RCBs do not differ much from those conducted by the PMDA. To pass product review, conformity with certification standards and essential principles is necessary. To pass the QMS inspection, a compliance certificate ensuring conformity with the QMS Ordinance is necessary.

The assessment of RCBs is conducted periodically by the PMDA. A RCB must first submit an application to the MHLW. The MHLW then conducts document review to see whether the necessary data are complete.

After the document review, the PMDA is requested to perform assessment. The PMDA conducts on-site assessment of the certification body to see if the certification body is capable of fulfilling its duties based on the PMD Act, ISO/IEC17021-1, and ISO/IEC17065. After completion of the assessment, PMDA reports the assessment results to the MHLW. Lastly, the MHLW performs document review again to determine whether to register the certification body or not.

30.4 Quality Management System

As well known, the Quality Management System (QMS) is defined as a formalization system that documents the processes, procedures, and responsibilities for achieving quality policies and objectives. In Japanese medical device regulations, QMS is based on ISO-13485.

30.4.1 QMS (Quality Management System) Ordinance (Art. 23-2-5-2-4)

In Japan, a MAH for medical devices except for limited general medical devices have to demonstrate compliance with the Ordinance on Standards for Manufacturing Control and Quality Control of Medical Devices and IVDs.[17] This ordinance is usually recognized as a QMS Ordinance. This Japanese QMS Ordinance is fundamentally referred to ISO 13485 but there are some additional requirements. These differences are including document control, record control, responsible engineering manager, infrastructure, etc. For example, there may be different requirements for the length of time required to keep copies of documents.

Table 30.7 Contents of the ordinance on standards for manufacturing control and quality control of medical devices and IVDs

Chapter 1	General Provisions	
Chapter 2	Basic Requirements Regarding Manufacturing Control and Quality Control of Medical Devices, etc.	
Section 1	General Requirements	ISO 13485 Compliant Requirements
Section 2	Quality Management System	
Section 3	Management Responsibility	
Section 4	Resource Management	
Section 5	Product Realization	
Section 6	Measurement, Analysis and Improvement	
Chapter 3	Additional Requirements Regarding Manufacturing Control and Quality Control of Medical Devices, etc.	Additional Requirements
Chapter 4	Manufacturing Control and Quality Control of Biological Medical Devices, etc.	
Chapter 5	Manufacturing Control and Quality Control of Radioactive In Vitro Diagnostic Reagents	
Chapter 6	Application *mutatis mutandis*, etc., to Manufacturers, etc., of Medical Devices, etc.	

MAHs for limited general medical devices have to demonstrate compliance with parts of the QMS Ordinance. Limited general medical devices are medical devices other than medical devices designated by the MHLW as requiring attention to manufacturing management and quality control of general medical devices. Limited general medical devices are

1. Medical devices which are designated by the MHLW and announced the list by Ministerial Notification, and Ministerial Notification No. 316 of the Ministry of Health, Labour and Welfare in 2014" General medical devices designated by the Minister of Health, Labour and Welfare as requiring attention to manufacturing management or quality control pursuant to the provisions of Article 6, paragraph (1) of the Ordinance of the Ministry of Health, Labour and Welfare regarding the standards for manufacturing management and quality control of medical devices and IVDs"

2. Medical devices which are sterile in the manufacturing process.

Items that are exempted from application include:

- requirements for documenting quality control supervision systems
- requirements related to quality control supervision system standards
- requirements for document management
- requirements related to the placement of personnel (management supervisors), etc.
- requirements related to staffing (management officers), etc.

30.4.1.1 QMS conformity as an essential requirement

QMS conformity is an essential requirement for product approval/certification and for retention of approval/certification under the PMD Act. As requirements for product approval/certification, during the examination of application documents for approval/certification, a QMS conformity inspection is performed by the relevant review body. Approval/certification is granted only after conformity is established. As requirements for retention of approval/certification, approval/certification holders are obliged to undergo a QMS inspection every five years. If the approval/certification holder does not fulfil this obligation or does not follow improvement instructions resulting from the QMS inspection, approval/certification may be withdrawn by the relevant authorities.

30.4.2 QMS Organizational Structure and Personnel Requirements (Art. 23-2-2-1, 69-1)

Organizational structure and personnel requirements for MAH are required by the Ordinance on Standards for Organizational structure and personnel requirements of Manufacturing Control and Quality Control of Medical Devices and IVDs.

1. Development of organizational structure

    MAH need to establish a system to comply with QMS Ordinance.

    Specifically, the following requirements must be met:

- Establishment of quality control supervision system
- Document and implementation of quality control supervision system and maintain its effectiveness
- Management of quality control supervision documents
- Management of quality control supervision records

2. The right person in the right place

    In order to comply with the provisions of the QMS Ordinance, MAH must appropriately assign the following personnel according to their respective qualification requirements.

- Administrative supervisor
- Responsible person for management
- General manufacturing and sales manager of medical equipment, etc.
- Domestic Quality Operations Manager

    MAHs for limited general medical devices are exempt from some of these requirements.

30.4.3 QMS Inspections (Art. 23-2-5-6)

A MAH who intends to receive approval of medical devices or IVDs must undergo a document-based or on-site investigation by the MHLW on whether the method to control manufacturing or the quality of the item complies with the standards specified by Order of the MHLW, Furthermore, a MAH who has already received approval of medical devices or IVDs in every period of not less than three years specified by Cabinet Order after obtaining the approval.

QMS inspections are conducted for medical device manufacturing facilities in Japan and in other countries based on the requirements set by the QMS Ordinance. The on-site and document-based inspections of manufacturing sites is performed by PMDA, applicable prefecture or the RCB (registered certification body), according to its application category or other issues.

Table 30.8 Type of QMS inspection

1	Pre-approval inspection	Requisition of the marketing approval. Conducted prior to obtain the approval.
2	Pre-partial change approval inspection	Conducted prior to obtain the partial change approval. e.g., main assembling site change, etc.
3	Periodic post-approval inspection	Requisition of maintaining existing marketing approval. Conducted every 5 years after obtaining the marketing approval.
4	Additional inspection	Conducted where appropriate. e.g., specialized inspection for biological products, micro machine and medical devices utilizing nanomaterials etc.

30.5 Post Market Safety Management

A MAH must continue controlling risks by conducting investigation/consideration of information related to post-marketing safety management, planning/implementation of safety measure, and provision of information appropriately and smoothly. MAH must make efforts to collect information related to the efficacy and safety of medical devices and provide it to Health care professionals/user.

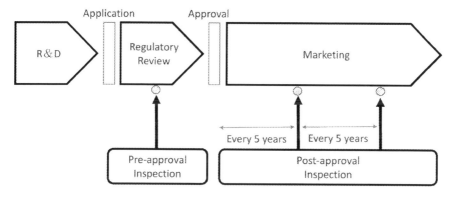

Figure 30.6 Timing of QMS Inspection.

When MAH learned that the use of medical devices might cause onset or spread of hazards, the necessary measures for safety should be taken, including recall, repair, discontinue distribution and provision of information to prevent such hazards. MAH of medical devices must employ a person "general marketing compliance officer" to undertake quality control and post-marketing safety control of medical devices. General marketing compliance officer must conduct the work concerning quality control and safety management impartially and appropriately. When it is confirmed to be necessary for impartial and appropriate conduct of the work, the general marketing compliance officer must present to MAH (head of the company) opinions in writing.

30.5.1 GVP Ordinance (MAH'S Obligations during Post-Market Phase) (Art. 23-2-2-2)

GVP Ministerial Ordinance refers to post-marketing safety management of medical devices, that is, methods related to quality/effectiveness and safety, collection of information necessary for proper use, examination, and necessary measures based on the results.

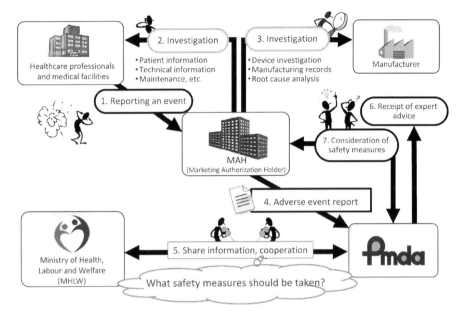

Figure 30.7 Flow of adverse event reporting–reports from MAHs.

Table 30.9 Responsibility of safety control manager

Requirements for safety control manager of medical devices of first class MAH	Requirements for safety control manager of type second and third class MAH
(1) Be in charge of the safety management control department. (2) A person who has been engaged in safety assurance work or other similar work for three years or more. (3) A person who has the ability to carry out safety assurance work properly and smoothly. (4) Being a person who does not belong to a department related to the sale of pharmaceuticals, etc., and who is not likely to interfere with the proper and smooth execution of safety assurance work.	(1) A person who has the ability to carry out safety assurance work properly and smoothly. (2) Being a person who does not belong to a department involved in the sale of pharmaceuticals, etc., and who is not likely to interfere with the proper and smooth execution of safety assurance work.

The requirements of GVP differ according to the type of MAH license, and there are provisions that are excluded for second- and third-class MAH license.

30.5.2 Requirements for MAH (Art. 12-2)

A MAH has also to comply with the Good Vigilance Practice (GVP) which is the standards of post-marketing safety management of a medical device (issued as MHLW ordinance No. 135 dated Sep. 22, 2004). In GVP requirements, MAH must continue controlling risks by conducting investigation/consideration of information related to post-marketing safety management, planning/implementation of safety measure, and provision of information appropriately and smoothly. GVP includes standards concerning organization and staffs engaged in post-market surveillance measurement, collection of safety information, internal audit and so on.

30.5.3 Collection, Analysis and Reporting of Post-marketing Safety Information (68-10)

It is very important to properly collect post market safety information and communicate it to the medical field. In Japanese medical device regulations, it is required to collect and analyze the following items.

1. Breakage, failure, malfunction, etc., of the device regardless of the occurrence of patient harm,
2. Problem with device specifications,
3. Defective products,
4. Patient adverse events related to use of the device and
5. Deficiency in information written in the package insert.

30.5.3.1 Obligations of MAH (Art. 68-10-1)

When MAHs learn of the occurrence of any disease, disability or death suspected to be caused by the adverse events use of the medical devices that they manufactured and sold, the occurrence of any infectious disease suspected to be caused by the use of such items, and other matters on the efficacy and safety of medical devices, MAH must report such information to PMDA within a certain period of time as stipulated in the relevant Ministerial Ordinance issued by the MHLW.

Table 30.10 shows the reporting period for adverse event reports. For serious health damage and for unknown cases, as well as foreign field safety corrective actions reports, the reporting period is 15 days, as opposed to 30 days for known cases. Known or unknown are determined by whether the adverse event is stated or can be anticipated from the contents of the Package Insert. This is the reporting period for a formal report, and that serious cases should be reported via phone, FAX or email as soon as possible to PMDA.

Table 30.10 Report due date: health damage

	Serious Health Damage		Unknown and Non-Serious	Foreign field safety corrective action (FSCA) reports	Reports of study results
	Death or Unknown	Known			
Reporting Period:	15 days*	30 days	Annual report	15 days*	30 days

Note: This is the reporting period for a formal report; but serious incidents should be notified to PMDA as soon as possible.

The criteria of what to report is determined by the seriousness of the adverse health effect, and the corresponding description in the package insert. Health damage that are seen as serious adverse events include:

1. Death
2. Disability or permanent damage
3. Life-threatening event
4. Hospitalization (initial or prolonged)
5. Congenital anomaly
6. Other important medical events

Non-serious adverse events are events other than those described above.

Adverse events listed in package inserts are referred to as known events. Events not listed are referred to as unknown events (Table 30.11).

Table 30.11 Types of adverse event report

1. Adverse event reports (domestic/overseas case reports)
2. Foreign field safety corrective action (FSCA) reports
   The purpose of the FSCA report is to consider what kind of action should be taken in Japan based on information about actions taken by overseas regulators.
   To rapidly collect information, information from manufacturers and information published on the websites of overseas regulatory agencies must be reported to PMDA by MAHs within 15 days.
3. Adverse infection reports (only designated biological products)
4. Annual reports (unknown and non-serious cases)
5. Reports of study results

30.5.4 Recall

When MAH undertake recalls/repairs of medical devices, MAH must report to the governor of the prefecture where the company is located.

MAH

- Article 68-10, Paragraph 1 of the PMD Act

When a marketing authorization holder (MAH) learns of the occurrence of adverse events or similar that are suspected to be related to the efficacy and/or safety of medical devices, MAH must report such information to PMDA within a certain period of time as stipulated in the relevant Ministerial Ordinance issued by the Ministry of Health, Labour and Welfare (MHLW).

Healthcare Professionals

- Article 68-10, Paragraph 2 of the PMD Act

When a healthcare professional learns of adverse events, etc. suspected to be related to medical devices and they confirm that it is necessary to take affirmative measures to prevent the onset or spread of risks to public health or safety, healthcare professional must report such information to PMDA.

Figure 30.8 Obligation of adverse event report.

30.5.4.1 Obligations of healthcare professional (Art. 68-10-2)

When a healthcare professional learns of adverse events, etc., suspected to be related to medical devices and they confirm that it is necessary to take affirmative measures to prevent the onset or spread of risks to public health or safety, that healthcare professional must report such information to PMDA.

Proprietors of pharmacies; proprietors of hospitals or clinics for humans or domesticated animals; or physicians, dentists, pharmacists, registered sales clerks, veterinarians and other medical industry professionals must, in cases where they learn of the occurrence of any disease, disability or death suspected to be caused by the side effects use of the pharmaceuticals, medical devices or regenerative medicine products, or the occurrence of any infectious disease suspected to be caused by the use of such items, and when it is found to be necessary in order to prevent the occurrence or spread of hazards in health and hygiene, report the same to the PMDA.

30.5.5 PMDA's Obligations during Post-Market Phase

PMDA collects safety information promptly and efficiently; safety staff receive reports from companies when cases of adverse events (AEs) and infections caused by drugs as well as malfunctions of medical devices are detected during the development and post-marketing periods.

Table 30.12 Types of adverse event report's evaluation

Problems of the device or MAH	• Quality defect, problem with durability, strength • Sophistication / complication of the device • Lack of the information provided by the MAH
Problems of use	• Inexperienced or unskilled in using the device • Poor maintenance • Insufficient environmental management • Inappropriate use (off label use, contraindications)
Other	• Patient condition (complication of diseases, allergy)

PMDA also consolidates all essential safety information reported by healthcare professionals, information from international sources, and conference papers and research reports. The collected information is then promptly compiled into a database and shared with the MHLW.

In adverse event reports, PMDA place particular focus on causes of adverse events for considering safety measures. In order to lead to appropriate safety measures, the causes of adverse events are divided into three categories. Keeping these in mind, PMDA consider what measures are adequate for prevention of recurrence of the event.

30.5.5.1 Various approaches

To improve and enhance safety measures, PMDA takes various approaches such as the introduction of data mining methods (which allow statistical analyses of information on adverse drug reactions as reported by companies or medical institutions and the detection of cases to be investigated), promotion of the application of electronic medical records to safety measures in order to establish evaluation methods of safety (known as the MIHARI Project), building of a system for evaluating medical device malfunctions, and building of a system for gathering and evaluating data from medical devices subject to tracking.

30.5.5.2 Information services

PMDA makes available a wide range of information on the quality, efficacy, and safety of drugs and medical devices, such as the package inserts for drug products and medical devices, recalls, and urgent safety information ("Dear Healthcare Professional" Letters), on the Medical Product Information page on its website (http://www.info.pmda.go.jp). All cases of adverse drug reactions and malfunctions reported by companies on or after April 1, 2004 (the date PMDA was established) are posted on the same web page.

PMDA provides information not only to healthcare professionals but also to the general public, such as the "Drug Guide for Patients," which is an easy-to-understand explanation for patients to teach them about prescription drugs with warnings labels, and the "Manuals for Management of Individual Serious Adverse Drug Reactions (for the general public)," which outline individual ADRs, initial symptoms, and key points for early detection and treatment in an easy-to-understand manner.

In addition, the Agency offers a free email information service called "PMDA medi-navi" (available in Japanese only), through which important safety information is distributed to healthcare professionals who subscribe to the service.

30.6 New Regulatory Challenge

To provide patients and healthcare professionals with rapid access to safer, more effective drugs, medical devices, and regenerative medical products, the PMDA is engaged in ensuring quality, efficacy, and safety from development to post-market stages. To achieve this, the MHLW and the PMDA took various approaches.

Innovative medical devices created by medical venture enterprises are expected to have extremely effective and safe profile; however, these medical devices tend to target extremely few patients. In that case, the development might be stagnated because of difficulties in collecting cases for clinical trial.

Table 30.13 Rapid approval scheme for medical devices that are particularly necessary for medical care

Category	Definition
Orphan medical devices	The number of patients who may use the medical devices should be less than 50,000 in Japan, or the medical devices should be indicated for difficult-to-treat diseases.
SAKIGAKE MDs	(1) The action mechanism is apparently different from that of products approved in Japan and foreign countries.
	(2) With regard to the use, they have particularly excellent value for use.
MDs for Special Use	(1) Medical devices are used for treatment of diseases belonging to the specific classification*.
	*Confirmation of efficacy and safety is not sufficient because the Directions for use for children is not established
	• Drug-resistant bacterial infection
	(2) The need is significantly insufficient for MDs related to the use.
	(3) With regard to the use, they have particularly excellent value for use.

Considering such a situation and administrator's mission to introduce innovative medical devices to the public, the government should construct the scheme which accelerate the approval the innovative medical devices by minimizing the burden regarding clinical trials and enhancing the post-market surveillance.

30.6.1 SAKIGAKE Designation System

The *SAKIGAKE* Designation System aims to promote the R&D of innovative medical devices and early clinical research/trials aiming at early practical application for innovative medical products in Japan by collaboration between developing manufacturers and the PMDA. An innovative MD/IVD for patients in urgent need of innovative therapy may be designated as a *SAKIGAKE* Product if;

1. its premarket application will be filed in Japan firstly or simultaneously in some countries including Japan, and
2. prominent effectiveness can be expected.

This system is based on effective usage of prioritized consultation to lead quicker approvals for innovative products. In short, this is a priority scheme for innovative MDs. It is like Breakthrough Devices Program in the US and PRIME in EU. The designated products can have priority consultation, priority review and other priority services. One professional staff in PMDA is designated as concierge and continuously support until the approval.

When mechanisms of and medical devices are clearly different from approved medical devices in Japan and foreign countries, they are identified as "SAKIGAKE designation medical devices" in the system.

Once an MD/IVD is designated, its developer can benefit following:

(A) Prioritized Consultation by PMDA
(B) Pre-application substantive review
(C) Prioritized Review (12 months → 6 months [MD])
(D) Review Concierge assigned by PMDA

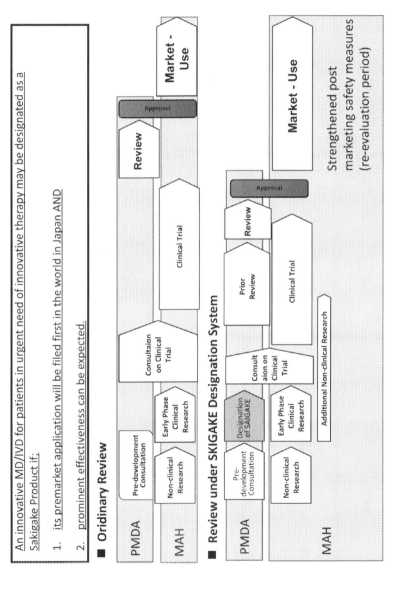

Figure 30.9 SAKIGAKE designation system.

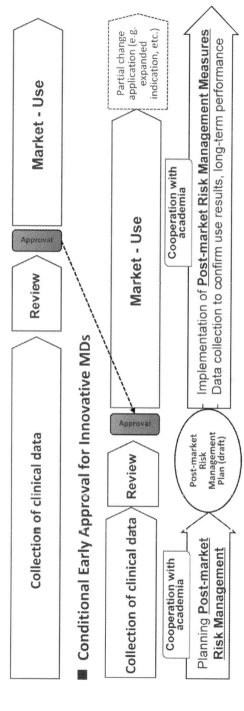

Figure 30.10 Conditional early approval for innovative medical devices.

30.6.2 Conditional Early Approval System

The objective of the Conditional Early Approval System is to accelerate the approval of highly needed medical devices where it is difficult to conduct clinical trials due to small patient populations.

By shifting the emphasis of clinical data from pre-market to post-market, devices can more readily obtain approval and reach patients more quickly. However, in order for manufacturers to achieve conditional approval, devices must meet a set of criteria to demonstrate safety and efficacy based on available clinical data while also appropriately addressing post-marketing risk management. Post-marketing risk management activities should be planned in collaboration with relevant academic societies. After the risk management measures are properly implemented, and the safety and efficacy of a device are confirmed, the device is approved without conducting new clinical trials.

30.6.3 Introduction of Approval System Based on the Characteristics of Medical Devices

For medical devices expected to be improved*, confirmation of the change plan in review processes and recognition of partial change of rapid approval issues within the planned range, <u>the approval review system can be introduced to make continuous improvement possible</u>. [Article 23-2-10-2]

Medical devices whose performance is consistently changed after marketing, including medical devices utilizing AI, the improvement of medical devices using real world data (RWD: actual clinical data) collected after marketing, and the addition of optional parts for improving usability.

Medical device regulations were enacted based on the concept of pharmaceutical regulation. However, it is important to remember that medical devices differ greatly from pharmaceuticals.

Medical devices have the following characteristics:

- various types (scalpels, pace makers, MRIs, etc.),
- repeated Post-marketing changes (software),

- continuous updates by improvement and refinement (devices with wide areas of specifications capable of confirming the performances) and
- large effect of operators (some medical devices such as cautery devices are considered the function to be applied to other regions).

There are some medical devices which are expected to be used for specific disease areas and developed and approved sequentially for each area, and the others are not limited to specific procedures and areas with the purpose of surgical invention to human bodies. Continuous improvement and refinement as medical device characteristics are handled with pharmaceutical procedures including partial change approvals and minor change notifications in response to the effect level of the change to the efficacy and safety. Additionally, a program including AI has been developed for continuously improving the performance from the initial approval. Considering the above, the following approval review system should be established for rapid access for patients.

- Addition of rapid application of medical devices having cauterization and irradiation functions which can be applied for other organs and body parts except by limiting facilities and operators and improving and strengthening post-marketing safety measures.
- Approval reviews capable of continuous improvement and refinement by confirmation of the improvement and refinement plan in the review process and recognition of partial change of rapid approvals in the area, for medical devices which are expected to be improved and refined immediately after approval.
- A measure capable of flexible modification of approval according to the performance change after marketing, for medical devices of which constant functions are changed after marketing, by evaluation of the improvement and refinement process of them.

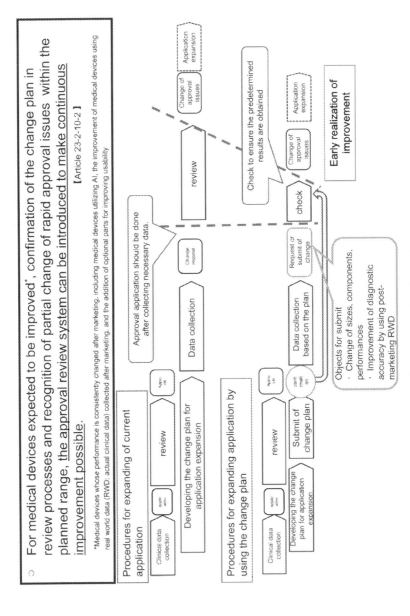

Figure 30.11 Improvement design within approval for timely evaluation and notice.

30.6.3.1 Early realization of change plan

Post-Approval Change Management Protocol will be introduced for medical devices to enable continuous improvements. This system is called PACMP internationally, and it was originally an idea that was originally invented at the International Conference on the Harmonization of Pharmaceutical Regulations (ICH), but Japan does not only introduce it into the pharmaceutical system as well as expanding it to medical devices.

30.6.3.2 Improvement design within approval for timely evaluation and notice

In addition to the requirements to the early realization of change plan are those related to the post-market change process for continuously improved devices, applying to devices that undergo continuous lifecycle improvements. The intention of Improvement Design within Approval for Timely Evaluation and Notice, IDAEN, is to promote early introduction of improved features by reducing regulatory burdens. The requirement will ensure that a risk management plan is in place for a device prior to any actual product changes. For example, emerging AI technology will need such a framework.

30.6.4 Electronic Labelling/Instructions for Use (eIFU)

The labelling had to be shipped with the product in the past, but it but it should be supplied electronically in current PMD Act. In addition to electronic methods of delivery, it is the responsibility of the MAH and seller to provide them in paper media at the time of the first delivery of medical devices, in cooperation with wholesale sellers as necessary. Furthermore, information that allows access to the latest labelling information is displayed in the outer box of the medical device or on the medical device itself, and when the information is revised, a mechanism has to be required to reliably deliver it to medical institutions by paper media. However, medical devices purchased directly by consumers, since it is necessary to ensure a state in which the contents of the attached document information can be immediately confirmed at the time of use, to bundle the paper medium as usual.

Package Insert (*Tenpu-Bunsho*)
Summary of important warnings, precautions, and instructions for use

Instructions For Use
More detailed instructions for use that are not included in the package insert

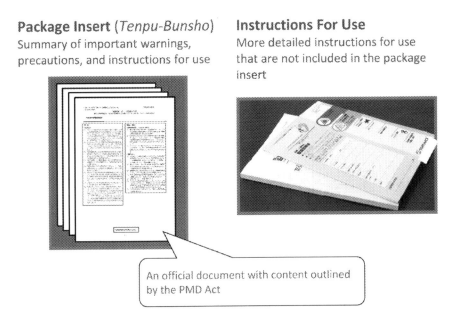

An official document with content outlined by the PMD Act

Figure 30.12 Basic requirement of labelling.

30.6.5 PMDA-ATC Medical Device Seminar

"Asia Training Center for Pharmaceuticals and Medical Devices Regulatory Affairs (PMDA-ATC)" was established on April 1, 2016. This Training Center will provide the training for regulators in Asia in response to the demands made by them by making use of the accumulated knowledge and experiences of PMDA. The content of the training will include basic lectures on information necessary to build regulatory capacity in each country/region, such as benefit/risk evaluation of the medical products and post-marketing safety measures.

Besides, the Center will provide the programs such as on-site mock inspection in cooperation with actual manufacturing facilities. Announcement of the Trainings/Seminars will be posted on the Symposia/Workshop site. PMDA will, through the Center, contribute to enhancement and mutual understanding of regulations, and strengthening of cooperation in Asia and other parts of the world.

30.6.6 Guidance for the Evaluation of Emerging Technology Medical Devices

30.6.6.1 PMDA Science Board

PMDA established the Science Board in 2012, as a high-level consultative body which discusses scientific aspects of pharmaceuticals and medical devices review. The purposes of the Science Board are, advancing regulatory science and evaluate products with advanced science and technology in appropriate manner by enhancing cooperation and communication with academia and medical institutions, based on PMDA's philosophy to deliver safe and effective drugs and medical devices to the people and further promotion of medical innovations. PMDA Science Board published various Guidance for the Evaluation of Emerging Technology Medical Devices.

Table 30.14 Example of outputs from PMDA Science Board

Regulatory Science on AI-based Medical Devices and Systems Adv Biomed Eng., 7: 118–123, 2018. https://www.pmda.go.jp/files/000224080.pdf
Report on the Use of Numerical Analysis for Strength Evaluation of Orthopedic Implants https://www.pmda.go.jp/files/000213002.pdf
Discussions on Evaluation of Medical Devices in Pediatric Use https://www.pmda.go.jp/files/000209947.pdf
Guidance for evaluation of artificial intelligence-assisted medical imaging systems for clinical diagnosis

30.6.6.2 Publication of the guidance for the evaluation of next-generation medical devices

The MHLW intends to establish a guidance for the evaluation of emerging technology medical devices, for which there is a high demand in clinical practice, by selecting the target areas to be studied in order to facilitate efficient product development and speed up the approval process through the advance preparation and publication of documents, such as a guidance for technical evaluation, to be used for the review.

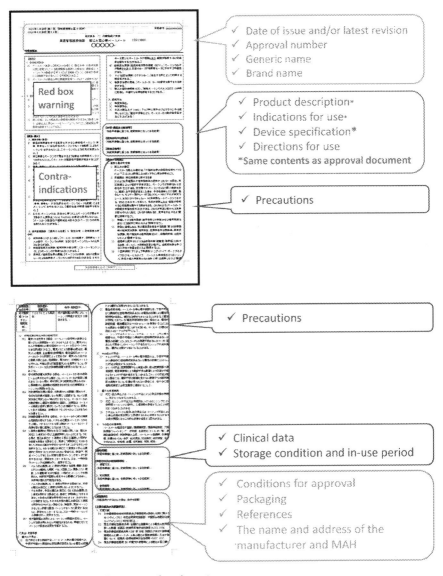

Figure 30.13 Contents of package insert.

1. The guidance makes recommendations regarding points that should be considered in product evaluation (evaluation items) from the viewpoints of collecting application documents for approval and accelerating the review

process. The guidance is not positioned as a legal standard but is designed simply to suggest currently recognized evaluation items for emerging technology medical devices. It has to be noted that other kinds of evaluation may be needed, or there may be some exceptions to this guidance depending on the product characteristics.

2. When collecting documents and data required for submission of an application for approval of individual products, it is recommended that the issues presented in the guidance be discussed in advance, and to use the consultation service of the Pharmaceutical and Medical Devices Agency at the earliest possible opportunity.

Table 30.15 Examples of typical outputs of next-generation medical device evaluation indicators[19]

Guidance on Evaluation of Customized Orthopedic Devices for Osteosynthesis
PFSB/ELD/OMDE (Yakushokuki) Notification 1215 No. 1, 2010.12.15
http://dmd.nihs.go.jp/jisedai/tsuuchi/customized_devices_for_osteosynthesis.pdf
Guidance on Evaluation of Orthopedic Customized Artificial Hip Joint Prosthesis
PFSB/ELD/OMDE (Yakushokuki) Notification 1207 No. 1, 2011.12.7
http://dmd.nihs.go.jp/jisedai/tsuuchi/customized_hip_prosthesis.pdf
Guidance on Evaluation of Autologous Induced Pluripotent Stem Cells-derived Retinal Pigment Epithelial Cells
PFSB/ELD/OMDE (Yakushokuki) Notification 0529 No. 1, 2013.05.29 (Annex 1)
http://dmd.nihs.go.jp/jisedai/tsuuchi/auto_iPS_derived_retinal.pdf
Guidance on Evaluation of the Devices for Physical Function Recovery
PFSB/ELD/OMDE (Yakushokuki) Notification 0529 No. 1, 2013.05.29 (Annex 2)
http://dmd.nihs.go.jp/jisedai/tsuuchi/devices_for_physical_function_recovery.pdf
Guidance on Evaluation of Accelerator Neutron Irradiation Device System for Boron Neutron Therapy
PSEHB/MDED (Yakuseikisinn) Notification 0523 No. 2, 2019.05.23 (Annex 5)
http://dmd.nihs.go.jp/jisedai/tsuuchi/accelerator_neutron_irradiation_device_system_for_BNCT.pdf
Guidance for Evaluation of Artificial Inteligence-Assisted Medical Imaging Systems For Clinical Diagnosis
PSEHB/MDED (Yakuseikisinn) Notification 0523 No. 2, 2019.05.23 (Annex 4)
http://dmd.nihs.go.jp/jisedai/tsuuchi/Guidance_for_evaluation_of_AI_assisted_systems.pdf

30.6.7 Traceability

From the viewpoint of ensuring medical safety, it is important to improve traceability by utilizing barcodes, such as management of medical device information, tracking of use records, and prevention of errors, in a series from manufacturing and distribution to the medical field. The display of barcodes based on international standards which is under Unique Device Identifier requirements is required for direct containers, packaging and body of medical devices. In addition, the MAH are required to register a database of product information.

30.6.8 Information Services

PMDA provides the following safety information regarding medical devices. You can reach all these information from the Safety Information regarding Medical Devices.[20]

The Yellow Letter/Blue Letter

PMDA makes available a wide range of information on the quality, efficacy, and safety of drugs and medical devices, such as the package inserts for drug products and medical devices, recalls, and urgent safety information ("Dear Healthcare Professional" Letters), on the Medical Product Information page on its website (http://www.info.pmda.go.jp). All cases of adverse drug reactions and malfunctions reported by companies on or after April 1, 2004 (the date PMDA was established) are posted on the same web site.

Safety Information announced by MHLW

Important safety information (e.g. press release) announced by MHLW regarding medical devices are also provided PMDA web site.

PMDA Risk Communications

This contains the most recent Risk Communications from PMDA including early communications or ongoing safety review. The webpage intends to provide the public with easy access to important medical devices safety information.

(https://www.pmda.go.jp/english/safety/info-services/devices/0006.html)

Revisions of PRECAUTIONS

This webpage contains the information announced by MHLW regarding medical devices on safety revisions of PRECAUTIONS in package inserts.

(https://www.pmda.go.jp/english/safety/info-services/devices/0002.html)

MHLW Pharmaceuticals and Medical Devices Safety Information

This Pharmaceuticals and Medical Devices Safety Information (PMDSI) is issued based on safety information collected by the MHLW. It is intended to facilitate safer use of pharmaceuticals and medical devices by healthcare professionals.

(https://www.pmda.go.jp/english/safety/info-services/drugs/medical-safety-information/0002.html)

Notification on self-check

This webpage contains the information announced by the MHLW regarding notification on self-check of medical devices. Marketing authorization holders perform self-check according to the notice, consequently they will revise the package inserts or user manuals.

(https://www.pmda.go.jp/english/int-activities/outline/0006.html)

PMDA Alert for Proper Use of Medical Devices

"PMDA Alert for Proper Use of Medical Devices" aims to communicate to healthcare providers with clear information. The information presented here includes such cases where the reporting frequencies of similar reports have not decreased despite relevant alerts provided in package inserts, among medical device failure/infection cases reported.

(https://www.pmda.go.jp/english/safety/info-services/devices/0005.html)

PMDA Alert for Proper Use of Medical Devices (for patients)

PMDA provides information not only to healthcare professionals but also to the general public, such as the "Drug Guide for Patients," which is an easy-to-understand explanation for patients to teach them about prescription drugs with warnings labels, and the

"Manuals for Management of Individual Serious Adverse Drug Reactions (for the general public)," which outline individual ADRs, initial symptoms, and key points for early detection and treatment in an easy-to-understand manner.
(https://www.pmda.go.jp/english/safety/info-services/devices/0008.html)

Notifications Related to Safety Measures

Notifications issued by the MHLW and other organizations regarding safety measures for medical devices are available here.
(https://www.pmda.go.jp/english/safety/info-services/devices/0007.html)

PMDA medi-navi

In addition, the Agency offers a free email information service called "PMDA medi-navi" (available in Japanese only), through which important safety information is distributed to healthcare professionals who subscribe to the service.

30.6.9 Harmonization by Doing

Through the U.S.–Japan Medical Device Harmonization by Doing (HBD), both countries' Regulators (MHLW/PMDA, US FDA), academia, and industry have discussed and addressed regulatory barriers that could delay the approval of medical devices between the two countries in a timely manner. Taking cardiovascular devices as an example, the implementation of international joint trials facilitated faster global development.

References

1. Population and demographics based on the Basic Resident Register [Online], Available: [09/24/2020]; https://www.soumu.go.jp/main_sosiki/jichi_gyousei/daityo/jinkou_jinkoudoutai-setaisuu.html.
2. *Yearbook of Statistics of Production by Pharmaceutical Industry* (in Japanese [Online], Available: [09/24/2020]; https://www.mhlw.go.jp/topics/yakuji/2018/nenpo/index.html.
3. Organization of the Ministry of Health, Labour and Welfare [Online], Available: [09/24/2020]; https://www.mhlw.go.jp/english/org/detail/dl/organigram.pdf.

4. Organization of the Pharmaceuticals and Medical Devices Agency [Online], Available: [09/24/2020]; https://www.pmda.go.jp/files/000234601.pdf.
5. Act on Securing Quality, Efficacy and Safety of Products Including Pharmaceuticals and Medical Devices [Online], Available: [09/24/2020]; http://www.japaneselawtranslation.go.jp/law/detail/?id=2766&vm=04&re=02&new=1.
6. Order for Enforcement of the Act on Securing Quality, Efficacy and Safety of Products Including Pharmaceuticals and Medical Devices [Online], Available: [09/24/2020]; http://www.japaneselawtranslation.go.jp/law/detail/?id=3214&vm=04&re=2&new=1.
7. Ministerial Ordinances [Online], Available: [09/24/2020]; https://www.pmda.go.jp/english/review-services/regulatory-info/0001.html.
8. Regulation for Enforcement of the Act on Securing Quality, Efficacy and Safety of Products Including Pharmaceuticals and Medical Devices [Online], Available: [08/17/2020]; https://www.pmda.go.jp/english/review-services/regulatory-info/0001.html.
9. Essential Principles [Online], Available: [09/24/2020]; https://www.std.pmda.go.jp/stdDB/Data_en/InfData/Infetc/MHLW_Notification_122_of_2005.pdf.
10. Certification Criteria [Online], Available: [09/24/2020]; https://www.std.pmda.go.jp/scripts/stdDB_en/refetc/stdDB_refetc_sum_absbttm.cgi?absdisp=2.
11. Approval Criteria_[Online], Available: [09/24/2020]; https://www.std.pmda.go.jp/scripts/stdDB_en/refetc/stdDB_refetc_sum_absbttm.cgi?absdisp=3.
12. Notifications and Administrative Notices [Online], Available: [09/24/2020]; https://www.pmda.go.jp/english/review-services/regulatory-info/0003.html.
13. Summary Technical Documentation for Demonstrating Conformity to the Essential Principles of Safety and Performance of Medical Devices (STED), [Online], Available: [09/24/2020]; http://www.ghtf.org/sg1/sg1-final.html.
14. Accreditation of Foreign Manufacturers [Online], Available: [09/24/2020]; https://www.pmda.go.jp/english/review-services/reviews/foreign-mfr/0001.html.
15. Accreditation Category [Online], Available: [09/24/2020]; https://www.pmda.go.jp/files/000153258.pdf.

16. Application for Accreditation of Foreign Manufacturers [Online], Available: [09/24/2020]; https://www.pmda.go.jp/files/000153619.pdf.
17. MHLW Ministerial Ordinance No. 169 in 2004 December 17. Ordinance on Standards for Manufacturing Control and Quality Control of Medical Devices and In Vitro Diagnostic Reagents [Online], Available: [09/24/2020]; https://www.pmda.go.jp/files/000207979.pdf.
18. K. Mori, M. Watanabe, N. Horiuchi, A. Tamura, H.Kutsumi, The Role of the Pharmaceuticals and Medical Devices Agency and Healthcare Professionals in Post-Marketing Safety, *Clin. J. Gastroenterol.* 7, 103–107 (2014) [Online], Available: [09/24/2020]; https://pubmed.ncbi.nlm.nih.gov/26183623/.
19. Output of the Next-Generation Medical Device Evaluation Indicators [Online partially in English], Available: [09/24/2020]; http://dmd.nihs.go.jp/jisedai/tsuuchi/.
20. Safety Information Regarding Medical Devices by PMDA [Online], Available: [09/24/2020]; https://www.pmda.go.jp/english/safety/info-services/devices/0001.html.
21. Tamura, H. Kutsumi. Multiregional Medical Device Development: Regulatory Perspective, *Clin. J. Gastroenterol* 7, 108–116 (2014) [Online], Available: [09/24/2020]; https://link.springer.com/article/10.1007/s12328-014-0478-2.
22. M. Krucoff, A. Tamura, Report from Co-Chairs: After Successful Japan–US HBD East 2013 Think Tank Meeting [Online], Available: [09/24/2020]; http://www.pmda.go.jp/hbd/meeting/report-think_tank_meeting13.html.

Chapter 31

Korea: Medical Device Regulatory System

Young Kim,[a] Soo Kyeong Shin,[a] and Jamie Noh[b]

[a]*Synex Consulting Ltd.*
[b]*Asia Regulatory Professional Association*
Jack.wong@arpaedu.com

31.1 General Market Overview

31.1.1 Key Healthcare Market Indicators of Korea

The Republic of Korea is located in the northeast part of Asia between China and Japan. It occupies the southern part of the Korean Peninsula. The area is around 100,000 km^2, around five times of that of Israel.

The country's GDP per capita reached US$27,195 in 2015, and it ranked the second richest country in East Asia following Japan. For reference, Israel reached US$35,343 of GDP per capita.

Medical Regulatory Affairs: An International Handbook for Medical Devices and Healthcare Products (Third Edition)
Edited by Jack Wong and Raymond K. Y. Tong
Copyright © 2022 Jenny Stanford Publishing Pte. Ltd.
ISBN 978-981-4877-86-2 (Hardcover), 978-1-003-20769-6 (eBook)
www.jennystanford.com

Figure 31.1 The Republic of Korea.

The population was estimated at 51 million in 2015 and is expected to grow very slowly to reach 52 million in 2030. For reference, the population was as large as around six times that of Israel in 2015.

The rapidly aging population and low fertility rate are the most serious societal problems in Korea. The total number of senior citizens 65 years old or older reached 13.1% in 2015. The pace of aging has been faster than any other OECD countries in recent years. The life expectancy was 81.3 years between 2010 and 2013, among the top in the world. In the meantime, the fertility rate was 1.23 between 2010 and 2014, representing one of the lowest rates in the world. For the past 10 years, the government has put the highest priority in its policy directions to raise fertility rates by offering various programs to support pregnancy, birth and child care for young working couples.

Table 31.1 Healthcare indicators

Description	Information
Population (2015)	51 million
Projected future population (2030)	52 million
Senior citizens aged 65 or older (2015)	13.1%
Projected future senior citizens (2030)	24.3%
Life expectancy at birth (2010–2014)	81.5 years
Fertility rate (2010–2014)	1.23

Country Statistics (General)

Population	51,446,201	2017
GDP	$1.404 trillion (KRW 1,615 trillion)	2016
GDP per capita	$ 27,633 (KRW 31,777,950)	2016
Inflation rate	1.2 %	2016
Unemployment rate	3.8 %	2017

Source: http://kosis.kr/. Exchange rate: 1,150 KRW/$.

Country Statistics (Healthcare)

	Korea	OECD average
Life expectancy (age)	82.2	80.8
Infant mortality rate (per 1,000 no. of birth)	3.0	4.0
No. of bed (per 1,000)	11.7	4.7
No. of physician (per 1,000)	2.2	3.3
No. of nurse (per 1,000)	5.6	9.6
No. of Kidney transplantation (per 100,000)	3.6	3.6
Healthcare spending % of GDP	7.1	9.1
Drug/Medical device expenditure among total healthcare spending (%)	20.6	15.9

Source: OECD Health Statistics 2016.

31.1.2 Medical Device Market

Korea is the eleventh largest market in the world and the third largest in Asia following Japan and China. In 2014, the market was estimated at about US$4.9 billion. The pace of growth is also impressive: It recorded a growth rate of 6.3% in compound annual growth rate (CAGR) for 2010–2014. The market is expected to advance by the similar rate for the years to come.

Table 31.2 Trends in medical device market in Korea (unit: USD 1 million, %)

Classification	2010	2011	2012	2013	2014	CAGR ('10~'14)
Domestic product	2,964	3,366	3,877	4,224	4,604	11.6
Export	1,681	1,853	2,216	2,580	2,714	12.7
Import	2,619	2,793	2,931	2,988	3,084	4.2
Trade surplus[1]	Δ938	Δ939	Δ714	Δ407	Δ370	Δ20.7
Market size[2]	3,902	4,306	4,592	4,631	4,974	6.3

[1]Trade surplus is Export – Import; Δ indicates deficit.
[2]Market size is Domestic Product – Export + Import.
Source: The Ministry of Food and Drug Safety (2015). *Food & Drug Statistical Year Book*.

Healthcare Industry

- Healthcare spending of GDP is 7.1%; it is lower than the OECD average (9.1%).
- Drug/medical device expenditure among total healthcare spending is 23.5% in 2009; it decreased 2.9% in 2014 (20.6%) but is still higher than the OECD average (15.9%).
- Total healthcare expenditure per person is US$2,361 PPP; it is lower than the OECD average (US$3,689 PPP).
- The import/export status of the healthcare industry (2011–2015) is as follows:

$ Million

Import	2011	2012	2013	2014	2015	YoY	CAGR ('11~'15)
Total	8,426	8,663	8,409	9,113	8,862	−2.8	1.3
Drugs	4,916	5,084	4,708	5,095	4,830	−5.2	−0.4
Medical devices	2,521	2,601	2,729	2,971	2,944	−0.9	4.0
Cosmetics	989	978	972	1,048	1,088	3.8	2.4

$ Million

Export	2011	2012	2013	2014	2015	YoY	CAGR ('11~'15)
Total	4,231	5,083	5,764	6,780	8,239	21.5	18.1
Drugs	1,754	2,049	2,117	2,403	2,940	22.3	13.8
Medical devices	1,673	1,967	2,357	2,577	2,711	5.2	12.8
Cosmetics	805	1067	1290	1800	2588	43.8	33.9

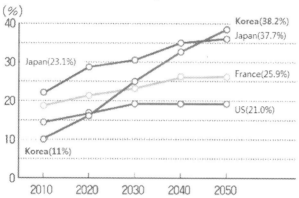

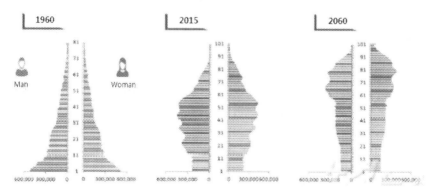

Changes in Population Structure

Source: OECD Health Statistics 2016, A major Performance of Health Industr_2015_ KHIDI, IMS Market Prognosis 2015–2019, South Korea, https://kostat.go.kr/.

- The National Health Insurance (NHI) reimbursement coverage continues to expand and there will be full reimbursement of treatment for cancer, cardiovascular, cerebrovascular, and rare diseases by 2017.

- The proportion of GDP spent on healthcare will keep growing driven by the ageing population. Korea is the fastest ageing society in the world and is expected to have the highest percentage of 65 years old in 2060.
- Total medical costs for elderly (>65 years old) patients are $1,912 million, which is 38.4% of total medical costs, even if the percentage of elderly patients is 12.3%.

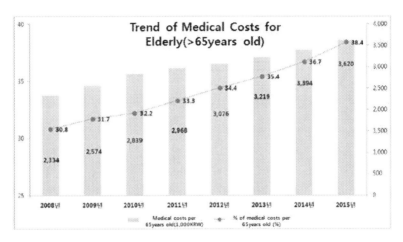

Source: OECD Health Statistics 2016, A major Performance of Health Industr_2015_KHIDI, IMS Market Prognosis 2015–2019, South Korea, https://kostat.go.kr/.

31.1.3 Import

Imported devices have been the major source of device supply for decades, representing the market shares of about 60%. In 2014, Korea spent around US$3 billion on the import of devices, which accounted 62% of the domestic market for the year.

Korea imports the majority of high-tech devices from advanced countries, such as, the United States, Europe, and Japan. In contrast, China has been emerging rapidly as the largest supplier of low-tech devices.

The United States has always been the largest supplier of medical devices to Korea. The country's import from the United States in 2014 amounted to around US$1.3 billion, or 44.4% of all imports. For the same year, other countries Korea imported devices from were Germany (US$486 million, 16.6%), Japan (US$309 million, 10.5%), and China (US$137 million, 4.7%), respectively.

Table 31.3 Medical device import by country (unit: $1,000, %)

Ranks	Importer	2012 Import	2012 Portion	2013 Import	2013 Portion	2014 Import	2014 Portion	Growth rate compared to previous Year
1	USA	1,171,751	45.1	1,230,563	45.1	1,299,241	44.4	5.6
2	Germany	378,123	14.5	402,387	14.7	486,803	16.6	21.0
3	Japan	336,597	12.9	331,435	12.1	308,220	10.5	-7.0
4	China	130,248	5.0	121,184	4.4	137,042	4.7	13.1
5	Swiss	115,178	4.4	149,298	5.5	131,686	4.5	-11.8
6	France	60,858	2.3	55,645	2.0	63,975	2.2	15.0
7	UK	39,711	1.5	48,185	1.8	55,792	1.9	15.8
8	Netherlands	32,030	1.2	35,565	1.3	45,800	1.6	28.8
9	Italy	40,012	1.5	41,532	1.5	42,373	1.4	2.0
10	Sweden	35,742	1.4	37,870	1.4	41,461	1.4	9.5
	Total of top 10 countries	2,340,250	90.0	2,453,664	89.9	2,612,393	89.2	6.5
	Import total	2,600,999	100.0	2,728,888	100.0	2,928,357	100.0	7.3

Source: The Ministry of Food and Drug Safety (2015). *Food & Drug Statistical Year Book*.

The imports of medical devices in Korea are heavily focused on large-volume items: The aggregate amount of those 30 products ranking at the top shares around a half of the total import. Among those high-demand items for imports are soft contact lenses, vascular stents, magnetic resonance imaging (MRI) equipment, dialysis equipment, computed tomography (CT) equipment, knee joint implants, etc.

Table 31.4 Medical device import of top 30 items (as of the end of 2013) (unit: $1,000, %)

Ranks	Items	2011	2012	2013	Rate of change
1	Soft contact lens	94,909	99,977	123,117	23.1
2	Stent	107,323	103,453	108,810	5.2
3	MRI system	81,823	92,356	72,963	−21.0
4	Water purification system, hemodialysis	57,756	60,941	64,461	5.8
5	CT system	86,953	70,329	63,383	−9.9
6	Knee prosthesis, internal, total	73,877	62,365	62,915	0.9
7	Accelerator system collimator electron applicator	35,812	40,632	57,782	42.2
8	Chemiluminescence assay (CLIA) analyzer, automated	—	14,480	54,143	273.9
9	Probe	43,624	53,581	54,122	1.0
10	Catheter, cannula and tubing, vascular	45,295	49,440	51,437	4.0
11	Graft/prosthesis	27,927	35,334	44,868	27.0
12	Sight corrective spectacles	40,913	60,268	43,753	−27.4
13	Orthopedic bone screw, non-biodegradable	30,711	42,016	43,048	2.5
14	Surgical Instrument	27,590	19,075	41,400	117.0
15	Staple	37,531	36,230	39,302	8.5
16	Intraocular lens	39,303	34,025	38,508	13.2
17	Ultrasound system, imaging, diagnostic	40,240	43,059	34,928	−18.9

Ranks	Items	2011	2012	2013	Rate of change
18	Dressing	35,907	35,973	34,767	−3.4
19	Endoscope Instrument	27,678	28,195	32,480	15.2
20	Laser, surgical	22,773	23,584	31,151	32.1
21	X-ray system, intravascular, angiography	25,396	23,357	31,102	33.2
22	Orthopedic fixation plate, non-biodegradable	15,742	15,885	29,825	87.8
23	Gloves for medical use	21,938	24,543	29,410	19.8
24	Orthopedics appliance	31,160	29,173	29,394	0.8
25	System, image processing device	28,983	31,704	27,532	−13.2
26	Hip prosthesis, internal, total	27,585	24,194	26,280	8.6
27	Clip for medical use	17,881	19,360	23,609	22.4
28	Clinical chemistry automated analyzer	23,073	38,577	22,916	−40.6
29	Hemodialysis system, high-permeability	19,617	22,709	22,654	−0.2
30	Stimulator, electrical, auditory, cochlear	19,240	17,348	21,593	24.5
Total of top 30 items (A)		1,188,561	1,252,163	1,361,732	8.8
Total of all items (B)		2,521,148	2,600,999	2,728,888	4.9
Proportion (A/B × 100)		47.1	48.1	49.9	—

Source: The Ministry of Food and Drug Safety (2014). Medical Product and Import and Export Results in 2013.

Korea Health Industry Development Institute (2014). Analysis Report of Medical Devices Industry in 2014.

The gigantic global device companies play major roles in the supply of quality devices. Those firms have local entities and staff in the country and promote their products directly to end-users. In 2014, Johnson & Johnson Medical was the largest importer, and Siemens, Roche, Covidien, Medtronic, and Abbott followed in that order.

Table 31.5 Medical device import of top 30 items

Ranks	Device category	Values in 2014 (US$ thousands)	% of all imports
1	Johnson and Johnson Medical	148,745	5.01
2	Siemens	142,858	4.81
3	Roche Diagnostics Korea	120,306	4.05
4	Covidien Korea	95,619	3.22
5	Medtronic Korea	94,726	3.19
6	Abbott Korea	84,414	2.84
7	Johnson and Johnson Korea	71,976	2.42
8	Stryker Korea	68,645	2.31
9	Phillips Electronics Korea	59,111	1.99
10	Fresenius Medical Care Korea	55,985	1.88

Source: Korea Medical Device Association.

31.1.4 Domestic Production

The medical device manufacturing industry of Korea has been unable to compete with imported products for decades. However, it has advanced significantly in recent years empowered by government policies visioning healthcare industry as one of the country's future growth engines, affluent public research funds and achievement of necessary technologies in the disciplines of information technology, biomedical engineering, electronics, biology, material science, etc. As a result, Korean manufacturers have successfully localized devices in a few specific sectors and have emerged as strong competitors to foreign suppliers in a few specific product markets in the domestic market. The Korean manufacturing industry is now in the process of stepping up its position from the manufacturer of low-tech, low-price product groups to the supplier of mid-tech, mid-price products.

The domestic production of medical device was valued at US$4.6 billion in 2014, summing up to 11.6% of CAGR for 2010–2014. Those sectors contributing significant volumes to total production were, in order of ranking, radiology imaging equipment (22%), dental implants and materials (19.8%), medical suppliers (15.1%), electrical surgical devices (7.3%), in vitro diagnostics (6.2%), and home-use medical devices (5.2%).

Table 31.6 Medical device production by category (unit: USD 1,000,000, %)

Category	2013 Production	2013 Portion	2014 Production	2014 Portion	Growth rate compared to previous year
Radiology imaging equipment	975	23.1	1,013	22.0	3.9
Biomedical measurement	152	3.6	163	3.6	7.8
IVD	157	3.7	161	3.5	2.7
Table, examination/treatment	104	2.5	109	2.4	5.3
Anesthesia and breath instrument	41	1.0	35	0.8	−15.1
Electrical surgical devices	298	7.1	334	7.3	11.9
Non-motorized medical instrument	134	3.2	139	3.0	3.4
Orthopedic materials	157	3.7	156	3.4	−0.9
Artificial internal organ	52	1.2	82	1.8	57.2
Speculums for medical use	30	0.7	26	0.6	−12.2
Medical suppliers	675	16.0	697	15.1	3.2
Dental instrument	59	1.4	76	1.7	28.7
Dental implants and materials	895	21.2	913	19.8	1.9
Home-use medical devices	252	6.0	239	5.2	−5.2
Medical device for orthopedics and restoration	150	3.6	169	3.7	12.1
In vitro diagnostics reagents	83	2.0	284	6.2	240.6
U-healthcare medical device	928	0.0	1,204	0.0	29.7
Total	4,224,169	100.0	4,604,814	100.0	9.0

Source: Korea Health Industry Development Institute (2015). Analysis Report of Medical Devices Industry in 2015.

Table 31.7 Medical device production of top 30 items (unit: USD 1,000,000, %)

Ranks	Category code	Items	Production 2011	Production 2012	Production 2013	Growth rate compared to 2012
1	C12050	Dental implants	249 (2)	448 (2)	556	24.0
2	A26380	Ultrasound imaging system	380 (1)	460 (1)	5127	11.2
3	C01020	Dental alloys	231 (3)	177 (3)	136	−23.2
4	A77030	Soft contact lenses	99 (5)	98 (8)	128	29.7
5	A77010	Lenses for eye glasses	177 (4)	143 (4)	121	−15.4
6	A26430	Software solutions for picture archiving and communication system (PACS)	47 (14)	113 (5)	112	−0.9
7	A58020	Probe for medical use	82 (9)	90 (9)	105	16.9
8	A11110	Digital x-ray imaging equipment	69 (11)	99 (7)	96	−3.3
9	A83060	Heating pad system under/overlay, electric, home use	94 (6)	106 (6)	91	−13.9
10	A83080	Combination stimulator	93 (7)	87 (10)	91	4.8
11	C12090	Spine implants	36 (21)	68 (12)	73	6.8
12	A54010	Syringes	60 (12)	62 (14)	69	11.5
13	B03160	Orthopedics appliance	89 (8)	66 (13)	69	3.6
14	A37010	Surgical laser equipment	72 (10)	73 (11)	68	−7.5
15	A11010	CT system	44 (16)	48 (17)	62	30.1

16	A78010	Hearing aid	43 (17)	49 (16)	53	8.6
17	B10010	Glucose strip	59 (13)	51 (15)	46	-9.7
18	B05010	Splint	38 (19)	44 (18)	45	2.9
19	A68010	Unit and chair, dental	44 (15)	39 (20)	41	3.8
20	A55030	Handpiece for medical use	37 (20)	39 (22)	40	3.8
21	A79160	Infusion instruments	23 (30)	39 (21)	38	-1.4
22	A22030	Glucose meter	32 (23)	33 (25)	38	14.3
23	A53010	Syringe needles	25 (27)	30 (26)	37	24.9
24	B03300	Stent	21 (34)	28 (29)	35	25.2
25	D06080	IVD reagents for infectious disease marker	—	—	35	—
26	A79030	Intravascular administration set	39 (18)	28 (28)	33	17.1
27	A79010	Infusion pump	33 (22)	37 (23)	33	-10.3
28	B07070	Wound dressing	118 (53)	21 (38)	32	52.8
29	B04230	Tissue fillers	5 (82)	18 (45)	31	70.9
30	C01030	Alloy, casting, base metal	12 (47)	28 (30)	31	9.6
Total of top 30 items (A)			2,259	2,635	2,870	8.9
Total of all medical devices (B)			3,366	3,877	4,224	8.9
Proportion (A/B × 100)			67.1	68.0	68.0	—

Note: The numbers within the parentheses indicate the rank for each year.
Source: The Ministry of Food and Drug Safety (2014). Medical Product and Import and Export Results in 2013.

Domestic manufacturing is limited to specific types of devices: the aggregate production value of those 30 highest-volume items represents a total of 68% of the total domestic production.

Dental implant is the number 1 product in terms of production value in recent years. It has made a noticeable growth in both domestic and oversea markets in recent years. Its production value for 2013 was estimated at US$550 million in 2013. Other high-volume items for the same year are digital x-ray imaging equipment, dental alloys, soft contact lenses, lenses for eye glasses, software solutions for picture archiving and communication system (PACS), syringes, spine implants, surgical laser equipment and CT equipment. In the meantime, those devices making high growth rate in 2013 from the previous year were tissue fillers (70%), wound dressing (52%), soft contact lenses (29%), CT equipment (30%), stents (25%), dental implants (24%), and syringe needles (24%).

Korea Health Industry Development Institute (2014). Analysis Report of Medical Devices Industry in 2014.

The medical device industry has received the highest level of attention from people and government in its history since the government selected the industry as one of the future growth engines of the country, especially in terms of export. Korea relies heavily on exports for its economic growth, so the continued growth in exports is crucial. It has since focused on improving the business climate to culture the device industry by increasing research funds, eliminating unreasonable regulation, and building networks of experts in various disciplines.

The government set the specific goal for the device industry to accomplish US$13.5 billion of export by the end of the year 2020. In an effort to realize such goal, it has laid down four specific policy directions:

(1) strategic investment for R&D focused on commercialization
(2) support for market entry through improving images of domestic products and de-regulation
(3) supporting exports to overseas markets where high values are added
(4) building open-innovation platforms

Table 31.8 Revenues of large device companies in Korea (unit: USD 1,000,000, %)

No.	Companies	2010	2011	2012	2013	2014	Growth rate compared to previous year
1	Samsung Medison Co., Ltd.	257	261	276	268	284	5.9
2	Carecamp Inc.	285	302	262	271	272	0.1
3	Johnson & Johnson Medical Korea Ltd.	252	267	258	241	268	11.2
4	GE Healthcare Korea	232	245	238	224	212	−5.1
5	Baxter Korea	138	153	158	165	174	5.0
6	VATECH. Co., Ltd.	126	124	138	153	172	11.9
7	OSSTEM IMPLANT Co., Ltd.	113	131	151	155	171	10.3
8	Fresenius Medical Care Korea	110	130	138	149	152	2.2
9	BODYFRIEND Co., Ltd	18	34	42	78	143	83.1
10	Standard Diagnostics, Inc.	84	113	91	117	135	15.9
11	SHIN HUNG Co., Ltd.	167	168	152	122	123	0.5
12	CHEMIGLAS Corp.	95	110	131	118	112	−5.3
13	Ceragem Co., Ltd.	67	64	69	86	105	22.2
14	PJ Electronics Co., Ltd.	89	102	111	113	104	−7.5
15	INSAN MTS Co., Ltd.	7	7	111	152	104	−31.7
16	ISM Co., Ltd.	122	125	116	102	101	−1.0
17	IIDX Corporation	64	70	74	106	99	−6.0
18	NUGA MEDICAL Co., Ltd.	58	75	95	104	94	−10.3
19	Gambro Korea Ltd	92	79	88	92	93	1.5
20	i-SENS, Inc	37	52	66	82	91	11.4

Source: Korea Health Industry Development Institute (2015). Analysis Report of Medical Devices Industry in 2015.

Table 31.9 Medical device export by country (as of the end of 2014) (unit: $1,000, %)

Ranks	Importer	2012 Import	2012 Portion	2013 Import	2013 Portion	2014 Import	2014 Portion	Growth rate compared to previous year
1	USA	346,673	17.6	424,595	18.0	406,157	15.8	−4.1
2	China	175,088	8.9	231,455	9.8	254,146	9.9	9.8
3	Germany	208,467	10.6	221,328	9.4	198,672	7.7	−10.2
4	Japan	183,131	9.3	170,988	7.3	187,898	7.3	9.9
5	Russia	159,600	8.1	156,200	6.6	150,280	5.8	−3.8
6	India	66,786	3.4	61,663	2.6	99,789	3.9	61.8
7	Brazil	56,031	2.8	105,778	4.5	76,861	3.0	−27.3
8	UK	25,840	1.3	34,967	1.5	65,802	2.6	88.2
9	Iran	46,908	2.4	54,573	2.3	65,056	2.5	19.2
10	Turkey	37,897	1.9	39,771	1.7	64,641	2.5	62.5
	Total of top 10 countries	1,309,421	66.4	1,501,319	63.7	1,569,303	60.9	4.5
	Export total	1,966,557	100.0	2,356,866	100.0	2,576,914	100.0	9.3

Source: Ministry of Food and Drug Safety (2015). Food & Drug Statistical Year Book.

31.1.5 Export

Korea has made a substantial growth in the export of medical devices in recent years. In 2014, the total amount of exports amounted to US$2.7 billion, leading to 27% of CAGR for the past five-year period from 2010.

The export of medical devices has been limited to a number of specific countries. The total amount of exports to the top 10 countries shared 63% of the total export of the country in 2014. Those top 10 countries are the United States, China, Germany, Japan, Russia, India, Brazil, the United Kingdom, Iran and Turkey.

31.2 Regulatory Approvals

31.2.1 Responsible Authorities

Medical devices are regulated by Medical Device Act. Under the jurisdiction of the Act, requirements for medical devices are provided in several different titles of regulations. Below is a list of the titles of the laws and major regulations:

- Medical Device Act
- Medical Device Enforcement Decree
- Medical Device Implementing Regulations
- Regulation for Medical Device Groups and Class by Group
- Regulation for Medical Device Approval, Notification, Review, etc.
- Standards for Medical Device Manufacturing and Quality Management
- Regulation for Clinical Investigation Plan on Medical Devices
- Regulation for Designating Medical Devices Subject to Tracking
- Regulation for Medical Device Re-examination
- Regulation for Medical Device Re-evaluation
- Regulation for Medical Device Safety Information Management, such as Adverse Event

In general, the definition of products regulated as medical device sounds very similar in wording to the ones in the United

States, the European Union and other countries. So, those products regulated as medical device in other countries are also regulated as medical device in Korea in most cases. However, there are some products regulated as medical device in other countries but as drug in Korea. It is related to the history of medical device regulation: medical devices in Korea used to be regulated under the Pharmaceutical Affair Act until the National Assembly legislated the Medical Device Act for the first time in its history in 2003. Some products were pre-occupied by drug regulations, which have remained unchanged even after the legislation of a separate device regulation. Foreign-made products in this type of category should be ready for submission as drug.

The Ministry of Food & Drug Safety (MFDS) is the government agency responsible for regulating medical devices. It has the main office located in the Life Science Park in the city of Osong, Chung-cheong-buk-do province, about 120 km south of Seoul and regional offices in six different cities including Seoul, Incheon, Daejeon, Daegu, Busan, and Gwangju. MFDS also has its subordinate agency called National Institute of Food & Drug Safety (NIFDS) for policy research and technical review of applications for pre-market approvals.

The Medical Device Safety Bureau of MFDS has overall responsibility for device regulation. It has three divisions in portfolios of responsibility as follows:

Medical Device Policy Division is the control tower of all policy and regulations primarily for pre-market requirements and issues. It initiates and coordinates with other divisions to make and change regulations.

Medical Device Management Division is responsible for post-market surveillance activities, such as promotional labeling, raids and destroys of incompliant products, etc.

Medical Device Safety Evaluation Division oversees safety administration of devices in markets, such as adverse event reporting, tracked devices, re-examination, re-evaluation, etc.

MFDS works with NIFDS for a review of pre-market applications by five different offices depending on product specialty as follows:

- Cardiovascular Device Division
- Orthopedic & Rehabilitation Device Division
- Oral & Digestive Device Division

- High-Tech Device Division
- In-Vitro Diagnostics Task Force

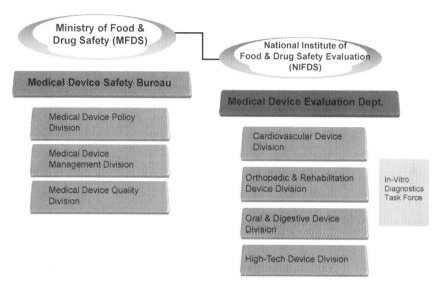

Figure 31.2 Office of MFDS and NIFDS for medical devices.

31.2.2 Qualifications for Medical Device Business

In order for a company to begin medical device business in Korea, the company is required to have three different licenses as follows:

- Medical Device Manufacturer/Importer Business License
- Certification to Medical Device Quality System Management called Korean Good Manufacturing Practice (KGMP)
- Medical Device Product Approval or Certification or Notification

31.2.2.1 Medical device business license

A company involved in one the following medical device businesses must obtain a business license for basic qualification: (i) manufacture, (ii) import, (iii) distribution, (iv) rental business, and (v) repairing company. This license must be completed before a company starts manufacturing, importing or distributing the first device product in its company.

Once a company obtains it, it is good for the entire organization. For registration, a company must have (i) a health checkup statement of the legal representative of the company, (ii) at least one product license for a medical device, and (iii) a full-time quality management system manager for medical device.

31.2.2.2 Certification to medical device quality system management (KGMP)

All Korean and foreign device manufacturers are required to have certification in compliance with quality management system standards or Korean Good Manufacturing Practice (KGMP) standards. For this purpose, MFDS has a total of 27 device categories. A company must be certified to a relevant category before it can be approved for a product belonging to the category. It is supposed to receive certification to as many categories as applicable to the products dealt by the company. Certification is valid for three years and must be renewed every three years. It is called Korean Good Manufacturing Practice (KGMP).

For KGMP certification, MFDS or third-party organizations inspect device manufacturers' quality systems on site. A certification on a foreign device manufacturer is provided in tandem with its Korean importer holding import product licenses in Korea. In case Korean license holder moves to a new company, the new importer must obtain a new KGMP for the manufacturer on its own name. If a foreign manufacturer has a multiple number of importers in Korea, each import must be certified for KGMP separately. When a Korean company imports device products from diverse companies, only one foreign manufacturer is inspected on site: The other firms are required to submit certain documents on the factory's quality management system for documentary review. In general, a foreign manufacturer providing the highest-class product is most likely to be chosen for on-site inspection.

31.2.2.3 Medical device product approval or certification or notification

31.2.2.3.1 Pre-market pathways

The MFDS classifies medical devices into four classes depending on the level of potential risk exposed to human body. The level of risk is assessed by the regulatory authority in consideration of duration of a device contacting the human body, level of invasiveness, delivery of drug or energy, and biological effects. In most cases, the Korean medical device classes match the ones in the medical device classification in the EU.

Class	level of potential risk to human body
1	Almost no potential risk
2	Low level of potential risk
3	Intermediate level of potential risk
4	High level of potential risk

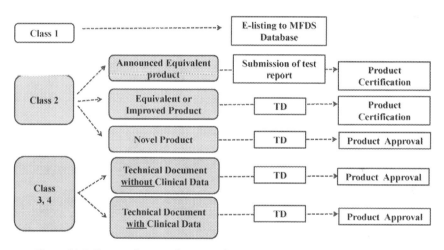

Figure 31.3 Pre-market regulatory pathways.

The regulatory system offers a few different pathways for pre-market clearance depending on medical device classes, level of innovativeness and availability of standards for finished products. The chart below illustrates pre-market pathways for different class and product groups. In general, MFDS reviews Technical Documents (TD) of class 3 and 4 devices and issues product licenses finally. MFDS also reviews certain class 2 devices for which clinical investigation report is required. For most of the class 2 devices, third-party organizations review TDs, and Medical Device Information and Technology Assistance Center (MDITAC) issues a product license. For most of the class 1 devices except for sterile or measuring devices, MDITAC operates an electronic system where companies submit information on products directly.

31.2.2.3.2 Review times

Regulations provide the maximum number of days that regulating authorities may take to review pre-market applications. MFDS or third-party organizations are required to complete a review of an application within such timeframes as follows. In the meantime, the timeframes include the number of days taken purely by MFDS reviewer: the review clock stops once the agency issued a letter of request for additional information. A review is conducted usually through the two cycles of questions and answers between the reviewer and the applicant. Extensions of timelines are available on applicants' request if adequate are not ready within a given timeframe. In aggregate, pre-market processes for approval may take from a few months to a year or more depending on the complexity of requirements and level of proficiency of a regulatory affairs professional.

Table 31.10 Official review times of technical documents of medical device pre-market applications

			Official process timeframes			Overall timeframes including company preparation
Class	Categories	Review body	Technical document review	Product license	Total	
Class 1	All class 1 devices except for sterile, measuring devices	No review	Not applicable	Not available	0 days	1–2 months
Class 2	Announced Equivalent Product	Test lab	Not applicable	10 days	10 days	2 months
	Equivalent or Improved Product	Third Party	25 days	10 days	35 days	3–5 months
	Novel Product	MFDS	70 days	10 days	80 days	9–12 months

Class	Categories	Review body	Official process timeframes			Overall timeframes including company preparation
			Technical document review	Product license	Total	
Class 2, 3, 4	Clinical Study Review Exempt	MFDS	55 days	10 days	65 days	9–12 months
	Clinical Study Review Mandated	MFDS	70 days	10 days	80 days	9–12 months

31.2.2.3.3 Key steps for medical device registration

Step 1: Classify your device

MFDS has a list of more than 2,500 device groups with each of a pre-determined class. A full list of products regulated as medical devices is available on the MFDS regulation titled "Regulation for Grouping Medical Devices and Class by Group." Companies first need to single out a device group which is applicable to a product for pre-market clearance. In case a suitable product group is not available from the list, companies can submit to MFDS a written inquiry asking for assignment of an appropriate product group and a class.

Step 2: Identify the scope of information necessary to TD review

MFDS determines the information required for a specific product based on how different it is from similar products previously approved by the agency: The more fundamental difference is, the more substantial information required. MFDS determines a product submitted to pre-market clearance to one of the three different categories depending on the nature of differences it introduces against the most similar products ("comparator") previously approved by the agency:

- **"Nobel Product"**: Not equivalent with the comparator primarily from the aspects of intended use, principles of mechanism, and raw materials;
- **"Incrementally Improved Product"**: Same as the comparator in intended use, principles of mechanism and raw materials but not equivalent in performance, test standards, and methods for use; or

- **"Equivalent Product"**: Equivalent with the comparator in intended use, principles of mechanism, raw materials, performance, test standards, and methods for use.

The information on the scope of requirements for a Technical Document (TD) of a product is available on the MFDS regulation titled "Regulation for Medical Device Approval, Notification and Review," provided in Table 7. This presents a series of matrixes indicating types of information necessary or unnecessary to TD by product category. A sample matrix is attached at the end of this document (Appendix 1). Starting with such matrix, companies can narrow down the scope of information specific to a device product.

Step 3: Prepare a TD

Companies prepare a TD application in the form called "Application for Technical Document Review," provided in Form No. 8., and submits it to the responsible organization. The application must be prepared in Korean. MFDS has released many guidance documents (in Korean) on how to prepare a TD in general or by type of product.

The required information for a TD application is itemized as follows:

- Information on manufacturer, manufacturing site and importer
- Product name and Model No.
- Shape, structure and dimensions
- Raw materials or components and amounts of contents
- Manufacturing Process
- Sterilization
- Performance and purpose of use
- How to use
- Precautions
- Packing units
- Storage conditions and period of validity
- Labeling
- Test standards

For a class 4 device, additional package of dossier called Summary Technical Documentation (STED) is required. The information necessary for STED is itemized as shown in Fig. 31.4.

1. Application	1.1 Application for Review
	1.2 Comparison chart with already approved products
2. STED (Summary Technical Documentation)	2.1 Table of Contents
	2.2 Device Descriptions & Product Specifications
	2.3 Labeling
	2.4 Design and Manufacturing Information
	2.5 Essential Principles (EP) Checklist
	2.6 Summary Risk Analysis and Management
	2.7 Summary of Product Verification and Validation
3. Attachments	3.1 Table of Contents
	3.2 Manufacturing Processed
	3.3 Risk Analysis Report
	3.4 Product Verification and Validation
	3.5 References

Figure 31.4 Table of contents for STED.

Applications are submitted to MFDS via electronic submission system. For imported devices, certain original documents provided by manufacturers must be submitted as part of an application package, so review bodies should know that important information on the product is originated from foreign manufacturers, not by discretion of importers.

Step 4: Receive a Certificate of Approval

As soon as MFDS or MDITAC has completed a review, a certificate of product license will be provided in the electronic submission system. Applicant companies pay a review fee electronically and then download it to print a hardcopy.

There are two types of product licenses: product manufacturing license and product import license. Both licenses are given to a business entity registered in Korea.

Product license has no expiration date: There is no need for periodic renewal. It must be updated whenever changes are made to the product.

31.2.2.3.4 Local laboratory testing

MFDS accepts test reports conducted in a foreign country, if the test was conducted in accordance with standards accepted by MFDS. No such full list of the standards recognized by MFDS exists. Instead, MFDS has also released review guidelines by type of product (in Korean): Those documents provide detailed information on standards applied to TD reviews. In general, MFDS accepts international standards frequently referenced to medical devices, such as, ISO/EN, IEC, ASTM, etc. The U.S. FDA's recognized standards are a good source of information when written guideline is not available from MFDS for a specific product: MFDS accepts most of the standards recognized by the U.S. device regulatory agency.

The information provided by such foreign test reports is also important. For example, foreign test reports on electrical safety or EMI should be prepared in the form of CB scheme. Biological safety test reports should indicate that the testing was conducted in compliance with GLP. Performance test reports conducted by foreign manufacturers are also acceptable when tests were conducted within quality management system, such as, ISO 13485.

Local test laboratories are available for conducting a test in Korea, in case foreign test reports do not provide adequate information for review. Local testing should be conducted in a laboratory registered with MFDS as a medical device test lab.

No.	Laboratories
1	Korean Testing Lab (KTL)
2	Korea Conformity Laboratories (KCL)
3	Korea Testing and Research Institute for the Chemical Industry (KOTRIC)
4	Korea Electric and Electric Testing Institute (KEETI)
5	Yonsei University College of Dentistry
6	Kunghee University College of Dentistry
7	Kyungbuk University College of Dentistry
8	Seoul University Clinical Research Institute
9	Yonsei University Medical Technology Quality Evaluation Center
10	Kunkook University Medical Device Center

Figure 31.5 Major test labs for medical device testing.

31.2.2.3.5 Clinical investigation

MFDS accepts foreign clinical study reports under certain conditions. Foreign manufacturers don't have to conduct a clinical investigation in Korea if they have clinical investigation reports readily available from outside of Korea. Among such important conditions are

- a report published in an academic journal listed on the Science Citation Index (SCI);
- study design, statistical power, and other information provided by the report should be adequate to the purpose of review.

MFDS does not accept clinical evaluation reports in place of clinical investigation report. If the agency requires a clinical report, it must be a report involving the specific product applied for product approval.

If foreign manufacturers need to conduct clinical investigations somewhere in the globe, Korea is a good candidate country for study conduct as Korea has an excellent infrastructure for medical device trials. Physicians practicing at university hospitals are highly receptive to new technologies and very much interested in participating clinical studies. Such hospitals already have a system in place to conduct a study according to Good Clinical Practices (GCP). People in Korea have positive attitude toward clinical studies, and patient enrollment could be much faster. Especially, large university hospitals located in Seoul or Gyonggi province have a relatively large size of patient pools that helps earlier enrollment of study subjects. Study funds for investigators or study sites are also reasonable.

31.2.3 Strategic Plan and Useful Tips for Efficient Registration

31.2.3.1 Face-to-face meeting with reviewer

Face-to-face meetings are most productive in resolving regulatory questions. The system in Korea offers device industry a very good access to reviewers. Meetings with reviewers are available at 1–2 week prior requests at any point of process, e.g., before submission of application or in the middle of review. Especially for complex products, pre-submission meetings with responsible reviewers are strongly recommended, so reviewers can have a chance to become familiar with a product for submission. Companies can also find out in advance the main questions the reviewers might have about the product and address those questions in applications. By doing this way, companies can reduce the overall review time significantly. There is no fee for this type of consultation meeting.

31.2.3.2 Respond to reviewer's questions with respect

In general, device reviewers in Korea are open to new information or technology. In the meantime, they also feel responsible strongly for protection of public safety and future medical device user through review of applications. So, reviewers try to stay with review criteria provided by relevant guidance documents. When companies find out gaps with reviewers in terms of requirements, e.g., reviewers request certain documents, but such documents are not available, explain your views politely and ask the reviewers about their perspectives asking for such information. They should have reasons to do so. You will then think about alternative ways to address their concerns on your product.

31.2.3.3 Retain experienced regulatory affairs professionals

Like in all other countries, good documentation of application is the basics to success. On top of that, timely management of communication during the review is equally important. To engage those essential competencies to your important project, retain experienced regulatory affairs professional for submission. Especially for an innovative product, the professional's competencies in setting up regulatory strategies and faithful relationships with the responsible reviewers are crucial as it is unlikely for MFDS to have review criteria for such product, so companies work with the reviewers from agreeing to review standards. Escalation of controversial issues to higher-level authorities in a timely manner is also important, so companies can move forward properly.

31.3 Reimbursement

31.3.1 Overview of Reimbursement Scheme

The healthcare system in Korea has two components: health insurance and medical aid. The national health insurance system provides coverage to all citizens, and it is managed comprehensively in the form of social insurance. It is funded by beneficiaries' contributions. The Medical Aid component provides support to lower income groups and is funded by the general revenue.

The Ministry of Health and Welfare (MOHW) oversees the national health insurance system. The National Health Insurance Service (NHIS) serves as the insurer and the Health Insurance Review and Assessment Service (HIRA) conducts reviews and assessment of medical fees.

"Fee for Service (FFS)" has been the traditional reimbursement system. Given that FFS' payment is based on individual visit or procedure, it encounters use of more services. In order to reduce the number of unnecessary services, the Diagnostic Related Group (DRG) system has been implemented since 2002. For certain illnesses, the DRG method pays a lump sum based on the patient's diagnosis.

The reimbursement process starts with the health institution filing a claim for medical fees to HIRA. After HIRA reviews the claim, it notifies the result to NHIS and the health institution for payment.

The insured pays a monthly premium according to his/her income level to the insurer and a co-payment (a fixed-rate) to the health institution after receiving medical treatments.

Under the public NHI system, MOHW maintains the three kinds of reimbursement price lists separately: (1) Medical Service Fee List, (2) Medical Supplies List, and (3) Drug Price List. All listed prices announced by MOHW mean the maximum reimbursement amounts available to the healthcare providers.

Medical supplies refer to disposable or implantable devices of therapeutic function and are covered by NHI. Approved medical supplies are announced by MOHW brand and listed on the reimbursement price list. For listing, device companies submit reimbursement applications to HIRA.

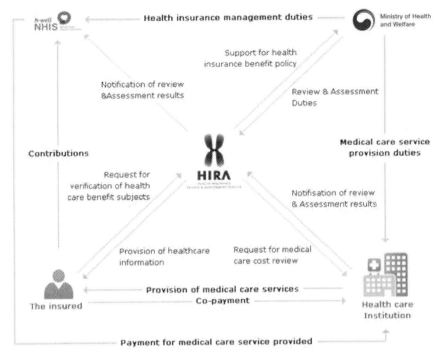

The costs of capital equipment and medical supply which is embedded in a medical service fee of a procedure involving the equipment. Service fee includes not only the costs of such medical supplies but also all other costs incurring to hospitals whenever treating a patient, such as salaries of doctors, nurses and other supporting staff, depreciation of facilities, cost of wards, etc. The submission of service fee application is the responsibility of healthcare service provider.

Some products fail to obtain reimbursement approvals as medical supplies although they look like worth reimbursement. Such products, called "products for No Separate Charge" or "Product Deemed Inclusive of Service Fee," may be used for patient treatment but should not be charged to either patients or the government for additional fee.

As of March 1, 2016, a total of more than 22,500 devices were listed for reimbursement as medical supplies. In addition, around 3,200 devices were approved for non-reimbursement/ full patient charge. Around 530 device products were listed on the list of no separate charge.

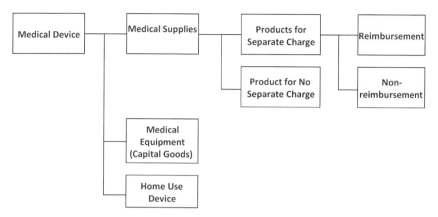

Figure 31.6 Categories of devices in NHI payment system.

Descriptions	Reimbursement or non-reimbursement	Number of products	%
Separate charge	Reimbursement	22,559	88%
	Non-reimbursement	3,271	12%
Sub-total of product for separate charge		25,830	100%
Products of no separate charges		530	31.4

Figure 31.7 Number of device products covered by NHI.

31.3.2 Medical Supplies

31.3.2.1 Submission

Medical supplies are required for reimbursement submission within 30 days of regulatory approval. Medical supplies must submit the application even if it does not pursue reimbursement by government but non-reimbursement or patient self-pay, so the device has a non-reimbursement code.

Once a company submitted the reimbursement application of a product within such timeframe, the company's customers as healthcare providers are granted by law rights to charge patients a full cost of the device until the government announces its decision officially.

31.3.2.2 Pricing options

The MOHW/HIRA determines the price of a new product based on its comparative analysis of features and benefits of a new product against the price of similar products already listed for reimbursement. If a new product is basically same as those products listed on the respective category in terms of purpose of use, clinical values, appearance, etc., the product is priced at the same amount with those existing products ("Standard Price").

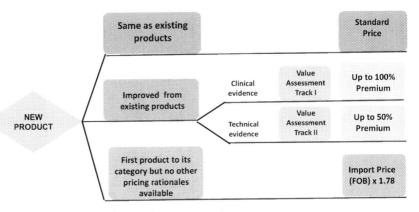

Figure 31.8 Options for reimbursement prices.

If a new product is submitted for a higher price ("Premium Price"), it takes much longer processes called "Value Assessment." The Value Assessment has two tracks depending on the type of evidence submitted by a device company. If the product submits evidence of clinical improvement, the product is referred to the process called Value Assessment Track I. The Track I open up the possibility of 100% premium over the price of the comparator product. If the product has made technical improvements, it goes through a shorter track called "Value Assessment Track II." The Track II allows the product to be approved at up to 50% higher than the price of the comparator product. If a new product introduces a totally new concept and has no comparator product, the government applies a cost-accounting calculation method for pricing in consideration of the costs of existing therapies for the similar disease state.

The normal percentage of government reimbursement is 70%, for the remaining 30% is paid by patients. The government introduced a new coverage option called "Selective Reimbursement" to allow a higher portion of amount to be covered by patient (50~80%) due to unclear clinical benefit or lack of cost-effectiveness.

31.3.2.3 Timeframes

The government is given a total of up to 100 days to make a decision on an application for medical supplies. HIRA accepts a reimbursement application from device companies and reviews it technically. And the application is forwarded to MOHW for a final decision and public announcement. In the meantime, depending on the level of price a product is proposed for, the evaluation process takes a shorter or longer period of time. The processes for a product applied to Standard Price are simple and takes an average of three months from filing to announcement of a result. For premium pricing, HIRA forms up a value assessment committee specific to the product and takes up 5-9 months in general.

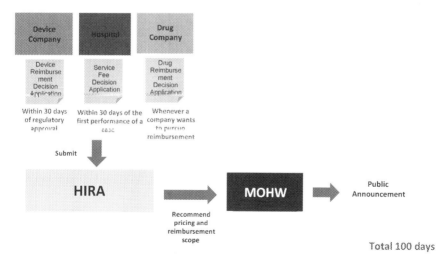

Figure 31.9 Reimbursement application parties and review times.

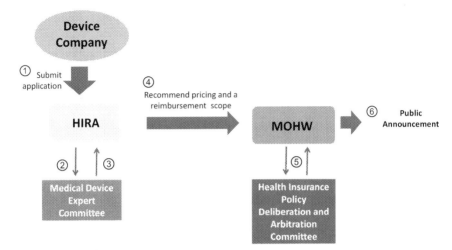

Figure 31.10 Evaluation process for a product applied to standard price.

31.3.2.4 Stakeholders

31.3.2.4.1 HIRA

In all review cases, HIRA plays the most important roles in medical supplies reimbursement decision. The HIRA reviewers mostly have nursing background and have a deep understanding of how a device is used in clinical setting. A responsible reviewer also collects as many information as possible on the submitted product from various parties, such as internal and external advisors, relevant academic societies and formulates a few price options for consideration of the Medical Supplies Expert Evaluation.

31.3.2.4.2 Medical supplies expert evaluation committee ("expert committee")

In the evaluation process, the Expert Committee provides expert opinion on clinical usefulness and safety as a physician.

The Expert Committee consists of about 300 expert individuals recommended by stakeholders in device reimbursement approvals as follows:

- Academic societies representing medical doctors, dentists, herbal medicine doctors, nurse and pharmaceutical affairs

Reimbursement | 569

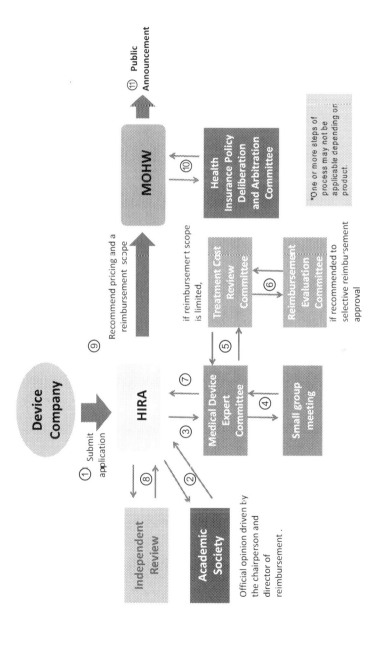

Figure 31.11 Evaluation process for a product applied to premium pricing.

- National Health Insurance Service
- Health Insurance Review and Assessment Service
- Consumer groups
- Experts working in relevant societies or experts
- Ministry of Food and Drug Safety
- Other organizations

31.3.3.4.3 Academic society

Academic society members routinely participate in a decision-making process to give comments from academic society perspectives. It is a routine process for HIRA to obtain comments from relevant academic societies on reimbursement, non-reimbursement, reimbursement prices, scope of reimbursement (reimbursement guideline), etc., based on clinical evidence available for the submitted product.

31.3.3 Health Technology Assessment (HTA)

In case there is no medical service code available for a new procedure and new medical supplies, it must go through HTA first before it applies for reimbursement request. And before a device can proceed to HTA, it must go to HIRA first to obtain the agency's official decision on the status of the procedure introduced by the device considered as a new procedure.

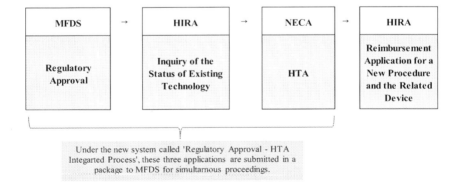

Figure 31.12 Processes for new devices introducing new procedures to NHI.

A product can be submitted for HTA following regulatory approval: the government introduced recently a new process called "Regulatory Approval—HTA Integrated Process," so products falling under certain criteria can apply for HTA at the same time with regulation submission to MFDS.

The National Evidence-based Collaborating Agency (NECA) is the responsible authority for HTA. It reviews applications submitted by device companies and conducts a systematic literature review with clinical evidence available for the medical technology involving the device. The timeframes for review are 280 days for general devices and 140 days for in vitro diagnostics.

HTA is the highest hurdle for device companies introducing new medical technologies. While HTA requires the highest level of clinical evidence, most of the device companies have limited clinical evidence at the time of introducing a new product. It has caused the substantial hurdle in the commercialization of new medical technologies in Korea. Presently, the Korean government is working on a new system to resolve this problem and to facilitate market access of medical innovations.

Chapter 32

Overview of Medical Device Regulation in Malaysia

Ir. Sasikala Devi Thangavelu

Policy Code & Standard, Medical Device Authority, Ministry of Health, Malaysia
Jack.wong@arpaedu.com

32.1 Medical Device Industry in Malaysia

The medical devices industry in Malaysia is a highly diversified industry that produces a broad range of products and equipment ranging from medical gloves, implantable devices, orthopedic devices and dialyzers to diagnostic imaging equipment and minimal invasive surgical equipment and other devices which can be used for medical, surgical, dental, optical and general health purposes. Malaysia recorded a Compounded Annual Growth Rate (CAGR) of 14.5% for medical devices export between 2014–2018, from RM13.4 billion in 2014 to RM23.0 billion in 2018. Malaysia's medical device export continues to grow steadily, with the composition of gloves and non gloves to be of almost equal proportion. Malaysia is a world leading producer and exporter for catheters and rubber gloves, supplying up to 80% of the world market for catheters and 60% for rubber gloves (including medical gloves). The growth of non-glove medical devices export outpaced the growth of glove medical devices export. In 2013–2018, the export growth rate of non-glove medical devices export stood at 15.2% as compared to glove medical devices export growth rate of 9.4%.

32.2 Medical Device Regulatory Program

On February 16, 2005, the Malaysian cabinet approved the development of Medical Device Regulatory Program to regulate the medical devices industry in Malaysia. Medical device Control Division (MDCD) was established within the Ministry of Health Malaysia in August 2005 to develop the medical device Bill, subsidiary legislation and to establish an organization to implement the medical device regulatory program. The objectives of the Medical Device Regulation are to address public health and safety issues and facilitate medical device trade and industry.

This first phase of medical device regulatory program started with the launch of the Medical Devices Voluntary Registration Scheme—MeDVER on January 12, 2006. This program was to encourage the medical device industry in Malaysia to voluntarily register their medical devices with MDCD.

It was during this phase, two laws, the Medical Device Act 2012 (Act 737) and Medical Device Authority Act 2012 (Act 738), were enacted and passed in the Malaysian Parliament. Act 737 is to regulate medical devices, the industry and to provide for matters thereto. Whereas Act 738 is to provide for the establishment of the Medical Device Authority with powers to control and regulate medical device, its industries and activities, and to enforce the medical device laws, and for related matters. By virtue of this Act, MDA was established on June 14, 2012 as a federal statutory agency under the Ministry of Health Malaysia to implement and enforce the Medical Device Act 2012 (Act 737).

The second phase, between 2012 and 2019, began with the establishment of Medical Device Authority (MDA), the enforcement of the Medical Device Act 2012, development of subsidiary legislation and the relevant documents in the implementation of the medical device regulatory program in Malaysia. The Medical Device Regulations (MDR) 2012 was made to prescribe certain activities under the provision of the Act, gazetted in December 2012 and came into operation on July 1, 2013. In this phase, the main activities such as registration, licensing, and conformity assessment bodies were prescribed in the Medical Device Regulation 2012.

Act 737 was effective July 1, 2013, with a 1-year transition period for establishment license and 2-year transition for product registration as provided for under savings and transition by virtue of Section 80 of Act 737. This transition period for product registration was extended until June 30, 2016.

In order to assist the industry, Malaysian Medical Device Authority launched the medical device online platform known as the Medical Device Centralised Application System (MeDC@St) on July 1, 2013. MeDC@St is a web-based Online Application System for Establishment Licensing, Medical Device Registration and Export Permit. It is a centralized system where only one account needs to be created by an applicant for all the activities. Then, MDA launched a newly upgraded database system, known as MeDC@St 2.0, on January 2, 2019, in order to assist the establishment and medical device industry. This system includes six modules for payment, establishment license, medical device registration Class A, medical device registration Class B, C, D, change notification, and Conformity Assessment Body registration (CAB).

The third phase of the regulatory program begins with the full transposition of the ASEAN Medical Device Directive into the national legislation in April 23, 2020. It also includes the implementation of the post market activities and advertisement regulation pursuant to Medical Device (Duties and Obligations of Establishments) Regulations 2019 and Medical Device (Advertising) Regulation 2019 which came into operation on July 1, 2020. The Medical Device Authority had decided to allow a transitional period of 18 months commencing from July 1, 2020, until December 31, 2021, for the implementation of Medical Device (Advertising) Regulation 2019.

32.3 Introduction to Medical Device Act 2012 (Act 737)

Any product that conforms with the medical device definition pursuant to section 2 of Act 737 and related activities shall be regulated based on the provisions of Act 737. To date, three Medical device Regulations and three Medical device Orders have been gazetted.

Pursuant to Subsection 79(2) of Act 737, three regulations have been gazetted as follows:

i. Medical Device Regulation 2016 (MDR 2016): prescribes the regulatory controls on premarket and placement of medical devices in the market

ii. Medical Device (Duties and Obligations of Establishments) Regulations 2019: regulatory control on post market surveillance and vigilance activities; and

iii. Medical Device (Advertising) Regulation 2019: prescribes the regulatory control on advertisement of registered medical devices.

Pursuant to Subsection 77(1) of Act 737, in the interest of public health and safety, three orders have been gazetted as follows:

i. Medical device (Exemption) Order 2015, which provides extension of the transitional period for medical device registration from July 1, 2015, until June 30, 2016, effective July 1, 2015,

ii. Medical device (Exemption) Order 2016, which exempts medical device for certain purposes from Section 5 and Section 15(1) of the Act and exemption of Class A medical device from conformity assessment procedures by a conformity assessment body under Section 7 of the Act effective April 18, 2016, and

iii. Medical device (Declaration) Order 2017, which has declared a non-corrective contact lens as a medical device effective on January 1, 2018.

32.3.1 Definition of Medical Device

Section 2 of Act 737 defines a "medical device" as follows:

(a) any instrument, apparatus, implement, machine, appliance, implant, in-vitro reagent or calibrator, software, material or other similar or related article intended by the manufacturer to be used, alone or in combination, for human beings for the purpose of—

i. diagnosis, prevention, monitoring, treatment or alleviation of disease;
ii. (ii) diagnosis, monitoring, treatment, alleviation of or compensation for an injury;
iii. (iii) investigation, replacement or modification, or support of the anatomy or of a physiological process;
iv. (iv) support or sustaining life;
v. (v) control of conception;
vi. (vi) disinfection of medical device; or
vii. (vii) providing information for medical or diagnostic purpose by means of in-vitro examination of specimens derived from the human body, which does not achieve its primary intended action in or on the human body by pharmacological, immunological or metabolic means, but that may be assisted in its intended function by such means; and

(b) any instrument, apparatus, implement, machine, appliance, implant, in-vitro reagent or calibrator, software, material or other similar or related article, to be used on the human body, which the Minister may, after taking into consideration issues of public safety, public health or public risk, declare to be a medical device by order published in the Gazette

32.3 Establishment license

The Act provides that any establishment dealing with import, export or place in the market any registered medical devices shall hold an establishment license granted under Act 737 by Medical Device Authority Malaysia. Establishment means a person who is a manufacturer, authorized representative, importer, or distributor but does not include a retailer. Such person being a person domiciled or resident in Malaysia or a firm or company registered in Malaysia and carrying on the business in Malaysia.

The Act defines the manufacturer as follows:

(a) a person who is responsible for:

  i. the design, production, fabrication, assembly, processing, packaging and labelling of a medical device whether or not it is the person, or a subcontractor acting on the person's behalf, who carries out these operations; and

  ii. (ii) assigning to the finished medical device under his own name, its intended purpose and ensuring the finished product meets the regulatory requirement; or

(b) any other person who:

  i. assembles, packages, processes, fully refurbishes, reprocesses or labels one or more ready-made medical devices; and

  ii. assigning to the ready-made medical device under his own name, its intended purpose and ensuring the finished product meets the regulatory requirement,

but **shall not include** the following persons:

(a) any person who assembles or adapts medical devices in the market that are intended for individual patients; and

(b) any person who assembles, packages or adapts medical devices in relation to which the assembling, packaging or adaptation does not change the purpose intended for the medical devices

Authorized representative

The authorized representative must be authorized by the foreign manufacturer. The authorized representative must be natural or legal person with business registration in Malaysia. It must maintain linkage with its foreign manufacturer and should be able to obtain the support of its foreign manufacturer whenever required.

Distributor

Distributor must be authorized by manufacturer or authorized representative to distribute devices on its behalf. Any natural or legal person in the supply chain authorized by the manufacturer or authorized representative to further the availability of medical devices to the end-user. In some circumstances, more than one distributor may be involved in this process.

Importer

Importer must be authorized by authorized representative to import devices on its behalf. Any natural or legal person authorized by authorized representative, who first makes a medical device manufactured in other countries, available in the Malaysian market.

The establishments shall comply with the license requirements as stated in Table 32.1. Every application for an establishment license shall be accompanied through MeDC@St 2.0 with application fee, licensing fee and documents as specified in Tables 32.2 and Table 32.3. The validity of the license is for a period of 3 years and the licensee may apply for renewal of its establishment license to MDA not later than 1 year before the expiry date of the license.

The related guidance documents are available in the MDA's portal, portal.mda.gov.my.

Table 32.1 Establishment licensing requirements

Requirements	Local manufacturer	Authorized rep	Importer	Distributor
• Establishment details	/	/	/	/
• Appropriate authorization		/	/	/
Procedures for; • Distribution records • Complaint handling • Adverse incident reporting • Field safety corrective action	/	/	/	/
• List of medical devices	/	/	/	/
• ISO 13485 or equivalent	/			
• Good Distribution Practice for Medical Devices (GDPMD)		/	/	/

Table 32.2 Establishment license fee for new and renewal application

Establishment license	New application fee (RM)	Renewal fee (RM)
Application fee	250.00	200.00
Manufacturer	4,000.00	2,000.00
Authorized representative	4,000.00	2,000.00
Importer	2,000.00	1,000.00
Distributor	2,000.00	1,000.00

Table 32.3 Establishment license fee for combined activities license

Establishment license	New application fee (RM)	Renewal fee (RM)
Application fee	250.00	200.00
Licensing Fee		
Manufacturer + distributor	4,000	2,000.00
Manufacturer + authorized representative	8,000	4,000.00
Authorized representative + distributor + Importer	4,000	2,000.00
Importer + Distributor	2,000	1,000.00

32.4 Medical Device Registration

All products that conform with the definition of "medical device" pursuant to Section 2 of Act 737 shall be registered prior to importation and placement into the market.

The persons responsible for registering a medical device under Act 737 are

i. the manufacturer of medical device as defined in Section 2 of Act 737; and
ii. in the case of a medical device manufactured in foreign country, the authorized representative of the foreign manufacturer, as defined in Section 2 of Act 737.

The manufacturer or the authorized representative shall classify the medical devices based on classification rules as specified in first schedule of MDR2012, group them based on second schedule of MDR2012, compile evidence to conformity element based on Schedule three of MDR2012 on quality management system, post market surveillance system, technical documentation, and declaration of conformity. The manufacturer or the authorized representative shall collect all these evidences of conformity and depending on the class of a medical device shall appoint conformity assessment body (CAB) registered by MDA under Section 10 of the Act to conduct conformity assessment on the said product. The CAB shall issue a report and certificate of conformity assessment to the manufacturer or authorized representative. Class A medical device is exempted from conformity assessment procedures by a conformity assessment body under Section 7 of Act 737 as stipulated in Medical device (Exemption) Order 2016 (P.U.(A)) 103 Gazetted on April 18, 2016.

An application for registration of medical device shall be made through MedCast2.0 online system. It shall be accompanied with application fee, registration fee, relevant documents on quality management system, post market surveillance system, technical documentation, declaration of conformity, CAB's report and certificate of conformity assessment.

Any medical devices without any approvals from the recognized countries shall go through full conformity assessment of the product prior to registration with MDA as indicated in Fig. 32.1 (Class A medical device) and Fig. 32.2 (Class B, C, D medical devices), whereas medical devices that have obtained approvals from the recognized countries shall go through verification route prior to registration with MDA. Based on circular letter No 2/2016; the recognized countries are European Union (EU), United States of America, Canada, Australia, and Japan.

32.4.1 Classification and Grouping for the Purpose of Medical Device Registration

All medical devices shall be classified into four classes, namely Class A, B, C and D a manufacturer based on the level of risk it poses, its intended use, duration of use and the vulnerability of the human body in accordance with the classification rules as specified in Appendix 1 for general medical devices and

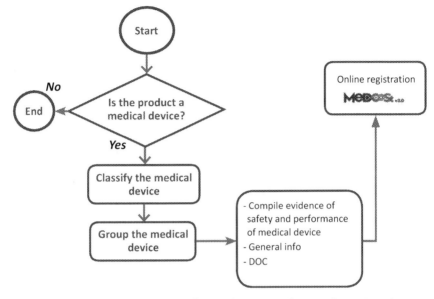

Figure 32.1 Steps to be taken before making an application for registration of a medical device for Class A.

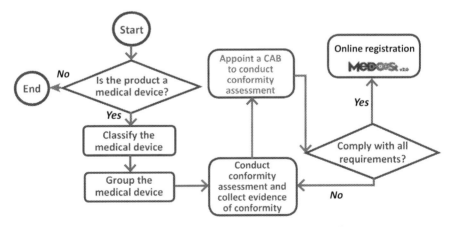

Figure 32.2 Steps to be taken before making an application for registration of a medical device for Class B, C, and D.

Appendix 2 for in vitro diagnostic medical device of the First Schedule of MDR2012. The classification of IVD is determined by taking into consideration both the intended purpose of the device and level of risk to the patient and public health. Samples of product classifications are shown in Table 32.4 and Table 32.5 for general medical devices and In Vitro Diagnostic Devices (IVD) respectively.

Table 32.4 Classification of general medical devices

Class	Risk level	Device examples
A	Low	Simple surgical instruments, tongue depressor, liquid-in-glass thermometer, examination light, simple wound dressing, oxygen mask, stethoscopes, walking aids
B	Low-moderate	Hypodermic needles, suction equipment, anesthetic breathing circuits, aspirator, external bone growth simulators, hearing aids, hydrogel dressings, patient controlled pain relief, phototherapy unit, x-ray films
C	High-moderate	Lung ventilator, orthopedic implants, baby incubator, blood oxygenator, blood bag, contact lens disinfecting/cleaning products, deep wound dressing, defibrillator, radiological therapy equipment, ventilator
D	High	Pacemakers and their leads, implantable defibrillators, implantable infusion pumps, heart valves, inter-uterine contraceptive devices, neurological catheters, vascular prostheses, stents

Table 32.5 Classification of in vitro diagnostic medical devices

Class	Risk level	Device examples
A	Low individual risk and Low public health risk	Clinical chemistry analyzer, prepared selective culture media
B	Low public health risk or moderate personal risk	Vitamin B12, pregnancy self testing, anti nuclear antibody, urine test strips
C	Moderate public health risk or high personal risk	Blood glucose self testing, HLA typing, PSA screening, rubella
D	High public health risk	HIV blood donor screening, HIV blood diagnostic

There are a wide range of medical devices, from a simple medical device to a highly complex and sophisticated medical device. The various components can be sold as a separate component, individual customized pack or group and can be categorized as SINGLE, FAMILY, SYSTEM, SET, IVD TEST KIT, and IVD CLUSTER.

Each of the categories mentioned can be submitted in the medical device registration application. Three basic rules that must all be fulfilled for the grouping are that the devices have one generic proprietary name, one manufacturer, and one common intended purpose. The medical devices are grouped using rules of grouping as specified in Second Schedule of MDR2012 for the purpose of registration of medical devices. The related guidance document is MDA/GD/0005 for general medical device and IVD.

32.4.2 Conformity Assessment

All medical devices except exempted medical device (Medical device (Exemption) Order 2016) shall be subjected to conformity assessment (CA) by a conformity assessment body registered with MDA. The CA for the purpose of registration of a medical device shall comprise the following elements:

(a) Conformity assessment of quality management system (QMS);
(b) Conformity assessment of post market surveillance system (PMS);
(c) Conformity assessment of technical documentation; and
(d) Declaration of conformity (DOC)

The level of CA is proportional to the risk associated with a medical device (risk-based classification) pursuant to Reg4 MDR 2012 as indicated in Table 32.6.

Table 32.6 Requirements for conformity assessment based on classification of medical devices

Class	QMS	PMS system	Summary technical documentation	DOC
Class A, Class A (S/M)	Establish and maintain QMS • Can exclude design and development control • May be audited for special cases	Establish and maintain adverse event reporting procedure for audit • Class A: may be audited to investigate specific safety or regulatory concerns	Prepare, make available upon request.	Prepare, sign and submit for review
Class B	Establish, maintain full QMS make available for audit		Prepare and submit for review	
Class C				
Class D				

32.4.6 Conformity Assessment through Verification Route

MDA circular 2/2014 sets the policy relating to conformity assessment for medical devices approved by recognized countries by MDA. The policy simplifies the process of conformity assessment and accelerate medical device registration under Act 737. Such medical devices need to undergo a simplified conformity assessment, through the process of verification of evidence obtained from the manufacturer (verification process). The eligible CAB shall conduct verification on evidence of conformity of the report and certificate. The recognized foreign regulatory authorities and notified bodies and the respective approval types eligible for conformity assessment by way of verification process are indicated in Table 32.7.

Conformity assessment elements and parameters to be verified by the eligible CAB

The parameters to be verified shall comprise the conformity assessment elements as stipulated in Third Schedule of MDR2012:

- Conformity assessment of QMS
- Conformity assessment of PMS system
- Conformity assessment of technical documentation
- Conformity assessment of declaration of conformity

The verification requirements for Class A (active, sterile or with measuring function) medical device (exempted from CAB) include:

- Basic medical device information
- Valid QMS certificate
- Technical documentation
- Declaration of conformity (DOC)

The verification requirements for Class B, C or D medical device will be verified by registered and eligible CAB include the following elements:

- Basic medical device information
- Valid QMS certificate
- PMS

- Technical documentation
- Declaration of conformity (DOC)

Medical devices which have not been subjected to conformity assessment and have not obtained any approval are required to undergo full conformity assessment in accordance with the requirements stipulated in Section 7(1)(a) of Act 737.

Table 32.7 Recognized regulatory authority and approval types

Recognized foreign regulatory authority or notified body	Approval type
Therapeutic Goods Administration (TGA) Australia	TGA license
Health Canada, Canada	Health Canada medical device license
Notified bodies listed in New Approach Notified and Designated Organisations (NANDO) database of European Union (EU)	For general medical device: • Annex II Section 3 or Annex V of MDD (for Class IIA) • Annex II Section 3 or Annex III coupled with Annex V of MDD (for Class IIB) • Annex II Section 3 and 4 of MDD (for Class III) • Annex II Section 3 and 4 of AIMDD (for active implantable medical device) For IVD medical device • Annex IV (Including Section 4 and 6) of IVDD (for List A IVD) • Annex IV (excluding Section 4 and 6) or Annex V coupled with Annex VII of IVDD (for List B and self-testing IVD) Annex III, EC declaration of conformity (section 1 to 5 of Annex III). Applicable for only class B IVD Medical Device in accordance with Medical Device Regulation 2012
Ministry of Health, Labour and Welfare (MHLW), Japan	• Pre Market Certification from a Japanese Registered Certification Body (RCB and PMDA) • Pre Market Approval from MHLW
US Food and Drug Administration (FDA)	• USFDA 510(k) clearance letter • USFDA pre-market approval (PMA) letter

32.5 Combinational Medical Devices

Drug-medical device/medical device-drug combination products are regulated according to the classification whether as drug or medical device based on the primary mode of action (PMOA).

Combination products regulated as drug by Drug Control Authority is in accordance with the requirements set forth in the Control of Drugs and Cosmetics Regulations 1984 which is promulgated under Sale of Drug Act 1952 and any other relevant documents published by National Pharmaceutical Regulatory Agency (NPRA).

Regulation of combination products as a medical device by the Medical Device Authority is in accordance with the requirements set forth in the Medical Device Act 2012 (Act 737) and its subsidiary legislations, and any other relevant documents published by MDA.

32.5.1 Definition of Combination Product

The term combination product includes:
   i. A product comprising two or more regulated components, i.e., drug/device, biological/device, or drug/device/biological, that are physically, chemically, or otherwise combined or mixed and produced as a single entity; OR
   ii. Two or more separate products packaged together (co-packaged) in a single package or as a unit and comprising drug and device products, device and biological products

Products that are excluded from the term combination product and will be regulated separately:
   i. A drug, device, or biological product packaged separately that according to its investigational plan or proposed labelling is intended for use only with an approved individually specified drug, device, or biological product where both are required to achieve the intended use, indication, or effect and where upon approval of the proposed product labelling of the approved product would need to be changed, e.g., to reflect a change in intended use, dosage form, strength, route of administration, or significant change in dose; or

ii. Any investigational drug or device packaged separately that according to its proposed labelling is use only with another individually specified investigational drug, device, or cosmetic product where both are required to achieve the intended use, indication, or effect.
iii. Convenience pack product (example: first aid kit consists of medical device and non-scheduled poison product)

Prior to registration, an applicant may apply for classification to NPRA through product classification form (NPRA 300.1) which is available at http://npra.moh.gov.my.

Drug-medical device combination product (DMDCP) is when the primary mode of action is based on pharmacological, immunological, or metabolic action in/on the body where NPRA is the primary agency of the combination product.

Medical Device-Drug Combination Product (MDDCP) is when the primary mode of action in or on the human body is not based on pharmacological, immunological or metabolic means, but that may be assisted in its intended function by such means where MDA is the primary agency of the combination product.

Primary agency is the agency with primary regulatory responsibility for a combination product which is determined by the primary mode of action of the product. The primary mode of action that provides the greatest contribution to the overall therapeutic effects of the combination product

Endorsement letter is issued by the secondary agency after satisfactory review of the ancillary component. The secondary agency is the agency that regulate the other part(s) included in the combination product.

32.5.2 Registration Process of Combination Product

The primary agency for registration of combination product is based on the primary mode of action/the principal mechanism of action by which the claimed effect or purpose of the product is achieved:

i. The drug is based on pharmacological, immunological, or metabolic action in/on the body; shall be regulated by the NPRA.

ii. The medical device does not achieve its primary mode of action in or on the human body by pharmacological, immunological or metabolic means, but that may be assisted in its intended function by such means; shall be regulated by the MDA.

The flow of registration process of combination product is illustrated in Fig. 32.3.

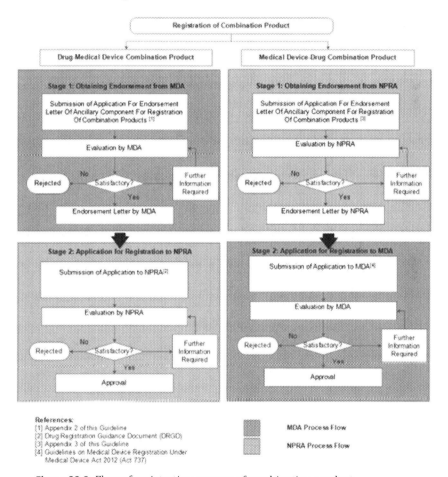

Figure 32.3 Flow of registration process of combination product.

The applicant may submit an application for the endorsement letter and registration of a combination product concurrently to both secondary and primary agency. However, the approval of combination product registration is subject to primary agency based on the fulfillment of registration requirements, as well as the receipt of endorsement letter from secondary agency.

Stage 1 Drug-medical device combination product is NOT mandatory for:
(a) low risk ancillary medical device components. They are not required to apply endorsement letter from MDA prior to their registration.
(b) Ancillary medical device components that have already obtain registration approval with MDA through a medical device registration application with MDA. Proof of medical device registration certificate is required to be presented to the NPRA when applying for drug-medical device combination product registration.

32.5.3 Drug-Medical Device Combination Product Registration Process (NPRA as Primary Agency)

The registration process of drug-medical device combination product shall undergo the following two stages:
i. Stage 1 – Obtaining Endorsement from MDA
ii. Stage 2 – Application for Registration to NPRA

32.5.4 Medical Device-Drug Combination Product Registration Process (MDA as Primary Agency)

The registration process of Medical Device-Drug combination product shall undergo the following 2 stages:
i. Stage 1 - Obtaining Endorsement from NPRA
ii. Stage 2 - Application for Registration to MDA

32.5.5 Changes/Variation to Particulars of a Registered Drug-Medical Device Combination Product

The application for changes to the particulars of a drug component for a registered drug-medical device combination product shall be required to comply with Section 5.2: Amendments to Particulars of a Registered Product of DRGD.

The application for changes to particulars of ancillary medical device component for a registered drug-medical device combination product shall be submitted to MDA manually according to Appendix 6 along with Appendix 4: Application Form for Change/Variation of Ancillary Components for Combination Products.

The MDA shall issue an approval letter upon satisfactory evaluation, and this approval letter shall then be sent to the NPRA for notification.

32.5.6 Changes/Variation to Particulars of a Registered Medical Device-Drug Combination Product

The application for changes to particulars of medical device component for a registered medical device-drug combination product shall be required to comply with the Guidance Document of Change Notification to Registered Medical Device.

The application for changes to particulars of ancillary drug component for a registered medical device-drug combination product shall be submitted to NPRA manually according to the categories and supporting documents in the current MVG or MVGB guidelines along with Appendix 4: Application Form for Change/Variation of Ancillary Components for Combination Products to the respective sections. Timelines for the evaluation of the application will follow as per the respective guidelines.

The NPRA shall issue an approval letter upon satisfactory evaluation, and this approval letter shall then be sent to the MDA for notification.

32.7 Medical Device Labelling

General Requirements

The labelling for all medical devices shall adhere to these general requirements pursuant to Regulation 16 of MDR2012. No person shall place any medical device in the market, use or operate any medical device to another person, use or operate any medical device to another person and used to any other person in any investigational testing unless it has been appropriately labelled.

The label shall include medical device registration number and this shall be carried out within 6 months from the date of registration of the medical device. The label shall not contain any statement to the effect, to indicate it is being promoted or endorsed by the Authority or the Ministry of Health or any of its organizational bodies. The label of a medical device shall be legible, permanent and prominent.

The medium, format based on international standard where applicable, content, readability and location of labelling should be appropriate to the particular device, its intended purpose and the technical knowledge, experience, education or training of the intended user(s).

In particular, instructions for use should be written in terms readily understood by the intended user and, where appropriate, supplemented with drawings and diagrams. Some devices may require separate information for the healthcare professional and the lay user. Paper versions of all labelling should accompany the product. Any residual risk identified in the risk analysis should be reflected as contraindications or warnings within the labelling.

Instructions for use (IFU) may not be needed or may be abbreviated for medical devices of low or moderate risk if they can be used safely and as intended by the manufacturer without any such instructions. Where relevant, for devices intended for home users, the IFU should contain a statement clearly directing the user not to make any decision of medical relevance without first consulting his or her health care provider.

Labelling activities to meet the Medical Device Act and regulations, may be conducted post importation or manufacturing, but prior to placing in the market. Contents of labelling shall be as per submitted to the authority during medical device registration. There shall be no over labelling on the lot/batch or serial number, date of manufacturing and date of expiry. The use of Bahasa Malaysia shall be required for home use medical devices, whereas English language shall be used on the labelling for other types of medical devices.

32.7.1 General Contents of Labelling

The label of a medical device shall contain the following information:

(a) details of medical device to enable user to identify it, which include name and model;
(b) An indication of either the batch code/lot number (e.g., on single use disposable medical devices or reagents) or the serial number (e.g., on electrically powered medical devices), where relevant, to allow appropriate actions to trace and recall the medical devices.
(c) name and contact details (address and/ or phone number and/ or fax number and/ or website address to obtain technical assistance) of the manufacturer and AR on the labelling;
(d) technical details concerning the medical device;
(e) description and intended use of the medical device;
(f) instructions for use of the medical device
(g) any undesirable side-effects, limitations, warnings and/or precautions on the safe use of the medical device;
(h) any necessary post-market servicing needs for the medical device; and
(i) any decommissioning or disposal information.

The IFU should include the date of issue or latest revision of the instructions for use and, where appropriate, an identification number. Where relevant, for devices intended for home users, the IFU should contain a statement clearly directing the user not to make any decision of medical relevance without first consulting his or her health care provider.

32.7.2 Additional Information for in vitro Diagnostic Medical Devices

For in vitro diagnostic medical devices, the following additional information shall be included in its label:

(a) Intended use/ purpose (e.g., monitoring, screening or diagnostic) including an indication that it is for in vitro diagnostic use

(b) test principle;
(c) specimen type, collection, handling and preparation;
(d) reagent description and any limitation (e.g., use with a dedicated instrument only);
(e) assay procedure including calculations and interpretation of results;
(f) information on interfering substances that may affect the performance of the assay;
(g) analytical performance characteristics, such as sensitivity, specificity, accuracy (trueness and precision);
(h) reference intervals; and
(i) use of drawings and diagrams.

32.8   Medical Device Order 2016

Medical device (Exemption) Order 2016 (P.U.(A) 103 Gazetted on April 18, 2016.

(a) Exemption from registration of medical devices

The exemption from Section 5 of Act 737 if the medical device is:

i. for the purpose of personal use
ii. for the purpose of demonstration for marketing
iii. for the purpose of education
iv. for the purpose of clinical research or performance evaluation of medical device
v. A custom made medical device; or
vi. A special access medical device

A person who imports or manufactures any medical devices of the above devices except for personal use shall notify the MDA and is exempted from establishment license.

(b) A class A medical device is exempted from conformity assessment procedures by a conformity assessment body under Section 7 of Act 737.

32.9 Post-Market Surveillance and Vigilance

The Medical device (Duties and Obligations of Establishments) Regulations 2019 prescribes the detail requirements of duties and obligation of the establishments regulated under Act 737. It is a regulatory tool to monitoring of safety and performance of medical device and to carry out post-market obligations.

As such the establishment shall put in place a post market surveillance and vigilance system which encompasses distribution records, complaint handling, mandatory problem reporting, Field corrective preventive action, recall to conform with this regulation to ensure the device is safe and effective throughout the life span of the device.

32.9.1 Distribution Records

An establishment shall ensure that a distribution record of each medical device manufactured, imported, exported and placed in the market. Record of distribution need to be established and maintained for tracking and traceability.

32.9.2 Records of Complaint Handling

The establishment shall ensure to maintain records of complaint handling. Procedure to handle complaint by the customer.

32.9.3 Mandatory Problem Reporting

An establishment shall submit an investigation report to the Authority within 30 days from the date of submission of mandatory report or within any extension of time given by the Authority.

An establishment shall submit mandatory report under Section 40 of the Act to the Authority relating to any incident that comes to the establishment's attention occurring inside or outside Malaysia. After submission of mandatory report, an establishment shall conduct investigation and if necessary, field corrective action to prevent recurrence of the incident. The establishment shall submit investigation report within 30 days from the date of submission of mandatory report. The requirement to submit a mandatory report shall not apply to any incident that occurs outside Malaysia if that incident has been reported by the establishment to the regulatory authority of the country.

32.9.4 Field Corrective Preventive Action

The establishment shall undertake corrective and preventive action to prevent recurrence of incident or problem. An establishment shall notify the Authority before undertaking any field corrective or preventive action.

32.9.5 Recall

The establishment shall undertake action to withdraw all defect medical device from the market. There are two types of recall, voluntary and mandatory recall.

32.9.6 Voluntary Recall

An establishment shall notify the Authority before undertaking any voluntary recall, notify MDA within the period in Table 32.8 and within 30 days after completion of a voluntary recall submit a report to MDA.

Table 32.8 Types of recall and requirements

Types of recall	Risk	Definition	Notification period to MDA
Class I recall	High risk	Possibility of serious health problem or death	Not less than 48 hours
Class II recall	Medium risk	Possibility it will cause temporary or reversible health problems or there is a remote possibility will cause serious problem	Not less than 3 days
Class III recall	Low risk	Probability it will cause health problem	Not less 5 days

32.9.7 Mandatory Recall

MDA may order an establishment to recall any medical device at any time due to patient safety and public health. The establishment shall report the result of the recall to MDA. If MDA is satisfied with the report, may close the matter or if it is not satisfied further order the establishment to take action to ensure that the device is not available in the market, cancel medical device registration or suspend or revoke the establishment license. MDA may publish the information of recall to the public.

32.10 Regulatory Action for Changes and Device Modifications

Policy on change of ownership for medical device registration with certain changes is allowed by MDA. The change ownership due to the reasons such as:

(a) A manufacturer outside Malaysia that has set up a company in Malaysia and intends to obtain the ownership of medical device registration from an authorized representative.

(b) Replacement of an existing authorized representative with new authorized representative by the manufacturer to place the medical device in the market

(c) Merger and acquisition activities

(d) The existing authorized representative closed its business

32.10.1 Change Notification for Registered Medical Device

Change to a registered medical device may be categorized into the following three categories:

1. Category 1: changes of medical devices that affect their safety and performance and require new registration of the medical device;
2. Category 2: changes that require evaluation and endorsement from the MDA prior to the implementation of the change and before placing on the market; and

3. Category 3: minor changes that may be implemented immediately.

For all categories of changes, prior to submission of change notification to the Authority the registration holder may submit a request for confirmation on change category.

32.10.2 Refurbishment of Medical Device

Based on circular 1/2016 (Rev 2), MDA will regulate all types of medical device refurbishment activities. The control methods for refurbishment activities are shown in Table 32.9.

Table 32.9 Regulatory control of refurbished medical device

Establishment that undertakes refurbishment	Type of medical device	Registration/notification requirements
Refurbishment by manufacturer	Unregistered refurbished medical device	Registration application shall be in accordance with Section 5
		The refurbished activities shall be included in the scope of quality management system for manufacturer of medical device
		Refurbished activities shall comply with GRPMD
	Registered medical device to be refurbished	Manufacturer/authorized representative shall provide Notification to the Authority
		The refurbished activities shall be included in the scope of quality management system for manufacturer of medical device
		Refurbished activities shall comply with GRPMD and provide technical details for the medical device
Refurbishment activities conducted by third party		Third party who wishes to carry out refurbishment activities shall obtain establishment license as Manufacturer and shall be responsible for registration of the medical devices
		Refurbished activities shall comply with GRPMD

32.11 Regulatory Control on Usage, Operation, and Maintenance of Medical Devices

The post market regulatory activities shall also include the regulatory control on usage, testing and commissioning, maintenance, and disposal of medical devices. The new regulation, which prescribes the manner to monitor and regulate these activities, is still in draft pending approval. This regulation also prescribes the competency requirement of person using, operating, installing, testing, commissioning, maintaining and disposing of medical devices.

To ensure proper use of medical device by competent operator, device which has high risk, patient safety, complex and public health concern may be designated as a designated medical device by an order published in the Gazette. Once such device is gazetted as designated medical device, any person who operates or uses this device shall be required to apply for a designated medical device permit. The new regulation to prescribe the requirements to control designated medical device is still in the draft stage pending approval.

References

1. Medical Device Act 2012.
2. Medical Device Authority Act 2012.
3. Medical device Regulation 2012.
4. Appointment of date of coming into operation: P.U.(B)73.
5. Appointment of date of coming into operation: P.U.(B) 126.
6. Medical device (Exemption) Order 2015.
7. Medical device (Exemption) Order 2016.
8. Medical device (Declaration) Order 2017.
9. Medical Device (Duties and Obligations of Establishments) Regulations 2019.
10. Medical Device (Advertising) Regulation 2019.
11. Medical device (Exemption) Order 2016 (P.U.(A) 103 Gazetted on April 18, 2016.
12. MDA annual report 2012/2013,2014,2015,2016.

13. MOH Annual Report 2006,2007, 2008,2009,2010,2011,2012.
14. http://www.matrade.gov.my/.
15. Guideline for registration of drug-medical device and medical device-drug combination products
16. Industry Mapping and Value Chain Analysis of Medical Devices Companies in Penang by Lee Siu Ming, Penang Institute.
17. Life Sciences Cluster in Selangor (Part II: Medical Devices), Invest Selangor 2013.
18. https://www.mida.gov.my.
19. https://portal.mda.gov.my.

Chapter 33

The Philippine Medical Device Regulatory System

Rhoel Laderas

Philippine Association for Medical Device Regulatory Professionals, Inc, The Philippines

Jack.wong@arpaedu.com

33.1 The Medical Device Market Profile

The Philippines has a population annual growth rate of 1.72% per year (2010–2015 data) with a life expectancy of 72 years.

The Department of Health statistics indicates that there are more than 1800 licensed hospitals, of which 60% are privately owned.

The medical device products are mostly imported, and 100% of medical equipment is imported. Local production limited to supplies such as masks, gloves, cotton, tube sets, and elastic bandage. There are manufacturers of syringes, stents however, these are for exports and not intended for local distribution.

The current demand for medical devices is based on the growing incidences of various diseases. There is a high incidence of diabetes, disease of the heart, disease of the vascular system, tuberculosis, cancer, pneumonia, and respiratory diseases. There is a high demand for different types of medical equipment to

---

Medical Regulatory Affairs: An International Handbook for Medical Devices and Healthcare Products (Third Edition)
Edited by Jack Wong and Raymond K. Y. Tong
Copyright © 2022 Jenny Stanford Publishing Pte. Ltd.
ISBN 978-981-4877-86-2 (Hardcover), 978-1-003-20769-6 (eBook)
www.jennystanford.com

diagnose these diseases, especially the radiation-emitting equipment such as x-ray machines, CT scans and MRIs. There is also an increased need for various in vitro diagnostic products or test kits for laboratory needs of patients.

The big percentage of procurement of medical device is done by the government through the public bidding.

33.2 Introduction to the Philippines Regulatory System

In 2009, Republic Act No. 9711 was approved to strengthen and rationalize the regulatory capacity of the Bureau of Food and Drugs and renaming it to the Food and Drug Administration.

The Food and Drug Administration of the Philippines is mandated to safeguard the health of the people. Primarily, the law has created four Centers established based on the product category that is regulated. All the Centers' primary function is to "regulate the manufacture, importation, exportation, distribution, sale, offer for sale, transfer, promotion, advertisement, sponsorship of, and/or, where appropriate, the use and testing of health products. The Centers shall likewise conduct research on the safety, efficacy, and quality of health products, and to institute standards for the same" [1].

The Food and Drug Administration has three regulatory tools:

1. Licensing of the Health Product Establishment. All establishments who intend to manufacture, import, distribute, and sell health products should secure first the license to operate (LTO) before the start of any business activities.
2. Authorization of Health Product. Once the LTO is granted, the health product establishment should secure the appropriate product authorization. Each regulated product needs to have the appropriate authorization before being allowed to be placed on the market. At present, the FDA is issuing the Certificate of Product Registration (CPR) and the Certificate of Product Notification (CPN) depending on the required regulatory control of the health product.

3. Post Market Surveillance Activities. The FDA conducts various PMS activities to ensure the continuous compliance of the regulated establishments and products to the FDA rules and regulations.

The **Center for Device Regulation, Radiation Health and Research (CDRRHR)** is one of the Centers that is established to regulate medical devices (to include the radiation emitting devices for medical application), health-related devices, ionizing and non-ionizing radiation facilities.

33.2 The Medical Device Regulatory System

The Medical Device Regulatory System employs the FDA's three regulatory tools: LTO, CPR/CPN, and PMS Activities.

33.2.1 The Licensing of Medical Device Establishment

There are two main classifications of the medical device establishments: Distributor and Manufacturer. Figure 33.1 shows the types of establishments under each classification.

Distributor	Manufacturer
• Importer • Wholesaler • Exporter • Retailer	• Manufacturer • Trader • Repacker

Figure 33.1 Classifications of medical device establishments.

Distributors include the importers, exporters, wholesalers, and retailers while manufacturers include the manufacturer, repackers and traders.

Figure 33.2 shows the distribution of the regulated establishments that have secured the license to operate as medical device establishment. The operational definition of TRADER in the context of this discussion is the owner of the formulation or design of the medical device. This figure shows that the Philippines is heavily reliant on the import of medical devices.

Medical Device Establishment

Figure 33.2 Distribution of regulated establishments as of June 2020.

The application for the licensing of the medical device establishments is filed online through the FDA e-portal. Each company has a qualified person who is the authorized person to transact business with the FDA. At present, retailers are not yet required to secure the LTO.

The process of filing the application is shown in Fig. 33.3.

The requirements to be uploaded in the e-portal for the LTO application is based on the Administrative Order No. 2016-003: "Guidelines on Unified Licensing Procedures of the Food and Drug Administration (FDA)," as shown in Fig. 33.4.

The manufacturers are inspected prior to the issuance of the LTO. Aside from the documentary requirements uploaded during the application, the manufacturing site is assessed and evaluated based on the submitted site masterfile. Areas considered includes the personnel (qualification and training), the storage area, the production area, the production process, the functionality of the different areas of the production, the sanitation, the ventilation, availability of required reference materials, documentation among others.

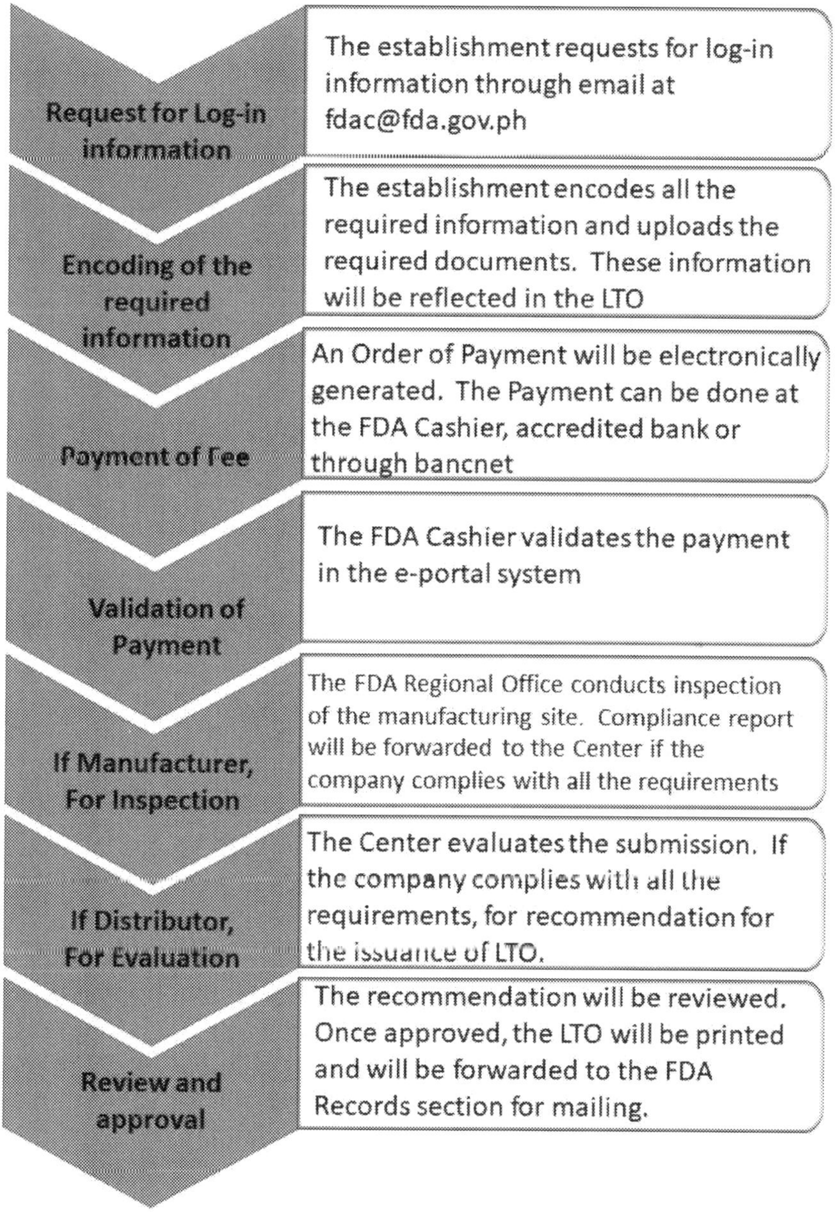

Step	Description
Request for Log-in information	The establishment requests for log-in information through email at fdac@fda.gov.ph
Encoding of the required information	The establishment encodes all the required information and uploads the required documents. These information will be reflected in the LTO
Payment of Fee	An Order of Payment will be electronically generated. The Payment can be done at the FDA Cashier, accredited bank or through bancnet
Validation of Payment	The FDA Cashier validates the payment in the e-portal system
If Manufacturer, For Inspection	The FDA Regional Office conducts inspection of the manufacturing site. Compliance report will be forwarded to the Center if the company complies with all the requirements
If Distributor, For Evaluation	The Center evaluates the submission. If the company complies with all the requirements, for recommendation for the issuance of LTO.
Review and approval	The recommendation will be reviewed. Once approved, the LTO will be printed and will be forwarded to the FDA Records section for mailing.

Figure 33.3 The licensing process.

> Declaration and undertaking of the responsibilities of the applicant as a condition for the processing and approval of the LTO. This is indicated in the electronic application and the applicant need to agree with the undertaking.

> The location plan and the global position system (CPS) coordinates of the establishment. This is provided electronically in the system and the company will select its location,

> The name of the qualified person and the relevant credentials which includes the PRC ID/Diploma. For medical device, the required profession of the qualified person includes the graduates of engineering (electrical, electronics, mechainical), nurses, medical technologist, chemists, and other medical allied courses appropriate for the product being distributed.

> For manufacturers, the credentials of the following staff should also be provided: Production Manager/Head,
> Quality Assurance Manager/Head, Quality Control Manager/Head

> Proof of Business Name Registration

> Site Master File (For manufacturers)

> Risk Management Plan

> Payment

Figure 33.4 Documentary requirements for LTO application based on Administrative Order 2016-003.

For the distributors, post-licensing inspection is conducted usually before the renewal of the LTO. The inspection covers all the physical structure of the office, ventilation, sanitation, storage area, historical records, documentations pertaining to the sources of the medical devices which includes the agreements, ISO certificates, issued CPRs, credentials of the qualified person among others.

The Administrative Order 2016-003 is already amended by Administrative Order 2020-017, which provides a wider coverage of the establishments to be licensed by the FDA. For medical device, retailers will now be licensed once this new AO is implemented. There will be another platform for this new AO where the request for the information login will be done online.

33.2.2 The Authorizations for Medical Device Products

The regulation for the medical device products is governed by the Administrative Order 2018-002: "Guidelines Governing the Issuance of an Authorization for a Medical Device based on ASEAN Harmonized Technical Requirements." This guidance document is the implementation of the ASEAN Medical Device Directive (AMDD). The definition and classification of medical devices in the said administrative order is based on the AMDD. In this guidance document, there are two sets of requirements, one is the legal requirements and the other one is the technical requirements. All classifications have the same legal requirements; however, the technical requirements differ for each classification. Class A has very minimum technical requirements, whereas Class D has full technical requirements based on the AMDD.

There are two authorizations issued for the medical device products depending on the classification:

1. *Certificate of Product Notification*: This is issued to all Class A medical devices that comply with the requirements.

2. *Certificate of Product Registration*: This certification is issued to Class B, C, and D medical devices that comply with the requirements.

The Certificate of Product Notification

The filing of application for notification is done online through the e-portal system. The login information is the same as the login information for the LTO application. The platform for the notification is designed in such a way that explanations are provided for every requirement. This is done to ensure that the applicants will submit the correct documents and to minimize the outright disapproval of the application. Upon application, there is a pre-evaluation if the company was able to provide the correct requirements. All incorrect or non-compliant applications are disapproved and a letter of rejection will be received by the applicant. At this stage, no fee is required. However, when the application meets all the requirements, the applicant will be issued an order of payment. The application will be forwarded for review, evaluation, and approval. Once the application is approved, the applicant will be notified by the system that the certificate is ready for printing to be done by the applicant.

Figures 33.5 and 33.6 show the legal and technical requirements for the application for notification of Class A medical devices:

Legal Requirements for Application for the Notification of Medical Devices under Class A and Registration of Medical Devices under Classes B, C and D

1. Notarized Application Form (Annex G or H)
2. Payment
3. Copy of Letter of Authorization. For imported medical devices, the copy of the Letter of Authorization shall be accompanied by an original copy of a notarized declaration from the legal manufacturer or product owner attesting that the authorization is true and correct.
4. A government – issued certificate attesting to the status of the Manufacturer with regard to the competence and reliability of the personnel and facilities, a Quality Systems Certificate of approval, or a compliance certificate for ISO 13485. For imported medical devices, the copy of the certificate shall be accompanied by an original copy of a notarized declaration from the legal manufacturer or product owner attesting that the certificate is true and correct.
5. For imported medical devices, the Certificate of Product Notification, Certificate of Product Registration, or any equivalent document attesting to the safety and effectiveness of the device issued by the regulatory agency or accredited notified body in the country of origin. The copy of the certificate shall be accompanied by an original copy of a notarized declaration from the legal manufacturer or product owner attesting that the certificate is true and correct.
6. Colored picture of the device from all sides. However, the CDRRHR can require a representative sample or commercial presentation for verification purposes.

Figure 33.5 Legal requirements for notification for Class A and registration of medical devices under Classes B, C and D in accordance with AO 2018-002.

> **Technical Requirements for Application for the Notification of Medical Devices under Class A**
>
> 1. Device description consisting of the following:
>   a. Intended use
>   b. Instruction for use
>   c. List of all raw materials
>   d. Technical specification of the finished product
>   e. List of reference codes, sizes, colors, models and variance, whichever is applicable.
> 2. Certificate of Conformity (issued by government agency dealing with metrology) on the aspect of manufacture relating to metrology for devices with measuring functions, if applicable
> 3. Declaration of Conformity (self declaration by the manufacturer) with product standards, if applicable
> 4. Clear and complete colored pictures of label from all sides of the packaging (loose label or artworks of all layers of packaging)
> 5. Declaration of shelf life

Figure 33.6 Technical requirements for notification for Class A in accordance with AO 2018-002.

The Certificate of Product Registration

The Certificate of Product Registration (CPR) is issued to medical device products under the classifications B, C and D. The manner of application is manually submitted at the FDA Action Center. However, the soft copies of the documents are required to be submitted. The copy is placed in a thumb drive and will be copied to the FDA System if the application is accepted. Pre-evaluation as to the completeness of the documents is done prior to payment If the documentary requirements are incomplete, the application will not be accepted. Only complete applications will be issued an order of payment and will be forwarded to the Center for review and evaluation.

The requirements for each classification are provided in the Administrative Order. This will provide clarity to the applicants on the specific requirement for each classification. Below is the list of the technical requirements per classification of medical device.

Technical Requirements for the Initial Registration of Medical Devices under Class B in accordance with AO 2018-002

1. Executive Summary. The executive summary shall include the following information:
    a. an overview, e.g., introductory descriptive information on the medical device, the intended uses and indications for use of the medical device, any novel features and a synopsis of the content of the CSDT;
    b. the commercial marketing history;
    c. the intended uses and indications in labeling;
    d. the list of regulatory approvals or marketing clearances obtained;
    e. the status of any pending request for market clearance; and
    f. the important safety/performance related information.
2. Relevant essential principles and method/s used to demonstrate conformity, if applicable.
3. Device description with the following information:
    a. Intended use
    b. Indications of use
    c. Instruction for use
    d. Contraindications
    e. Warnings
    f. Precautions
    g. Potential adverse effects
    h. Alternative therapy (practices and procedures)
    i. Materials. A description of the materials of the device and their physical properties to the extent necessary to demonstrate conformity with the relevant Essential Principles. The information shall include complete chemical, biological and physical characterization of the materials of the device.
    j. Other Relevant Specifications to include the following:

        j.1 The functional characteristics and technical performance specifications of the device including, as relevant: accuracy, sensitivity, specificity of measuring and diagnostic medical devices, reliability, and other factors

j.2 If applicable, other specifications including chemical, physical, electrical, mechanical, biological, software, sterility, stability, storage and transport, and packaging.

k. Other Descriptive Information to demonstrate conformity with the relevant Essential Principles (e.g., biocompatibility category for the finished medical device

4. Summary of Design Verification and Validation Documents: The validation documents shall consist of the following:

   a. Declaration/Certificates of Conformity to the product standards issued by the manufacturer;
   b. Summaries or reports of tests and evaluation based on other standards, manufacturer methods and tests, or alternative ways of demonstrating compliance, such as a listing of and conclusions drawn from published reports that concern the safety and performance of aspects of the medical device with reference to the Essential Principles;
   c. Data summaries or tests reports and evaluations covering the following appropriate test reports, whichever is applicable:

      c.1 Engineering test
      c.2 Laboratory test
      c.3 Biocompatibility test
      c.4 Animal Test
      c.5 Simulated Use
      c.6 Software Validation
      c.7 Pre-clinical studies

5. Clear and complete colored pictures of label in all angles of the packaging (loose label or artworks of all layers of packaging)
6. Risk to include the results
7. Physical Manufacturer information

a. Manufacturing process, including quality assurance measures. This should include the manufacturing methods and procedures, manufacturing environment or conditions, facilities and controls. The information may be presented in the form of a process flow chart showing an overview of production, controls, assembly, final product testing, and packaging of finished medical device.

b. A brief summary of the sterilization method should be included.

Technical Requirements for the Initial Registration of Medical Devices under Class C in accordance with AO 2018-002

1. Executive Summary. The executive summary shall include the following information:
   a. an overview, e.g., introductory descriptive information on the medical device, the intended uses and indications for use of the medical device, any novel features and a synopsis of the content of the CSDT;
   b. the commercial marketing history;
   c. the intended uses and indications in labeling;
   d. the list of regulatory approvals or marketing clearances obtained;
   e. the status of any pending request for market clearance; and
   f. the important safety/performance related information.
2. Relevant essential principles and method/s used to demonstrate conformity, if applicable.
3. Device description with the following information:
   a. Intended use
   b. Indications of use
   c. Instruction for use
   d. Contraindications
   e. Warnings
   f. Precautions
   g. Potential adverse effects
   h. Alternative therapy (practices and procedures)

i. Materials. A description of the materials of the device and their physical properties to the extent necessary to demonstrate conformity with the relevant Essential Principles. The information shall include complete chemical, biological and physical characterization of the materials of the device.

j. Other Relevant Specifications to include the following:

j.1 The functional characteristics and technical performance specifications of the device including, as relevant: accuracy, sensitivity, specificity of measuring and diagnostic medical devices, reliability, and other factors

j.2 If applicable, other specifications including chemical, physical, electrical, mechanical, biological, software, sterility, stability, storage and transport, and packaging.

k. Other Descriptive Information to demonstrate conformity with the relevant Essential Principles (e.g., biocompatibility category for the finished medical device)

4. Summary of Design Verification and Validation Documents: The validation documents shall consist of the following:

a. Declaration/Certificates of Conformity to the product standards issued by the manufacturer

b. Summaries or reports of tests and evaluation based on other standards, manufacturer methods and tests, or alternative ways of demonstrating compliance, such as a listing of and conclusion drawn from published reports that concerns the safety and performance aspects of the medical device with reference to the Essential Principles

c. data summaries or tests reports and evaluations covering the following appropriate test reports, whichever is applicable:

c.1 Engineering test
c.2 Laboratory test
c.3 Biocompatibility test
c.4 Animal Test
c.5 Simulated use

d. Clinical evidence for the following:

d.1 implantable devices

d.2 newly introduced devices

d.3 devices incorporating new materials coming into contact with the patients

d.4 Existing materials applied in a body part not previously expose to that material, and for which no prior chemical experience exists.

d.5 Existing device that is modified and the modification might affect safety and effectiveness

e. Software validation studies, if applicable

f. Biological evaluation, if applicable

5. Clear and complete colored pictures of label from all sides of the packaging (loose label or artworks of all layers of packaging)

6. Risk assessment consisting of risk analysis, evaluation, and reduction measures.

7. Physical Manufacturer's Information:

   a. Manufacturing process, including quality assurance measures. This should include the manufacturing methods and procedures, manufacturing environment or conditions, facilities and controls. The information may be presented in the form of a process flow chart showing an overview of production, controls, assembly, final product testing, and packaging of finished medical device.

   b. A brief summary of the sterilization method should be included.

Technical Requirements for the Initial Registration of Medical Devices under Class D in accordance with AO 2018-002

1. Executive Summary. The executive summary shall include the following information:

   a. an overview, e.g., introductory descriptive information on the medical device, the intended uses and indications for use of the medical device, any novel features and a synopsis of the content of the CSDT;

b. the commercial marketing history;
c. the intended uses and indications in labeling;
d. the list of regulatory approvals or marketing clearances obtained;
e. the status of any pending request for market clearance; and
f. the important safety/performance related information.

2. Relevant essential principles and method/s used to demonstrate conformity, if Applicable
3. Device description with the following information:
   a. Intended use
   b. Indications of use
   c. Instruction for use
   d. Contraindications
   e. Warnings
   f. Precautions
   g. Potential adverse effects
   h. Alternative therapy (practices and procedures)
   i. Materials. A description of the materials of the device and their physical properties to the extent necessary to demonstrate conformity with the relevant Essential Principles. The information shall include complete chemical, biological and physical characterization of the materials of the device.
   j. Other Relevant Specifications to include the following:

   j.1 The functional characteristics and technical performance specifications of the device including, as relevant: accuracy, sensitivity, specificity of measuring and diagnostic medical devices, reliability, and other factors

   j.2 If applicable, other specifications including chemical, physical, electrical, mechanical, biological, software, sterility, stability, storage and transport, and packaging.

   k. Other Descriptive Information to demonstrate conformity with the relevant Essential Principles (e.g., biocompatibility category for the finished medical device)

4. Summary of Design Verification and Validation Documents: The validation documents shall consist of the following:

   a. Declaration/Certificates of Conformity to the product standards issued by the manufacturer
   b. Summaries or reports of tests and evaluation based on other standards, manufacturer methods and tests, or alternative ways of demonstrating compliance, such as a listing of and conclusions drawn from published reports that concern the safety and performance of aspects of the medical device with reference to the Essential Principles;
   c. Data summaries or tests reports and evaluations covering the following appropriate test

      c.1 Engineering test
      c.2 Laboratory tes
      c.3 Biocompatibility test
      c.4 Animal Test
      c.5 Simulated Use

   d. Clinical evidence
   e. Software validation studies, if applicable.
   f. Biological evaluation, if applicable.
   g. A bibliography of all published reports dealing with the use, safety, and effectiveness of the device.

5. Clear and complete colored pictures of label from all sides of the packaging (loose label or artworks of all layers of packaging)
6. Risk assessment consisting of risk analysis, evaluation and reduction measures.
7. Physical Manufacturer information

   a. Manufacturing process, including quality assurance measures. This should include the manufacturing methods and procedures, manufacturing environment or conditions, facilities and controls. The information may be presented in the form of a process flow chart showing an overview of production, controls, assembly, final product testing, and packaging of finished medical device.

b. A brief summary of the sterilization method should be included.

References

1. Republic Act 9711: Strengthening of the FDA.
2. Administrative Order 2018-002.
3. DOH Website.

Chapter 34

Singapore Medical Device Regulation

May Ng, Ray Soh, Trish, Beatrice, Bing Kang, Yiyu, Xinyu, Ivy Lim, and Tiffany Hu

ARQon

Jack.wong@arpaedu.com

Enterprise Singapore reported that the number of home-grown medtech companies has increased from 100 in 2014 to more than 250 2019, and over half of these were start-ups. The Asia-Pacific medtech scene is expected to skyrocket to US$ 133 billion by 2021, from US$ 8 billion in 2015. It is set to overtake the European Union as the second-largest market the world. Local start-ups are involved in a wide range of technologies and devices such as medical implants, surgical robotics and AI-powered medical solutions, largely in the areas of in vitro diagnostics, cardiology and ophthalmology. Apart from the ageing population and the increasing prevalence of chronic disease, Singapore is a hotbed for the medtech sector because companies are getting more support from investors and the public healthcare clusters.

Over 30 medical technology companies, including global leaders Baxter International, Becton Dickinson, Hoya Surgical Optics, Medtronic, and Sivantos, have set up manufacturing, R&D

and /or headquarter functions in Singapore. In 2015, Singapore's medical technology sector contributed about SGD 10 billion in output and about 16,000 jobs across manufacturing, R&D, and headquarters functions.

The Health Sciences Authority (HSA) was formed in 2001. A medical device must be registered in the SMDR under the Health Products (Medical Devices) Regulation 2010 and the Health Product Act in order to be imported, sold and/or supplied in Singapore, unless otherwise exempted under the regulations. The regulatory controls are in place to safeguard public health and safety and assuring access to good quality, safe and effective products.

34.1 Definition of Medical Device

A "medical device" means

(a) any instrument, apparatus, implement, machine, appliance, implant, reagent for in vitro use, software, material or other similar or related article that is intended by its manufacturer to be used, whether alone or in combination, for humans for one or more of the specific purposes of

  i. diagnosis, prevention, monitoring, treatment or alleviation of disease;
  ii. diagnosis, monitoring, treatment or alleviation of, or compensation for, an injury;
  iii. investigation, replacement, modification or support of the anatomy or of a physiological process, mainly for medical purposes;
  iv. supporting or sustaining life;
  v. control of conception;
  vi. disinfection of medical devices; or
  vii. providing information by means of in vitro examination of specimens derived from the human body, for medical or diagnostic purposes,

and which does not achieve its primary intended action in or on the human body by pharmacological, immunological or metabolic means, but which may be assisted in its intended function by such means; and

(b) the following articles:
   i. any implant for the modification or fixation of any body part;
   ii. any injectable dermal filler or mucous membrane filler;
   iii. any instrument, apparatus, implement, machine or appliance intended to be used for the removal or degradation of fat by invasive means.

Telehealth medical devices that are intended for medical purposes by the Product Owner, will be classified as medical device and are regulated by HSA.

34.2 Classification of Medical Device

A rule-based system is used in the determination of medical device risk classification. It is dependent on the intended purpose and indications for use claimed by the product owner. If the medical device may be assigned into two or more classes of medical devices, the highest risk class must be assigned. The inherent risk depends on its intended purpose and the effectiveness of the risk management techniques applied during design, manufacture, and use. Additional considerations include its intended user(s), its mode of operation, and/or the technologies used. Among other factors, the duration of medical device contact with the body, the degree of invasiveness, whether it delivers medicinal products or energy to the patient, whether it is intended to have a biological effect on the patient and local versus systemic effects affect the risk classification.

Risk classification	Risk level	Examples
A	Low risk	Wheelchairs, tongue depressors, bandages, walking aid, gauze dressings
B	Low-moderate risk	Hypodermic needles, single-use catheters, contact lenses, digital blood pressure monitors, hearing aids
C	Moderate-high risk	Lung ventilator, orthopaedic implant, baby incubators, blood bags
D	High risk	Heart valves, implantable defibrillator, IUDs, absorbable sutures

For In Vitro Diagnostic (IVD) medical device, consideration is also being given to the importance of the information to the diagnosis and the impact of the results to the individual and/or public health.

Risk classification	Risk level	Examples
A (IVD)	Low individual risk and Low public health risk	Specimen collection tube, prepared general culture media
B (IVD)	Moderate individual risk and/or Low public health risk	Pregnancy self-test, urine test strip, cholesterol assay
C (IVD)	High individual risk and/or Moderate public health risk	Blood glucose self-test, HLA typing test, cyclosporine assay
D (IVD)	High individual risk and High public health risk	HIV diagnostic tests, ABO blood grouping tests

For telehealth medical device, classification consideration is based on the following:

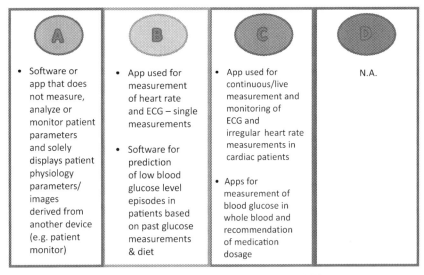

A	B	C	D
• Software or app that does not measure, analyze or monitor patient parameters and solely displays patient physiology parameters/ images derived from another device (e.g. patient monitor)	• App used for measurement of heart rate and ECG – single measurements • Software for prediction of low blood glucose level episodes in patients based on past glucose measurements & diet	• App used for continuous/live measurement and monitoring of ECG and irregular heart rate measurements in cardiac patients • Apps for measurement of blood glucose in whole blood and recommendation of medication dosage	N.A.

The following guidance documents are available on the HSA website:
- GN-13 Guidance on the Risk Classification of General Medical Devices

- GN-14 Guidance on the Risk Classification of In Vitro Diagnostic Medical Devices

34.3 Singapore Medical Device Notification, Registration and Other Authorisations

34.3.1 Steps to Access MEDICS

In Singapore, applications to HSA must be submitted online via Medical Device Information and Communication System (MEDICS).

The company that submits the application to HSA must be a local entity company based in Singapore. An overseas company without local entity can appoint ARQ on consultancy with local entity in the country to be the product license holder, a country-caretaker, also known as Local Authorised Representative or Registrant in Singapore.

The local entity company must complete the following steps to access MEDICS:

Steps	Purpose (defined by ARQon)	Processing time
CorpPass	Company Director creates GovTech's CorpPass ID for their staff and grant access for them to transact into the selected Government eservice such as HSA eservices.	Immediate
CRIS Administrator	Company Director to appoint the HSA's Client Registration & Identification Service (CRIS) administrator(s), for their key staff to access MEDICS, a HSA e-service.	3 to 4 days
CRIS User	If required at any time, CRIS administrator(s) can add other staff and service provider as CRIS user to access MEDICS, using the HSA CRIS management module e-service	Immediate

Note: If any change to your staff, company to remember to update for removal of old staff and add new staff.

If required, the Company Director also can apply HSA PIN for an overseas employee to draft the application. However, submission cannot be done by HSA Pin holder.

34.3.2 Establishment or Dealer Licenses

The Local Manufacturer, Importer and Wholesaler dealing with medical devices activities must have a Dealer License with Quality Management System (QMS) SS620 GDPMDS or ISO 13485 compliance. Dealers are to comply with the licensing obligations and must establish and maintain a QMS File throughout the supply chain to ensure the quality of the products is not adversely affected during manufacture, import, and supply. Among other things, dealers must ensure that procedure is in place with respect to management of supply records, complaint, product recall, field safety corrective action, adverse event reporting, handling, storage, delivery, installation and servicing of the products they deal in.

Class A, B, C and D medical devices must be submitted to HSA before it can be supplied in Singapore.

For Class A medical device, product notification is not submitted by Registrant, but by the Local Importer or Local Manufacturer dealer license company.

For Class B, C and D medical device,

- The local entity company to apply as Registrant in Dealer License application in MEDICS.
- Then, the Registrant/Local Authorised Representative company can submit the Product registration in MEDICS. At any time, Registrant can appoint the Local Importer(s) and Wholesaler(s) for the approved products.

The scope of ISO 13485 certificate must include import, storage and/or distribution of the categories of medical devices and the categories of medical devices and the activities performed at the facility, where applicable. Declaration of conformity is applicable for licensee companies who manufacture, import or wholesale Class A medical devices only.

SS 620 GDPMDS certification can include secondary assembly in its scope.

	Manufacturer license	Importer license	Wholesaler license	Registrant
Purpose (defined by ARQon)	Dealer license to conduct medical device manufacturing activity in Singapore	Dealer license to conduct medical device importing activity in Singapore	Dealer license to conduct medical device supply activity in Singapore	Registrant account to register Class B, C and/or D medical device
Types of QMS certification	ISO 13485 or Declaration of conformity to a QMS	ISO 13485 or SS 620 GDPMDS or Declaration of exemption from SS 620 GDPMDS or Declaration of conformity to a QMS	ISO 13485 or SS 620 GDPMDS or Declaration of exemption from SS620 GDPMDS or Declaration of conformity to a QMS	Nil
Other document where applicable	Declaration letter of non-dealing in Class A medical devices	Declaration letter of non-dealing in Class A medical devices	Nil	Nil

The following guidance documents are available on the HSA website:

- GN-02 Guidance on Licensing of Manufacturers, Importers and Wholesalers of Medical Device
- GN-01 Guidance on the Application of Good Distribution Practice for Medical Devices in Singapore
- GN-33 Guidance on the Application of Singapore Standard Good Distribution Practice for Medical Devices
- TS-01 Good Distribution Practice for Medical Devices in Singapore – Requirements
- Singapore Standard for Good Distribution Practice for Medical Devices – Requirements (SS 620)

Lead Time and Fees

Application type	Fees for all medical devices	Turnaround time (in working days)
New application	$1,030	10
Annual renewal	$1,030	10

34.3.3 Class A Notification

1. Information required

No documents are required in the submission. All class A products can be submitted in one application with information on device description, product owner, country of manufacturing, sterile or non-sterile. The manufacturer has to ensure that the product has documentation in place to demonstrate compliance to safety and performance requirements in the essential requirements. Evidence is to be provided if any clarification or input is requested by the HSA.

Submitted Class A product will be listed in the **'Class A Medical Device Register'**.

The following guidance document is available on the HSA website:

- GN-22 Guidance for Dealers on Class A Medical Devices Exempted from Product Registration

2. Lead time and fees

For lead time and fees, refer to Importer or Manufacturer License.

34.3.4 Class B, C and D Pre-Market Product registration

1. Evaluation Routes and Priority Review Scheme

Class B, C, D medical device pre-market product registration can be submitted under following Evaluation routes:

- **Immediate Evaluation:** Class B MDs and Standalone Medical Mobile Applications

  The medical device will be registered immediately and listed on the Singapore Medical Device Register (SMDR) after submission. Product can be placed in the market. An input request will be sent to the applicant if any clarification or additional information is required by HSA during the verification and evaluation. The registration will be cancelled if the application fails to fulfil ALL the eligibility criteria and/or the information submitted is inadequate.

- **Expedited Evaluation:** Class C and D MDs

  Upon submission, the application will be verified/screened for the completeness before it is accepted for evaluation. An input request will be sent to the applicant if clarification or additional information is required. The medical device will be listed on the Singapore Medical Device Register (SMDR) if the regulatory outcome is positive.

- **Abridged Evaluation:** Class B, C and D MDs. One Reference agency approval.

- **Full Evaluation:** Class B, C and D MDs. No Reference agency approval. See the figure on the next page.

- **Priority Review Scheme**

  This scheme only applies for devices under full evaluation route, and not applicable for Class D with a registrable drug. The turnaround time is 25% shorter compared to the actual full route. There are 2 Routes under the Priority Review Scheme:

Route 1	Route 2
Criteria 1: Fall under one of the focused areas ○ Cancer ○ Diabetes ○ Ophthalmic Diseases ○ Cardiovascular Diseases ○ Infectious Diseases	Does not meet the two criteria in first route.
Criteria 2. Designed & validated to meet unmet clinical needs ○ Intended for a medical purpose with no existing alternative treatment or means of diagnosis ○ Represents a breakthrough technology that provides a clinically meaningful advantage over existing legally marketed technology	

Evaluation Route	Immediate (Class B MDs and Standalone Medical Mobile Applications)	Expedited (Class C and D MDs)	Abridged (Class B, C and D MDs)	Full (Class B, C and D MDs)
Criteria	Class B MDs • 2 Reference agencies approval • No prior rejection/withdrawal by/from any Reference agency or HSA OR • 1 Reference agency approval • 3 years marketing history • No major safety issues globally • No prior rejection/withdrawal by/from any Reference agency or HSA Standalone Medical Mobile application • 1 Reference agency approval • No major safety issues globally • No prior rejection/withdrawal by/from any Reference agency or HSA	Class C and D MDs • 2 Reference agencies approval • No prior rejection/withdrawal by/from any Reference agency or HSA OR Class C MDs • 1 Reference agency approval • 3 years marketing history • No major safety issues globally • No prior rejection/withdrawal by/from any Reference agency or HSA	1 Reference agency approval	No Reference agency approval

Note: HSA's reference agencies: Health Canada, US FDA, Australian Therapeutic Goods Administration, European Union and Japan Ministry of Health, Labour and Welfare. Please refer to the corresponding approvals indicated in GN-15

2. Grouping of Medical Device

Product registration can be submitted in one application based on the product grouping. The classification of the medical device for product registration is based on the highest risk class found in the product grouping.

Medical device covers a vast range from simple tongue depressor to complex programmable patient monitoring system that comprises multiple components. The components can be sold individually or in various combinations as replacement parts. Additionally, medical devices are usually available in wide assortment of configurations including lengths, diameters, volume etc.

Medical devices that can be grouped into one of the following categories: Single, Family, System, Group, IVD Test kit, IVD Cluster. Refer also how some products are grouped in the Device Specific Grouping.

- **Single**

    A SINGLE medical device is a medical device from a product owner identified by a medical device proprietary name with a specific intended purpose. It is sold as a distinct packaged entity and may also be available in a range of pack sizes. One example is condoms that are sold in pack of 3, 12, and 24.

- **Family**

    A medical device FAMILY is a collection of medical devices and each medical device FAMILY member

    o is from the same product owner;
    o is of the same risk classification;
    o has a common intended purpose;
    o has a common design and manufacturing process; and
    o has variations that are within the scope of the permissible variants.

    Example of a FAMILY is IV administrative sets that differ in features such as safety wings and length of tubing but are manufactured from the same material and common manufacturing process and share a common intended purpose.

- **System**

  A medical device SYSTEM comprises a number of medical devices and/or accessories that are
  - from the same product owner;
  - intended to be used in combination to achieve a common intended purpose;
  - compatible when used as a SYSTEM; and
  - sold under a single SYSTEM name or the labelling, instruction for use, brochures or catalogues

  for each constituent component states that it is intended to be used together or for use with the SYSTEM. An example will be a hip replacement system comprising femoral and acetabular components. The components must be used in combination to achieve a common intended purpose of total hip replacement. The size of the components may vary.

- **Group**

  A medical device GROUP is a collection of two or more medical devices that is labelled and supplied in a single packaged unit by a product owner. The medical device GROUP comprises the following:
  - a single proprietary GROUP name
  - a common intended purpose

  The product owner is able to customize for supply, in a single packaged unit, from the closed list of devices while maintaining the same GROUP name and intended purpose. Any other single packaged unit combination (permutations of devices within the closed list) of devices in that GROUP can be supplied on the market. For example, a first aid kit consisting of medical devices such as bandages, gauzes, drapes, and thermometers, when assembled together as one package for a common medical purpose by a product owner.

- **IVD Test Kit**

  An IVD TEST KIT consists of reagents or articles that are
  - from the same product owner;
  - intended to be used in combination to achieve a specific intended purpose;

- compatible when used as an IVD TEST KIT; and
- sold under a single IVD TEST KIT name or labelling, instructions for use, brochures, or catalogues for each reagent or article states that it is intended for use with the IVD TEST KIT.

IVD TEST KIT excludes instruments such as analysers that are required to perform the test. IVD Medical Device SYSTEM may include test kits and instruments. For example, a glucose-monitoring system comprising a glucose meter, test strips, control solutions, and linearity solutions can be grouped as a SYSTEM.

- **IVD Cluster**

An IVD CLUSTER comprises a number of in vitro diagnostic reagents or articles that are

- from the same product owner;
- is of the same risk classification (either Class A only or Class B only);
- of a common test methodology as listed in the guidance document; and
- of the same IVD CLUSTER category as listed in the guidance document.

The IVD CLUSTER may include analysers that are designed for use with the reagent in the IVD CLUSTER.

- **Device Specific Groupings details are available for:**
  - Dental Grouping Terms (DGT)
  - Hearing Aids
  - ImmunoHistoChemistry (IHC) IVD reagents
  - Fluorescence In Situ Hybridisation (FISH) Probes IVD reagents
  - IVF Media
  - IVD Analysers

The following guidance documents are available on the HSA website:

- GN-12-1 Guidance on Grouping of Medical Devices for Product Registration (General Grouping Criteria)
- GN-12-2 Guidance on Grouping of Medical Devices for Product Registration (Device Specific Grouping Criteria)

3. Document requirements in the format of ASEAN CSDT

Product registration dossier must be prepared for submission accordance to ASEAN Common Submission Dossier Template (CSDT).

	Documentary Requirements	A	B	C	D
1	Letter of authorization		✓	✓	✓
2	Annex 2 List of configurations		✓	✓	✓
3	Executive summary		✓	✓	✓
4	Essential principles checklist & declaration of conformity		✓	✓	✓
5	Device description		✓	✓	✓
6	Design verification and validation, e.g. • Functional test • Biocompatibility studies • Software V&V • Sterilisation validation • Shelf-life studies	Not applicable Essential safety & performance requirements shall be met	✓	✓	✓
7	Clinical evidence		If applicable	✓	✓
8	Proposed device labelling		✓	✓	✓
9	Risk analysis		✓	✓	✓
10	Manufacturing information • Site's name & address • Proof of QMS				

Device labelling refers to any written, printed, or graphic representation that appears on or is attached to the medical device or any part of its packaging and includes any informational sheet or leaflet that accompanies the product when it is being supplied.

The following guidance document is available on the HSA website:

- GN-15 Guidance on Medical Device Product Registration
- GN-16 Guidance on Essential Principles for Safety and Performance of Medical Devices
- GN-17 Guidance on Preparation of a Product Registration Submission for General Medical Devices using the ASEAN CSDT
- GN-18 Guidance on Preparation of a Product Registration Submission for IVD Medical Devices using the ASEAN CSDT

Upon approval, Class B, C and D product will be listed in the **'Singapore Medical Device Register'**.

4. Product registration lead time and fees

Registration route	Class B	Class C	Class D	Class D with a registrable drug
Immediate route	Immediate upon submission	Immediate upon submission	N.A.	N.A.
Expedited route	N.A.	120	180	N.A.
Abridged route	100	160	220	N.A.
Full route	160	220	310	310
Full route (Priority Review Scheme Route 1)	120	165	235	N.A.
Full route (Priority Review Scheme Route 2)	120	165	235	N.A.

Fees	Class B	Class C	Class D	Class D with a registrable drug
Application fee	$515	$515	$515	$515
Immediate route fee	$925	$3,090	N.A.	N.A.
Expedited route fee	N.A.	$3,090	$5,560	N.A.
Abridged route fee	$1,850	$3,605	$5,870	$10,200
Full route	$3,605	$5,870	$11,600	$75,200
Full route (Priority Review Scheme Route 1)	$4,100	$6,600	$13,200	N.A.
Full route (Priority Review Scheme Route 2)	$5,300	$8,600	$17,100	N.A.
Annual retention fee for SMDR listing	$36	$62	$124	$124

34.3.5 Special Authorisation Routes (SAR) in Singapore

There are others or Special Authorisation routes allowing the import and supply of unregistered medical devices in specific situations. Applicant may require certain pre-requisite HSA dealer license and Quality Management System (QMS) prior submission.

S/No.	Types of Special Authorisation Routes	Purpose (defined by ARQon)	Applicant	Validity	No. of import consignment allowed
1	Import and supply of Unregistered devices requested by licensed qualified practitioner (GN-26):	To import and supply unregistered devices under a qualified practitioners for use on their patients, to meet *special clinical needs in the case of emergency or where all conventional therapies have failed.	Applicant with a valid HSA Importer license or Manufacturer license (with QMS certified)	12 months	Multiple
2	Import and Supply of Unregistered devices requested by licensed PHMC (GN-27):	To import and supply unregistered devices under a Private Hospitals and Medical Clinics (PHMC) Act for use on their patients, to meet *special clinical needs in the case of emergency or where all conventional therapies have failed.	Applicant with a valid HSA Importer license or Manufacturer license (with QMS certified)	12 months	Multiple
3	Import for re-export of unregistered devices (GN-28):	To import medical devices for export or re-export requires to apply for approval	Applicant with a valid HSA Importer license or Manufacturer license (with QMS exemption)	12 months	Multiple
4	Import and Supply of Unregistered devices for non-clinical purposes (GN-29)	To import and supply unregistered medical devices for non-clinical purpose. Non-clinical use medical devices refer to: – Training equipment – Devices for us on animals – IVD for research-use only	Applicant with a valid HSA Importer license or Manufacturer license (with QMS exemption)	12 months	Multiple
5	Import of Registered medical devices on consignment basis (GN-30)	To import registered medical devices on a consignment basis, by dealers not authorised by the registrant.	Applicant with a valid HSA Importer license. Wholesaler license is applicable.	12 months	Single

Singapore Medical Device Notification, Registration and Other Authorisations | 635

S/No.	Types of Special Authorisation Routes	Purpose (defined by ARQon)	Applicant	Validity	No. of import consignment allowed
6	Import of unregistered medical devices for exhibition (GN-32)	To import unregistered medical devices for display, require approval. These devices are not allowed to be: -Used on humans, including demonstration -Supplied locally, including free samples distribution	Applicant with local company entity (for cargo) or any individual (for hand carry)	Authorised period.	Multiple
7	Import and Supply of Custom-made medical devices	To import or supply Custom-made device	Applicant with a valid HSA Importer license or Manufacturer license. Wholesaler license is applicable.	~	Multiple
8	Import of unregistered MD after refurbishment	To export and import unregistered medical devices which will be returned to the original PHMC after its refurbishment.	Applicant with a valid HSA Importer license.	Authorised period.	Multiple
9	Import of unregistered Specified MD under HSA Exemption order 2020	To import the Specified medical device including Surgical Mask, Thermometer, Particulate Respirator and Protective Gear under Health Products (Import, Wholesale and Supply of MDs - Exemption) Order 2021	Any Applicant. Post-market compliance requirements still applicable.	Authorised period.	Single
10	Import and Supply of Unregistered MD under Provisional Authorisation	To import and supply of COVID-19 Diagnostic Tests, Ventilators, Decontamination devices for respirators via Provisional Authorisation	Any Applicant. Post-market compliance requirements still applicable.	Authorised period.	Multiple

Note: Definition of *special clinical needs under GN-26 and GN-27.

Medical devices on compassionate use basis

- Absence of alternative treatment option; or
- Available alternative treatments failed or deemed ineffective or unsuitable for the patient according to the doctor's or the dentist's clinical judgement;

<u>and</u>

- Patient's health will be clinically compromised without the requested treatment

Alleviation of stock-out situation

- The unregistered medical device is needed to minimise disruption to the continued supply of a similar registered medical device

Established medical devices with history of use

- The unregistered medical device has been used
  - before 1 January 2012
  - in a licensed private hospital as approved by the relevant authority of that healthcare institution; or
  - in a licensed medical clinic as required by the doctor or dentist,

<u>and</u>

- There are no known safety issues related to the use of the device

Novel or established medical device or upgraded version of established medical devices (new models/ new features)

- Absence of registered alternatives or lack of a specific feature in registered medical device; or
- User's (doctor or dentist) familiarity or expertise in terms of device technology, design and/or operation that is likely to support or enhance the safety outcomes of the procedure or treatment for the patient;

<u>and</u>

- Patient's health will be clinically compromised without the requested medical device.

Lead time and Fees

Special access route	Fees	Turnaround time (in working days)
Import for re-export of unregistered devices	$258	14
Unregistered devices for non-clinical purposes	$258	14
Unregistered devices requested by qualified practitioners	$155	14
Unregistered devices requested by a licensed PHMC	$360	14

34.4 Post-Market Surveillance

Post-market surveillance ensures that an approved medical device in use continues to be safe and effective. It facilitates early detection of product design and/or usage problems and allows timely intervention to prevent further harm. Post-market surveillance serves to complement pre-market review and may drive the manufacturers to further improve its processes and/or product design.

The HSA adopts a number of post-market measures to ensure the continued safe use of medical devices. These measures include reporting from healthcare professionals, mandatory reporting from dealers, and exchange of regulatory information with other medical device regulatory agencies.

34.4.1 Adverse Event

All adverse event (AE) reports received by the HSA are reviewed and subject to trend detection analysis. Unexpected adverse events might be detected, and such findings may lead to refinement of the product labels such as introduction of specific warnings. When a hazard is considered unacceptable, withdrawal of the product may be warranted.

Registrants, manufacturers, importers, and suppliers of medical devices in Singapore have an obligation to report AE to the HSA and there should a pre-disposition to report rather than not to report. Any AE that fulfils the three basic reporting criteria below is considered reportable AE:

- An AE has occurred;
- The medical device is associated with the AE;
- The AE led to one of the following outcomes;
    - A serious threat to public health;
    - Death of a patient, user or other person;
    - Serious deterioration in state of health, user or other person;
    - No death or serious injury occurred but the event might lead to death or serious injury of a patient, user, or other person if the event recurs.

All AEs should be reported immediately and not later than 48 hours for events that represents a serious threat to public health; not later than 10 days for events that has led to the death, or a serious deterioration in the state of health, of a patient, a user of the medical device or any other person; not later than 30 days for events where a recurrence of which might lead to the death, or a serious deterioration in the state of health, of a patient, a user of the medical device or any other person.

34.4.2 Field Safety Corrective Action

A Field Safety Corrective Action (FSCA) is required when it becomes necessary for the product owner of the medical device to take action (including recall of the device) to eliminate or reduce the risk of, the hazards identified. Consideration must be given to devices that are no longer on the market or has been withdrawn but may still possibly be in use.

A notification of FSCA must be submitted to the HSA within 24 hours of having made the decision to initiate the FSCA and a preliminary report containing full details submitted within 24 hours from commencement of the FSCA. The final FSCA report must be submitted within 21 days from commencement of the FSCA. To ease the entire process, the HSA has introduced Online Safety, Compliance Application and Registration System (OSCAR).

The following guidance documents are available on the HSA website:

- GN-04 Guidance on Medical Device Recall
- GN-05 Guidance on the Reporting of Adverse Events for Medical Devices
- GN-10 Guidance on Medical Device Field Safety Corrective Action

34.5 Medical Device Advertisement

Advertisement means the publication, dissemination, or conveyance of any information for the purpose of promoting, whether directly or indirectly, the sale or use of the medical device by any means or in any form, including the following:

- publication in a newspaper, magazine, journal or other periodical
- display of posters or notices
- circulars, handbills, brochures, pamphlets, books or other documents
- letters addressed to individuals or bodies corporate or unincorporate
- photographs or cinematograph films
- sound broadcasting, television, the Internet or other media
- public demonstration of the use of the health product
- offer of trials of the health product to members of the public

It includes sales promotions, sales campaign, exhibition, competition, or any other activities meant to introduce, publicize, or raise the profile or public awareness or visibility of the product.

Advertisement and sales promotion of medical device do not require prior approval from the HSA. The onus is on the advertiser to ensure compliance with applicable regulations and guidelines including the Singapore Code of Advertising Practice (SCAP). The SCAP is a guidance code promulgated by the Advertising Standards Authority of Singapore (ASAS) and it sets out further guidelines on advertising. Advertisement of certain medical device such as hearing aids and condoms must comply with specific requirements stipulated in the SCAP.

Advertisement of medical device for research use only is prohibited. Advertisement of medical device approved for professional use only is prohibited unless the advertisement is intended for distribution to qualified healthcare professionals only. Advertisement of medical device must not expressly or implicitly claim, indicate, or suggest that the product will prevent, alleviate, or cure any disease or condition listed in the Second Schedule of the Health Products (Medical Devices) Regulations, unless the advertisement is intended for distribution to qualified healthcare professionals only.

In general, advertisements must be truthful, accurate, and substantiated with scientific studies. It must not encourage indiscriminate use, discourage public from seeking advice of a medical professional, and arouse public fear. Testimonials, recommendations, and endorsement by healthcare professionals are prohibited.

The following guidance documents are available on the HSA and ASAS websites:

- GN-08 Guidance on Medical Device Advertisements and Sales Promotion
- Singapore Code of Advertising Practice

34.6 Clinical Trial in Singapore

Clinical trials on medical devices in Singapore are not regulated. Clinical trial sponsors are exempted from the requirement to submit a clinical trial notification or obtain a clinical trial authorisation or clinical trial certificate prior to commencement of the trial. The notice of import (by the importer) or supply (by the local manufacturer) must be made to the HSA before the import or supply of medical device as clinical research material for use in clinical trial in Singapore. The sponsor, the manufacturer, the importer, and the local supplier must comply with the following:

- Maintain record of manufacture, receipt and supply
- Comply with labelling requirement

- Report defect and adverse event
- Maintain record of complaints
- Notify the HSA concerning FSCA and recall
- Ensure CRM is for clinical research purpose only
- Ensure CRM is used only in an IRB approved clinical research (applicable to sponsor only)
- Ensure CRM is disposed/exported within 6 months of completion/termination of research and records of disposal/export is maintained (applicable to sponsor only)

The following guidance documents are available on the HSA website:

- GN-CTB-2-00lC-001 Clinical Research Materials (CRM)

34.7 Price Control on Medical Device

Prices of medical devices are not regulated at present.

References

1. www.hsa.gov.sg.
2. www.arqon.com.

Chapter 35

Taiwan: Medical Device Regulatory System Introduction

Pei-Weng Tu

Division of Medical Devices and Cosmetics,
Taiwan Food and Drug Administration, Taiwan
pwtu23@fda.gov.tw

Disclaimer

The regulatory information contained in this report is intended for market planning and is subject to change frequently. Translations of laws and regulations are unofficial.

35.1 Market Overview

35.1.1 Overview of Structure and Funding of Local Healthcare System

The population of Taiwan is about 23 million. Its medical device industry consists of approximately 1,157 firms and 46,953 workers. The export value of Taiwan in 2019 was NT$ 71.8 billion and the import value was NT$ 85.16 billion (Fig. 35.1). The USA is its

---

Medical Regulatory Affairs: An International Handbook for Medical Devices and Healthcare Products (Third Edition)
Edited by Jack Wong and Raymond K. Y. Tong
Copyright © 2022 Jenny Stanford Publishing Pte. Ltd.
ISBN 978-981-4877-86-2 (Hardcover), 978-1-003-20769-6 (eBook)
www.jennystanford.com

largest exporting country (Fig. 35.2). Contact lenses, diabetes test strips, and medical scooters made in Taiwan rank among the top five of their respective global market shares.

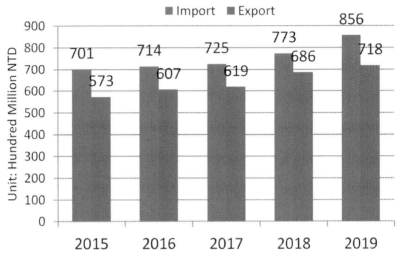

Figure 35.1 Taiwan's medical device market. *Source*: R.O.C. Customs Import/Export Statistics Data; ISTI, ITRI, Taiwan (April 2020).

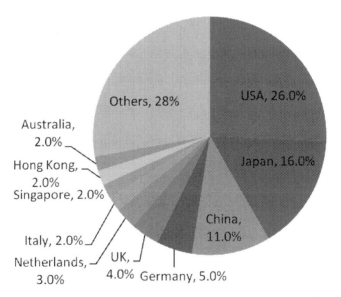

Figure 35.2 Taiwan's major exporting countries. *Source*: R.O.C. Customs Import/Export Statistics Data; ISTI, ITRI, Taiwan (April 2020).

35.1.2 Overview of Regulatory Environment and Laws/Regulations Governing Medical Devices

The regulation of medical devices in Taiwan dates back to 1970, when the Pharmaceutical Affairs Act was promulgated. Enforcement rules of the act were subsequently enacted in 1973.

Figure 35.3 Administration of medical device life cycle.

Regulatory changes in the premarket approval process and post-market surveillance have continued to take place over the years in order to safeguard public health and harmonize with international regulations. The major ones are as follows:

(1) Pharmaceutical Affairs Act (1970) (last amended 2019)
(2) Good Manufacturing Practice (1998) (last amended 2013)
(3) Reclassification of Medical Devices (2000)
(4) Regulations for Governing the Management of Medical Device (2004) (last amended 2019)
(5) Regulation for Registration of Medical Devices (2004) (last amended 2017)
(6) Good Laboratory Practice for Nonclinical Laboratory Studies (2006) (amended 2019)
(7) Medical Device Good Clinical Practice (2007) (amended 2015)

(8) Guidance for pre-clinical testing (2009-2020)
(9) Guidelines for Registration of In Vitro Diagnostic Medical Device (2010)
(10) Guidance for Medical Device Good Vigilance Practice (2011)
(11) Good Submission Practice (2014)
(12) Good Distribution Practice (2015)

On January 15, 2020, the Medical Devices Act was established and promulgated. Once its effective date for enforcement is determined, provisions governing medical devices in the Pharmaceutical Affairs Act shall no longer be applicable.

35.1.3 Detail of Key Regulator(s)

To assure the quality and safety of medical products and to promote industry development, the Taiwan government has strived to reform and renovate the organization of the Department of Health (DOH). By integrating the original Bureau of Food Safety, Bureau of Pharmaceutical Affairs, Bureau of Food and Drug Analysis, and Bureau of Controlled Drugs under the DOH into the newly established Food and Drug Administration (FDA), the Taiwan FDA (TFDA) was inaugurated on January 1, 2010. Later in 2013, the Ministry of Health and Welfare (MOHW) was established to supersede DOH. The TFDA had then become an agency under the new ministry.

The Division of Medical Devices and Cosmetics in the TFDA handles the administration of regulations concerning medical devices. All medical devices are required to be registered by TFDA before they can be manufactured domestically or imported from abroad.

35.2 Regulatory Overview

35.2.1 Definition of Medical Device

According to the definition in the Pharmaceutical Affairs Act, the term "medical devices" refers to instruments, machines, apparatus, materials, software, reagent for in vitro use, and other similar or related articles, which are used in diagnosing, curing, alleviating, or directly preventing human diseases, regulating fertility, or which may affect the body structure or functions of human beings, and do not achieve its primary intended function by pharmacological, immunological or metabolic means in or on the human body. The in vitro diagnostic devices (IVDs) are to comply with Guidelines for Registration of IVD Medical Device and include diagnostic reagents, instruments or systems used to collect, prepare, and test specimens from human body in order to diagnose disease or other conditions (such as status of health). Veterinary products are not under the scope of medical devices in Taiwan.

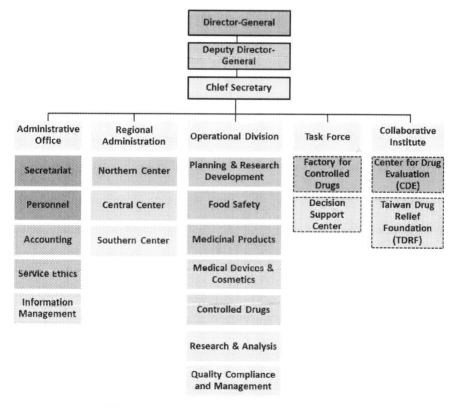

Figure 35.4 TFDA organization chart.

35.2.2 Classification of Medical Device

Medical devices are categorized by risk level and are divided into Class 1 (low risk), Class 2 (medium risk) and Class 3 (high risk). Another parallel classification system emphasizes the intended use and mode of action, as listed as follows:

A. clinical chemistry and clinical toxicology devices
B. hematology and pathology devices
C. immunology and microbiology devices
D. anesthesiology devices
E. cardiovascular devices
F. dental devices
G. ear, nose, and throat devices
H. gastroenterology-urology devices
I. general and plastic surgery devices
J. general hospital and personal use devices
K. neurological devices
L. obstetrical and gynecological devices
M. ophthalmic devices
N. orthopedic devices
O. physical medicine devices
P. radiology devices

For example, a resorbable calcium bone void filler would be under the scope of (N) and Class 2 medical device.

For detailed and updated information, please refer to the TFDA medical device databases website (http://www.fda.gov.tw/MLMS/H0001.aspx).

35.2.3 Role of Distributors or Local Subsidiaries

The TFDA issues product licenses only to domestic firms, and all foreign companies must make submissions through authorized domestic distributors or subsidiaries. The local firms must have licenses for wholesaling, retailing, importing or exporting medical devices. Also, these local firms are responsible for adverse events reporting and recalls.

The Good Distribution Practice (GDP) has been announced to enhance quality and management of local distributors.

35.2.4 Product Registration, Technical Material Requirement, and Time Required

Device evaluation is based on the risk management concept. For Class 1 products, submission requirements are: (1) affidavit, (2) GMP/QSD certificate for sterile devices or devices with measuring function, and (3) medical device business permit. For Class 2 and 3 products, technical documents are required for full review, including preclinical testing, quality control, performance testing, sterilization, etc. However, these data can be substituted by relevant documents if the product belongs to a Class 2 device and has a predicate device already approved for marketing. For products that do not have substantial equivalent devices, clinical evaluation reports are mandated.

As for the time required, Class 1 products are approved by on-site registration. The average review time in 2019 for new medical devices is about 151 days. The times for substantial equivalent products of Class 2 and Class 3 are about 89 days and 121 days, respectively. The aforementioned statistics does not include the supplemental time for companies. The risk-based approach to registration documents is shown in Fig. 35.5.

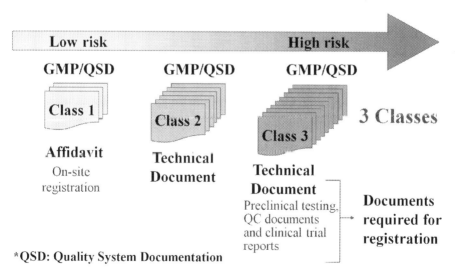

*QSD: Quality System Documentation

Figure 35.5 Risk-based regulation.

35.3 Quality System Regulation

The quality system of medical device manufacturers is audited to the requirements of Good Manufacturing Practice (GMP) using conformity assessment standard based on ISO 13485. Newly established manufacturers and new applications for device registration must comply with GMP beginning on February 10, 1999. Full compliance with GMP for all devices on the market became mandatory effective February 10, 2004.

35.4 Combined Device–Drug Product

The registration pathway depends on the primary mode of action. For example, drug-eluting stents are considered as medical devices since the stent provides the primary mode of function, and thus medical device requirements would be followed. The lead review division may still request pharmaceutical experts internally to review the drug component of the device.

35.5 Registration Fee

For new medical devices, the registration fee is NT$ 65,000; for those with predicates, the fees are NT$ 38,000 for Class 3 devices and NT$ 25,000 for Class 2 devices.

For Class 3 IVDs without predicates, the fee is NT$ 70,000; for those with predicates, the fee is NT$ 40,000.

For Class 1 devices, the fee is NT$ 10,000.

35.6 Labeling Requirements of Medical Devices

Detailed requirements are listed in the Regulation for Registration of Medical Devices. For imported products, labeling in traditional Chinese is necessary, including product names, license number, manufacturer and distributor names and addresses, and shelf life.

35.7 Post-Marketing Surveillance Requirement

Medical facilities, device firms, and pharmacies are required to report any serious adverse reactions caused by medical devices.

For certain devices, TFDA may designate a specific period of time to monitor their safety after they have been approved for marketing. Periodic safety update reports (PSUR) from the license holders shall be submitted accordingly in the designated period. License holders of the devices that have been designated are usually required to submit a PSUR every six months for three years.

35.8 Manufacturing-Related Regulation

The manufacturing of medical devices shall comply with the relevant parts of the Standards for Medicament Factory Establishments, which is formulated according to Article 57 of the Pharmaceutical Affairs Act. A manufacturer in compliance with Part 2 of the Standards, i.e., basic requirements for the factory establishments, will be issued a factory registration certificate. Such manufacturer may then apply for compliance with Part 3 of the Pharmaceutical Good Manufacturing Practice Regulations, i.e., Good Manufacturing Practices for Medical Devices, which is also formulated according to Article 57 of Pharmaceutical Affairs Act and is the requirement for medical device GMP certification.

35.9 Clinical Trial–Related Regulation

Preapproval of clinical trials are required for new medical devices claiming new design, new materials, new combination, etc. The process of submitting clinical trials protocols is shown in Fig. 35.6. Both approvals from the TFDA and the IRB (Institutional Review Board) are required prior to conducting clinical trials.

35.10 International Cooperation

Taiwan has signed medical device cooperative and confidentiality arrangements with several countries, including the United States, the European Union, Switzerland, Australia, China, the United Kingdom, and Japan.

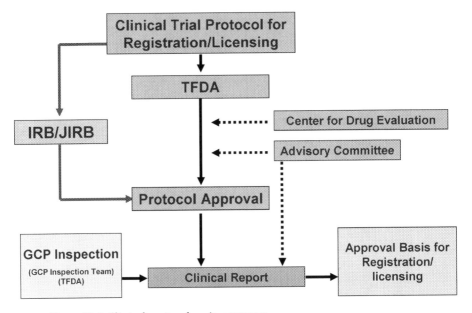

Figure 35.6 Clinical protocol review process.

Since 2011, the TFDA has been serving as the champion to lead the APEC 2020 Roadmap for Good Review Practices on Medical Products for the Regulatory Harmonization Steering Committee (RHSC) of the APEC Life Sciences Innovation Forum (LSIF). Later in 2020, the TFDA was endorsed as a formal Center of Excellence in Medical Devices and signed a memorandum of understanding with APEC LSIF to promote regulatory convergence in the APEC region.

During 2011 to 2014, the TFDA served as the vice-chair of the Asian Harmonization Working Party (AHWP) as well as the chair of the in vitro diagnostic device (IVDD) work group under the AHWP Technical Committee. The chairmanship of IVDD work group was twice reappointed in 2014 and 2017 after election. At the end of 2019, a total of 14 international guidances for IVDD led by the TFDA have been accepted as formal documents of the AHWP.

With years of accumulated experience on in vitro diagnostic regulation, the TFDA became a representative for AHWP to join the principles of IVD medical devices classification working group established by the International Medical Device Regulators Forum (IMDRF) in 2019 and collaborated with other IMDRF regulators for contribution to the global harmonization efforts.

35.11 Commercial Aspects

35.11.1 Price Control of Medical Device

The national health insurance system is implemented in Taiwan under the National Health Insurance Act. For some diseases, the payment of medical devices is covered in DRG (Diagnosis Related Groups). For those which are not under DRG programs, the insurer should pay the same amount for devices with the same functional type.

35.11.2 Parallel Imports

In order to ensure the quality of products, only the license holders or consignors can import devices. The authorization letter to the importer from the foreign original manufacturer can be one of the required documents.

35.11.3 Advertisement Regulation of Medical Devices

Persons other than medical device dealers are not allowed to make advertisements. For publishing or broadcasting advertisement, the dealers shall, before publishing or broadcasting, submit all the written or spoken words and/or drawings or pictures constituting the advertisement to the competent central or municipal health authority for approval. Interviews, news reports, or propaganda containing information implying or suggesting medical efficacy shall be regarded as medical advertisements.

35.12 Upcoming Events

The Taiwan FDA continues to enhance the administration of product life cycle to reach the goals of protecting public health and facilitating the medical device industry. Following the promulgation of Medical Devices Act in 2020 and pending the determination of its enforcement date, a set of subsidiary regulations are being developed to complement the subsequent implementation of this new act. Each of the drafted regulations would be fully communicated with the industry and comments from all sectors would be collected. Sufficient transition periods would also be given to minimize potential regulatory impact.

Acknowledgment

The author would like to thank Mr. Yung-Chuan Lee for assistance in editing this chapter.

Reference

Division of Medical Devices and Cosmetics,
Food and Drug Administration, Ministry of Health and Welfare
Telephone: +886-2-2787-8000
Address: No. 161-2, Kunyang St., Nangang District,
Taipei City 115-61, Taiwan (R.O.C.)
Website: http://www.fda.gov.tw/EN

Chapter 36

Thailand: Medical Device Control and Regulation

Kanokorn Pulsiri, Sirinmas Katchamart, Sansanee Pinthong, and Korrapat Trisansri

*Thai Food and Drug Administration,
Ministry of Public Health, Thailand*

pkanokorn@fda.moph.go.th

Disclaimer

The regulatory information contained in this handbook is intended for market planning and is subject to change frequently and may be not up to date since the regulatory scenario in Thailand is dynamic and moving fast. Translations of laws and regulations are unofficial.

36.1 Market Overview

Due to the Thailand 20-Year National Strategy, the government has an objective to make Thailand as the ASEAN medical hub by 2020. The healthcare sector including medical device is one of

---

Medical Regulatory Affairs: An International Handbook for Medical Devices and Healthcare Products (Third Edition)
Edited by Jack Wong and Raymond K. Y. Tong
Copyright © 2022 Jenny Stanford Publishing Pte. Ltd.
ISBN 978-981-4877-86-2 (Hardcover), 978-1-003-20769-6 (eBook)
www.jennystanford.com

the New S-curve industries. Therefore, the government decided to support in medical device industrials investment. As a research result from Krungsri Research showed that the value of medical device market in Thailand rose to 176 billion baht (US$ 6 billion), in 2019, an estimated around 70% arise over the last decade with the expected annual growth rate of 8–10% in 2019–2021. And the statistic of medical devices' market database from Plastics Institute of Thailand presented the semiannual data in 2020, the value of medical device importation market in Thailand was 43,805.26 million baht. Many of importers import 43% of disposable medical devices, 38% of durable medical devices, and 19% of reagent and test kits, lens, syringes, catheters (canula), prosthesis, devices, and appliances used in medical science, surgery, dental and diagnostic reagents as such. On the other hand, the data showed the value of medical device exportation market in Thailand was 73,124.62 million baht. Many of the exporters export 55% of disposable medical devices, 35% of durable medical devices and 10% of reagent and test kits; rubber gloves and examination gloves, lens, surgical gloves, devices and appliances used in medical science, surgery, dental and diagnostic reagents. The government support and encouragement through the Board of Investment Office (BOI), large capability of the labor force, the rise in the number of cases of non-communicable disease and a gradual move to societal aging in Thailand attract investors from around the world. Therefore, international manufacturers of medical devices should see the opportunity of investment in the medical devices industry in Thailand.

36.2 Medical Device Regulations

Prior to 1988, there was no specific law related to control over the manufacturing, import, export, sale, or handling of medical device. During that time, if problems related to medical devices occurred, they were normally solved by referring to the provisions of the Drug Acts. The Ministry of Public Health enacted the Medical Device Act, B.E. 2531 (1988) and had been in force since May 23, 1988, requiring the registration of all medical devices, including condoms and surgical gloves. And with the emergence of extensive research and development, rapidly advancing medical technology, the Medical Device Act, B.E. 2551 (2008) was promulgated to supersede the 1988 Medical Device Act. The new Medical Device Act added new sections on the registration of establishment licenses, medical device assessments, civil liabilities, revision of penalties and fees for greater suitability. To align the regulation of medical devices in Thailand with the ASEAN Medical Device Directive (AMDD) and to comply with them more appropriately under the current circumstances where there are rapid developments in technology and medical innovations in the country, the latest amendment of Medical Device Act, the Medical Device Act, B.E. 2562 (2019), has been legislated in 2019. There has been a revision to the definition of a medical device, and the definitions of "accessory" and "listing" have been added. Medical devices have been categorized based on the level of risk that they may pose to individual and public health. The amendment also provides updates for medical device assessments, civil liabilities, and revision of penalties and fees.

36.3 Definition of Medical Device

1. A "medical device" shall mean any instrument, apparatus, machine, appliance, implant, in vitro reagent used in or outside of laboratory, product, software, material or other similar or related article, intended by the product owner or manufacturer to be used, alone or in combination for human beings or animals for one or more of the specific purpose(s) of:

    i. diagnosis, prevention, monitoring, treatment or alleviation of disease,

    ii. diagnosis, monitoring, treatment or alleviation of an injury,

    iii. investigation, replacement, modification or support of the anatomy or of a physiological process,

iv. supporting or sustaining life,
v. control of conception or fertility development,
vi. aid or compensate for disabled/handicapped
vii. providing information for medical or diagnostic purposes by means of in vitro examination of specimens derived from body
viii. disinfection of medical device

2. Accessory used in combination with medical device in (1)
3. Apparatus, machine, product or other thing specified as medical device by the Ministry of Public health which does not achieve its primary intended action in or on the human or animal body by pharmacological, immunological or metabolic means.

36.4 Medical Device Classification

Following the Acts, there are three levels of Medical Device Control for the risk- based medical devices classification namely,

- Licensed Medical Devices,
- Notified Medical Devices,
- Listing Medical Devices

from highest to lowest risk-based, respectively.

As stated in the Notification of the Ministry of Public Health Re: Risk classification of Medical Devices, effective since November 14, 2019, in harmonizing with AMDD and the international regulations. The medical devices are categorized into two categories as in vitro diagnostic medical device (IVD) and non-in vitro diagnostic medical device (non-IVD).

In vitro diagnostic medical devices (IVD) are classified by the level of risk to individual and public health from lowest to highest as follows:

In vitro diagnostic medical devices (IVD) risk classification

Class 1   Low risk to individual and public health
Class 2   Moderate risk to individual or low risk to public health

Class 3   High risk to individual or moderate risk to public health
Class 4   High risk to individual and public health

Non-in vitro diagnostic medical devices (non-IVD) are classified by the level of risk that may arise from lowest to highest as follows:

Non-in vitro diagnostic medical devices (non-IVD) risk classification

Class 1   Low risk
Class 2   Low to moderate risk
Class 3   Moderate to high risk
Class 4   High risk

36.5 Pre-Marketing Control

36.5.1 Documents Required for Medical Device Registration

36.5.1.1 Licensed medical device and notified medical device

Based on the medical device classification and to ensure effective control of safety and performance of medical device, it is necessary to submit the required documentation in the Common Submission Dossier Template (CSDT) format for Licensed Medical Device (which will be subject to an in-depth revision) and Notified Medical Device. And there are country-specific documents required to be submitted together with CSDT.

Common Submission Dossier Template (CSDT)

Thailand requires Common Submission Dossier Template (CSDT) format to compatible with the ASEAN Common Submission Dossier Template (CSDT), an important element of the AMDD. Use of the CSDT is to least the burden for document preparation of the medical devices registration among the different regulatory authorities of the ASEAN member states.

The list of Common Submission Dossier Template (CSDT) is as follows:

1. Executive summary

2. Elements of the Common Submission Dossier Template

2.1 Relevant Essential Principles and Method used to demonstrate conformity

2.2 Medical device description

    2.2.1   Medical device description and features

    2.2.2   Intended purpose

    2.2.3   Indications

    2.2.4   Instructions for use

    2.2.5   Contraindications

    2.2.6   Warning

    2.2.7   Precautions

    2.2.8   Potential adverse effects

    2.2.9   Alternative therapy

    2.2.10 Materials

    2.2.11 Other relevant specification

    2.2.12 Other descriptive information

2.3 Summary of design verification and validation documents;

    2.3.1   Pre-clinical studies including biocompatibility tests, pre-clinical physical tests, pre-clinical animal studies

- Software verification and validation studies (if applicable)
- Medical device containing biological material

    2.3.2   Clinical evidence

- Use of existing bibliography
- Clinical evaluation report
- Clinical investigation

2.4 Medical device labeling

2.5 Risk analysis

2.6 Physical manufacturer information

    2.6.1   Manufacturing process

    2.6.2   Name and address of physical manufacturer

2.6.3 Certificate of quality management system for physical manufacturer

2.7 Destruction, termination or disposal of waste occurred after consumption

Following are the country-specific documents requirements to be submitted together with CSDT:

1. Establishment license number
2. Power of attorney
3. Declaration letter for the medical device quality; intended use, indication, packaging, labeling, instructions for use for supply in Thailand from product owner or physical manufacturer
4. Declaration of conformity from product owner or physical manufacturer
5. Declaration of marketing history from product owner or physical manufacturer
6. Declaration of safety from product owner or physical manufacturer
7. Reference of market approval from regulatory authority in reference countries
8. Letter of authorization for authorized representative from product owner (in case of importation)

The country-specific documents requirements are subject to change. The safety of individuals, public health, and performance of medical device are the major concern.

36.5.1.2 Listing

Listing Medical Devices are almost low risk medical devices. The manufacturers are responsible for ensuring that their products comply with all the relevant Essential Performances. Therefore, the authorized representative has to submit documents of the product to notify to the Thai FDA.

Following is the list of documents for listing medical devices:

1. Establishment license number
2. Power of attorney

3. Documents of product description, labeling, technical requirements, manufacturing information, including instructions for use (if instructions for use exist)
4. Documents of product registration history in foreign countries (if any)
5. Documents of product sterilization test for sterile medical devices
6. Documents of product calibration test for medical devices with a measuring function
7. Declaration of conformity from product owner or physical manufacturer
8. Letter of authorization for authorized representative from product owner (in case of importation)

36.5.2 Labeling

- All classes of medical devices have to be labeled in Thai. (The other languages are also allowed to be printed provided that the above labeling is identical to the one in Thai language.)
- Labels should include
  (a) the product name,
  (b) device description, such as active component, mechanism, etc.,
  (c) indication or objective, amount, quantity or quantity that can be packed according to the situation,
  (d) name, address and country of origin of the manufacturer as well as the importer according to the situation,
  (e) license number,
  (f) name, address and contact number or other contact channels for further information and complaint,
  (g) lot number or serial number,
  (h) month and date of manufacturing or expiration if any,
  (i) storage condition,
  (j) warning, contraindication, precaution.

As of July 2020, the Thai FDA is drafting the Notification of the Ministry of Public Health Re: Labeling and instruction for use of Medical Devices; thus, the information above may change.

36.5.3 Advertising

The Thai FDA must grant approval of any advertisement with a commercial purpose for a medical device. It is prohibited to falsely advertise a medical device.

36.6 Post-Market Controls

The licensee of the establishment, license notified or listing medical devices shall perform control and supervise the operation of the manufacture, import, or distribution of a medical device to comply with the quality standard of manufacture, import, or distribution of a medical device.

Adverse event and Field safety corrective action

The Notification of the Ministry of Public Health Re: Criteria, procedures and requirements on reporting medical device defects or adverse effects occurring to consumers and reporting of field safety corrective actions come into force in November 2016. The importers and manufacturers of licensed medical devices, notified medical device, and the establishment licenser have to submit reports on device defect or adverse reactions and its field safety corrective action of the medical device to the Secretary General of the Food and Drug Administration according to the medical devices defect or adverse reactions and its field safety corrective action.

Enforcement

The penalties (ranging from monetary fines to imprisonment) vary depending on the type of violation under the Medical Device Act and its related.

Summary

Medical Device Control and regulation in Thailand

Acts	The Medical Device Act, B.E. 2551 (2008) amendment with the Medical Device Act, B.E. 2562 (2019) and its related
Establishment	Establishment registration
Level of medical device control	Licensed Medical Device		Notified Medical Device		Listing Medical Device	
Risk classification	IVD risk Class 4	Non-IVD risk Class 4	IVD risk Class 2,3	Non-IVD risk Class 2,3	IVD risk Class 1	Non-IVD risk Class 1
Pre-marketing controls	Common Submission Dossier Template (CSDT)				Listing	
	All classes of medical devices have to do the labeling.					
	Grant approval of any advertisement with a commercial purpose for a medical device.					
Post-marketing controls	Control and supervise the operation of production, import, or distribution of a medical device to comply with the quality standard of manufacture, import, or distribution of a medical device.					
	Report device defects or adverse reactions and its field safety corrective action of the medical device					
	The penalties vary depending on the type of violation under the Medical Device Act and its related.					

References

1. The Ministry of Public Health enacted the Medical Device Act, B.E. 2531 (1988).
2. The Medical Device Act, B.E. 2551 (2008).
3. The Medical Device Act, B.E. 2562 (2019).
4. The Notification of the Ministry of Public Health Re: Risk classification of Medical Devices.
5. The Notification of the Ministry of Public Health Re: Labeling and instruction for use of Medical Devices.
6. The Notification of the Ministry of Public Health Re: Group of Licensed Medical Devices or Licensed Medical Devices.
7. The Notification of the Ministry of Public Health Re: Group of Notified Medical Devices or Notified Medical Devices.
8. Notification of the Ministry of Public Health Re: Rules, procedures and conditions for preparing the malfunction report of medical device or adverse event that occur to consumers and field safety corrective actions reports.
9. Thailand Races Ahead as Global Healthcare Hub, Investors bet on medical device market as health tourism surges, Thailand Board of Investment (BOI).
10. The statistics of medical devices market database, Plastics Institute of Thailand.
11. Thailand industry outlook 2019-2021 : MEDICAL DEVICES, Krungsri Research.

Chapter 37

Vietnam

Nguyen Minh Tuan

*Department of Medical Equipment & Construction,
Ministry of Health, S. R. Vietnam*

Ngmtuan@ymail.com

37.1 Market Environment

With the open-door policy, Vietnam is increasingly integrated into the regional and the world markets. The opening up of extensive integration together with efforts to simplify administrative procedures and integration in the field of management has created opportunities for people and health workers to have access to technologies, new and modern curative care methods and medical device, thereby reducing risks and adverse impacts on health and the environment.

In recent years, the total investment in the medical device market has grown strongly, if the total investment by 2010 was estimated at USD 515 million, in 2016 it was estimated at USD 950 million and reached the estimation of USD 1.1 billion in 2017 and USD 1.680 billion in 2019.

---

*Medical Regulatory Affairs: An International Handbook for Medical Devices
and Healthcare Products* (Third Edition)
Edited by Jack Wong and Raymond K. Y. Tong
Copyright © 2022 Jenny Stanford Publishing Pte. Ltd.
ISBN 978-981-4877-86-2 (Hardcover), 978-1-003-20769-6 (eBook)
www.jennystanford.com

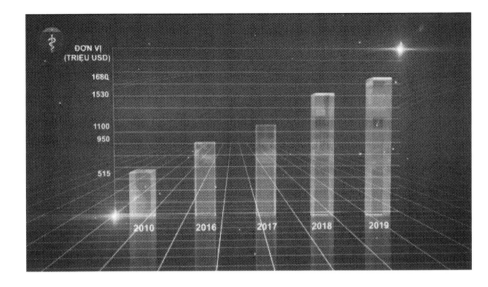

37.2 Roadmap for the Implementation of the Decree on Medical Device Management

Decree 36/2016/ND-CP of the Government on the management of medical device (MD) was issued and effective from July 1, 2016, with all regulations on classification, production and circulation of medical device, management of MD procurement and sale, services, information, labelling as well as management and use of medical device. At the same time, the health sector also regulates the online publicizing and registration, facilitating administrative procedures for businesses operating in the medical device field. Businesses have already implemented and responded to the requirements of the new regulations. The Ministry of Health has also issued the forms for announcement of eligibility to manufacture, classification, purchase and sale as well as the form of announcement of standards applicable to class A medical device.

After 18 months of implementation and evaluation, the government issued Decree 169/2018/ND-CP to amend and supplement a number of articles of Decree 36/2016/ND-CP to make it more suitable with practical conditions. At the same time, the Decree also allowed the automatic renewal of a number of medical device import permits and circulation permits of class B, C, and D. The Ministry of Health has also improved and finalized the online procedure system in order for businesses to proceed with administrative procedures more smoothly.

On January 1, 2020, the Government issued Decree 03/2020/ND-CP, which specifies the validity of licenses for a number of groups of medical device as follows:

1. The permit to import class B, C and D medical devices and the permit to import in vitro diagnostic bio-products, issued in 2018, 2019, 2020, and 2021, shall remain valid until the end of December 31, 2021, and the customs office will not control the imported quantity in this case.

2. For class A medical device, which has the receipt of applicable standard declaration dossier issued by the Province Department of Health, it can be imported according to need without quantity restriction and with no requirement for classification and certification of MD by the Ministry of Health when carrying out customs clearance procedures.

3. For a medical device that does not require import permits and is classified as class B, C, or D by an eligible classification entity and publicized on the Ministry of Health's official website, it is allowed to continue being imported until the end of December 31, 2021, according to demand without quantity restrictions and a written certification of the Ministry of Health when carrying out customs clearance.

4. If the medical device is an in vitro diagnostic bio-product being granted a circulation registration certificate under the provisions of the 2005 Pharmacy Law and its guiding documents, the granted registration number will be valid until the end of the period indicated on the registration certificate. Particularly for the circulation registration numbers of in vitro bio-products that expire after January 1, 2019, but before December 31, 2021, the issued registration number will be valid until the end of January 31, 2021.

5. Medical device being in vitro diagnostic bio-products for which registration dossiers were submitted under the provisions of the 2005 Pharmacy Law before January 1, 2019 shall be settled in accordance with the 2005 Pharmacy Law.

6. Imported medical devices being in vitro diagnostic biologicals whose dossiers submitted from January 1, 2019, to the end of December 31, 2021, shall be issued with the import license in accordance with the 2005 Pharmacy Law and valid until December 31, 2021.
7. Reception of dossiers and granting circulation registration numbers for domestically manufactured medical devices started from January 1, 2019. The circulation registration numbers take effect from the date of issuance.
8. For household and medically used insecticidal and germicidal chemicals/preparations, which solely aim at disinfecting a medical device, and which have been granted the certificate of free sale, if the certificate expires after the July 1, 2016, and before December 31, 2020, it shall remain valid until December 31, 2020.
9. Reception of dossiers and granting circulation registration numbers for insecticidal and germicidal chemicals/preparations for household and medical use, which solely aim at disinfecting a medical device, started from January 1, 2019. The registration number is effective from the date of issue.
10. For domestically produced medical device that has been granted a free sale registration certificate, validation shall remain until the end of the time stated in the registration certificate. Particularly for the circulation registration certificates which expire after July 1, 2016, and before December 31, 2020, the certificates will remain valid until December 31, 2020.

At the same time, the Decree also stipulates the roadmap to apply the ASEAN Common Submission Dossier Template (CSDT) from January 1, 2022.

The Ministry of Health (Department of Medical device and Constructions) has set up an electronic information portal (in Vietnamese) to manage medical device (in term of announcement of eligibility for production and sale; announcement of class A medical device; registration circulation for class B, C, and D medical devices; issuance of import permits in special cases...) and publicize prices of medical device (https://dmec.moh.gov.vn/).

37.3 License Fees

Along with the implementation process of Decree 36/2016/ND-CP, the Ministry of Health has reviewed the fee rates and adjusted them in accordance with the new regulations. Accordingly, the Ministry of Health coordinates with the Ministry of Finance to promulgate Circular 278/2016/TT-BYT regulating the fee rates, collection and payment modes, management, and use of fees in the health sector. The Circular stipulates different fee rates for appraisal and issuance of circulation certificates (in which different fees are set for class A, B, C, and D medical devices according to the principle that examination fee will be higher for dossiers of medical devices with a higher level of risk), appraisal for extension; re-appraisal and re-issuance of import and export permits; assessment of business conditions related to medical device field.

37.4 Labelling According to New Rules

The Ministry of Health has reviewed the mandatory contents required for medical device and included them in the Decree on goods labelling (Government Decree No. 43/2017/ND-CP dated April 14, 2017). In addition to the mandatory information required on the label like all other goods (including: name, responsible organization, origin), the MD label should contain other information, including circulation number or import permit number, batch or serial number, manufacture date or expiry date, warnings, instructions for use, instructions for storage.

In addition, the condition for the medical device to be allowed to circulate on the market is to have instructions for use in Vietnamese.

The Ministry of Health issued Circular No. 09/2015/TT-BYT dated May 25, 2020, stipulating requirements on advertisement content accuracy for special products, goods, and services under its management.

37.5 Technical Requirements for Raw Materials

The Decree includes requirements for material dossiers. For reagents, calibrators, in vitro control materials, material documentation is a mandatory component of the circulation registration application.

For class B, C, and D medical devices, according to the roadmap to January 1, 2022, the ASEAN Common Submission Dossier Template (CSDT) will be applied in which the material documentation is also a required component.

The current Decree also contains special provisions for the import and export of medical device and materials for production of medical device containing drugs and precursors.

37.6 Domestic Production

The Decree specifies the conditions for personnel, infrastructure, and equipment required for medical device manufacturing facilities. At the same time, it also regulates the quality management system of these facilities. They should have a quality management system that meets ISO 13485 before January 1, 2020. They should also declare eligibility to manufacturing medical device prior to manufacturing.

The Government also has preferential policies for the production of medical device to encourage the development of domestically produced medical device.

37.7 Clinical Trial Evaluation

The new Decree on the management of medical device requires in the registration dossier a summary of clinical data and results of clinical trial for class C and D medical devices which infiltrate the human body, except for the followings: medical device which is manufactured or processed in Vietnam solely for the purpose of export but the importing country does not require clinical trial; medical device which have been circulated and granted certificate for free sale by one of the countries or organizations, i.e., EU member states (including UK, Switzerland), Japan, Canada, TGA of Australia, FDA of America.

The Ministry of Health is also drafting a general regulation on clinical trial of medical device, including contents of different appraisal stages, appraisal of medical device for circulation, medical device subject to appraisal, medical device fully or partially exempted from evaluation stages; requirements for establishments eligible for appraisal as well as required dossiers and procedures for appraisal.

Part 5

Hot Topics

Chapter 38

A Strong Regulatory Strategy Is a Competitive Advantage to a Medical Device Company

Jacky Devergne

Asia Pacific, Renal Marketing, Baxter Healthcare Inc.

jacky.devergne@gmail.com

Asia-Pacific is a widely diverse region with a complex and rapidly changing regulatory environment. For medical device companies, this is a double-edge sword as compliance with ever-changing regulatory standards is quite costly and adds stress to an organization. The way to overcome these hurdles is to incorporate the regulatory strategy into the business planning and turn what was perceived as a burden on the organization into a strategic advantage.

By investing in an effective regulatory affairs (RA) organization and integrating it with the business planning process, a company will benefit from a competitive advantage in four areas: time to market, barriers of entry, continuity of business, and best use of resources.

Medical Regulatory Affairs: An International Handbook for Medical Devices and Healthcare Products (Third Edition)
Edited by Jack Wong and Raymond K. Y. Tong
Copyright © 2022 Jenny Stanford Publishing Pte. Ltd.
ISBN 978-981-4877-86-2 (Hardcover), 978-1-003-20769-6 (eBook)
www.jennystanford.com

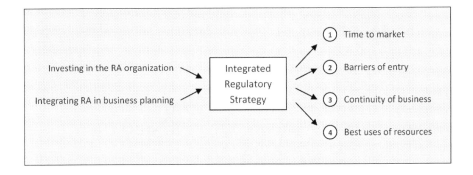

38.1 Competitive Advantage

38.1.1 Time to Market

When one thinks about the regulatory impact on medical device companies, the first thing that usually comes to mind is the impact on new product launches. It is undeniable that a fast and well-executed product launch will provide a company with the advantages of being the "first-mover" on the market: relatively easy acquisition of market share, higher initial profitability, positive impact on brand image, and opportunities to shape the market.

However, another equally important aspect to bear in mind is the impact on a multinational or global launch plan. Too often, multinational companies will choose to launch a medical device first in the USA or in Europe and then release it in Asia-Pacific over a long time span. "Second-mover" companies, benefiting from lower investments in R&D and sometimes lower costs of manufacturing, can quickly develop a copy of the product and launch it in Asian markets shortly after the innovative company, thus removing the first-mover advantage. There are even multiple examples of second-movers globally which became first-movers locally by launching their copy first in Asian countries.

Optimizing time to market in a global product launch is a complex endeavor, and companies that can design and execute such plans will be ahead of their competition. Doing so requires balancing the commercial objectives with the regulatory requirements of each country, taking into account

- countries that require registration;
- countries that do not require registration;
- countries that require local adjustments with impact on R&D (e.g., screen in local language) or labeling (e.g., manual in local language);
- countries that require precedence in registration (e.g., precedence of 510(k) or CE Marking);
- anticipated regulatory changes that will happen within the timeline of launch plan.

A well-executed global launch plan will optimize the chances of an innovative company to benefit from the first-mover advantage in as many markets as possible.

38.1.2 Barriers of Entry

A regulatory strategy can do more than supporting the entry of a product into a market. A stronger strategy will simultaneously plan to make it more difficult for competitors to follow with their products on the market. The following variables can be used to that effect:

- intended use: indications for use, targeted patient population
- claims: unique claims requiring or reinforced by regulatory approval
- manufacturing site: compliance requirements (e.g., Good Manufacturing Practices)
- unique therapeutic benefits: can allow the product to qualify for expedited review or humanitarian use designation
- Localization: documentation, labeling, or parts of the product adapted to the local language or practice
- differences in reimbursement: seeking a specific reimbursement category can favor the adoption of the device by medical professionals

38.1.3 Continuity of Business

Ensuring the continuity of business, or in other words ensuring that a company can continue offering an existing product to its customers, can provide a definite advantage compared to another company that will not.

This is true of all companies as the continuity of business is considered part of the service provided, but it is critical for medical devices companies—since patients depend on their products—and particularly true in Asia, where relationships between a manufacturer and its customer are built in the long term on trust and excellence of service. Examples abound of a medical manufacturer that had to interrupt the supply of a medical device to its customers and patients after losing its license; this provided an opportunity to a competitor to prove their higher level of reliability by responding quickly and providing an impeccable level of service in a time of crisis.

Continuity of business is best achieved through anticipation and planning: anticipation of a change of regulatory status of an existing product (e.g., expiry of registration), or anticipation of a change in the regulatory landscape. In the latter case, the effective RA professional will know when the change will happen, what it will entail, and how best to prepare for it.

Another characteristic of the effective RA professional will be their ability to not only anticipate but also influence a change in regulation. Members of government bodies responsible for writing and releasing medical device regulations cannot have expertise in all medical devices; so informing and educating them can help avoid unnecessary challenges. In all Asian countries, opportunities exist for the interactions with the regulatory bodies to change from a compliance mindset to an advocacy and educational approach.

38.1.4 Best Use of Resources

Finally, the integration of the regulatory function in the business planning will result in system efficiencies that will allow for better quality of the support provided by the RA team, reduce the resource consumption and allow for cost reductions, or free up resources to support other businesses within the organization.

The regulatory function plays a critical role in supporting a company's value chain, and an optimal use of resources will translate in cost savings or increased profits.

38.2 The Right Organization

38.2.1 Investing in an Effective RA Organization

The RA team is always considered a key partner to the business within medical device companies, due to its responsibility for ensuring appropriate and fast product registration. Regulatory professionals are expected to have thorough regulatory knowledge of the area they are responsible for and to be able to work efficiently managing regulatory work.

Beyond this critical primary responsibility lie the tasks of building and maintaining (internally) a relationship with their commercial partners and (externally) with government regulators. This additional role, strategically critical to the company, requires the regulatory team to develop, possess, or acquire additional skills. The highly functioning regulatory professional

- masters internal communication channels allowing for seamless work flow between global, regional, and local RA teams as well as between the regulatory function and its commercial partners;
- is able to transform regulatory knowledge into intelligence to inform the regulatory strategy;
- is comfortable working with government regulators and learns to navigate the sometimes complex and changing government organizations.

By developing and investing in a highly functioning RA organization, a medical device company will equip itself with an important competitive advantage over competitors that fail to see the potential that such a team can provide.

38.2.2 Integrating RA in Business Planning

Integrating regulatory affairs in the business planning requires rethinking the role of the RA professional, moving from a reactive model to a consultative approach. In this model, the RA professional collaborates with their commercial partners in a continuous manner, providing education and advice as well as regulatory guidance.

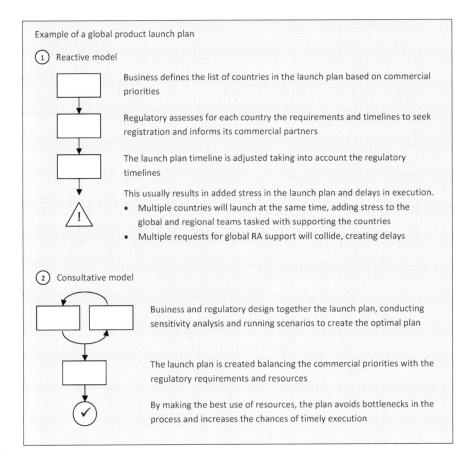

38.3 Conclusion

The role of the RA professional in medical device companies is evolving. Now more than ever, it is important for the RA professional to increase their understanding of the business and become a strong partner to their commercial counterparts. By changing from a reactive to a consultative model internally, and from a compliance to an advocacy mindset externally, the RA professional becomes a critical element in providing a competitive edge to their organization on the marketplace.

Chapter 39

Regulatory Strategy: An Overview

Pakhi Rusia

RAC-US, Regulatory Affairs Professional

Pakhi.rusia@gmail.com

Disclaimer: The views expressed herein represent those of the author and do not necessarily represent the views or practices of the author's employers or any other party.

Readers may expect to gain an understanding on the following:

- Definition of regulatory strategy
- The significance of regulatory strategy
- The development criteria to be considered for regulatory strategy
- Key factors to be considered for developing regulatory strategy
- Importance of successful implementation of regulatory strategy

---

Medical Regulatory Affairs: An International Handbook for Medical Devices and Healthcare Products (Third Edition)
Edited by Jack Wong and Raymond K. Y. Tong
Copyright © 2022 Jenny Stanford Publishing Pte. Ltd.
ISBN 978-981-4877-86-2 (Hardcover), 978-1-003-20769-6 (eBook)
www.jennystanford.com

39.1 Introduction

Among the numerous definitions of "strategy" available on the Internet and in books and articles, in my view strategy, in short, is the bridge between the gap of "where we are," "where we want to be," and "how to get there."

The knowledge of the goal, the order of various events, uncertainty around those events and possible plan to mitigate or deal with the uncertainty are essential for the development of any strategy. It is the blueprint of decisions by considering the objective, the key policies, plans for achieving the objectives for the company. The strategy provides the path to deliver the economic growth and the contribution it plans to bring to its patients, its shareholders, customers, and society at large.

Broadly, strategy is a holistic and proactive plan developed to prepare for and perform a project or to address and resolve a problem, so it achieves the predefined desired outcome successfully. Developing a strategy involve analyzing problems or projects from a broad perspective and identifying solution and plan to accomplish them, while at the same time creating opportunities and taking full advantage of the possibilities and options that present themselves.

Planning a good strategy is significant because it is not possible to foresee the future. Without a perfect foresight, teams would not be ready to deal with the uncertain events which constitute the business environment. It also enables development of long-term plans with probability of Innovations or new products, exploration of new markets to be developed for business in future. A brilliant strategy will be created considering the probable behavior of customers and competitors. It will articulate a credible vision of how to achieve a future state and anticipates consequences, risk, and trends accurately, while still being realistic and solidly grounded. Readers should note that strategy is not merely "How" "What," and "When" of completing a task, but a rather holistic plan of tactical maneuvers to achieve the predefined desired outcome.

Regulatory strategy is a plan of action designed to achieve a specific regulatory goal, such as to obtain marketing approval for drugs, biologics, and medical devices (will be referred to as products) and ensuring the approval of products complies with the laws and regulations of the countries where they are marketed. Since regulatory requirements continue to evolve and become increasingly complex, solid regulatory strategy and execution not only can maximize the opportunity to achieve the regulatory goal but also may provide competitive advantage.

Regulatory strategy provides a thought-out coordinate of regulatory affairs to launch the product in defined by commercial or marketing strategy. It is of immense importance to understand that regulatory strategy is not merely making the list of required R&D data, documents, intended product specifications, preclinical plan, clinical plan, and other stability data, etc., required for product registration. All these factors are important and indeed are part of regulatory strategy, but these are needed to meet the regulatory strategy objectives successfully rather than making a whole regulatory strategy.

The best approach to strategy development begins by involving internal stakeholder to identify product attributes that defines regulatory parameters such as classification of the product, then conducting regulatory intelligence on possible precedents, verifying the viability, and documenting it. Whether it is developed to address a transient problem or as a comprehensive plan tailored to deal with a specific challenge, in either case the steps involved in developing a strategy are essentially the same.

39.2 The Significance of Regulatory Strategy

A regulatory strategy as a formal document helps to align the regulatory activities, together with the business strategy, to bring a new or modified product to market. It identifies the important regulatory elements to be addressed and provides overall definition and direction to the project team. It should outline which path to take and the rationale of why a specific path is selected or recommended.

A well-defined and executed regulatory strategy is critical as it stipulates the pathway from conceptualization of product idea to fulfill the unmet need, its development, registration, and then finally launch, to facilitate faster access of the products to the patients; while providing assurance that the product meets regulatory criteria for quality, safety, and efficacy when used according to labels.

It ensures the internal stakeholder's alignment with the company's vision prospective; the path reach to the market with maximum potential and with optimized internal efficiencies. It also helps to build the trust and credibility with regulatory agencies by demonstrating the company has evaluated product risk carefully, has the proper risk mitigation strategies in place.

Regulatory strategy is mostly considered as a live document and is aimed to document discussions on crucial issues with regulatory authorities, evolution driven via those discussions, learnings for future, and pathways adopted to fulfill the ad hoc data needs that arise at the time of project execution. This documentation works as a perfect catalyst to offer a comprehensive elucidation of the project, develop intelligence and keeps the team from reinventing the negotiation strategies, path to market, mitigation plans for the risk and challenges which have already been faced in the past and benefit the overall efficiency.

39.3 Key Steps to Develop Regulatory Strategy

Although the regulatory strategy template provides the scaffold for the steps to be followed for its development, following are the general steps which could be helpful while forming a regulatory strategy:

39.3.1 Defining Project Scope

A deep understanding of what is being requested to define the parameters and scope is critical. A well-defined scope is a necessity to ensure the success of any project. Without it, no matter how efficient, how effective, and how perfect the drafted strategy is, a successful execution would be a challenge. Clear identification of these needs of the project is more likely to set a sound benchmark from the beginning.

Understanding the "what and why" of a project will enable the team to set specific goals and objectives and will set the groundwork to develop and deliver the regulatory strategy. It is of most significance to ask the right questions, challenge the assumptions, and identify the issues, problems, and their root cause.

39.3.2 Defining Timelines and Milestones

Timelines and milestones provide a benchmark to measure progress periodically and are a way of knowing how the project is advancing, and tasks being executed. Flawless execution of the strategy is the key, and no amount of planning can make up for weak execution and implementation. Therefore, creating a dashboard for all milestones and timelines that have been met and those that are lagging could be extremely helpful to manage the expectations from those who are involved in execution of strategy.

Timelines and milestones are the soul of the regulatory strategy as the time associated with each regulatory activity mostly constitutes the majority of overall timeline for the project.

39.3.3 Stakeholder Alignment

The stakeholder's alignment would ensure the strategy is accepted, funded, and will support its successful implementation. Gain alignment on the draft strategy for the defined scope and incorporate their feedback as needed. Ultimately get confirmation from decision makers, as they are critical to the strategy success and communicate the decision to team responsible for executing it. Defining the project scope and the stakeholder alignment is the key step to assess the need for proper allocation of resources as well.

39.4 Key Factors to Be Considered for Developing Regulatory Strategy

A few key factors to be considered for developing a successful regulatory strategy are discussed below. It is important to note that these are just a few factors; many more factors should be considered contingent upon the nature of the project and geography.

39.4.1 Identification of Project Attributes

The nature of the product and its attributes are the main factors which will typically define the regulatory strategy. A proper regulatory strategy must take all the desired product attributes; but most importantly the attributes which defines the labeling concept of the products, which is also referred as "Targeted Product Profile" (TPP). The sponsor uses these attributes to demonstrate product claims or communications and by working backwards, can describe the data needed to obtain regulatory approval successfully. It is to be noted that although TPP is not really required by regulatory authorities, it can be used by sponsor to initiate product development discussion with reviewers and vital tool to communicate the regulatory strategy for products. Various health agencies have defined meetings types to discuss specific stages of development of the product (e.g., sharing the TPP with FDA in preparation for the End-of-Phase 2 (EOP2) meetings), the product attributes and stages of development also aid the regulatory strategist to strategize the communication strategy, optimization of key opinion leaders, timelines and buffer time to be considered for these meetings.

Additionally, the formulation of open-ended questions will help to ensure that appropriate factors are given consideration for regulatory strategy development. A few key aspects about product attributes must be considered while developing a regulatory strategy are as follows:

- Which unmet medical need product is intended to be used for (intended use)
- Mechanism of action, design, or feature
- Proposed indication or claims
- Pharmacology
- Dosage and administration
- Contraindications, warnings, and precautions
- Drug abuse and dependence
- Overdose
- Toxicology
- Storage and handling

39.4.2 Geography and Regulatory Landscape

The regulatory landscape and geography of the intended market for product launch would be the second most important factor while developing the successfully executable regulatory strategy. In broad terms, it describes the laws, regulations, requirements, timelines, competitive intelligence relevant to the product. The regulatory landscape helps the strategist to focus on the evolution of regulations and the environment and regulatory situations to be leveraged.

Key drivers for consideration in regulatory landscapes are harmonization, transparency, simplicity, traceability, and collaboration. The first consideration must be given to the current regulatory harmonization initiatives around the world which could be leveraged while developing the regulatory strategy. Harmonization has significantly enabled the reach of new advanced product's reach to the patients for better outcome.

It is important to consider the current regulatory environment but also the future trend, as the regulatory process usually progresses for few years. Most of the regulatory agencies works very transparently and updates the information related to approvals, rejections, minutes of meetings for major evolution, warning letters etc. in their websites. This provides ample details to regulatory strategist to use online resources to research competitors' inspections and compliance documents. If this information is not readily available, then strategies must adopt alternative approaches to gather this information via developing the relationships with regulators and liaising with key opinion leaders working closely with regulatory agencies.

The online resources also enable the team and the regulatory strategist to perform a thorough assessment of publicly available knowledge on the set regulatory precedents to gain valuable information for, e.g., summary basis of approval or product characteristics (SmPC), Chemistry manufacturing and control (CMC) information, EU public assessment report (EPARs), etc. Research of this information provides value of health authority expectations on the topic and facilitates the future communication strategy.

Following are consolidated points on regulatory landscape to be considered for developing regulatory strategy:

- Competitor product profile or similar products available in the intended geography
- Upcoming changes in legislations, law, and regulations, etc
- Any foreseen change in political scenario and possible impact on regulatory landscape
- Jurisdiction concerns pre-market and post-market
- Data requirements to support the product
- Openness of regulatory authority for meetings and communications
- Regulatory precedence in the intended geography

39.5 Implementing the Regulatory Strategy

As presented above, implementation of the regulatory strategy is the most important aspect. Strategy is of no use without a successful implementation plan, therefore it is important that substantial amount of efforts are made to keep the strategy updated vis-à-vis the regulatory environment (external and internal) and keep the regulatory intelligence tools abreast of all landscape changes, specially which may impact the strategy. Along with this, a routine discussion with stakeholders to keep them informed with the changes is extremely important. This step will ensure no derailing of project milestones and mitigation plan readiness by the team.

It is often prudent to seek input from regulatory authority on the key components of the regulatory strategy, typically in form of pre-submission. This step is mainly recommended for fairly novel products where there are ambiguity and lack regulatory precedence. Agency feedback can give the confidence that the proposed strategy is likely to be effective or provide insight into areas where the strategy should be revised for successful implementation.

39.6 Conclusion

It is evident that a well-proportioned regulatory strategy and its successful implementation play a key role in boosting the opportunity of timely approval and helps in meeting the objectives.

A well-planned regulatory strategy should be balanced, realistic, and achievable to support the organization's mission and vision. It should also be easily understood, as it will need to be communicated to the development team.

Additionally, a good regulatory strategy is based upon a solid understanding of your product and its use, which drives the regulatory review process and data requirements. It drives the stakeholder's alignment and identify options and tradeoffs team be willing to take a different approach when appropriate. It is important not only during the initial development stage of the project but also for a complete life cycle because it facilitates optimized study design to meet regulatory end points and increases work process efficiencies and compliance with the law.

It is important to continually monitor the regulatory environment and internal drivers for change and adapt your strategy when needed so that it remains viable, achievable, and likely to address company goals.

References

1. Babatola M, and Kabir N. *Fundamentals of US Regulatory Affairs*, 9th ed, Chapter 12.
2. Deloitte Center for Health Solutions. A bold future for life science regulations prediction 2025, November 2018.
3. Magee JF. Decision trees for decision making. *Harvard Business Review* July 1964.
4. Kramer MD. Five steps of regulatory strategy development. March 11, 2014.

Chapter 40

Leading the New Normal by Accelerating Digital Transformation

Virginia Chan

Digital Health Chapter, Asia Regulatory Professionals Association
Jack.wong@arpaedu.com

40.1 China Medical Technology Market in 2025

Public health challenges call for safe, effective use of software and digitalization

By 2025, China will be the largest medical technology market with the projected size of USD 328 billion. The Chinese government has been actively investing to upgrade China medical device innovation and manufacturing to advance public health in China, accelerate "made in China 2025" and "Healthy China 2030," while controlling the cost of healthcare with access and affordability.

The investment focus according to the Healthy China 2030 Vision is as follows:

- Cultivation of large-scale, key research and medical technology development enterprises

---

Medical Regulatory Affairs: An International Handbook for Medical Devices and Healthcare Products (Third Edition)
Edited by Jack Wong and Raymond K. Y. Tong
Copyright © 2022 Jenny Stanford Publishing Pte. Ltd.
ISBN 978-981-4877-86-2 (Hardcover), 978-1-003-20769-6 (eBook)
www.jennystanford.com

- Focus on stem cells and regenerative medicine, new cutting-edge medical technologies such as type vaccines and biological treatments, robotic, additive manufacturing and strengthening of the prevention and control of chronic diseases, precision medicine, smart medicine and preventive health technology
- Strengthening of source integration and data exchange and deployment of national biomedical big data, to build medical data demonstrations on the heart, blood vessels, tumors, and geriatric diseases
- The new generation of traditional Chinese medicine to enhance major diseases treatment

The global medtech reached USD500 billion in 2019.
In 2025, China will be the largest medtech market
Covid accelerates digitalization

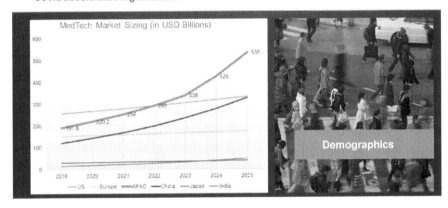

40.2 Patient-Centric Digital Healthcare Model

As Chinese gradually enter an aging society, China's medical information industry is developing rapidly. According to industry research data, the size of China's medical information market reached USD 8 billion in 2018, with an annual growth rate of more than 15%. In recent years, the continuous development of artificial intelligence, 5G, Internet of Things, cloud computing, big data, and other technologies has provided technical support for digital healthcare.

COVID-19 accelerates the digital delivery of delivery services, from remote patient monitoring, online booking, telemedicine and digitally assisted service with AI, machine learning for diagnostic and treatment, AR and VR for physician training and predictive maintenance service. For example, Ping An Good Doctor—China's largest healthcare platform—experienced a 900% surge in new users and 800% surge in online consultations from December 2019 to January 2020.

Medtech companies with software as a service (SaaS) are well positioned to transit into a patient-centric business model, and provide holistic and comprehensive care. Non-medtech companies in electronics, automobile, and consumer goods and insurance, such as Foxconn and Huawei, are entering the healthcare eco-system.

40.3 Accelerate Digital Transformation for Agility, Flexibility, Efficiency, Reimagine the New Business Model

Medtech companies can lead in the new normal with agility by accelerating digital transformation, across the product life cycle to accelerate new business model, shorten time-to-market, improve top- and bottom-line with operational efficiency, enable digital supply chain and post-market surveillance.

For this chapter, we focus on connected care, intelligent design control, digital innovation to shorten your time-to-market with compliance, lowering development time and cost, to achieve the key milestones for investors to secure funding rounds for the start-ups.

40.4 Digital Health Regulation: Focus on Quality Across the Product Life Cycle

The top three major reasons for recalls of medical devices are software-related, design controls, and production controls. Medical technology regulatory systems aim to protect and advance public health and safety while support innovation and access.

Public trust and confidence in these systems depend on the safety and performance of medical devices throughout the product life cycle. Software update and cyber security are the major challenges with the latest trend of software-based innovation, with the 5G and IoMT (Intelligent of Medical Things), SaMD (Software as Medical Devices), and the use of AI.

In July 2020, China's State Council released the "Key Tasks for Deepening the Reform of Medical and Health System in the Second Half of 2002" which highlights the accelerated development of "IoMT" platform, health management service models and promotes Internet and medical security with the application of new generation of information technology in the healthcare industry. With the "Law of the PRC on Basic Medical Care and Health Promotion" coming into effect in June 2020, medical and health institutions were urged to establish and improve their systems for medical information exchange and cyber security.

40.5 Connected Care

It is important for innovators to explore how to architect complex software solutions, balance agility with the current regulatory frameworks in digital health applications, how low-code development catalyzes DevOps.

- Agile software development to cover traceability of standards, integrating risk management
- Low-code cloud app development for fast, collaborative, controlled application
- IoT digital health platform with security, regulatory compliant, certified
- Designed for patient data privacy
- Using IoT and low-code to support real-world evidence and clinical studies

40.6 Intelligent Design Control to Shorten Time-to-Market with Compliance

Complexity in the development of medical devices, especially for smart medical devices that include software and electronics, demands software solutions to establish design controls that orchestrate the design process across inter-disciplinary engineering teams.

What is intelligent design control? A data-driven, model-based, and document smart product and software development process, with four key capabilities:

- User needs and design inputs as data
- Risk assessments and mitigations as data
- V&V test management and traceability
- Dashboards and reporting on "live" data

Intelligent design control increases engineering efficiency and speed across the device lifecycle while avoiding costly errors, quality issues, and recalls.

40.7 Digital Innovation for Design Excellence: Augment Virtual Evidence to Reduce Design Time and Cost with in silico Clinical Trials

Design excellence combines multi-disciplinary design collaboration with advanced design tools and multi-physics simulations to achieve competitively differentiated, premium-value devices. Unlock your design potential to a distinct competitive advantage.

Five key areas to achieve design excellence maturity:

- Advanced design and evidence re-use across products and programs. Data based—not document-based—for intelligent platforming
- Robust, concurrent design to enable quick domain-specific progress with highly capable design tools for efficient cross-domain workflow merges
- Design file integrity: accessible, integrated, high-quality data throughout the product lifecycle for trusted compliance and product maintenance
- Comprehensive risk management: traceability, application, and decision support through requirements, design, and verification and validation
- Digital evidence to reduce costly physical tests and iterate quickly through verification and validation

Design Excellence Maturity

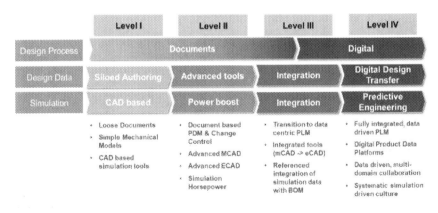

Digitalization facilitates real-time data across the product life cycle and post-market usage, which provides a closed loop for continuous quality improvement, which benefits patients and supports national health initiatives.

The FDA acknowledges the benefits to public health provided by modeling and simulation and advises the medical device industry on the use of simulation methodologies so that safe and effective new therapeutics can advance more efficiently, from design to preclinical studies through clinical trials to market. You can hear directly from the FDA's Dr. Tina Morrison, Deputy Director, Division of Applied Mechanics, Center for Devices and Radiological Health, provides an overview the possibility for in silico clinical trials to be used for evaluating medical products: https://www.plm.automation.siemens.com/global/en/webinar/fda-perspectives-on-modeling-and-simulation/33661.

40.8 Turning Complexity to Your Competitive Advantage: Gain Digital Dividend with Vision 2025

To summarize, as we are formulating Vision 2025 strategically, it is important to remember that China will be the largest market when addressing the user needs and requirement, with digital health represents one of the largest potential areas of growth in the healthcare sector.

To lead in the new normal, we all need to accelerate our digitalization journey across the enterprise, with data-driven decision making, turning complexity to your competitive advantage, and re-imagine healthcare with new business model with agility, flexibility, efficiency and quality, across the healthcare eco-system and value-chain. The finance and investment world starts to recognize the benefits digitalization brings and here we can achieve "digital dividend" for the shareholders with both top-line and bottom-line improvement, while improving safety, quality, affordability and accessibility of healthcare.

Chapter 41

An Overview of the Herbal Product Regulatory Classification in Asia and General Guidelines for Health Product Development

Jacob Cheong

Asia Regulatory Professionals Association
Jack.wong@arpaedu.com

Herbal products could fall under a wide range of regulatory classification in Asia. In general, the herbal product could be classified under seven categories. The definitions of the seven possible regulatory categories are as follows:

Prescription medicines: Medicines/drugs which can only be purchased with a prescription or a physician's order. Herbal medicines categorized as prescription medicines are sold in pharmacies and/or by licensed providers.

Non-prescription medicines: Medicines/drugs which can be purchased without a prescription or a physician's order, often at a pharmacy. The definition of "non-prescription medicines" may also often include the terms "self-medication" and/or "over-the-counter (OTC)" medicines.

Medical Regulatory Affairs: An International Handbook for Medical Devices and Healthcare Products (Third Edition)
Edited by Jack Wong and Raymond K. Y. Tong
Copyright © 2022 Jenny Stanford Publishing Pte. Ltd.
ISBN 978-981-4877-86-2 (Hardcover), 978-1-003-20769-6 (eBook)
www.jennystanford.com

Dietary supplements (including traditional medicines and health supplements, food supplements, Nutraceutical product, and food in capsule and tablet format): A dietary supplement could be intended to supplement the diet and the herbal product could combine with other active compounds, for instance, vitamin, mineral, amino acid and nutritional compounds such as CoQ10, glutathione, etc. A dietary supplement is intended to supplement the diet by increasing the total daily intake of a concentrate, a metabolite, a constituent, an extract, or a combination of these ingredients.

Health foods (including functional foods): Any natural food popularly believed to promote or sustain good health by containing vital nutrients. Functional foods also include any foodstuff enhanced by additives and marketed as beneficial to health or longevity. Health food/functional food could be marketed with specific health claims and may therefore be regulated differently from conventional foods.

Chinese proprietary medicine refers to a medicinal product that contains one or more active ingredients from any plant, animal or mineral, or any combination of sources. All of the active ingredients have to be documented for use in traditional Chinese medicine.

General food products: Food products consisting essentially of protein, carbohydrate, fat, dietary fiber, and other nutrients used in the body of an organism to sustain growth and vital processes and to furnish energy.

Others classification: Such as medication cream using herbal or cosmetic, or other topical product with herbal compound, etc.

The criteria for the classification of a product depend on four key factors:

1. Formulation
   - Name and the quantities of ingredient used
   - Scientific names and plant parts of botanical ingredients
   - Extraction ratio for botanical ingredient so that it could convert into crude form
   - Solvent use for extraction

2. Dosage forms and recommended dosing
   - Pharmaceutical dosage form (capsule, tablet, effervescent, etc.) or food format (gummy, confectionary, beverage, etc.).
   - Recommended daily dose, e.g., two capsules a day.
3. Product indication (the intensity—specific or non-specific claim)
   - General claim/nutritional claim
     - Nourishes the body
     - Maintains health and general wellbeing
     - Relieves general tiredness, weakness
     - etc.
   - Functional claim
     - Maintain/supports immunity
     - Promotes healthy skin
     - Aids in digestion
     - etc.
   - Disease risk reduction/ relief of symptom linked to a named condition/nutritional supplementation claim linked to a specific therapeutic effect
     - Helps to reduce the risk of dyslipidemia
     - Helps to reduce insomnia
     - Helps to reduce risk of osteoporosis by strengthening bone
     - etc.
4. Experience and history of use
   - Product classification in country of origin
   - Product knowledge, e.g., safety data and record of approval in the FDA database
   - FDA permitted list/prohibited list

Besides product classification, it is very crucial to understand the registration lead time and requirements during product development. The enclosed table summarizes the key information pertaining to Asia markets.

Countries	Governing Body	Classification	Likely time to register	English Label Allowed?	Claim Substantiation (exclude Drug and CPM)	Is stability studies required by authority?	Climate zone	Special registration requirements for import/export
1	DOH or VFA	Food Supplement/Health Supplement/Functional Food (FS/HS/FF) or Drug	FS/HS/FF - 2 months dossier prep + up to 9 months approval. Drug - 6-12 months dossier prep + 3 years approval.	Yes but will require oversticker in Vietnamese	Can Follow ANNEX VII ASEAN Guidelines On Claims & Claims Substantiation For Health Supplements.	Only for shelf life extension and drug.	Zone 4B	Imported product requires Certificates of Pharmaceutical Product (CPP), Certificates of Listed Product (CLP) or Certificate of Free Sales (CFS)
2	HSA	Health Supplement, Chinese Proprietary Medicine (CPM) or Drug	Health Supplement - No registration require; CPM - 60 working days for new product listing, Drug; 6-12 months dossier prep + 60 to 270 working days for approval.	Yes	Can Follow ANNEX VII ASEAN Guidelines On Claims & Claims Substantiation For Health Supplement.	No except drug	Zone 4B	Imported drug requires Certificates of Pharmaceutical Product (CPP) and Statement of Licensing Status (SLS); CPM product for export requires Certificate of Free Sales (CFS) and product listing.
3	DOH for Drug and CFS for Food	Food or Chinese Proprietary Medicine (CPM) or Drug	No register require for Food. Drug: 6-12 months dossier prep + 12 - 16months for approval.	Yes	Food: No specific guidelines. Practise similar like ASEAN. Look at market comparitive. More claims allowable compared to other markets.	No except drug	Zone 4A	None
4	FSSAI or CDSCO	Health Supplement or Nutraceutical or Food for Special Dietary Use or Drug	HS/ Nutraceutical/ FSDU- Self regulated if acceptable for direct import. Otherwise registration approval required prior to import. Dossier prep 3 months + 6 months approval. Drug: Dossier prep 6-12 months + 9 months approval	Yes	Local health authority has own set of guidelines for label claim. The claim requirement compatible with ANNEX VII ASEAN Guidelines On Claims & Claims Substantiation For Health Supplements.	No except drug	Zone 4B	Can be direct imported if ingredients are in the approved list and meeting product classification requirements. Otherwise registration required prior to import.
5	BPOM	Traditional Medicines (TM) or Health Supplement (HS) or Drug	Traditional Medicines/Health Supplement: 3 months dossier prep + 9 - 12 months for approval.	No	Can Follow ANNEX VII ASEAN Guidelines On Claims & Claims Substantiation For Health Supplements.	Yes	Zone 4B	Imported product requires Certificates of Pharmaceutical Product (CPP)/Certificate of Free Sales (CFS) and these certificates would need legalisation at Indonesia embassy which takes time.
6	NPRA	Food or Health Supplement or Traditional Medicines or OTC	Food: 1.5 months dossier prep + 2 months for approval, Health Supplement/Traditional Medicines: 3 months dossier prep + 18 months for approval, OTC: 6-12 months dossier prep + 18 to 24 months for approval.	Yes	Can Follow ANNEX VII ASEAN Guidelines On Claims & Claims Substantiation For Health Supplements.	Yes	Zone 4B	Imported product requires Certificates of Pharmaceutical Product (CPP)/Certificate of Free Sales (CFS).
7	KFDA	Health Functional Food	1 month dossier prep + 0.5 month for approval	No	Local health authority assess own evidence for label claim (permitted ingredient and claims).	Only for new formula that develop in domestic market with no record in KFDA.	Zone 2	May require products to be sent ahead to get custom clearance ahead of product arrival in Korea.

Herbal Product Regulatory Classification in Asia

#		Authority	Category	Lead time	Clinical trial required?	Claims guideline	GMP required?	Zone	Other requirements
8		Taiwan FDA	Conventional Food, Food in Capsule & Tablet Form, Health Food, Special dietary Food, Drug	No registration require for Conventional Food; Food in Capsule & Tablet Form: 1 month dossier prep + 4 to 6 months for approval; Special dietary Food: 3 months dossier prep + 8 to 10 months for approval; Health Food: 6-12 months dossier prep + 6 to 24 months for approval depends on the certification category; OTC Drug: 6-12 months dossier prep + 10 to 12 months for approval	No	Local Health authority assess own evidence for label claim (permitted ingredient and claims). Health Food would need to go through pre-clinical/clinical trial to prove for product efficacy and claim.	No except Health Food and Drug. Not needed most of the time.	Zone 4	Plasticizers required to meet the buyer's requirements during product listing on pharmacy/retail channel
9		TH FDA	Food (Dietary Supplement Product, confectionary, special purposed food, etc.) or Drug (such as dangerous drug, non dangerous drug, etc.) or Herbal Product (Herbal Supplement Product, Herbal Medicine)	Food, Confectionary: No dossier preparation require + 1 week for approval. Dietary Supplement Product: 1-2 months dossier prep + 3-4 months for approval. Special purposed food: 3-4 months dossier prep + 8-10 months for approval. Drug: 6-12 months dossier prep + 1-2 years for approval. Herbal Product (new category - Pending the announcement of new regulation)	No	Can Follow ANNEX VII ASEAN Guidelines: On Claims, & Claims Substantiation for food category.	No except Drug. Drug is 50% of the time.	Zone 1B	None
10	WFOE	SMAR and GACC	Food	0.5 months (For custom clearance only)	Yes, but will require oversticker in Chinese on mandatory info.	No function claim allow.	No	Zone 2	None
		SMAR and GACC	Health Food/Blue cap	5 years	Yes but mandatory info must be in Chinese	Local health authority assess own evidence for label claim (permitted ingredient and claims) based on domestic clinical trial.	Yes, to be conducted in CN lab.	Zone 2	Must have 1 year sales history prior to registration and animal toxicity is mandatory
		GACC		0.5 months (For custom clearance only)	Yes	Follow countries of origin's practice.	No	Zone 2	None
11	FTZ	Serviços de Saúde de Macau (SSM)	Food or Drug	Required to submit actual product for classification before importation. -No registration require for Food. Drug: 3 months dossier prep + 2 months for approval (CPP needed).	Yes	No specific guidelines. Practise similar like ASEAN. Look at market comparative. More claims allowable compared to other markets.	No	Zone 4A	None

Note: Lead time for dossier preparation is just for indication purpose.

References

1. Act/Regulation/Advisories. Retrieved October 26, 2020, from https://foodlicensing.fssai.gov.in/index.aspx.
2. ANNEX VII ASEAN Guidelines on Claims & Claims Substantiation for Health Supplements. Retrieved October 26, 2020, from https://asean.org/wp-content/uploads/2017/09/ASEAN-Guidelines-on-Claims-Claims-Substantiation-HS-V2.0-with-discla....pdf.
3. Drug. Retrieved October 26, 2020, from https://cdsco.gov.in/opencms/opencms/en/Drugs/New-Drugs/.
4. Food Legislation/Guidelines. Retrieved October 26, 2020, from https://www.cfs.gov.hk/english/food_leg/food_leg.html.
5. Guidance, Law & Regulations. Retrieved October 26, 2020, from https://www.fda.gov.tw/ENG/index.aspx.
6. Industry & Quest3 + System Main Page: Product Registration & Licensing System. Retrieved October 26, 2020, from https://npra.gov.my/index.php/en/industry.html.
7. Law & Regulations. Retrieved October 26, 2020, from https://www.mohw.gov.tw/np-120-2.html.
8. Links within the FDA. Retrieved October 26, 2020, from https://www.fda.moph.go.th/Pages/HomeP_D2.aspx.
9. Law & Regulations. Retrieved October 26, 2020, from http://food.fda.moph.go.th/law/index.php.
10. Peraturan/JDIH. Retrieved October 26, 2020, from https://www.pom.go.id/new/.
11. Product Regulation. Retrieved October 26, 2020, from https://www.hsa.gov.sg/.
12. Regulatory Affairs. Retrieved October 26, 2020, from https://www.dh.gov.hk/english/main/main_ra/main_ra.html.
13. Regulations. Retrieved October 26, 2020, from https://www.mfds.go.kr/eng/brd/m_15/list.do.

Chapter 42

Overview of Health Supplements: Singapore

Srilatha Sreepathy, Geeta Pradeep, and A. V. Rukmini

Vitamin, Mineral and Supplement (VMS) Committee,
Asia Regulatory Professionals Association

srilatha_bala1999@yahoo.com.au

42.1 Introduction

Increasing life expectancies and greater focus on lifestyle modifications towards improving the quality of life has led to an explosion of demand for products that can be used to achieve this. Food/food-like-substances that are taken alongside of pharmaceuticals, and which claim to provide physiological and medical benefits to *support* or *maintain* the healthy functions of the human body are generally known as nutraceuticals/health supplements. These products are used to supplement a diet with benefits beyond those of normal nutrients and also thought to contribute to the prevention and treatment of disease. Nearly 40% of the population of the ASEAN region, and about 50–70%

Medical Regulatory Affairs: An International Handbook for Medical Devices and Healthcare Products (Third Edition)
Edited by Jack Wong and Raymond K. Y. Tong
Copyright © 2022 Jenny Stanford Publishing Pte. Ltd.
ISBN 978-981-4877-86-2 (Hardcover), 978-1-003-20769-6 (eBook)
www.jennystanford.com

of the population of developed nations use these nutraceuticals with an annual increase of about 8.4–10%. The global nutraceutical industry accounts for more than US$ 80,700 million in 2019. Of this market, vitamins, minerals, and supplements amount to about 40% of the market. The regulation of nutraceuticals varies widely between different countries regarding these health supplements. Probiotics, prebiotics, food additives, and health supplements all fall under the gamut of nutraceuticals.

42.2 What Are Health Supplements?

Health supplements are often defined as substances used either individually or in combination intending to provide the body with micronutrients when required. They are presented in dosage forms such as the capsules, softgels, tablets, liquids, and syrups. Of these, softgels remain the most popular form, generating nearly one-fourth of the global revenue in this market. The market for liquids, capsules and tablets is growing steadily. In Singapore, health supplements can be imported and sold without a license from the Health Science Authority (HSA). This regulatory approach for health supplements has similarities to that adopted by USA, EU, and Japan.

All these different types of health supplements can contain one or more, or a combination of the following ingredients:

- Vitamins, minerals, or amino acids (both natural and synthetic). Examples: calcium, iron, Vit A, L-Lysine. The products containing only vitamins and minerals are classified as quasi medicine (see under advertisement). Different countries have different allowable limits set for vitamins and minerals and products must comply with these limits (see Tables 42.3 and 42.4).
- Substances derived from natural sources, including non-human, animal, and botanical materials in the forms of extracts, isolates, concentrates. Example: melatonin, fish oil products, curcumin, green tea extracts, Ginkgo biloba, Chinese ginseng.

42.3 What Are Not Health Supplements?

When attempting to define health supplements, it is useful to define also those products which cannot be categorized as health supplements. Any product as a sole item of meat or in the form of food and beverages, like biscuits, bread, coffee, and juice, cannot be considered as health supplements, even if they are fortified with micronutrients. Similarly, injectables or preparations that need to be sterile, for example, injections and eye drops, are also not health supplements.

42.4 How Are Health Supplements Regulated in Singapore?

In Singapore, the Health Science Authority (HSA) regulates health products to meet standards of safety, quality, and efficacy. Health supplements are however not subject to pre-market evaluation and approval by the Health Sciences Authority (HSA). Also, licensing by HSA is not needed for importation or manufacture before they can be sold locally. In the following sections, we provide an overview of some important aspects of health supplement regulation in Singapore.

42.5 Safety and Quality Standards

Dealers (which include importers, manufacture, wholesale dealers/distributors, repacker, and retailer) are responsible for the safety and quality of their health supplements. They also have the responsibility to ensure that their products meet HSA specified safety and quality standards.

The dealer in health supplements must consider the following facts/required actions to ensure the safety of health supplements.

- Health supplements must contain low-risk ingredients, such as vitamins, minerals and natural source substances, which are safe to consume.

- Products should not contain any harmful or undeclared active substances.
- Post-market surveillance system: Regular sampling and testing will ensure the products under safety control throughout the whole shelf life of such products and to initiate the timely product recall when necessary. The program includes a random sampling of products in the market and adverse reaction monitoring, which draws on the HSA's network of healthcare professionals and international regulatory partners to pick up signals of adverse reactions to products.
- Health supplement must not contain substances listed in Poisons Act (Chapter 234) and Poison Rules. (Example Steroid etc),or do not contain prohibited ingredients ex Lithium salts etc Ingredients which are derived from human part or may affect human health , active ingredients that are not stated in the label or use of active ingredients for a medicinal purpose to treat or prevent diseases or disorders.

42.6 Quality Standards

A dealer of health supplements must also ensure the products conform to HSA safety and quality standards, including heavy metals and microbial limits (Tables 42.1 and 42.2).

Table 42.1 Heavy metal limits—must comply for new and existing products

Heavy metal	Quantity (by weight)
Arsenic	5 ppm
Cadmium	0.3 ppm
Lead	10 ppm
Mercury	0.5 ppm

Table 42.2 Microbial limits

Microbe	Quantity CFU per G or ml of product
Total aerobic microbial count	Not more than 10^5
Yeast and mould count	Not more than 5×10^2
E. coli, Salmonellae, and *S. aureus*	Absent

Table 42.3 Vitamin limits for general adult population

Nutrient	Maximum daily limit
Biotin	0.9 mg
Folic acid	0.9 mg
Nicotinamide	450 mg
Nicotinic acid	15 mg
Vitamin A (Retinol)	1.5 mg (5000 IU)
Vitamin B1	100 mg
Vitamin B2	40 mg
Vitamin B5 (pantothenic acid)	200 mg
Vitamin B6	100 mg
Vitamin B12	0.6 mg
Vitamin C	1000 mg
Vitamin D	0.025 mg (1000 IU)
Vitamin E	536 mg (800 IU)
Vitamin K1/K2	0.12 mg

Table 42.4 Mineral limits for general adult population

Nutrient	Maximum daily limit
Boron	6.4 mg
Calcium	1200 mg
Chromium	0.5 mg
Copper	2 mg
Iodine	0.15 mg
Iron	15 mg*
Magnesium	350 mg
Manganese	3.5 mg
Molybdenum	0.36 mg
Phosphorus	800 mg
Selenium	0.2 mg
Zinc	15 mg

*For multivitamin and mineral supplements for pregnant women, a higher iron limit of 30 mg/day may be considered.

42.7 Stability Study and Shelf-Life of Health Supplements

Stability is an essential factor of quality in health supplements (HS) to ensure maintenance of the specifications of the finished product when packed in its specified packaging material and stored at the established storage condition within the specified shelf life.

For health supplements, stability data from at least two batches is required. It could be either from pilot scale, primary scale, production scale or their combination. The testing should be conducted on the product packaged in the primary container closure system proposed for marketing including, any secondary packaging.

The testing frequency for accelerated and real time is captured in Table 42.5.

Table 42.5 Testing frequency

Storage condition	Testing frequency
Real time	0, 3, 6, 9, 12, 18, 24 months and every 6 months over the second year and annually thereafter through the proposed shelf life
Accelerated	0, 3, and 6 months

42.8 Storage Condition

Table 42.6 Common storage conditions

Type of container closure system/study	Storage condition
Products in primary containers permeable to water vapor	30°C ± 2°C/75% RH ± 5% RH
Products in primary containers impermeable to water vapor	30°C ± 2°C
Accelerated studies	40°C ± 2°C/75% RH ± 5% RH

Note: Other storage conditions are allowable, if justified.

A tabulated list of stability indicating parameters for health supplement are captured in Table 42.7.

Table 42.7 Health supplement dosage form

HS Dosage Form	Organoleptic characteristics	Assay	Hardness/Friability	Dissolution/Disintegration	Water Content	Viscosity	PH	Microbial Content	Granules/Particle Size	Resuspendability
Oral powder	√	√			√			√		
Hard Capsule	√	√		√	√			√		
Soft Capsule	√	√			√			√		
Coated and Uncoated Tablet	√	√	√	√	√			√		
Coated and Uncoated Pill/Pellet	√	√		√	√			√		
Suspension	√	√				√	√	√	√	√
Solution	√	√				√	√	√		
Emulsion	√	√				√	√	√		
Granules	√	√			√			√	√	

42.9 Product Labelling Requirement

The product label should be prominent and conspicuous. The information must be in English and must be printed in a clear and legible manner. In case of small containers, the product name, batch and expiry date must be on the product (examples: blister strip, unit dose of sachet). The full product information should be on the outer container (example: carton).

The label must contain the following information:

- Product name
- Names and quantities of all active ingredients
- Product indications/Intended purpose
- Daily dosage
- Directions of use
- Pack size
- Batch number
- Expiry date (or "Use by," "Use before" or words with similar meaning)
- Cautionary label or statement, where necessary
- Name and address of the local manufacturer or local importer
- Name of the country of manufacture for imported products

Note

- Ingredient naming should be based on its internationally accepted, scientific names. Example: Malpighia glabra (acerola) fruit extract dry conc.
- The use of the common name of the ingredient name is optional. Example: Vitamin C (ascorbic acid).
- Common or chemical names should be used for Minerals. Example: Zinc.

42.10 Health Supplement Claims

The claims must be consistent with the definition of health supplements. Health supplements must not be labelled, advertised, or promoted for any specific medicinal purpose including treatment or prevention of any disease, disorder. The claims must be substantiated by good quality evidence that is relevant to claim. The evidence should be based on authoritative references, history of use, scientific opinion, and quality scientific evidence from human studies. The dealers should hold evidence to support the claims and provide it to Authority when requested.

Health supplement may make general or functional health claims.

Example

General Claim: Support good health and growth, Relieves general tiredness and weakness.

Functional claim: Assist in maintaining joint mobility, manage mild discomfort associated with menopausal symptoms.

Note: The general claims for vitamins and minerals are permitted only when the relevant amount of vitamins and minerals is more than 30% RDA value in the finished product.

42.11 Advertisements and Promotions

In general, health supplements are not subject to permit approval. However, vitamins and minerals (which are classified as quasi-medicinal products) as well as some other health supplements are subject to medical advertisements and sales promotion control (example: St John's wort, melatonin products).

The advertisement must comply with the guide on health supplement claims. Advertisement and promotional messages must not contain misleading claims and do not make claims for treatment or prevention of diseases.

42.12 Other Important Aspects

An important point to note is that health supplement regulation differs from country to country, even within an interconnected region such as the ASEAN region. These differences range from the minimum and maximum levels of micronutrients prescribed in each country to product labelling requirements, timelines, and application processes. While we have attempted to provide an overview of the vitamins, minerals and supplement regulation in Singapore, as with any medical device or pharmaceutical regulation, our concern is ultimately the consumer, i.e., the patient or the clinician. Here we provide a few brief points that would affect the consumers.

Although health supplements, vitamins, and minerals are increasingly consumed, there is not much evidence showing the benefits of these products in those people who are well nourished.

Different individuals and even different populations may differ in their requirements for these supplements based on their health status, dietary intakes, climate, and other environmental factors and these may also play a role in the perceived benefit derived from supplements. Some products, such as fat-soluble vitamins, may in fact produce toxicity when consumed in large quantities. The availability of these drugs over the counter or online leads to a less well-informed public consuming supplements that they may not need, and which may in fact lead to adverse effects. Clinicians face problems when prescribing supplements and other medicines because these drugs are available over the counter and patients may not reveal that they use any supplements. Some prescription drugs may potentially lead to adverse drug interactions with supplements as well. Despite these concerns, there is an ever-increasing demand for "more natural therapies" to ensure long-term health benefits which necessitates more evidence based research to fully elucidate their beneficial properties.

References

1. ASEAN guidelines on stability study and shelf-life of health supplements Version 1.0.
2. ASEAN-Guidelines-on-Claims-Claims-Substantiation-HS-V2.0.
3. DeFelice SL. FIM Rationale and Proposed Guidelines for the Nutraceutical Research & Education Act-NREA, November 10, 2002. Foundation for Innovation in Medicine. Available at: http://www.fimdefelice.org/archives/arc.researchact.html.
4. Dwyer et al, Dietary Supplements: Regulatory Challenges and Research Resources, Nutrients. January 2018.
5. Health Supplement Guidelines Revised September 2019 Health Science Authority.
6. Kantor ED, et al. Trends in dietary supplement use among US adults from 1999–2012. JAMA. 2016;316(14):1464–1474. doi:10.1001/jama.2016.14403.
7. Marik PE, et al. Do dietary supplements have beneficial health effects in industrialized nations: what is the evidence? *JPEN J Parenter Enteral Nutr*, 2012, 36(2), 159–168.

8. Tripathi C. et al. Nutraceutical regulations: an opportunity in ASEAN countries, *Nutrition*, 2020, 74, 110729. doi: 10.1016/j.nut.2020.110729.
9. https://www.hsa.gov.sg/consumer-safety/articles/how-health-supplements-are-regulated-by-hsa, accessed on September 4, 2020.
10. https://www.globenewswire.com/news-release/2018/12/18/1668804/0/en/Herbal-Nutraceutical-Supplements-in-Vogue-as-Consumer-Preference-for-Natural-Soars-Fact-MR.html, accessed on September 4, 2020.

Chapter 43

International Medical Device School Experience

Encey Yao

Chairperson for IMDS Alumni

Jack.wong@arpaedu.com

The International Medical Device School (IMDS) provides a comprehensive syllabus which covers different considerations involved in bringing a medical device into commercialization. Organized by Temasek Polytechnic, ARQ on (Asia Regulatory & Quality Consultancy), MedtechBOSS and ARPA (Asia Regulatory Professionals Association), IMDS was launched in October 2019. In a 4-day program, industry professionals shared and discussed important factors for medical device stakeholders should be aware of throughout the lifecycle of the medical device: from conceptualization, design, manufacturing, market access, regulations, clinical use, post-market responsibilities, even to decommissioning of medical devices. By inviting a carefully selected line-up of speakers with involved experiences in the handling and studying of the medical devices industry, the IMDS

Medical Regulatory Affairs: An International Handbook for Medical Devices and Healthcare Products (Third Edition)
Edited by Jack Wong and Raymond K. Y. Tong
Copyright © 2022 Jenny Stanford Publishing Pte. Ltd.
ISBN 978-981-4877-86-2 (Hardcover), 978-1-003-20769-6 (eBook)
www.jennystanford.com

attracts professionals of diverse backgrounds who benefited from the syllabus contents.

The IMDS adopts a holistic coverage in content allowing for individuals to appreciate the depth of the syllabus even without prior experience in the industry. Each speaker in the IMDS prepared a concise lecture which covers their topics in reasonable breadth and with minimal overlaps between each presentation and lecture to ensure less repetition of content and materials. The syllabus provides sufficient coverage in considerations for any participant in a way that topics are introduced at an entry level, and for participants who are more experienced, it still provides great value through the broad coverage of topics and networking potential throughout the program. For continuous improvement, a panel committee reviews industry trends, regulatory environment and participants' feedback to ensure the course stays relevant and fits of the needs of medtech professionals from engineers to management.

The regulatory updates of different economic spaces and their regulatory bodies are also broadly discussed. Regulatory models, their differences and focus are also discussed. Stakeholders being familiar with these regulatory models opens the understanding of how medical devices are to be prepared prior to market entry and how to be managed post-market as well.

The program also features a site visit whereby participants get to witness and interact with hosting partners on the technology they would like to showcase and for a more immersive learning in medical device technologies. This experience adds to the already enriching program in another dimension where there are more opportunities for learning through observing, instead of lecture style presentations.

The IMDS program was also densely packed, yet allowing for small breaks between sessions and time for thoughtful rumination of the discussed content. However, the accompanying materials and online resources made available for download allow students to review the syllabus and revisit the contents after the course for reference. Participants were also encouraged to engage and interact with the speakers during refreshment breaks. Networking opportunities were organized for the participants and partners during the course.

The multidimensional experience with the IMDS provides the participants a complete perspective on the lifecycle of medical devices. Such considerations proved to be valuable to any professional in the Medtech space.

Currently, IMDS has also attracted various overseas institutions' efforts to collaborate in order to to fund and train the local medtech professionals. In 2020, IMDS is available in South Korea and in China, with additional countries under discussion.

Chapter 44

Medtech Start-Up: Journey to First Product Approval

Sing Wee, Joel Tan, and Trish, May Ng

ARQon
Jack.wong@arpaedu.com

44.1 Introduction

This chapter guides medtech start-ups on key considerations, especially with regard to the bulk of regulatory and quality activities and expected outcomes in the product design and development phases of a medical device.

Briefly to recap, Global Regulatory Strategy [1] for medical devices is an important planning step that helps the medical device manufacturer, especially the medtech start-up in their journey to obtain first product approval in the country of interest, and all subsequent markets. A typical regulatory strategy consists of the following 10 key information points:

1. device name
2. device description

Medical Regulatory Affairs: An International Handbook for Medical Devices and Healthcare Products (Third Edition)
Edited by Jack Wong and Raymond K. Y. Tong
Copyright © 2022 Jenny Stanford Publishing Pte. Ltd.
ISBN 978-981-4877-86-2 (Hardcover), 978-1-003-20769-6 (eBook)
www.jennystanford.com

3. intended use
4. country of interest
5. device classification
6. product registration or conformity assessment route and its approval timeline
7. technical documentation
8. quality management system
9. clinical evaluation
10. reimbursement assessment

Medtech start-up companies face the problems of common errors, gaps, and misconception during the main activities related to the technical documentation, quality management system (QMS) and clinical evaluation:

- **Missing Technical documentation?** This is different from the QMS documentation. Medtech start-up assumed that having QMS documentation and certification means the technical documentation on the product is ready.
- **Overly complicated quality management system?** Do not unnecessarily complicate your company procedure or to go for an irrelevant certification.
- **Clinical evaluation or Clinical trial?** Often medtech start-ups confuse clinical evaluation versus clinical trial. Regulator currently requiring clinical evaluation that can be based on different clinical validation sources.

For medtech start-up companies new to the industry, it may take a while to understand relevant guidelines, and as such, comprehensive training and exposure to common industry standards is required. Ensure responsible personnel has sufficient resource and expertise for regulatory work during the product development, building the technical documentation, including also the clinical evidence based on clinical evaluation, and proper quality management system in place for the appropriate scope from design, development, manufacturing, storage to distribution.

44.2 Intended Use

The intended use of the product has a significant influence in determining the product classification, which in turn defines the regulatory route and associated requirements needed before the product may be marketed. Crafting the intended use that suits the product [2] is one of the tripping points of many start-ups.

Some start-ups look to take over the market by pioneering a new wave of disruptive technology. This zealousness may sometimes prove to be a double-edged sword, especially when the product includes many claims on the intended use. On one hand, this may seem good because it shows that the product is all encompassing and may claim a larger market share. However, more claims also mean more validation is required to support these claims, thus requiring more effort, time and money. And in the quest to validate all these claims, the primary intent for the device is also not met, which brings to mind the Russian proverb, "if you chase two rabbits, you will not catch either one." Companies, especially start-ups, should weigh the claims for the product intended use to effectively catch the rabbit.

44.3 Technical Documentation

Having identified the intended use of the product, the general regulatory route and requirements can be charted out. Noting the technical documentation required at the start of product development will help ensure compliance to relevant regulations and standards. Depending on the complexity of the medical device, the medical device authority/certification body will determine the need to review and audit the QMS, technical documentation and/or clinical trial.

Medtech start-ups often misunderstand or underestimate the quality management system documentation, product testing and manufacturing information sufficient to prepare for submission dossier to the authority or the conformity assessment body. When starting to develop a medical device, according to the quality guidelines and the device class, the manufacturer is required to build Technical documentation in the design control phases. If this was not performed, the manufacturer is required to replan and redo the documentation.

Would it not be frustrating if the company had to circle back and make changes to a product because the regulatory requirements were not taken into consideration when developing the product?

Using the regulatory framework of the European Union (EU) as an example, with the upcoming Medical Device Regulation in May 2021 and In-Vitro Diagnostic Device Regulation [3], these regulations will introduce a requirement for manufacturers to assign Unique Device Identifiers (UDI) to their products and respective packaging/labels. A company looking to attain first product approval in the EU will have to rework the packaging and labels if this requirement was not included as part of the technical documentation.

Such unnecessary detours can be avoided if attention is paid to the finer details of the regulatory approval requirements.

44.4 Quality Management System

Quality management system (QMS) is defined as the organisational structure, responsibilities, procedures, processes, and resources needed to implement quality management. In the cases of missteps, on one end of the spectrum are start-ups which enter product development without having a QMS in place. On the other end of the spectrum are companies with an overly complicated quality system and an organization unable to differentiate between a mandatory procedure versus extreme over-engineering.

Typical reasons why companies skip the initial step of setting up a QMS range from its cost, to the duration required, to "I don't need it now because product development is going to take a long time." The absence of a QMS during product development usually means process and documentation to ensure safety and efficacy of the product (also to comply to regulatory requirements) would probably be lacking or non-existent by the time verification or validation is required. Having to backtrack and retroactively take steps to implement a QMS, then close crucial process gaps (usually risk management) resultantly takes up more time and resource compared to if was set-up initially as part of the core foundation. Company should have in-place and in-use QMS procedures, however it is neither mandatory nor efficient to prematurely go for third-party certification. In early stages, external audit surveillance will incurred additional cost, resources and time to manage the audit, as well as follow-up actions from the audit. Nevertheless, it is understandable as the start-up is often requested, by the investors or the funding bodies, to achieve milestone such as the ISO 13485 QMS certification. QMS certification is normally advisable when the technical documentation is almost ready.

Medtech start-ups also need to realize that although QMS ISO13485 is an internationally recognised medical device manufacturing standard, some countries enforce their own local QMS requirements. For example, the US Quality System Regulation states the local requirements impacting companies intending to market their product in the United States. In this case then, the medtech start-up will need to include additional compliance elements of US QSR requirements in their QMS.

44.5 Clinical Evaluation

Clinical evaluation and pre-clinical testing are important sections in the technical documentation. Medtech start-up needs to properly assess available clinical data sources, with consideration of the state of the art standard-of-care or best practice, in order to provide clinical evidence to substantiate the product can be used safely and effectively based on the product's intended use.

The clinical safety and performance of a product is evaluated and verified through assessing and analysing relevant clinical data. Clinical data may be sourced from [4]

- literature search;
- clinical experience;
- clinical investigation.

Usually done during the validation phase, this activity is both time consuming and resource intensive. Budget and time allocation estimation in a regulatory strategy is often overly idealistic, and typically only based on the initial intended uses for the product. As the product was developed, more claims may be added which, as mentioned previously, leads to more validation requirements. Many start-ups fail to update or may have misconstrued the extent and amount of resources now required resulting in an overrun in time and budget.

44.6 Conclusion

A well planned-out regulatory strategy which is executed with precision will facilitate the start-up's journey to enter markets faster. This is especially important to start-ups where resources are limited, milestones are highly scrutinised, and timelines always pressing.

Some common hazards encountered by start-ups which hamper their efforts in achieving their first product approval have been identified. Companies may use this as a reminder when looking to embark on their journey, to learn from the mistakes of others because there is not enough time to make them all yourself.

Wishing all medtech start-ups a smooth sailing on their journey to their first product (medical device) approval.

References

1. *Global Regulatory Strategy for Product development and Product Registration*, Handbook for Regulatory Affairs, Edition 2.
2. *Content of a 510(k)*, U.S. Food and Drug Administration.
3. EU: Medical Device Directive (MDD 93/42/EEC), Active Implantable Medical Devices Directive (90/385/EEC) and in vitro Diagnostic Device Directive (IVDD 98/79/EC), NEW EU—Medical Devices Regulation (MDR 2017/745) and in vitro Diagnostic Medical Devices Regulation (IVDR 2017/746).
4. *Clinical Evaluation: A Guide for Manufacturers and Notified Bodies*, European Commission.

Chapter 45

Digital Transformation of Healthcare and Venture Capital's Role in It

Mark Wang

Pureland Global Venture, Singapore

Jack.wong@arpaedu.com

The Investment Thesis into the Digitization of Health

Venture capital is a form of private equity and a type of financing that investors provide to earlier-stage companies that are believed to have long-term growth potential. This normally involves a form of direct equity injection or convertible instrument injection to a company at its R&D stage to benefit business growth through capital appreciation. Given a VC does not require a company to mature before investing, it played a critical role at funding healthcare digitalization in the past two decades. As the size of venture capital AUM grow from millions into billions, the new approach of VC funding a company from the idea stage all the way to the IPO stage, bypassing traditional players, becomes more popular. A popular example would be Sequoia Capital, who would

---

Medical Regulatory Affairs: An International Handbook for Medical Devices and Healthcare Products (Third Edition)
Edited by Jack Wong and Raymond K. Y. Tong
Copyright © 2022 Jenny Stanford Publishing Pte. Ltd.
ISBN 978-981-4877-86-2 (Hardcover), 978-1-003-20769-6 (eBook)
www.jennystanford.com

inject into the cap-table from the earlier stage to the late growth stage as shown from many examples of its famous investments. This has ensured a fund to push out a new disruptive process or technology into the market with less interruption and resistance, such as the case of Uber. From this perspective, it can be argued that the role of VC is equally important as the role of regulatory bodies in the innovation of healthcare technologies.

Traditionally when venture capitalists look at the healthcare sector, they are often faced with a bit of an internal dilemma. On one hand, it represents one of the largest cyclic proofs market and growth opportunities in the world. The sector holds a healthy compound annual growth rate (CAGR), *and has many outdated technologies, with a large demand-supply gap from the current global aging population crisis. According to United Nations' 2019 World Population Prospects, the current global population structure consists of 9% of people over the age of 65. By 2050, this number will grow to 16%.* On the other hand, healthcare is one of the slowest sectors to adapt to the digital transformation, in contrast with other consumer centric sectors such as ecommerce, banking, and other services. With the number of different stakeholders in healthcare settings, any digitalization adoption requires the initiator to peel through various conflicts of interests between governments, payers, providers, healthcare organizations and pharmaceutical/device conglomerates before anything can be implemented for patients in real life. Regardless of the amount of capital invested into technology, *healthcare still remains a physically encounter-based service model in the past,* which is not built for the current aging population structure. This complexity and modest approach to digitization results in the reluctance of deployment of sufficient money into healthcare by venture capital funds and impact funds, despite healthcare being a natural fit for impact investors. Estimates from the WHO suggest that achieving SDG 3 health targets requires new investments increasing from an initial US$ 134 billion annually to US$ 274–371 billion annually by 2030. According to their study, private capital is not being invested at nearly the level needed.

Lack of technology itself is not the main problem as many healthcare tasks have already been digitally enhanced in the last 5 years. The amount of PE and VC funding for healthcare technology and solutions started to grow from 2014 and it was accelerated from 2017 onwards. Instead, the barriers to a digital transformation are often non-technological such as culture and mindset, organization structure and governance. According to a recent study by McKinsey & Company, six conditions are required to accelerate digital healthcare transformation, including "(1) government instigating and support; (2) payers and standard bodies adopting reimbursement guidelines for technology; (3) promote open innovation culture from Healthcare organization; (4) healthcare providers to focus on tangible value to consumers; (5) healthcare stakeholders to invest in the right mix of skills such as technology know-how; (6) build a long-term investment plan from the payers and providers." These points highlighted a few concrete considerations that drive healthcare's digital transformation. Unfortunately, such a matrix is inherently hard to achieve in reality from a venture capitalists' perspective. Therefore, this trait of digital healthcare increases the intrinsic risks of a technology investment, because of its longer investment horizon and lower revenue growth. As a result, most successful venture capital funds around the world partially invest in healthcare, but not many are solely focused on it. For example, B2C and B2B consumer centric technologies still account for the majority of the capital deployments in technology-focused funds. To combat this slow progression, nature has granted healthcare systems a serendipitous catalyst—a global pandemic. What we have experienced in 2020 inarguably has changed our ways of living and reshaped the global healthcare system forever.

Back in Q1 2020, we have seen how fast China adopted digitalization in healthcare and pushed it into a norm of everyday people's lives to combat COVID-19. China's largest online healthcare platform, Pin An Health, saw a 10-fold growth in new users' registration. The same trend was followed by other countries globally, including the United States, India, and Singapore. Although such changes are expected when governments mandate people to stay at home and hospitals turn away non-COVID-19-related ailments to cope spikes in capacity shortage. Only a pandemic has the kind of power to force all stakeholders to adopt digital healthcare solutions simultaneously and quickly change policy. What would normally take all stakeholders to buy-in and reach agreements for years of time was achieved in a matter of months. This shift will also incentivize healthcare investments from private capital in the years to come as the trend has already shown. As expected, Global healthcare funding to private companies reached a record high of US$ 18 billion+ in Q2 2020 from a record high of US$ 14.7 billion+ in Q1 2020. This was driven by digital health deals such as healthcare AI and telehealth deals. To date in 2020, there are already 46 healthcare Unicorns valued at US$ 116.8 billion led by the US and China.

When discussing the investment thesis for the next decade, one has to first look at the demand and supply gap of healthcare, as well as taking a bird's eye view to observe the shift in healthcare's focus over the past few years. The old era of healthcare and associated investments was focused on hardware equipment and medical consumables in hospitals. The most successful investments theses were hospital infrastructure and Capex heavy machineries. This hospital centric system deals well with serious health problems and ill people. However, the design is not intended to deal with large numbers of people whose conditions are chronic, complex and require long-term care plans. As increases in population lifespan and the associated chronic conditions that come with an aging population burden the health care system and exacerbate existing resources, the digital era of healthcare was required to begin. It can be clearly seen that by solely adding the number of caregiving professionals alone will not be sufficient anymore in developed markets. Back in 2008, researchers at Institutes for Healthcare Improvement (IHI) described the Triple Aim as simultaneously "improving the individual of care; improving the health of population; and reducing the per capita costs of care for population." In 2014, increasing clinician productivity was added. It is clear that these goals are meant for technology innovation and digitalization to achieve and they are still not fully addressed as at 2020. Examples of interesting investment areas during this transition period were wearable devices, IOT, big data analytics, clinical flow management tools, and AI to drive healthcare towards an outcome-based care model.

New advancements in hardware and the eruption of consumer wearable devices in the past 10 years has laid the foundation for the next stage of digitalization—Data availability and abundance. In order for healthcare digitalization tools such as Artificial Intelligence to become accepted and standardized, data abundance is one of the paramount drivers and conditions. This past decade has seen an exponential increase in the amount of healthcare data and newly available information. The volume of healthcare related data will reach yottabytes in the next decade with 80% of that data being unstructured. *These volumes are beyond the scope of any human to handle both individually and manually.*

Last but not least, patients are now more active than ever thanks to wellness devices and Dr. Google. Healthcare is no longer about plainly listening to doctors' instructions. The availability of data and information start to bring patients into the center of healthcare treatment processes and the patients start to take ownership on the decisions they have made regarding their health.

The shift in the healthcare trend into digitalization strengthens the investment thesis for the future of healthcare: **on-demand patient care, backed by AI/ML-driven decision-making supporting systems, and blockchain-backed electronic patient information flow, run on dedicated medical devices to deliver a seamless patient experience.**

Let us take a look at each one of the six sub-theses quickly:

1. *On-demand healthcare*

With the advancement of mobile phone and digital innovation, consumers now have the opportunities to book appointments at their own convenience, with the choice of their preferred doctors, from the comfort of their home. As more investment is injected into telehealth and primary healthcare providers, the following experience is no longer in the science fiction:

During this process, all the patients' health history will be recorded onto the healthcare management's platform, making it accessible for physicians to review. Subsectors include technologies in online-offline care, virtual care, pharma supply chain, health plans and benefit management, and telehealth.

Good technologies in these subsectors will keep getting venture capital funding until each component of this process integrated with each other and deliver this service seamlessly.

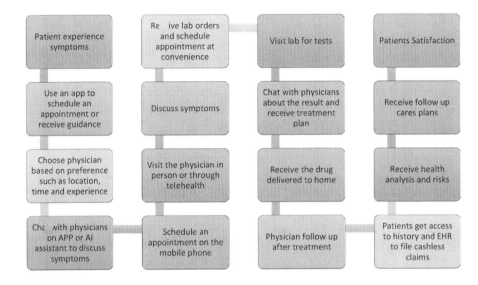

2. Big data-driven decision supporting system

Big data-driven clinical support systems will still be a major focus for investors with the following goals: (1) Reducing medication errors; (2) Facilitating preventive care and reducing unnecessary visits to the ER; (3) Improve staffing for hospitals and clinics to reduce inefficiencies. Interesting technologies of this sector are focused on administrative automation and digitization, diseases management, screening and diagnostics, and finally clinical intelligence and enablement.

3. Wearable medical devices targeting niche segments

When people look at the transition to digital technology, they may often overlook medical devices. With the help of brands such as Fitbit and Apple, smart watches and wearables that continuously collect data and provide up-to-date monitoring of a person's health are mature in the market. There is still growth potential for consumer wearable devices such as ECG sensors, smart watches, and sweat diabetic monitors. The most extreme example of a wearable device has always been Elon Musk's Neuralink. The company is now planning to implant medical wearable trackers directly into our brain. Before body implant technology becomes a trend or even the norm, there will be more remote monitoring devices. These technologies will be used to target specific high-risk individuals. Some examples including children asthma management post discharge, or breast cancer monitoring that can be prescribed by physicians for patients to use at home. These devices are specially made to monitor diseases/conditions and can capture data that regular consumer wearables, such as smart watches, cannot.

4. Predictive and preventive solutions

Given the aging population, preventive solutions are required to better manage the chronic conditions for patients as they become older. Combining predictive analytics, machine learning and redesigning the primary healthcare process will enable preventive platforms to be able to analyze a patients' health history and provide a personalized healthcare experience. For instance, after an online consultation and an annual physical checkup, a patient will be automatically warned by his digital twin about the high-risk areas of their health and will be directed to do additional checkups or meet with a specialist to take preemptive measures.

5. Artificial intelligence (AI) and machine learning

AI is more than a buzzword or a digital transformation trend in healthcare. It represents the future. Despite billions of dollars already invested in the sector globally, a good solution has yet to get investors tired. In 2019 alone, more than US$ 4 billion were poured into AI healthcare startups. Nevertheless, the sheer quantity of healthcare data alone is already impossible for any human to intercept without machine learning support. Hence, AI funding can only continue to grow. It is projected by different market analysis that the AI healthcare market can potentially exceed US$ 30 billion in the next 5 years. Without diving into details, the major subsectors of healthcare AI include: Imaging Diagnostics, Predictive analytics and risk scoring, hospital decision support, remote monitoring, drug discovery, genomics, virtual assistant, clinical trial support, compliance, mental health and wellness.

6. Blockchain-backed EHR

Last but not least, blockchains will naturally fit in the electronic health record storage given its advantage in privacy, safety and cost efficiency.

In the end, a quick note as a venture capitalist. For the fund, it is not about identifying unicorns, because luck also plays an important role in the investment journey to determine which company can succeed. Venture capital is not private equity, which invests in the mature companies at the growth stage and benefits from business growth and EBITDA increase. Venture capital invests in much earlier staged companies and many of them are still at the research and development phase. There are normally two fundamental questions: (1) Does the technology has a high chance to succeed in the next 5 years from investment. (2) What does it take to make it a successful business and does the VC have the resource to assist the company in the right amount of time. In the end, having technologies before its time normally will not work. For example, telehealth was first brought up in the 20th century. Since the early 1960, there are already examples of telehealth, but the popularity for investors did not grow until the 21st century and peak until lately. Hence, given VC funds have a responsibility of generating returns for LPs during a finite amount of years, picking the right company requires both skills and luck. Once the technology becomes a common thesis within the VCs, the VCs will help to push the technology to become a norm with its resources. For example, during the R&D stage, VCs would need to keep funding the company or look for alternative funds to ensure the company does not run out of cash flow during the long healthcare R&D cycle. Post R&D, the VCs will push the technology out to more commercial partners such as hospitals. This is the role of VC in healthcare technology.

Chapter 46

A Regulatory Career in Asia

Ambrose Chan

Integrity, Singapore

ambrose@integritypl.com

You have probably picked up this book to help you get started on your next career endeavor in regulatory affairs. Whether you are preparing for your new role or searching for an opportunity in Asia, this chapter will provide an introduction of the hiring potential in Asia, the skill sets that are required and ways to get ahead in your search.

46.1 Hiring Landscape in Asia

There has never been a better time to explore a career in regulatory in Asia. The growth of Asian markets coupled with its increasing complexity has created a plethora of job opportunities stemming from many of the largest medical device giants as well as the rising number of MedTech start-ups. It is no surprise that emerging markets in China and India are expected for strong growth, traditionally riding on the backs of

---

Medical Regulatory Affairs: An International Handbook for Medical Devices and Healthcare Products (Third Edition)
Edited by Jack Wong and Raymond K. Y. Tong
Copyright © 2022 Jenny Stanford Publishing Pte. Ltd.
ISBN 978-981-4877-86-2 (Hardcover), 978-1-003-20769-6 (eBook)
www.jennystanford.com

international players. However, with national initiatives such as "Made in China 2025" looming on the horizon, the industry is also evolving with more local start-ups joining the race for talent as they reap the benefits from robust government support.

Japan, South Korea and Taiwan are commonly acknowledged as the "reimbursement markets," characterized as having well-developed healthcare systems with considerable public expenditure. These countries have always been anchor markets for medical device companies and consequently, salaries are comparatively higher amongst the countries in Asia.

Against the backdrop of these established markets, the ASEAN cluster (Association of Southeast Asia Nations) consisting of countries such as Indonesia, Vietnam, Thailand and Myanmar continue to lure investment from multinationals given its huge untapped potential within their healthcare sectors. For many of these multinationals, that could be further expanding their distribution network in the region and thereby bolstering the headcount for local representatives, or going direct sales and setting up a local office.

At the centre of it all is Singapore, a city-state designated by many multinationals as Asia's regional hub for its pro-business environment and political stability. With international connectivity as a competitive advantage, there is a consistent demand for top talent with inter-cultural skills.

46.2 Opportunities across Markets

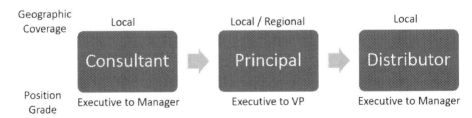

In order to understand what regulatory opportunities are available, we look at where the prospective employer lies in the value chain.

- Are they a principal and innovate their own portfolio of medical devices?
- Are they a regulatory consultancy providing third-party support to principals?
- Or are they a distributor managing an extensive product portfolio from multiple principals?

Depending on whether the company is a principal, consultancy or distributor, this will determine the difference in job scope, geographical coverage and progression opportunities. It is important to note that there will always be outliers but as a rule of thumb, the medical device industry across Asia generally adheres to these characteristics.

Principal companies consist of both start-ups and multinationals.

MedTech start-ups typically have a headcount of between five to twenty people, with regulatory affairs (RA) playing an integral role throughout the product lifecycle from conceptualization to commercialization.

The role of the RA is to work closely with cross-functions across R&D, engineering and commercial, commonly with the objective of achieving certification for Europe CE Mark or US FDA 510K. Following which, the objective will shift to local registration and thereafter, regional registrations if the office also doubles as the regional headquarters.

In a start-up, resources in the form of SOPs (standard operation procedures), databases and systems may not be readily available. To thrive well in this setting, the RA must be comfortable with ambiguity and also exhibit an independent and resilient mindset.

As for career progression in a nimble setup, the RA typically starts off as an Executive and can go up to a Manager role.

For multinationals, headcount varies in accordance to the size of business operations – is it a commercial office with a mid-sized headcount or does it encompass R&D and manufacturing?

In a commercial office, the headcount can range from 10 people all the way to 200. Here the role of the RA is frequently occupied with sales and marketing stakeholders, with the key objective of achieving fast registration approvals so that products can be monetized at the earliest.

The RA person could also be part of a larger regulatory team, handling a particular range of product, geographical scope or a specific process in the lifecycle e.g., artwork and labelling. Structure in the form of SOPs, databases and systems would also be in-place, though not necessarily translating to good order as process overlaps is a common challenge in the modern era of mergers and acquisitions.

To thrive well in this environment, there is a heavy emphasis on collaboration as getting your decision through requires the buy-in from multiple layers of stakeholders ranging from your immediate department, global headquarters as well as the commercial team.

For multinationals that encompass R&D and manufacturing, headcount can go up to 2000 or more people. In these organizations, an RA is part of a larger team and is allocated to a specific process as a subject matter expert. The role of the RA would be to liaise frequently with the various engineering functions to further refine processes.

Finally, a career with multinationals entails more layers in the hierarchy, allowing for progression from executive, manager, director and in some cases, vice-president.

In a consultancy or distributor, the RA acts as a vendor to different principals and is responsible for providing pre- and post-market support for local registrations. Depending on the size of the organization, the RA can operate as a standalone or part of a small team of around 5 headcounts. Working culture is therefore similar to that of a start-up or mid-sized where the RA is expected to be hands-on with many of the SOPs, database and processes.

Consultancies are generally nimble in size with an average headcount of 10, while distributors can employ up to a mid-sized headcount. In terms of career path, executive and manager positions are most commonly available in these organizations.

46.3   Getting Ahead in your Search

Many of us have heard of the old adage, "Your network is your net worth." Other than applying for active job postings, how do you better your chances of securing regulatory opportunities abroad?

The answer is effective networking in the digital age and that can be performed on a number of avenues. When it comes to cultivating professional relationships remotely, LinkedIn naturally comes to mind as the ideal platform for professional networking. However, most professionals limit its use to connecting with colleagues and updating themselves on industry trends.

The key to LinkedIn lies with its accessibility to professionals and particularly, prospective hiring managers, who would otherwise be out of your physical network. Make it a practice to craft a personalized introductory message with people you connect with, as it could serve as a catalyst for meaningful conversations that could be beneficial. There are a handful of medical device and regulatory groups where you can also play an active part in the community by sharing your perspective when circumstances are appropriate.

The COVID-19 pandemic has also changed the face of conferences. Iconic events such as APACMed and Medlab Asia are pivoting their formats to virtual events, facilitating networking opportunities that were previously only made possible by hopping onto the next flight.

Lastly, one additional platform is Asia Regulatory Professionals Association (ARPA), where there are online curriculums designed to improve your knowledge on regulatory frameworks. Networking opportunities are abound with a member base consisting mainly of regulatory professionals from various medical device companies across Asia.

The regulatory landscape in Asia is teeming with challenges arising from the rapid growth of the healthcare sectors across the region. With so much innovation on the horizon, every day regulatory health agencies are tasked with a greater responsibility on patient safety, and this inevitably brings along significant regulatory changes that would impact several business units in any medical device organization. This continual challenge also translates to continual opportunities for regulatory professionals with prolific knowledge and expertise.

Chapter 47

A Former FDA Investigator's Views on Compliance with the Medical Device Regulations

Ken Miles

Former US FDA Management/US FDA Investigator
Jack.wong@arpaedu.com

The U.S. Food and Drug Administration (FDA) and International Standards Organization (ISO)s are working with industry to protect lives by assisting manufacturers to improve their manufacturing and marketing processes so that only safe and effective medical devices are introduced into the marketplace. American and foreign manufacturers should have a positive attitude about the U.S. FDA. By following FDA regulations and International Standards Organization requirements allows them to make safe and effective products which is good for customers, as well as for the medical device manufacturing companies and own-label distributors. Good-quality products which are safe and effective for use will result in far fewer customer complaints and recall issues.

Medical Regulatory Affairs: An International Handbook for Medical Devices and Healthcare Products (Third Edition)
Edited by Jack Wong and Raymond K. Y. Tong
Copyright © 2022 Jenny Stanford Publishing Pte. Ltd.
ISBN 978-981-4877-86-2 (Hardcover), 978-1-003-20769-6 (eBook)
www.jennystanford.com

I joined the U.S. FDA in 1975 after receiving my degree in biology and having worked in research and technical field with the University of California at Berkeley and General Electric Company. I was assigned to the FDA's San Francisco District Office, which covers the Western United States and Pacific Islands. My specialty from the beginning was with medical devices and radiological health fields. The FDA provided extensive training to its investigators so that they could implement the new regulatory programs to regulate various products including medical and radiological devices.

During the late 1970s, the FDA developed the Medical Device Good Manufacturing Practices (GMPs) regulations and an inspection checklist which focused primarily on American manufacturing companies. Several years later, I was both selected to serve as a trainer for new FDA recruits and was assigned to the Foreign Inspection program to inspect medical device manufacturers in Canada, Europe, and Asia. My work included numerous month-long trips to Japan, China, Europe, and Canada. Each inspection took 2 or 3 days on average. During the early 1990s, I received additional training in the FDA Center of Devices and Radiological Health's (CDRH) Current Good Manufacturing Practices (GMPs) regulations program, along with the FDA's QSIT (Quality System Inspection Technique) program as the FDA was broadening its approach to regulating medical device manufacturers worldwide who were producing medical devices for the American market. I soon became regarded as one of top FDA device experts in the field as having inspected hundreds of medical device manufacturers around the world and served as a CDRH trainer for medical device inspections and FDA foreign inspection program requirements. I also provided medical device training to Japanese, Canadian and German ministry of health inspectors while assigned to conduct foreign inspections.

The FDA's initial regulation for medical device manufacturers was termed the "Medical Device Good Manufacturing Practices (GMPs)," which was enforced during the mid-1970s to early 1990s. The GMPs were good, but in retrospect were not good enough to assure manufacturers would produce safe and reliable medical device products. The GMPs focused primarily on American medical device manufacturers that produced finished medical device products only, and were NOT concerned about component quality, management responsibilities, design controls, improving production or product controls, etc. The FDA's *Current* Good Manufacturing Practices (CGMPS) developed during the early 1990s and during 1996 became the Quality System Regulations (QSRs), which was a big improvement in how the FDA regulates and works with the device industry. The QSRs were codified under Title 21 Code of Federal Regulations (CFR), Subpart 820 (https://www.fda.gov/media/88407/download). The key requirements of the CGMPs/QSRs include management review and responsibilities, planning, procedures, component control production records, quality manuals, device master records, design controls, validation and verification records, traceability, corrective and preventative actions (CAPA; also considered as Continual Improvement), strong vendor/supply chain controls requiring notification of planned changes, and written documentation. The FDA investigators' motto is, "If it is not written, it has not been done."

The International Standards Organization (ISO) developed medical and risk management requirements like the U.S. Food and Drug Administration's medical device regulations during the late 1990s and were adopted by the European Union. ISO 13485, "Medical Devices Quality Management Systems," was published in 1996. It, like the FDA's CGMPs and QSRs, represents the essential requirements for a comprehensive quality management system that covers the development of a device from design to manufacturing to corrective repair. ISO 14971, "Risk Management, Medical Devices," updated during the late 1990s, is both a product safety standard and a process for identifying, evaluating, and reducing risks associated with harm to people and damage to property or the environment. Risk Management, an integral part of medical device design and development, production processes and evaluation of field experience, is applicable to all classes and types of medical devices. The evidence of its application is required by most regulatory bodies such as the U.S. FDA. The standard provides a process framework and associated requirements for management responsibilities, risk analysis and evaluation, risk controls and lifecycle risk management. Canada, and later other countries, followed suit by developing their own medical device regulations, guidelines and existing standards such as those developed by ISO or the International Electrotechnical Commission (IEC).

The FDA inspects medical device manufacturing companies that come in all sizes and stages of development, from one-man startups developing a single device to global corporations employing thousands of people with factories in multiple countries developing numerous devices. The FDA/CDRH provides directives for trained investigators to inspect both domestic and foreign manufacturers and distributors responsible for introducing medical device products for use in the United States. The highly trained investigator will review manufacturing operations from the incoming shipping dock to finished product testing, the equality manual and procedures, production records, complaint files and Correction and Preventative Action files, and interview staff and management, and more. They know the "red flags" to look for and will document all findings. The investigator may also collect samples of labeling and copies of records which may or may not be related to a product or production deficiency. At the close of the inspections, the investigator will hand an FDA form 483, List of Observations, to the top manager present, usually the president of the company, and ask for his or her response to each listed observation. The FDA investigator cannot provide recommendations to the company. The best practice is to be honest and willing to correct the deficiencies/observations as soon as it is practical to do so. The investigator provides a written report, called an Establishment Inspection Report, to his or her supervisor for further review and classification as No Action Indicated (NAI), Voluntary Action Indicated (VAI) or Official Action Indicated (OAI). Firms that have serious QSR deviations (classified as OAI) are likely to receive a Warning Letter or Information Letter from the FDA, which the company top manager must reply to within 15 days addressing how each issue will be corrected and planned dates of correction. It is important to know that Warning Letters issued by the FDA are available to the public and posted on the FDA's website. Firms that do not comply with QSRs are considered by the FDA to be producing "adulterated products." Companies may be required to hold all "adulterated products" from shipment until corrections are made, and/or recall certain lots, or in the case of some foreign manufacturers, may not be able to ship products to the U.S. due to the FDA's Automatic Detention declaration.

The FDA was, and still is, finding it difficult to inspect the thousands of medical device firms in the U.S. and in other parts of the world. Congress long ago mandated that the FDA inspect their regulated manufacturers every other year. Over the years, the number and complexity of the manufacturing industries has greatly increased while the FDA has been unable to hire enough qualified investigators and inspectors to inspect each firm as often as required. Slowly, it became impossible for the agency's inspectors to meet the requirement set by Congress. FDA insiders and industry leaders were saying in effect, "There are not enough FDA inspectors to do the job." For example, 2012 was the FDA's most active inspection year in history, but only 5% of overseas manufacturers registered to sell in the United States were inspected, which was one-tenth the number of foreign inspections that are needed to be inspected each year. The FDA is trying to find ways to meet the demands of Congress and assure the manufacturers are adequately inspected as frequency as needed.

The FDA's Accredited Person (AP) program was established as an amendment to the Medical Device User Fee and Modernization Act (MDUFMA) of 2002. During early 2014, the FDA introduced a third-party inspection program, titled "Medical Device Single Audit Program" (MDSAP), to free up resources for it to better evaluate high-risk facilities and encourage combining inspections for multiple nations. The FDA collaborated with international partners to participate in a pilot program to conduct regulatory audits of medical device manufacturers that covers the relevant requirements of the regulatory authorities participating in the program. It approved the MDSAP program and 15 firms, termed as Accredited Persons, to perform third-party inspections of certain Class II and III medical device manufacturing facilities to assess whether their quality systems are suitable for the production of Class II and III devices under 21 CFR Part 820. Their findings are submitted to the FDA, which determines the firms' compliance. The manufacturer must notify the FDA of which AP it wants to use, and the FDA must sign off on it. The manufacturer must market a device in the United States and market or intend to market it in at least one foreign country. Another requirement is that every third inspection of a facility must be done by the FDA.

I was assigned to the FDA's Pacific Region Office during the later phase of my career with the FDA. One of the Region Office's responsibilities was to review, and monitor inspections and field tests conducted under contract with several states. The quality of the work performance and reporting was often problematic. Eventually, the contracts had to be cancelled. Later, during my career as a consultant, I also observed third-party audits at several of clients' facilities that, in my opinion, were not nearly as effective nor up to the high quality of the FDA's trained and experienced field investigators.

I retired from the FDA during 2004 and became a quality systems consultant for the medical device industry, and once again traveled throughout the U.S. and to Europe and Asia to work with medical device manufacturers. I conducted audits or "mock FDA inspections" to spot the "red flags" and to assess their levels of noncompliance and compliance with FDA regulations and ISO requirements. I also provided written reports covering my observations recommendations for improvement to upper management. Another service I provided was quality management systems training to the clients' QA and QC staff. On occasion, I have worked as a subcontractor for large quality system consulting companies to help their clients address major issues they had with their quality systems. Generally, the larger consulting firms' fees were extremely expensive as they hired numerous subcontractors to work on different quality management systems, sometimes taking many months to achieve partial corrections. The clients could have saved a vast amount in expenses and time by hiring one or two highly trained specialists to conduct an FDA-type inspection/audit of their entire quality management system and find the root causes of their problems, and to advise and/or assist them in making corrective changes.

Whether it is a small or a large company, the root causes of an ineffective quality management system are generally the same:

- Top management failed to adequately provide the funds, training, planning, QA programs, production procedures,

production material resources, design control reports, vendor audits, validation studies, CAPA, vendor contracts and audits, or utilize other important components of an effective quality management system.
- Top management failed to attend the quarterly management review meetings or review the reports, or the management review meetings were seldom held.
- Middle managers failed to monitor and collect important metrics such as consumer complaints or processing equipment malfunctions or did not provide these metrics to top management.

Whether you are a small startup or a global giant manufacturing company, it is a good practice to have some of the key members of your staff join a medical device quality association such as the American Society for Quality (ASQ), Medical Device Manufacturers Association (MDMA), Advanced Medical Technology Association (AdvaMed), or other industry associations that may meet your needs. You should also encourage members of quality management to develop a relationship with the FDA's Center of Devices and Radiological Health (CDRH) by visiting their user-friendly website, and/or request information from their Office of Ombudsman (https://www.fda.gov/about-fda/center-devices-and-radiological-health/cdrh-ombudsman). The CDRH Ombudsman investigates complaints from outside the FDA and facilitates the resolution of disputes between CDRH and the industry it regulates. The CDRH Ombudsman is a good starting point if you have a complaint, question, or dispute of a scientific, regulatory, or procedural nature. The Ombudsman can answer questions, follow up on a complaint, discuss appeal and dispute resolution options, or mediate a dispute. While providing this assistance, they maintain impartiality and neutrality. The Ombudsman also advises the CDRH Director, to whom they report, on ways to assure that CDHR procedures, policies, and decisions are of the highest quality and are fair and equitable.

Lastly, the FDA as a science-based regulatory agency whose mission is to protect the public's health and safety is undergoing incredible stress during the COVID-19 pandemic. Pressure is being applied to allow manufacturing companies that produce products that are not medical devices, such as automobiles, to rapidly adjust and produce human-use ventilators, Class III critical devices. The agency approved COVID-19 test kits (in vitro diagnostic devices) for the Center of Disease Control that had faulty components. Currently, the agency is being pressured to rapidly approve several types COVID-19 vaccines. There will probably be critical need for the FDA to assign numerous investigators to inspect device, biologic and drug manufacturing facilities, clinical trials, and related facilities. Will the FDA be able to develop and maintain a cadre of well-trained inspectors and investigators to do carry out its mission?

Index

acceptable risks 195, 309
active implantable medical device
   (AIMD) 273, 285, 298-299,
   313, 318, 391, 394-395, 586
adverse events (AE) 12, 34, 115,
   135-140, 250, 315, 347, 357,
   362, 377-379, 406, 409, 428,
   463, 467, 478, 512-517,
   551-552, 584, 624,
   637-639, 641, 648
AE, see adverse events
AHWP, see Asian Harmonization
   Working Party
AIMD, see active implantable
   medical device
AMDD, see ASEAN Medical Device
   Directive
ancillary medicinal substance 286,
   322, 327-333
Argentina 336-337, 339-342,
   344-345
   medical device system 342-343,
   345, 347, 349, 351
ARGMD, see Australian Regulatory
   Guidelines for Medical
   Devices
ARPA, see Asia Regulatory
   Professionals Association
ARTG, see Australian Register of
   Therapeutic Goods
ASEAN Medical Device Directive
   (AMDD) 575, 607, 657-659
Asia 20-21, 97-98, 114, 175, 218,
   433, 526, 535, 538, 680,
   701-706, 737-741, 744, 748

ARPA, see Asia Regulatory
   Professionals Association
Asia Regulatory Professionals
   Association (ARPA) 2, 719,
   741
Asian Harmonization Working
   Party (AHWP) 5, 113, 652
Asian markets 2, 20-21, 26, 97,
   678, 737
   countries 146, 678, 680
   herbal product regulatory
      classification 702-706
   RA team 20-22, 24, 26
   regulatory career in 737-740
Asia-Pacific 63, 70, 123-124, 126,
   128, 130, 132-134, 136, 138,
   140-142, 445, 677
Asia Regulatory Professionals
   Association (ARPA) 2, 719,
   741
Australia 104, 114, 117, 131, 133,
   174, 278, 382, 387-389,
   392-393, 395-397, 400, 402,
   434, 440, 581, 586, 651, 673
ARGMD, see Australian Regulatory
   Guidelines for Medical
   Devices
ARTG, see Australian Register of
   Therapeutic Goods
Australian Declaration of
   Conformity 393, 395
Australian medical device
   regulations 387-402
Australian Register of Therapeutic
   Goods (ARTG) 133-134,
   392-398, 400, 402

Australian Regulatory Guidelines
  for Medical Devices (ARGMD)
  389–391, 393–394, 398–399
  medical device in 392–393, 395
  Technical File Review (TFR)
    396–397
TFR, *see* Technical File Review
TGA, *see* Therapeutic Goods
  Administration
TGA Conformity Assessment
  certificates 394–395, 400
Therapeutic Goods
  Administration (TGA) 132,
  134–135, 370, 388–398,
  400–401, 586
Australian Register of Therapeutic
  Goods (ARTG) 133–134,
  392–398, 400, 402
Australian Regulatory Guidelines
  for Medical Devices (ARGMD)
  389–391, 393–394, 398–399

biological products 238, 264–267,
  269, 493, 510, 587
biomedical devices 95–96, 98, 100,
  102, 104–106
Brazil 335–342, 347–349, 352,
  551
  medical device system 342–343,
  345, 347, 349, 351
business planning 677, 680–681
businesses, continuity of 677–680

CA, *see* conformity assessment
CAB, *see* conformity assessment
  bodies
CDRRHR, *see* Center for Device
  Regulation, Radiation Health
  and Research (Philippines)

CDSCO, *see* Central Drugs Standard
  Control Organization (India)
Center for Device Regulation,
  Radiation Health and
  Research (Philippines)
  (CDRRHR) 603
Center for Food and Drug
  Inspection (China) (CFDI)
  412, 423
Center of Medical Device Evaluation
  (China) (CMDE) 132, 135,
  412, 427
Central Drugs Standard Control
  Organization (India) (CDSCO)
  448, 453–455, 459–461,
  463–464, 468, 704
Central License Approving
  Authority (India) (CLAA)
  450, 465
CEP, *see* clinical evaluation plan
CFDI, *see* Center for Food and Drug
  Inspection (China)
CFR, *see* Code of Federal
  Regulations
China 2–4, 11, 40, 97, 112, 114,
  119, 131, 135–136, 145,
  403–412, 414–418, 420, 422,
  424, 426–428, 430, 432, 535,
  538, 540, 551, 651, 693–695,
  699, 721, 731–732, 737, 744
  Center for Food and Drug
    Inspection (CFDI) 412, 423
  Center of Medical Device
    Evaluation (CMDE) 132, 135,
    412, 427
  CFDI, *see* Center for Food and
    Drug Inspection
  CMDE, *see* Center of Medical
    Device Evaluation
  National Medical Products
    Administration (NMPA) 112,
    132, 135–136, 183, 403,

Index | 753

405–407, 409–413, 415, 422–424, 427–428, 431
NMPA, *see* National Medical Products Administration
CII, *see* Confederation of Indian Industries
CLAA, *see* Central License Approving Authority (India)
clinical evaluation plan (CEP)   307, 316, 374
CMDE, *see* Center of Medical Device Evaluation (China)
Code of Federal Regulations (CFR)   112, 174, 193, 240, 249–250, 255, 258, 260, 264, 266–267, 269, 745
Colombia   336–337, 339, 341, 352–355, 357
   Instituto Nacional de Vigilancia de Medicamentos y Alimento (INVIMA)   352, 354–355, 357
   INVIMA, *see* Instituto Nacional de Vigilancia de Medicamentos y Alimento
   medical device system   352–353, 355
combination products   125, 263–271, 321–322, 324–327, 329, 331–332, 460, 472, 499, 587–589, 591
   registered drug-medical device   590–591
   regulation of   263–264, 266, 268, 270, 321–322, 324, 326, 328, 330, 332, 587
Common Submission Dossier Template (CSDT)   610, 612, 614, 632, 659, 661, 664, 670, 672
competitive advantage   20, 30, 43, 46, 48, 50, 55, 677–680, 682, 685, 699, 738

complaint handling   171, 185–186, 579, 595
compliance requirements   635, 679
Confederation of Indian Industries (CII)   451, 468
conformity assessment (CA)   191, 290, 294, 299, 301, 305, 370, 400, 581, 584–586, 650
conformity assessment bodies (CAB)   293, 382, 439–440, 575–576, 581–582, 584–585, 594, 725
conformity assessment of technical documentation   584–585
conformity assessment procedures   290, 300–301, 303, 311, 321, 382, 452, 458–459, 464, 467, 576, 581, 594
conformity assessment services   209–210, 212, 214, 216
conventional food   704
COVID-19   81, 695
COVID-19 Impact   1–2, 4
CSDT, *see* Common Submission Dossier Template

DCGI, *see* Drugs Controller General of India
device classification   4, 140, 300, 437, 724
device combination products   321, 323
device modifications   346, 351, 356, 361, 475, 597
device products, medical   20, 503, 601, 607, 609, 746
devices
   diagnostic   102, 117, 148, 248, 388, 448, 485, 647, 652, 750
   electrical surgical   544–545

highest risk 244, 247
imported 540, 559
invasive 104, 147, 159, 161, 163, 297
single-use 288–289
wearable 732–735
diagnosis 1, 77, 79–80, 82, 98, 125, 150, 210, 259, 265, 283, 289, 323, 343, 353, 358, 381, 389–390, 408, 436–437, 450, 456, 471, 497, 577, 602, 620, 622, 627, 653, 657
diagnostic tests 77–78, 80–81, 622, 635
diagnostics 63, 73, 76–77, 79, 81–83, 102, 146, 148, 283, 285, 299, 459, 563, 734
companion 85, 303
dietary supplements 702, 705
DMDCP, see drug-medical device combination product (Malaysia)
drug-medical device combination product (Malaysia) (DMDCP) 588, 590–591
drugs, registrable 627, 633
Drugs and Cosmetics Act 446–447, 450, 452, 456, 460, 462–463, 467
Drugs Controller General of India (DCGI) 132, 136, 448–449, 455–456, 459, 466

EEA, see European Economic Area
EFTA, see European Free Trade Association
EMA, see European Medicines Agency
EP, see essential principles

essential principles (EP) 275, 392, 400–401, 440, 479, 486, 498, 506, 559, 611, 613, 616, 632
ETSI, see European Telecommunications Standards Institute
European Economic Area (EEA) 274, 276, 318–319
European Free Trade Association (EFTA) 274, 276
European Medicines Agency (EMA) 2, 74–75, 330, 333
European Telecommunications Standards Institute (ETSI) 274, 281
European Union 174, 191, 235, 275–276, 280, 290, 304, 313, 339, 353, 408, 434, 440, 552, 581, 619, 651, 726, 745
medical device regulatory system 273–274, 276, 278, 280, 282, 284, 286, 288, 290, 292, 294, 296, 298, 300, 302, 304, 306, 308, 310, 312, 314, 316, 318
regulation of combination products 321–322, 324, 326, 328, 330

FDA, see Food and Drug Administration, US
field safety corrective actions (FSCA) 274, 312, 345, 377–378, 380, 514–515, 638, 641, 663
field safety notice (FSN) 312, 380–381, 441
Food and Drug Administration, US (FDA) 101, 111–112, 237–238, 240–242, 244, 247, 249–254, 256–258, 260, 264, 266–267, 269–271, 338

food and drug safety 538, 541, 543, 547, 550
free sale certificate (FSC) 140, 345, 349, 354, 477, 670
FSC, *see* free sale certificate
FSCA, *see* field safety corrective actions
FSN, *see* field safety notice

GCP, *see* Good Clinical Practices
GDPMD, *see* Good Distribution Practice for Medical Devices
General Safety and Performance Requirements (GSPR) 289–290, 294, 304, 306, 314, 332
General Safety and Performance Requirements (GSPRS) 289–290, 294–295, 304, 306, 314, 332
GMPs, *see* Good Manufacturing Practices
Good Clinical Practices (GCP) 127–128, 131–132, 137, 140, 142, 241, 258, 416, 422, 498, 561
Good Distribution Practice 646, 648
Good Distribution Practice for Medical Devices (GDPMD) 579, 625
Good Manufacturing Practices (GMPs) 140, 250, 344, 457, 460, 463, 645, 650, 679, 744
Good Review Practice (GRevP) 217, 219–220, 234, 652
Good Submission Practice (GSubP) 217–218, 220–222, 224, 228, 230, 232, 234, 646

GRevP, *see* Good Review Practice
GSPR, *see* General Safety and Performance Requirements
GSPRS, *see* General Safety and Performance Requirements
GsubP, *see* Good Submission Practice

Health Insurance Review and Assessment Service (Korea) (HIRA) 563, 567–570
Health Sciences Authority (Singapore) (HSA) 114, 132, 138, 620–621, 624, 626, 637–641, 704, 708–709
health supplement regulation 709, 715
health supplements 702, 704, 707–710, 712–716
health technology assessments (HTA) 76, 570–571
healthcare 1–2, 20, 87, 211, 273, 283–284, 292, 336, 446, 470, 516, 537, 540, 592, 693, 699, 729–736
healthcare facilities 376–377
healthcare industry 538, 696
healthcare organizations 730–731
healthcare professionals 319, 516–518, 531–532, 637, 640, 710
healthcare providers 21, 125, 366–367, 376–377, 531, 563, 565, 731
HIRA, *see* Health Insurance Review and Assessment Service (Korea)
HIV 77, 79–80, 153, 448, 622
Hong Kong 1, 5, 112, 114, 136, 191, 433–436, 438, 440, 442–443

MDACS, see Medical Device Administrative Control System
MDCO, see Medical Device Control Office
MDD, see Medical Device Division
Medical Device Administrative Control System (MDACS) 5, 114, 435, 437–439, 443
Medical Device Control Office (MDCO) 435, 440
Medical Device Division (MDD) 136, 435–436, 442–443

HSA, see Health Sciences Authority (Singapore)
HTA, see health technology assessments
human body 99, 127, 135, 147, 150, 289–290, 298, 307, 323, 328, 381–383, 390–391, 405, 408, 416, 419–420, 422, 472, 481, 523, 554–555, 577, 581, 588–589, 620, 647, 672, 707

IDE, see Investigational Device Exemption
IEC, see International Electrotechnical Commission
IMDRF, see International Medical Device Regulators Forum
IMDS, see International Medical Device School
in vitro diagnostics (IVDs) 73–74, 76, 78, 80, 82, 117, 150, 285, 318, 344, 364–366, 368, 388, 391, 398–399, 459, 462, 495–496, 501–502, 504–510, 544, 571, 619, 622, 647, 650

India 11, 40, 132, 136, 145, 445–458, 460, 462–466, 468, 551, 731, 737
CDSCO, see Central Drugs Standard Control Organization
Central Drugs Standard Control Organization (CDSCO) 448, 453–455, 459–461, 463–464, 468, 704
Central License Approving Authority (CLAA) 450, 465
CII, see Confederation of Indian Industries
CLAA, see Central License Approving Authority
Confederation of Indian Industries (CII) 451, 468
DCGI, see Drugs Controller General of India
Drugs and Cosmetics Act 446–447, 450, 452, 456, 460, 462–463, 467
Drugs Controller General of India (DCGI) 132, 136, 448–449, 455–456, 459, 466
drugs inspectors 454–455
medical devices 452, 462–463
Indonesia 471, 475
medical device market 470
information technology 46–47, 106, 210, 263, 544, 696
injury 138, 203, 239, 243–244, 247, 250–251, 256, 264, 289, 323, 353, 357, 379, 381, 390, 408, 428–429, 436–437, 471, 577, 620, 638, 657
Institutional Review Board (IRB) 128, 131, 136, 139–140, 258, 641, 651

Instituto Nacional de Vigilancia de
Medicamentos y Alimento
(Colombia) (INVIMA) 352,
354–355, 357
intended purpose 196, 199, 242,
265, 297, 299, 306, 309–310,
329, 374–375, 382–383, 390,
399, 578, 582, 584, 592, 621,
629–630, 660
International Electrotechnical
Commission (IEC) 121, 192,
210–215, 274, 281, 349, 407,
479, 560, 746
International Medical Device
Regulators Forum (IMDRF)
10–11, 70, 76, 223, 234, 274,
491, 653
International Medical Device School
(IMDS) 719–721
International Standards
Organization (ISO) 2, 47, 99,
119–120, 169–178, 180–182,
184–188, 191–194, 196, 198,
200, 202–204, 206, 212–213,
215, 250, 274, 278, 280–281,
305, 307–309, 345, 350, 354,
369, 400, 407, 415, 440–441,
452, 460, 462, 464, 479, 507,
560, 579, 624–625, 650, 672,
727, 743, 745–746
Investigational Device Exemption
(IDE) 12, 131, 248, 250, 258
INVIMA, see Instituto Nacional de
Vigilancia de Medicamentos y
Alimento (Colombia)
IRB, see Institutional Review Board
ISO, see International Standards
Organization
ISO 14971 181, 191–194, 196,
198, 200, 202
IVDs, see in vitro diagnostics

Japan 97–98, 130–131, 145, 217,
358, 382, 434, 440, 491–500,
502, 504–508, 510, 512,
514–516, 518–520, 522,
524–526, 528, 530, 532, 535,
540, 551, 581, 586, 651, 673,
708, 738, 744
Japanese market 499–501
Japanese medical device
regulations 506, 513
legislation of medical devices
497–499, 501, 503, 505
Pharmaceuticals and Medical
Devices Agency (PMDA)
217–218, 492–496, 500–501,
503, 506, 510, 514–519,
526–527, 529–531, 586
PMDA, see Pharmaceuticals and
Medical Devices Agency
PMDA alert for proper use of
medical devices 531
regulatory agency 492–493, 495

KFDA, see Korean Food and Drug
Administration
KGMP, see Korean Good
Manufacturing Practice
Korea 112, 114, 119, 131, 139,
535–540, 542, 544, 546,
548–554, 556, 558–564, 566,
568, 570–571, 704–705, 721
Health Insurance Review and
Assessment Service (HIRA)
563, 567–570
HIRA, see Health Insurance
Review and Assessment
Service
KFDA, see Korean Food and Drug
Administration

KGMP, *see* Korean Good Manufacturing Practice
Korea Health Industry Development Institute 543, 545, 548–549
Korean Food and Drug Administration (KFDA) 114, 132, 139–140, 704–705
Korean Good Manufacturing Practice (KGMP) 553–554
MDITAC, *see* Medical Device Information and Technology Assistance Center
Medical Device Information and Technology Assistance Center (MDITAC) 555–556, 559
MFDS, *see* Ministry of Food & Drug Safety
Ministry of Food & Drug Safety (MFDS) 552–562, 570–571
Korean Food and Drug Administration (KFDA) 114, 132, 139–140, 704–705
Korean Good Manufacturing Practice (KGMP) 553–554
Korea Health Industry Development Institute 543, 545, 548–549
KSA market 370, 376, 380

labeling 109–118, 120–121, 128, 147, 225, 246, 248, 250, 254, 290, 295, 297, 305, 311, 315, 345, 350, 372, 383, 420, 447, 462–463, 467, 475, 478, 482, 487, 489, 525–526, 558–559, 578, 587–588, 591–593, 610, 612, 615, 630–631, 650, 661–662, 664, 668, 679, 740, 746
 definition of 111–112
 elements of 111, 113
 labeling content 110, 112, 116, 119, 121
 labeling requirements 114, 246, 344–345, 349–350, 357, 359, 373, 441, 462, 499, 640
Latin America 76, 235, 335–342, 344, 346, 348, 350, 352, 354, 356, 358, 360, 362
 market analysis 336–337
 medical device regulatory affairs in 335–336, 338, 340, 342, 344, 346, 348, 350, 352, 354, 356, 358, 360, 362
license to operate (LTO) 602–604, 606
local responsible person (LRP) 5, 438–443, 457
LRP, *see* local responsible person
LTO, *see* license to operate

MAH, *see* marketing authorization holder
Malaysia 133, 137, 218, 573–574, 576–578, 580, 582, 584, 586, 588, 590, 592, 594–598
 drug-medical device combination product (DMDCP) 588, 590–591
DMDCP, *see* drug-medical device combination product
MDA, *see* Medical Device Authority
Medical Device Authority (MDA) 573–575, 579, 581, 584–585, 587–591, 594, 596–598
 medical device industry 573–574
 manufacturer's evidence 393, 395, 397

market clearance 610, 612, 615
marketing approval 134, 440, 510, 684
marketing authorization 220, 249, 253, 255, 257, 365–366, 369, 372
marketing authorization holder (MAH) 428–429, 499–501, 503–505, 509–516, 525, 530–531
marketing clearances 610, 612, 615
materials of animal origin 393–395
MDA, see Medical Device Authority (Malaysia)
MDACS, see Medical Device Administrative Control System (Hong Kong)
MDCG, see medical device coordination group
MDCO, see Medical Device Control Office (Hong Kong)
MDD, see Medical Device Division (Hong Kong)
MDDCP, see medical device-drug combination product
MDITAC, see Medical Device Information and Technology Assistance Center (Korea)
MDNR, see Medical Devices National Registry
MDR, see medical device regulation
MDSAP, see medical device single audits programme
medical device
  labelling 591, 593
  reporting 246, 250, 378, 383
Medical Device Administrative Control System (Hong Kong) (MDACS) 5, 114, 435, 437–439, 443

Medical Device Authority (Malaysia) (MDA) 573–575, 579, 581, 584–585, 587–591, 594, 596–598
Medical Device Control Office (Hong Kong) (MDCO) 435, 440
medical device coordination group (MDCG) 274, 282, 287–289, 292, 294, 317, 326
Medical Device Division (Hong Kong) (MDD) 136, 435–436, 442–443
medical device-drug combination product (MDDCP) 588, 590
Medical Device Information and Communication System (Singapore) (MEDICS) 623–624
Medical Device Information and Technology Assistance Center (Korea) (MDITAC) 555–556, 559
medical device regulation (MDR) 1–2, 22, 112, 130–131, 171, 174, 240, 246, 251, 274, 278, 282, 285–290, 294–298, 302, 304–306, 313–317, 321–323, 326, 329, 332, 338–341, 343, 345, 352, 357, 363–364, 382, 387–388, 390, 392, 394, 396, 398, 400, 402, 409, 434, 443, 457–458, 495, 506, 513, 522, 552, 573–576, 578, 580, 582, 584, 586, 588, 590, 592, 594, 596, 598, 619–620, 622, 624, 626, 628, 630, 632, 634, 636, 638, 640, 656, 680, 726, 743–750
medical device regulatory system 235, 385, 403, 433, 445, 469, 491, 535, 603, 605, 607, 609, 611, 613, 615

medical device single audits
  programme (MDSAP) 76,
  747
medical devices 438, 579
  active 147, 149, 298, 358
  active implantable 285, 298, 313,
    318, 391, 586
  advanced 97, 446
  advertisement regulation of 443,
    466, 653
  approval of 510, 532, 551, 558
  approved 519, 637
  categories of 370, 447, 624
  classification 353, 438, 653
  clinical use of 416, 419
  combinational 587, 589
  commercialization 339, 352
  conformity assessment for
    214–215, 585
  design 3, 745
  designated controlled 498,
    500–501, 503, 505
  development of 95, 106, 697
  diagnostic 150, 153, 285, 299,
    303, 318, 582–583, 593, 611,
    613, 615, 658–659
  directives 2, 68–69, 86, 193, 282
  disinfection of 381, 472, 577,
    620, 658
  disposable 593, 656
  distribution of 367, 382
  distributors 362, 428
  finished 415, 578, 611–616
  foreign manufacturer of 469,
    502, 504
  high-risk 406, 422
  implementation of 133, 575
  import/ imported 136, 410–411,
    417, 428, 462, 469–470,
    541–542, 544, 604, 668, 670
  industry 10–11, 21, 47, 91, 97,
    170, 173, 239, 446, 543, 545,
    548–549, 573–575, 643, 654,
    656, 719, 739, 748
  innovation 251, 693
  labeling of 114, 305, 660
  labeling requirement for
    372–373
  laws 286, 366, 574
  licensed 658–659, 663
  local 2, 130
  manufacture of 250, 406, 464
  manufacturers 98, 170, 194, 336,
    365, 369, 429–432, 580, 598,
    650, 723, 744, 747–748
  price control of 443, 465, 653
  product approval 553–554
  product registration 462, 632
  recall 429, 639
  registration 345, 354, 409, 411,
    415, 422, 557, 576, 580–581,
    583, 585, 592, 597, 659
  registration application 414, 584,
    590
  registration certificate 411, 424,
    426, 590
  registration number 114, 592
  registration process 344,
    347–348, 474
  registration submission 2, 4, 346
  regulatory approval of 346, 361
  risk classification of 621, 658
  safety and performance of 595,
    659, 696
  specially-controlled 501, 504
  trials 125, 138, 561
  types 240, 372
  unapproved 134, 141, 402
  unregistered 138–139, 633, 635
  in vitro diagnostic 292, 593, 623
Medical Devices National Registry
  (MDNR) 365, 370, 383
medicinal substances 148, 303,
  307, 324, 327–333, 400–401,
  441

Index | 761

medicines 73-74, 88, 160-161, 238, 293, 326, 329, 374, 390, 393-395, 400-401, 437, 465, 701, 704-705, 716
MEDICS, see Medical Device Information and Communication System (Singapore)
medtech 67, 69-70, 86, 723-724, 726-728
medtech regulations 67-68
Mexico 114, 218, 335-339, 341, 357-360
  medical device system 357, 359
MFDS, see Ministry of Food & Drug Safety (Korea)
Ministry of Food & Drug Safety (Korea) (MFDS) 552-562, 570-571

National Medical Products Administration (China) (NMPA) 112, 132, 135-136, 183, 403, 405-407, 409-413, 415, 422-424, 427-428, 431
NMPA, see National Medical Products Administration (China)

periodic safety update reports (PSUR) 307, 316, 651
pharmaceutical regulations 522, 525, 715
Pharmaceuticals and Medical Devices Agency (Japan) (PMDA) 217-218, 492-496, 500-501, 503, 506, 510, 514-519, 526-527, 529-531, 586

Philippines 218, 601-602, 604
CDRRHR, see Center for Device Regulation, Radiation Health and Research
Center for Device Regulation, Radiation Health and Research (CDRRHR) 603
certificate of product notification (CPN) 602, 607
certificate of product registration (CPR) 602, 607, 609
CPN, see certificate of product notification
CPR, see certificate of product registration
medical device regulatory system 601-602, 604, 608, 610, 612, 614, 616
PMA, see pre-market approval
PMDA, see Pharmaceuticals and Medical Devices Agency (Japan)
PMOA, see primary mode of action
PMS, see post-market surveillance
PMSP, see post-market surveillance plan
PMSR, see post-marketing safety reporting
PMSV, see post-market surveillance and vigilance
post-market review 396-397
post-market safety management 510-511, 513, 515, 517
post-market surveillance (PMS) 116, 275, 287, 292, 312, 316, 346, 352, 357, 361, 372, 439-440, 452, 463, 476, 519, 584-585, 595, 637, 645, 695
post-market surveillance and vigilance (PMSV) 292, 595
post-market surveillance plan (PMSP) 307, 312

post-market surveillance
  requirement 376–377, 379
post-market surveillance system
  312, 476, 581, 584, 710
post-marketing safety 269, 493
post-marketing safety reporting
  (PMSR) 269, 307
post-submission stage 228–229
pre-market approval (PMA)
  242–244, 247–248, 253–257,
  260, 439, 501, 552, 586
pre-market assessment 388–389
pre-market notification process
  244, 253
pre-market notifications 243, 246,
  252–255
pre-market submission 243, 257,
  259
pre-marketing control 659, 661
predicate device 253–255, 649
premarket submission 252–253,
  255, 257, 259
primary mode of action (PMOA)
  266–268, 587
product labelling requirement
  713, 715
PSUR, see periodic safety update
  reports
public clinical trial registry
  128–129

QC, see quality check
QMS, see quality management
  system
QP, see qualified person
qualified person (QP) 314, 604,
  606
quality check (QC) 227–228, 234
quality management system
  (QMS) 169–170, 172–175,
  177–181, 183–188, 194, 250,
  301–302, 307–308, 364, 369,
  423, 441, 464, 482, 484, 495,
  503, 506–507, 509, 554, 560,
  581, 584–585, 598, 624–625,
  632–634, 660, 672, 724–727,
  745, 748–749

regulatory activities 133, 270, 599,
  685, 687
regulatory affairs professionals
  63–64, 124, 129–130, 141,
  335, 562, 683
  roles 63–65
risk control measures 4, 197–199,
  205–206
risk management 2, 110, 115, 118,
  171, 181, 191–196, 199, 202,
  213, 215, 278, 308–309, 313,
  316, 345, 350, 354, 372, 488,
  697, 726, 745

Saudi Arabia 235, 363–368, 370,
  372–374, 376, 378, 380–384
Saudi Food and Drug Authority
  (SFDA) 364, 366–371, 373,
  376–382, 409
  MDEL, see Medical Device
    Establishment Licensing
  MDMA, see Medical Device
    Marketing Authorization
  Medical Device Establishment
    Licensing (MDEL) 368
  Medical Device Marketing
    Authorization (MDMA)
    369–370, 749
SFDA, see Saudi Food and Drug
  Authority

Saudi Food and Drug Authority
   (SFDA) 364, 366–371, 373,
   376–382, 409
SCAP, see Singapore Code of
   Advertising Practice
sham device/procedure 125–126
SFDA, see Saudi Food and Drug
   Authority
Singapore 11, 19, 27, 114, 131,
   133, 138–139, 145–146, 150,
   153, 218, 619–620, 623–625,
   633, 637, 640, 707–709, 715,
   729, 731, 737–738
   Health Sciences Authority (HSA)
      114, 132, 138, 620–621, 624,
      626, 637–641, 704, 708–709
   HSA, see Health Sciences
      Authority
   Medical Device Information and
      Communication System
      (MEDICS) 623–624
   medical device regulation
      619–620, 622, 624, 626, 628,
      630, 632, 634, 636, 638, 640
   MEDICS, see Medical Device
      Information and
      Communication System
   SCAP, see Singapore Code of
      Advertising Practice
   Singapore Code of Advertising
      Practice (SCAP) 639–640
   Singapore Medical Device
      Notification 623, 625, 627,
      629, 631, 633, 635, 637
   Singapore Medical Device Register
      (SMDR) 620, 626–627, 632
   SMDR, see Singapore Medical
      Device Register
   Singapore Code of Advertising
      Practice (SCAP) 639–640
   Singapore Medical Device
      Notification 623, 625, 627,
      629, 631, 633, 635, 637

Singapore Medical Device Register
   (SMDR) 620, 626–627, 632
Single Registration Number (SRN)
   288, 317
SMDR, see Singapore Medical
   Device Register
SOPs, see Standard Operating
   Procedures
SRN, see Single Registration
   Number
Standard Operating Procedures
   (SOPs) 140, 219, 226–227,
   233–234, 490, 739–740
STED, see Summary Technical
   Documentation
Summary Technical Documentation
   (STED) 275, 401, 501,
   558–559

Taiwan 130–131, 140, 174,
   643–644, 646–648, 650–654,
   738
   Division of Medical Devices and
      Cosmetics 140, 643, 646,
      654
   Taiwan Food and Drug
      Administration (TFDA) 132,
      140–141, 643, 646, 648,
      651–652
   TFDA, see Taiwan Food and Drug
      Administration
Taiwan Food and Drug
   Administration (TFDA) 132,
   140–141, 643, 646, 648,
   651–652
technical documentation 286, 288,
   291, 296, 301–303, 305–306,
   318, 349, 354, 372, 376,
   400–401, 441, 501, 559, 581,
   584–586, 724–727

technical documents 224, 232, 365, 405–406, 555–556, 558, 649
review 556–558
TFDA, see Taiwan Food and Drug Administration
TGA, see Therapeutic Goods Administration (Australia)
Thai FDA, see Thailand Food and Drug Administration
Thailand 141, 218, 655–664, 738
Thailand Food and Drug Administration (Thai FDA) 141, 218, 661–662
The Organisation for Professionals in Regulatory Affairs (TOPRA) 85–87, 89–90, 92
Therapeutic Goods Administration (Australia) (TGA) 132, 134–135, 370, 388–398, 400–401, 586
TOPRA, see The Organisation for Professionals in Regulatory Affairs

UDI, see unique device identifier
UK, see United Kingdom
unacceptable risk 204–205
unique device identifier (UDI) 111–112, 251, 275, 287–288, 292, 306, 317, 530, 726
United Kingdom (UK) 73, 90, 112, 116, 275–276, 293, 317–318, 330, 387, 551, 651, 673
market 318
responsible person 318
UKCA mark 318–319
United States (US) 3–4, 40, 79, 96–98, 104, 112, 120, 131, 145, 193, 235, 238–239, 241, 251, 260–261, 263–266, 268, 270, 275, 337–340, 358, 424, 519, 540, 551, 651, 727, 731–732, 746–747
CDRH, see Center for Devices and Radiological Health
Center for Devices and Radiological Health (CDRH) 237–239, 241, 243–244, 246, 260, 266, 699, 744, 749
FDA, see Food and Drug Administration
FD&C Act, see US Federal Food, Drug, and Cosmetic Act
Food and Drug Administration (FDA) 101, 111–112, 237–238, 240–242, 244, 247, 249–254, 256–258, 260, 264, 266–267, 269–271, 338
Health and Human Services (HHS) 237
HHS, see Health and Human Services
MDUFMA, see Medical Device User Fee and Modernization Act
medical device regulatory framework 237–238, 240, 242, 244, 246, 248, 250, 252, 254, 256, 258, 260
Medical Device User Fee and Modernization Act (MDUFMA) 252, 747
Q-Submission Program 258–259
Q-Submissions 258, 260
Quality System Regulations (QSRs) 171, 174–175, 745–746
Request for Designation (RFDs) 267–268, 270

RFDs, *see* request for designation
US Federal Food, Drug, and
 Cosmetic Act (FD&C Act)
 239–243, 246, 248, 253, 255,
 257, 260, 265–267
unregistered devices 634, 637
unregistered devices for non-
 clinical purposes 634, 637
US, *see* United States

VCs 736
venture capital's role 729–736
Vietnam 667–668, 670, 672, 738
Vietnamese 670–671, 704

WHO, *see* World Health
 Organization
World Health Organization (WHO)
 75, 80–82, 220, 234, 341, 730

Printed in the United States
by Baker & Taylor Publisher Services